T0188945

Digital Logic Design Using Verilog

Vaibbhav Taraate

Digital Logic Design Using Verilog

Coding and RTL Synthesis

Second Edition

Vaibbhav Taraate
1 Rupee S T (Semiconductor
Training @ Rs.1)
Pune, Maharashtra, India

ISBN 978-981-16-3201-3 ISBN 978-981-16-3199-3 (eBook)
https://doi.org/10.1007/978-981-16-3199-3

This Springer imprint is published by the registered company Springer Nature Singapore Pte Ltd.
The registered company address is: 152 Beach Road, #21-01/04 Gateway East, Singapore 189721, Singapore

Dedicated to my Inspiration

Bharat Ratna Lata Mangeshkar
&
Bharat Ratna Sachin Tendulkar

For giving me real happiness and many happy moments!

Preface

I am delighted to have the second edition of the *Digital Logic Design Using Verilog* book. During the past five years, the first edition has more than 80K downloads and then I thought to work on the second edition of the book.

This edition includes the Verilog RTL design and verification using the Verilog-2005 style constructs. Throughout this book, I have used the constructs from the stable release of Verilog, that is, IEEE 1364-2005. The keywords are highlighted using bold blue color, and this book is useful to RTL design engineers who wish to pursue their career in RTL design, FPGA design, and ASIC design. Even the performance improvement of the design and overall design improvement techniques are included in this edition!

For the synthesis of the RTL designs, I have used Xilinx Vivado and ISE 14.7. The readers can go to www.xilinx.com to download the EDA tool, and even they can purchase the Xilinx FPGA boards and tools to implement the products and ideas.

The book has 25 chapters and is mainly useful to understand about the RTL design concepts, synthesizable and non-synthesizable Verilog constructs, and basics of testbenches to check for the functional correctness of the design.

The book even covers the advanced concepts used in the ASIC design synthesis, with the low power and multiple clock domain design concepts.

Chapter 1 "Introduction" describes about the evolution of logic design, design methodology, and the basics of Verilog. The chapter discusses basics of Verilog Simulation and synthesis flow.

Chapter 2 "Concept of Concurrency and Verilog Operators", for any language, the operator plays an important role. The Verilog supports various operators, and the chapter discusses the use of these operators in the RTL design.

Chapter 3 "Verilog Constructs and Combinational Design-I" discusses the combinational logic design using the synthesizable Verilog constructs. Also, it discusses the practical and real-life scenarios, useful while implementing combinational designs.

Chapter 4 "Verilog Constructs and Combinational Design-II" discusses RTL design for few of the arithmetic resources and the code converters.

Chapter 5 “Multiplexers as Universal Logic” discusses the efficient RTL coding for multiplexers and parallel versus priority logic.

Chapter 6 “Decoders and Encoders” discusses the efficient RTL coding for decoders and encoders. The RTL design strategies for these combinational design elements are discussed using the synthesizable constructs.

Chapter 7 “Event Queue and Design Guidelines” discusses event queue and few important design and coding guidelines for the combinational logic design.

Chapter 8 “Basics of Sequential Design Using Verilog” is useful to understand about the RTL design for the latches and flip-flop. The concept of the synchronous and asynchronous reset is also discussed.

Chapter 9 “Synchronous Counter Design Using Synthesizable Constructs”, the RTL design of various synchronous counters using the synthesizable constructs is discussed. The chapter discusses the RTL design, simulation, and synthesis concepts.

Chapter 10 “RTL Design of Registers and Memories” is useful to understand the RTL design techniques and strategies to code the RTL for registers, shift registers, and memories.

Chapter 11 “Sequential Circuit Design Guidelines” discusses the sequential design guidelines which need to be followed while coding an efficient RTL using synthesizable Verilog constructs.

Chapter 12 “RTL Design Strategies for Complex Designs” discusses the use of synthesizable Verilog constructs to implement the complex designs for the desired functionality.

Chapter 13 “RTL Tweaks and Performance Improvement Techniques” discusses the area, speed, and power improvement basics and is useful during the RTL design and synthesis stage to improve the design performance.

Chapter 14 “Finite State Machines Using Verilog”, the RTL design for the Moore and Mealy machine is discussed. The FSM encoding styles are binary, gray, and one-hot encoding and are discussed in this chapter.

Chapter 15 “Non-synthesizable Verilog Constructs and Testbenches” discusses the inter-, intra-delay assignments and other non-synthesizable constructs useful during the testbenches. The chapter is useful to understand about the non-synthesizable constructs and how to check for the functional correctness of the design.

Chapter 16 “FPGA Architecture and Design Flow” discusses the FPGA architecture, design flow, and the simulation using the FPGA.

Chapter 17 “FPGA Design and Guidelines” discusses the design guidelines for FPGA-based designs. How to use the design guidelines is explained with the RTL designs coded using the synthesizable Verilog constructs.

Chapter 18 “ASIC Design” discusses the ASIC types and basics of ASIC design flow.

Chapter 19 “ASIC Synthesis and SDC Commands” discusses the ASIC synthesis and important SDC commands used during synthesis.

Chapter 20 “Static Timing Analysis” discusses the STA concepts useful during the timing analysis and during the timing closure.

Chapter 21 "Design Constraints And Optimization" discusses the design constraints and optimization using Synopsys DC.

Chapter 22 "Multiple Clock Domain Design" discusses the multiple clock domain design techniques and the control and data path synchronizers and their use!

Chapter 23 "Case Study: FIFO Design" is useful to understand the FIFO depth calculations and discusses the FIFO design, simulation of FIFO, and synthesis.

Chapter 24 "Low Power Design" discusses the low power design techniques and the need of Unified Power Format. This chapter is also useful to understand about the UPF concept and its use.

Chapter 25 "System-On-Chip (SOC) Design", the SOC consists of many complex blocks like processors, arbiters, memories, and peripherals. These blocks are discussed in this chapter. This chapter even focuses on the generalized SOC architecture and the SOC design flow.

The book includes many practical examples to understand how to code an efficient RTL using Verilog. The book is also useful to understand the synthesizable designs and frequent issues in the RTL design and how to overcome them. The book even covers the performance improvement using RTL tweaks and the ASIC and FPGA synthesis for a better understanding.

This book is useful to the engineering students, VLSI beginners, and professionals who wish to implement synthesizable design using Verilog!

Pune, India Vaibhav Taraate

Acknowledgements

The second edition of this book is originated due to my extensive work in FPGA and ASIC design from the year 2000. The journey to design the architectures will continue in the future also and will be helpful to many professionals and engineers.

This book is possible due to the help of many people. I am thankful to all the participants to whom I taught the subject digital design and RTL design using Verilog in various multinational corporations. I am thankful to all those entrepreneurs, design/verification engineers, and managers with whom I worked in the past almost around 20 years.

I am thankful to my dearest friends for their constant support. I am especially thankful to my friends and well-wishers and family members. Special thanks to Neeraj and Deepesh for their best wishes and for their valuable help during the manuscript work.

The book is possible due to the best wishes and constant encouragement of Raghu, Satya, Venky, Srinath, Rohit, Acharya, Suresh, Nitin, Rohit, Amit, Anil, Ashok, Deepak, Shrinivas, Sunil, Madhuseth, Sanjuseth, and Rahul. I am thankful to my dearest sister Manisha, Dhanashree, Sharmistha, Neha, Annu, Anjali, Vaishali, Anjani, and Esha for their faith and belief on me and for best wishes.

Finally, I am thankful to the Springer Nature staff, especially Swati Meherishi, Muskan Jaiswal, Ashok Kumar, Silky, Rini Christy, and Umamagesh for their great support during various phases of the manuscript.

Special thanks in advance to all the readers and engineers for buying, reading, and enjoying this book!

Contents

About the Author

Vaibhav Taraate is an entrepreneur and mentor at “1 Rupee S T”. He holds B.E. (Electronics) degree from Shivaji University, Kolhapur (1995) and received a Gold Medal for standing first in all engineering branches. He completed his M.Tech. (Aerospace Control and Guidance) at the Indian Institute of Technology (IIT) Bombay, India, in 1999. He has over 18 years of experience in semi-custom ASIC and FPGA design, primarily using HDL languages such as Verilog, VHDL and SystemVerilog. He has worked with multinational corporations as a consultant, senior design engineer, and technical manager. His areas of expertise include RTL design using VHDL, RTL design using Verilog, complex FPGA-based design, low power design, synthesis and optimization, static timing analysis, system design using microprocessors, high-speed VLSI designs, and architecture design of complex SOCs.

Chapter 1
Introduction

RTL Design engineer should have the strong logic design fundamentals. This chapter describes about the evolution of logic design, design methodology and the basics of Verilog. The chapter discusses about basics of Verilog Simulation and Synthesis flow.

The Verilog is one of the powerful hardware description languages and supports the concurrent and sequential constructs. The language is popular in the industry as it supports the use of synthesizable and non-synthesizable constructs. In the design and verification context, it is essential to understand the basic design flow, and the chapter discusses about the basics of hardware description language and the design flow! We will use the stable release of Verilog, that is, IEEE 1364-2005, which is Verilog-2005 coding style.

1.1 Evolution of Logic Design

The actual invention of the prototype transistor model during the year 1946–1947 at Bell Labs by William Shockley, John Bardeen, and Walter Brattain had revolutionized the use of semiconductor in switching theory and in the design of chip. The working transistor used in the design was the biggest contribution by the Tanenbaum during the year 1954-1956.

During the year 1958, Jack Kilby, young electrical engineer at Texas Instrument, figured out how to place the circuit elements transistors, resistors, and capacitors on small piece of Germanium! But prior to the year 1958, many more revolutionized ideas were published and conceptualized!

The invention of CMOS logic during 1963 has made integration of logic cells very easy, and it was predicted by Intel's co-founder Gordon Moore that the 'Number of transistors in a dense integrated circuit (IC) doubles about every two years' What we call as Moore's law! The Moore's law is just observation and is used for the overall planning of the chip design by considering doubling rate of the

V. Taraate, *Digital Logic Design Using Verilog*,
https://doi.org/10.1007/978-981-16-3199-3_1

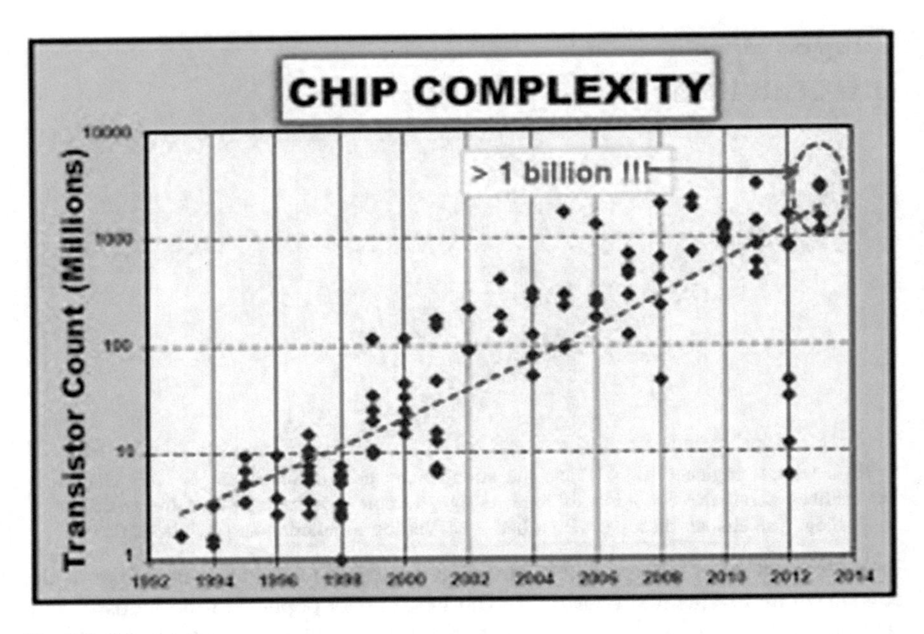

Fig. 1.1 Moore's law

transistors. The Moore's law with the Rock's law (second Moore's law), that is, 'The cost of semiconductor chip fabrication doubles for every four years,' is used to plan the overall design cycle and investments required during the chip design and fabrication.

Both laws are used in the overall planning including the financials during the chip design cycle. The industry has witnessed the doubling of transistors according to Moore's law till almost 2014. Below 10 nm, the doubling of transistor during two-year time span has failed, and now, we can say that to double the transistors at lower process nodes it needs almost 30 to 36 months of the time (Fig. 1.1).

But still how Moore's prediction was right that experience engineers can get with the complex VLSI-based ASIC chip designs. In the present decade, the chip area has shrunk enough and process technology node on which foundries are working is below 10 nm and chip has billions of transistors with small silicon die size. With the evolutions in the algorithmic design and manufacturing processes, most of the designs are implemented by using **V**ery **H**igh **S**peed **I**ntegrated **C**ircuit **H**ardware **D**escription **L**anguage (**V$_{\mathbf{HSIC}}$HDL**) or using Verilog. During the last decade, SystemVerilog has become popular for hardware description and hardware verification.

In this manuscript, we are focusing on the Verilog as hardware description language. The book is useful to understand Verilog constructs and their use during RTL design and verification. Even the book is useful to understand the performance improvement techniques and optimization of the RTL design using Verilog synthesizable constructs.

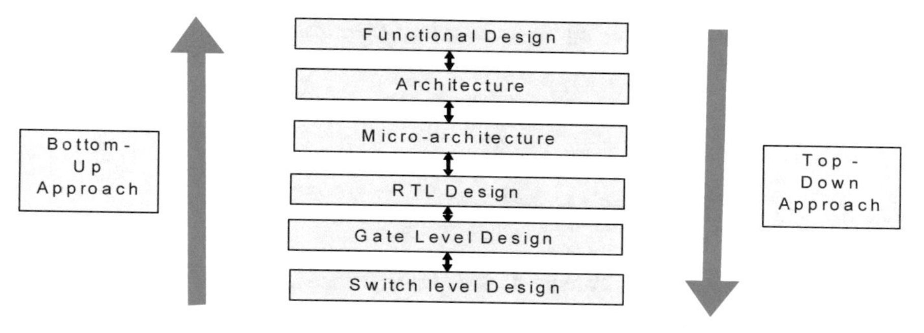

Fig. 1.2 Design abstraction

1.2 System and Logic Design Abstractions

As shown in Fig. 1.2, most of the designs have various abstraction levels. The design approach can be top-down or bottom-up. The implementation team takes decision about the right approach depending on the design complexity and availability of design resources. Most of the complex chips are also designed by using the top-down approach instead of bottom-up approach.

1.2.1 Architecture Design

The design is described as functional model initially, and the architecture and micro-architecture of the design is described by understanding the functional design specifications. Architecture design involves the estimation and role of the functional blocks, memory, processor logic, and throughput with associative glue logic with reference to the functional design requirements. Architecture design is in the form of functional blocks and represents the functionality of design in the block diagram form. In simple words, we can consider the architecture design as block-level representation which has evolved from the functional design specifications.

1.2.2 Micro-architecture Design

The micro-architecture is detail representation of every functional block which is specified in the architecture document. It describes the block- and sub-block-level details like interfaces and pin connections and hierarchical design details. The information about synchronous or asynchronous designs and clock and reset trees should be described in the micro-architecture document. Even the detail timing information and the overall data flow can be included in the micro-architecture document.

1.2.3 RTL Design and Synthesis

RTL stands for Register Transfer Level. RTL design uses micro-architecture as reference design document, and the main strategy is to code the design using synthesizable Verilog constructs to meet the required design functionality. The efficient design and coding guidelines at this stage play important role, and efficient RTL reduces the overall time requirement during the implementation phase. The RTL design is used as one of the inputs by synthesis tool to get the gate-level netlist. Gate-level netlist is representation of the functional design in the form of combinational and sequential logic cells.

1.2.4 Switch Level Design

Finally, the switch-level design is the abstraction used at the layout to represent the design in the form of standard cells and macros of ASIC. Or for the FPGA-based design, the switch-level design can be representation of the design implementation on FPGA fabric using the dedicated FPGA architecture resources.

1.3 Integrated Circuit Design and Methodologies

With the evolution of VLSI design algorithms and with the shrinking process node, the designs are becoming more complex and SOC-based designs are feasible in shorter time span. The demand of the customers to deliver product in the shorter span of time is possible by using efficient design flow which is used during frontend and backend design. The design needs to be evolved from specification stage to final layout. The use of EDA tools with the suitable features has made it possible to have the bug-free designs with proven functionality. The design flow is shown in Fig. 1.3, and it consists of the three major steps to get the netlist.

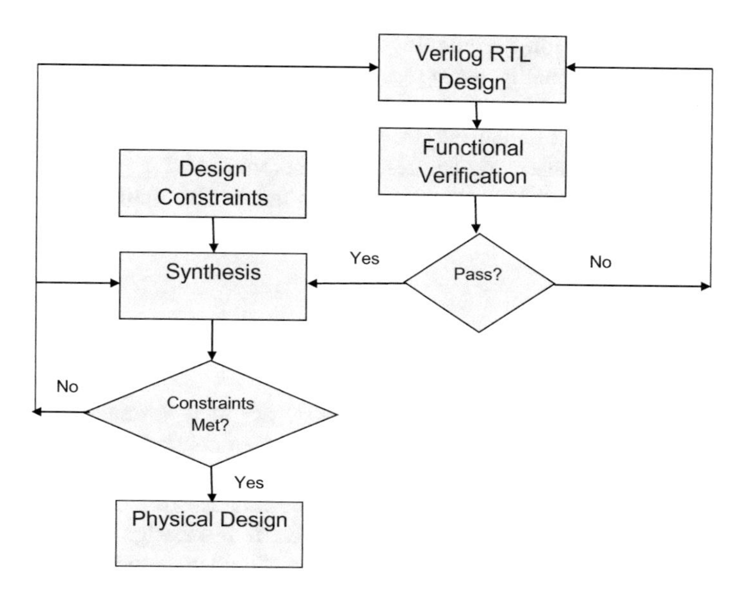

Fig. 1.3 Simulation and synthesis flow

1.3.1 RTL Design

As discussed in previous section, the functional design is described in the form of document using the architecture and micro-architecture. The RTL design using Verilog uses the micro-architecture document as a reference to code the design. The engineers need to spend more time to complete the RTL design at block and top levels. RTL designer uses the recommended design and coding guidelines while implementing the RTL design. An efficient RTL design always plays important role during implementation cycle. During this, designer describes the block-level and top-level functionality using the synthesizable Verilog constructs.

1.3.2 Functional Verification

After completion of an efficient RTL design using synthesizable Verilog constructs, the design functionality needs to be verified by using simulator. The intentional is to check for the functional correctness of design. Functional simulation is without considering any delays, and during this stage, the main intent is to verify the functionality of design. But common practice in the industry is to verify the functionality by using the testbench. In simple words, the testbench uses a driver which is used to drive the signals to the design under test and to monitor the output from the design under test. In the present scenario, automation in the verification flow and new verification methodologies has evolved and is used to verify the complex design functionality during the shorter span of time using the dedicated resources. The role of verification team is to test the functional mismatches between the expected output and actual output. If functional mismatches are found during simulation, then it need to be resolved before moving to the synthesis step. Functional verification is iterative process unless and until design meets the required functionality and various coverage goals.

1.3.3 Synthesis

When the functional requirements of the design are met, the next step is design synthesis. Synthesis tool uses the RTL design, design constraints, and libraries as inputs, and the goal is to get the gate-level netlist as an output. Synthesis is iterative process until the design and optimization constraints are met. The primary optimization constraints are area, speed, and power. If the design and optimization constraints are not met, then the synthesis tool should be used to perform optimization of the design after RTL or architecture tweaks. Again, after the optimization, if the constraints are not met, then it becomes compulsory to tweak the RTL or micro-architecture. The synthesis tool is also used to generate the area, speed, and power reports with the gate-level netlist.

1.3.4 Physical Design

The physical design stage involves the floor planning of design, power planning, clock tree synthesis, place and route, post-layout verification, static timing analysis, and final outcome is GDSII for any kind of ASIC design. This step is out of scope for the subsequent discussions!

1.4 Verilog as Hardware Description Language

Verilog is standardized as IEEE 1364 and is used to describe digital electronic circuits. Verilog is used mainly as hardware description language and is even popular during the verification. Verilog was created by Prabhu Goel, Phil Moorby, Chi-Lai Huang, and Douglas Warmke during 1983–1984 at Gateway design automations. The language supports synthesizable and non-synthesizable constructs and is useful in the design and verification of digital designs. Verilog IEEE standards are Verilog-95 (IEEE 1364-1995), Verilog-2001 (IEEE 1364-2001), and Verilog-2005 (IEEE 1364-2005).

Few of **the** important points to understand about the Verilog are listed below

1. *Verilog is case sensitive language, and we will recommend to use the lowercase letters while coding the RTL and testbenches.*
2. *For single line comment we can use //*
   a. *For example: // The Verilog RTL for processor*
3. *For block comment we can use /* ...... */*
   a. *For example : /* The block comments*
      assignment 1;
      *assignment 2; */*
4. *Verilog supports the declaration of input and output ports and keywords are* ***input, output,*** *respectively.*
5. *Verilog supports the declaration of bidirectional port and keyword to declare is* ***inout****.*
6. *Verilog has main net data types as* ***wire*** *and* ***reg****.*
7. *Verilog includes the concurrent constructs and sequential constructs.*
8. *Verilog includes the synthesizable and non-synthesizable constructs which are useful during design and verification, respectively.*
9. *Verilog Supports the various arithmetic, logical, bit wise, shift, equality and conditional operators which are useful to model the designs.*
10. *Verilog supports the various time constructs, blocking, non-blocking assignments, edge sensitive constructs and other various delay constructs (intra and Inter delay).*
11. *Verilog includes the function, task, loops, and other compiler directives.*

The stable release of Verilog standard is IEEE 1364-2005, and before we proceed further to discuss about the Verilog constructs, it is essential to have the basic understanding of the Verilog code structure. Throughout this book, we will use the coding style Verilog-2005 during RTL design and verification. Throughout this book, all the keywords in the Verilog RTL Design and testbenches are documented using the bold text have blue color.

As shown in the Verilog code structure template (Fig. 1.4).

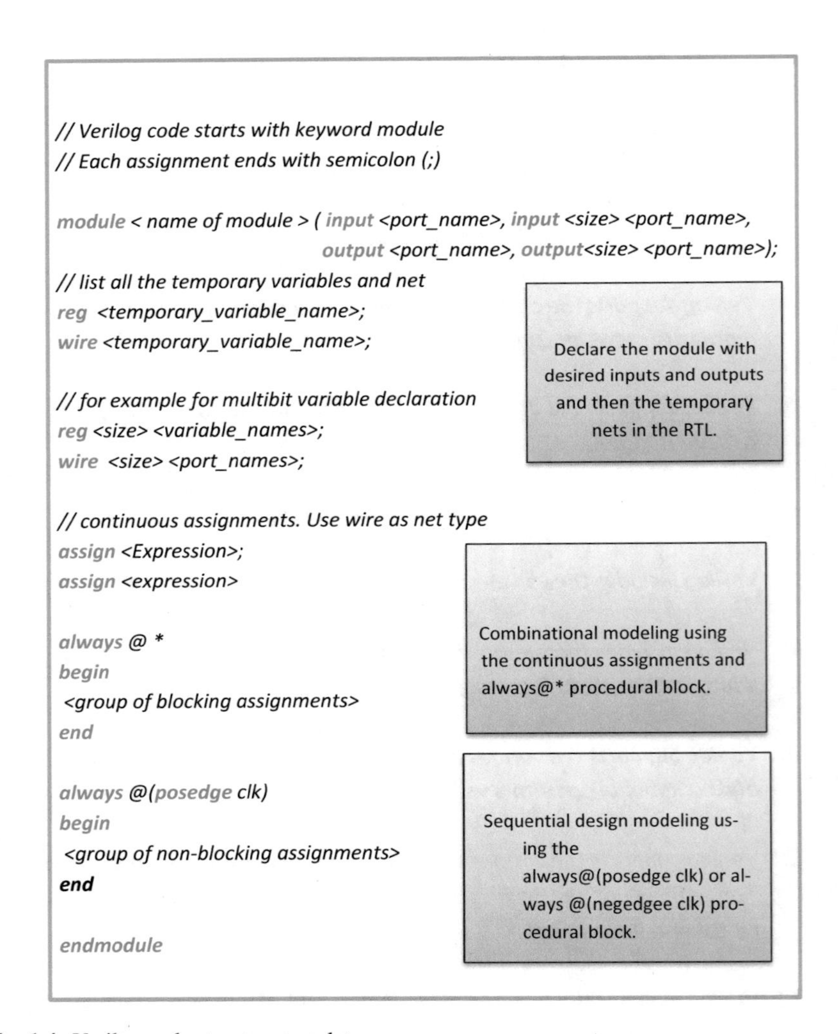

Fig. 1.4 Verilog code structure template

// indicates the comment line

< module_name > is the name of **module** *. It is recommended to have some meaningful name while declaring*

< port_name > is the name of **input** *or* **output** *or* **inout** *port.*

<size> is the width of the input port, output port or net

wire *and* **reg** *are net types,* **wire** *doesn't hold any data and used in continuous assignment.* **reg** *is used to hold data and used for the procedural assignments.*

<net_name> is the name of net declared

always *and* **assign** *are keyword and used to describe the design functionality.*

assign *statements are continuous assignments and executes concurrently .*

always *block is procedural block and all the statements inside* **always** *block are executed sequentially if they are within the* **begin ...end***. Multiple* **always** *procedural blocks executes concurrently.*

endmodule *is key word and indicates the end of the design module!*

*Every Verilog code starts with the '***module***' keyword and ends with '***endmodule***'. Module consists of the port declaration, net declaration and the functionality of design.*

1.5 Verilog Design Description

In the practical scenarios, the Verilog is categorized into three different kinds of coding descriptions. The different styles of coding descriptions are structural, behavioral, and RTL design. Consider the design structure of half adder which is shown in Fig. 1.5, and let us get familiar with the different coding styles. Figure 1.5 is useful to understand about the truth table, schematic, and logic diagram of half adder.

1.5.1 Structural Design

As name indicates, the structural design is used to describe the overall structure of the design. The main intention is to describe the logic in the form of gate or block-level design using the net connections. Structural design is mainly the instantiation of different small complexity digital functional blocks. It is basically design connection of small modules to realize moderate complex logic. Example 1 describes the structural code style to infer the 'half_adder.'

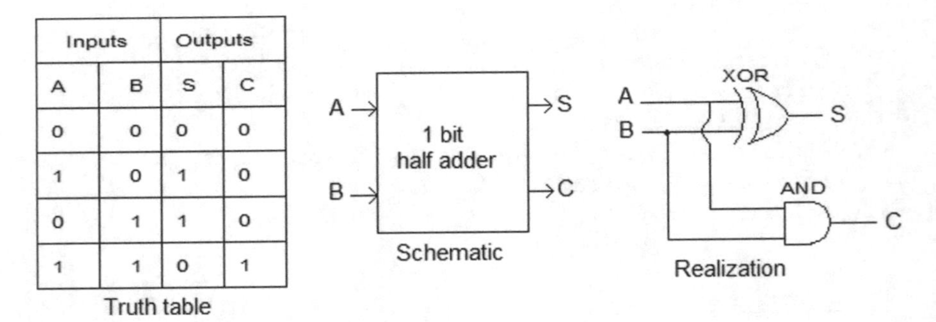

Inputs		Outputs	
A	B	S	C
0	0	0	0
1	0	1	0
0	1	1	0
1	1	0	1

Fig. 1.5 Logic structure of half adder

```
//////////////////////////////////////////////////////////////
module half_adder( input wire A, B, output wire S,C);

// design functionality

xor_gate U1 ( .A, .B, .S);
and_gate  U2 ( .A, .B, .C);

endmodule

module xor_gate( input wire A, B, output wire S);

// design functionality

assign S= A ^ B;

endmodule

module and_gate( input wire A, B, output wire C);

// design functionality

assign C= A & B;

endmodule

//////////////////////////////////////////////////////////////
```

Example 1 Structural Style to infer half-adder

As described in Example 1, the description of half_adder uses the instance of other two modules, and those are xor_gate and and_gate. The schematic is shown in Fig. 1.6.

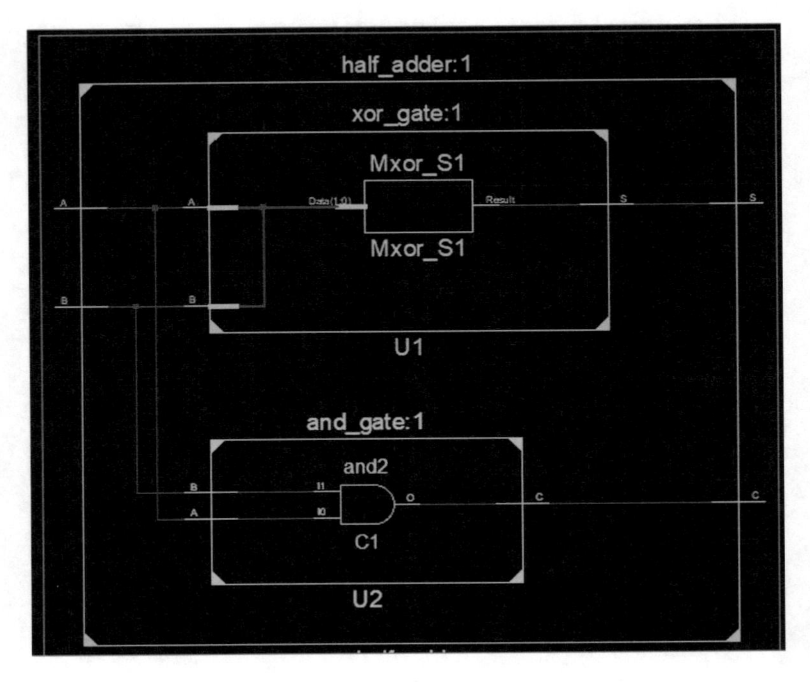

Fig. 1.6 Top-level schematic of Example 1

1.5.2 Behavior Design

Name itself indicates the nature of coding style is to describe the behavior to perform the addition. In the behavior style of Verilog code, the functionality is coded using the relationship between the inputs and the outputs to get the intended design functionality. It is assumed that the design is black box with the inputs and outputs. The main intention of designer is to map the functionality at output according to the required set of inputs. The description to implement the half adder is shown in Example 2.

```
/////////////////////////////////////////////////////////////////
module half_adder( input wire A, B, output reg S,C);

// design functionality to generate sum (S) output

always@*
begin

if ( A==B)
 S= 0;
else
S=1;
end

// design functionality to generate carry(C) output


always@*
begin

if ( A==1 && B==1)
 C= 1;
else
C=0;
end
endmodule
/////////////////////////////////////////////////////////////////
```

Example 2 Behavior style to implement half-adder

As coded in Example 2 to implement half adder, the description of half_adder uses the ***if-else*** constructs within the multiple ***always*** procedural blocks. It infers the comparator AND gate due to concurrent execution of the multiple procedural blocks. The schematic is shown in Fig. 1.6 (Fig. 1.7).

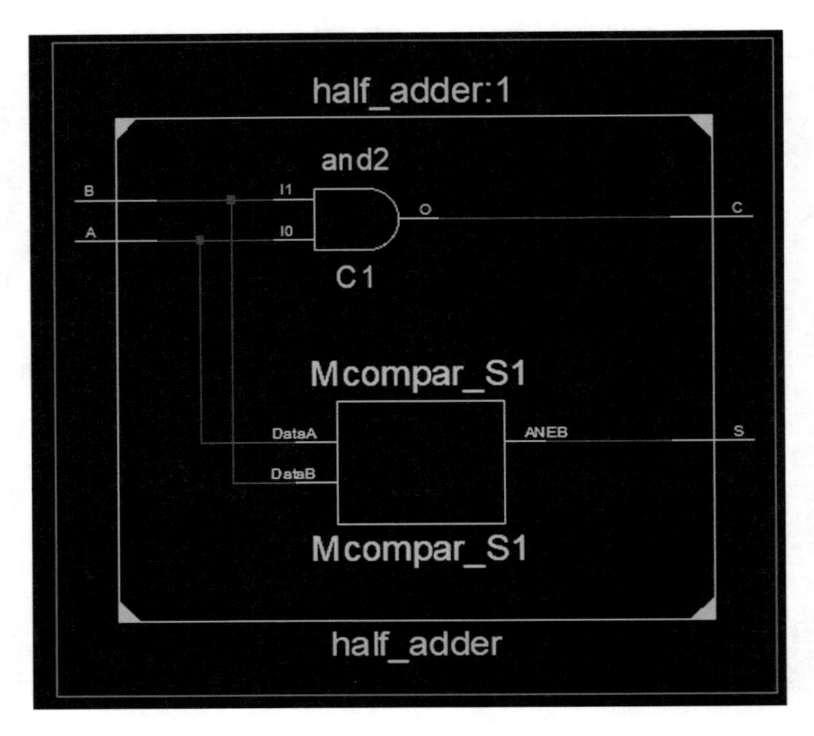

Fig. 1.7 Top-level schematic of Example 2

1.5.3 Synthesizable Design

In the practical environment to describe the functionality of design using Verilog, we always use the synthesizable constructs. The RTL code style is higher-level description of functionality using synthesizable Verilog constructs. Many times, the RTL coding style may resemble as the structural or behavioral model. The main intent of the designer is to infer the intended logic by using synthesizable Verilog constructs. Any Verilog code which infers the gate-level structure during synthesis we can treat as RTL design. Consider Example 3 (Fig. 1.8).

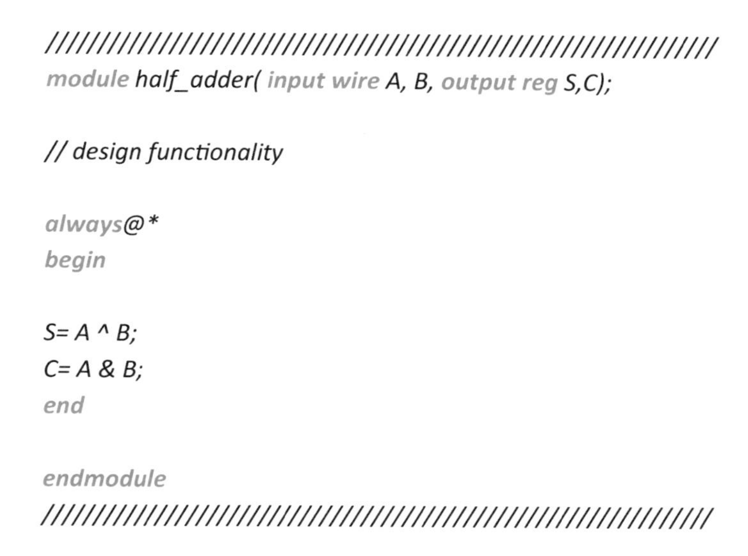

```
////////////////////////////////////////////////////////////////////
module half_adder( input wire A, B, output reg S,C);

// design functionality

always@*
begin

S= A ^ B;
C= A & B;
end

endmodule
////////////////////////////////////////////////////////////////////
```

Example 3 RTL Design to infer the half-adder

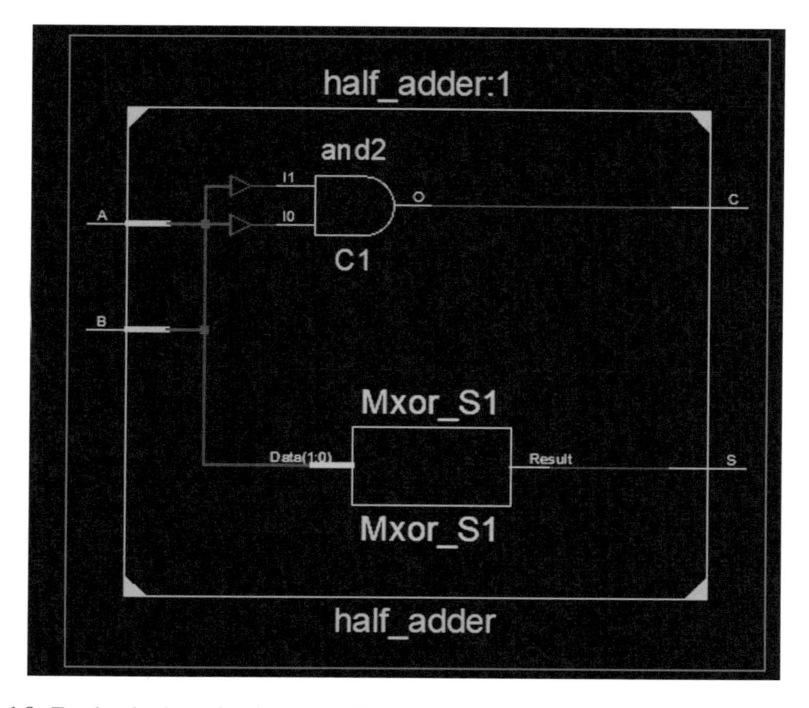

Fig. 1.8 Top-level schematic of Example 3

1.6 Few Important Verilog Terminologies

Before the subsequent discussion on Verilog constructs, it is essential to understand how Verilog works? Why it is popular hardware description language?

- Verilog is different from the software languages as it is used to describe the hardware. Verilog supports time constructs and delays.
- Verilog supports concurrent (parallel) execution of statements and even sequential execution of statements.
- Verilog supports blocking (=) assignments and non-blocking assignments (<=). Blocking assignments are used to describe combinational logic, and non-blocking assignments are used to describe sequential logic. These assignments will be discussed in subsequent chapters.
- Verilog supports the declaration of input, output, and bidirectional (inout) ports.
- Verilog supports definition of constants and parameters. Verilog supports file handling.
- Verilog supports four value logic logical '0', logical '1', high impedance 'z', and unknown 'X'.
- Verilog has procedural blocks '***always***' and '***initial.***' Procedural block with keyword '***always***' indicates free running process and executes always, and procedural block with '**initial**' keyword indicates the execution of block only once. Both procedural blocks execute at simulator time '0'. These blocks will be discussed in the subsequent chapters.
- Verilog supports synthesizable constructs as well as non-synthesizable constructs.
- Synthesizable constructs are used during the RTL design.
- Non-synthesizable constructs are used during the RTL verification.
- Verilog supports use of tasks and functions for recursive use.
- Verilog supports Program Language Interface (PLI) to transfer control form Verilog to functions written in 'C' language.

The template shown (Example 4) describes few of the important Verilog constructs which are useful to describe most of the combinational logic design

1. Continues assignment using 'assign'

```
// consider 'y' is dedared as output port and 'a' , 'b' as input port

assign y = a;          // assigns input 'a' to output 'y'

assign y =a ^ b;       // assigns XOR of 'a', 'b' to output 'y'

assign y =~a;          // assigns NOT of 'a' to output 'y'
```

2. Use of 'always' procedural block

```
// Consider input ports are dedared as 'a', 'b', 's' and output port is dedared as
'y'

 always (a or b or s)

 if (s)

  y= a;

else

 y=b;
```

Procedural block 'always' executes when there is an event of either 'a', 'b' or's'. If s is equal to logical '1' then 'a' is assigned to output 'y'.

If 's' is equal to logical '0' then 'b' is assigned to output 'y'

If-else is used for assigning the output value depending on the true or false condition.

3. Declaration using 'wire'

```
// 'wire' and 'reg' are used to dedare the nets and reg variable respectively.

wire y = a ^ b;
```

4. Declaration using 'reg'

```
//Consider 'y' is dedared as an output port and used in 'always' block

reg y;      // used as variable and to be declared in 'always' block

always @ ( a or b or s)

y = (s) ? a : b;
```

Example 4 Basic Verilog definitions and descriptions

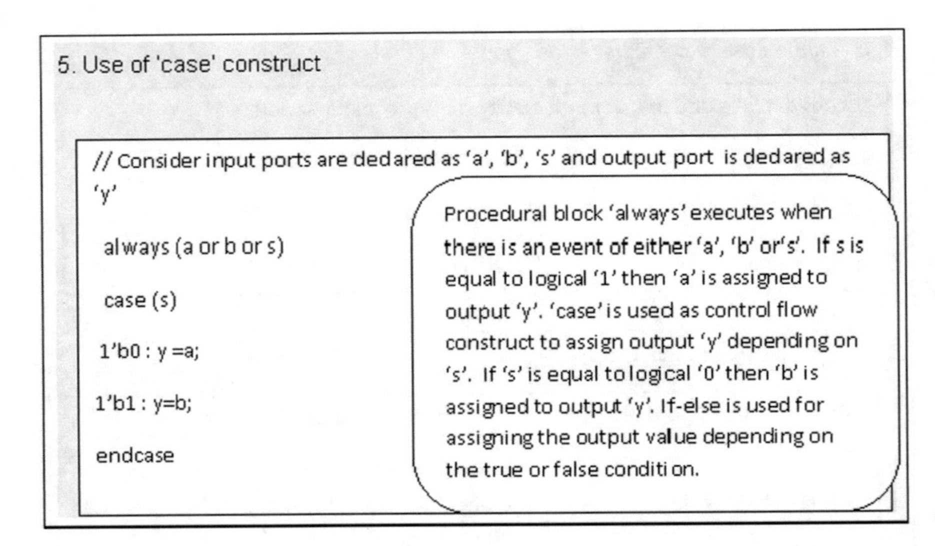
5. Use of 'case' construct
// Consider input ports are dedared as 'a', 'b', 's' and output port is dedared as 'y'
always (a or b or s)
case (s)
1'b0 : y =a;
1'b1 : y=b;
endcase
Procedural block 'always' executes when there is an event of either 'a', 'b' or's'. If s is equal to logical '1' then 'a' is assigned to output 'y'. 'case' is used as control flow construct to assign output 'y' depending on 's'. If 's' is equal to logical '0' then 'b' is assigned to output 'y'. If-else is used for assigning the output value depending on the true or false condition.

1.7 Exercises

For better understanding of the subsequent chapters, complete the following exercises on basics of digital design.

1. Using 2-input NAND as universal gate implement the 2-input XOR gate. Use minimum number of logic gates.
2. Implement the half subtractor using the minimum number of logic gates.
3. Implement the full adder using minimum number of half adders and additional minimum number of combinational gates.
4. Implement design of 4-bit binary to gray code converter.
5. Implement design of 4-bit gray to binary code converter.

1.8 Summary

As discussed earlier, Verilog is case-sensitive language and is used for design and verification of logic circuits. The following are few of the important points to summarize this chapter.

1. Verilog is efficient hardware description language and is used to describe the design functionality.
2. Although there are different description styles, practically designer uses the RTL coding style. Verilog supports concurrent and sequential design constructs.

3. Verilog is used as an efficient HDL and supports four value logic, logic '0', logic '1', high impedance 'z', and unknown 'x'.
4. Verilog uses concurrent and sequential constructs. Verilog HDL supports different operators to perform logical and arithmetic operations.
5. Verilog supports the synthesizable and non-synthesizable constructs.
6. Verilog supports the delay assignments and constructs to specify the positive and negative edge.
7. Verilog supports the port definition as input, output, and inout.
8. Verilog is used for both design and verification of digital logic.
9. Verilog supports the declaration of ***input***, ***output,*** and ***inout*** ports.
10. Verilog supports the time constructs and delays and other non-synthesizable constructs which are useful during verification.
11. Verilog supports edge-sensitive design modeling using ***always*** procedural block sensitive to posedge or negedge.

Chapter 2
Concept of Concurrency and Verilog Operators

For any language, the operator plays important role. The Verilog supports various operators, and the chapter discusses about the use of these operators in the RTL design.

For the better understanding of the use and application of Verilog, we need to focus on various operators supported by the language. As discussed in the previous chapter, the Verilog supports various operators and is useful during the design. What exactly we need is the arithmetic, logical, bitwise, shift, reduction, and equality operators to infer the intended logic. The logic may be combinational or sequential design; the powerful understanding of the operators can be useful during the design and verification of the digital circuit; the chapter discusses various operators used during the design. Even the chapter discusses the continuous assignments and always procedural block used to model the combinational design.

2.1 Use of Continuous Assignment to Model Design

The continuous assignment is used to model the combinational logic. In combinational design, an output is function of the present input. The ***assign*** keyword is used to model the combinational design with the logic expression on right-hand side. The continuous assignment is neither blocking nor non-blocking and executes when there is event on any one of the inputs or intermediate net. These assignments are updated in the active event queue. For more details, refer Chap. 7.

In the RTL, if multiple continuous assignments are there, then all the assignments are executed concurrently and mainly the ***assign*** construct is used to model the glue logic.

The logic inferred for Example 1 is shown in Fig. 2.1.

V. Taraate, *Digital Logic Design Using Verilog*,
https://doi.org/10.1007/978-981-16-3199-3_2

```
////////////////////////////////////////////////////////////////////////////////

module half_adder( input a_in, b_in, output sum_out, carry_out);

//concurrent execution of multiple assign constructs

assign sum_out = a_in ^ b_in;

assign carry_out = a_in & b_in;

endmodule

////////////////////////////////////////////////////////////////////////////////
```

Example 1 Half adder using continuous assignment

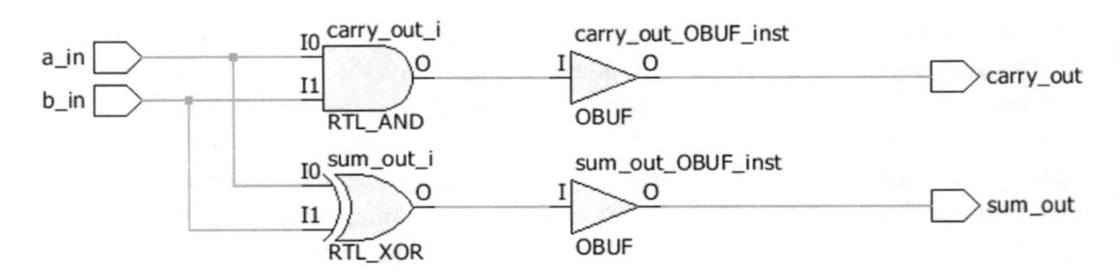

Fig. 2.1 RTL schematic of Example 1

2.2 Use of *always Procedural Block* to Implement Combinational Design

The real beauty of Verilog is the powerful synthesizable construct ***always***. The procedural block is used to model the combinational logic if specified as always @ (//sensitivity list).

For example, if we consider the ***always*** @ (a-in, b_in), then the procedural always block invokes when there is event on any one of the inputs a_in, b_in. Event indicates the transition from 0 to 1 or 1 to 0.

Now let us consider the design of half subtractor, where inputs are a_in, b_in and outputs are diff_out, borrow_out (Table 2.1).

Table 2.1 Truth table of half subtractor

a_in	b_in	diff_out	borrow_out
0	0	0	0
0	1	1	1
1	0	1	0
1	1	0	0

So, the RTL design should have the diff_out = a_in XOR b_in, borrow_out = NOT(a_in) AND b_in functionality. The RTL is coded using the multiple always procedural block. The first always procedural block is used to code the functionality of difference output and another procedural block for the functionality of borrow output.

```
////////////////////////////////////////////////////////////////////////
    module combo_design(input a_in, b_in, output reg diff_out,
    borrow_out);

// Functionality of half subtractor diff_out is XOR of a_in , b_in

always @ ( a_in, b_in)

if ( a_in==b_in)

    diff_out = 0;

else

    diff_out =1;

// Functional description of the logic for borrow_out that is ~a_in &
    b_in

always @(a_in , b_in)

if ( a_in ==0 && b_in==1)

    borrow_out = 1;

else

    borrow_out = 0;

endmodule

////////////////////////////////////////////////////////////////////////////
```

Example 2 Half subtractor synthesizable design

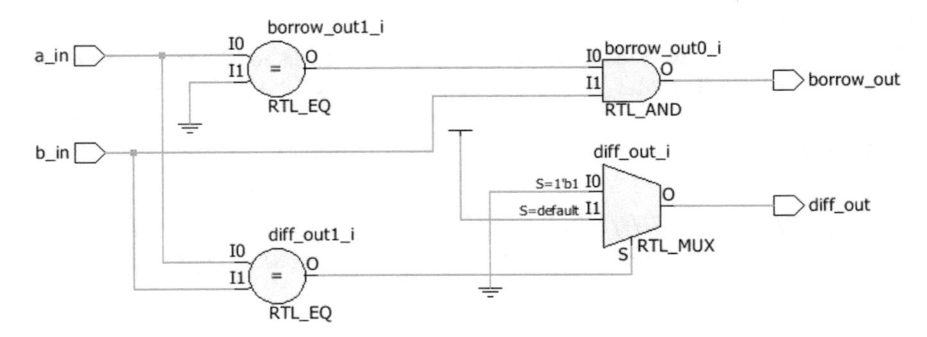

Fig. 2.2 RTL schematic of half subtractor

The multiple ***always*** procedural blocks execute concurrently and infer the logic shown in Fig. 2.2.

2.3 Concept of Concurrency

The powerful feature of the Verilog is the concurrent execution, and the multiple ***always*** procedural blocks and continuous assignments execute concurrently. Consider Example 3; as described, the always procedural block and assign execute concurrently and infer the parallel combinational logic.

```
//////////////////////////////////////////////////////////////////////////

module combo_design(input a_in, b_in, output reg diff_out, output
borrow_out);

// Functionality of half subtractor diff_out is XOR of a_in , b_in

always @ ( a_in, b_in)

 diff_out = a_in ^ b_in;

 // Functional description of the logic for borrow_out that is ~a_in &
b_in

assign borrow_out = (~a_in) & b_in;

endmodule

//////////////////////////////////////////////////////////////////////////
```

Example 3 RTL to understand concept of concurrency

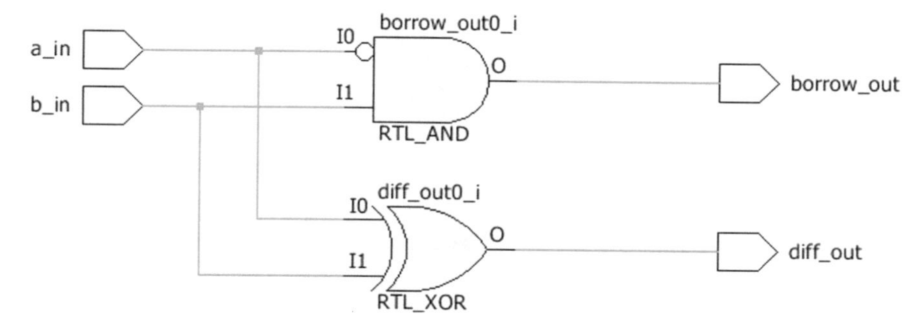

Fig. 2.3 Logic inferred for Example 3

The logic inferred is XOR gate due to the assignment *diff_out = a_in ^ b_in; within* ***always*** procedural block and the AND gate with one of the inputs controlled by NOT of a_in due to assignment ***assign*** *borrow_out = (~a_in) & b_in;* (Fig. 2.3).

2.4 Verilog Arithmetic Operators

Verilog supports addition, subtraction, multiplication, and division and modulus operators to perform arithmetic operations. Table 2.2 describes the arithmetic operators.

The logic inferred is shown in Fig. 2.4, and as shown, it consists of the logic to perform arithmetic operations.

Table 2.2 Verilog arithmetic operator

Operator	Name	Functionality
+	Binary addition	To perform addition of two binary operands
−	Binary minus	To perform subtraction of two binary operands
*	Multiplication	To perform multiplication of two binary operands
/	Division	To perform division of two binary operands
%	Modulus	To find modulus from division of two operands

```
//////////////////////////////////////////////////////////////////////////////

module arithmetic_operators (
                          input [3:0] a_in, b_in,
                          output reg [4:0] y1_out,
                          output reg [7:0] y3_out,
                          output reg [3:0] y2_out, y4_out, y5_out
                          );

always@ (a_in, b_in)
begin
  y1_out = a_in + b_in;
  y2_out = a_in -b_in;
  y3_out = a_in * b_in;
  y4_out = a_in / b_in;
  y5_out = a_in % b_in;
 end
endmodule

//////////////////////////////////////////////////////////////////////////////
```

Example 4 Use of Verilog operators in RTL design

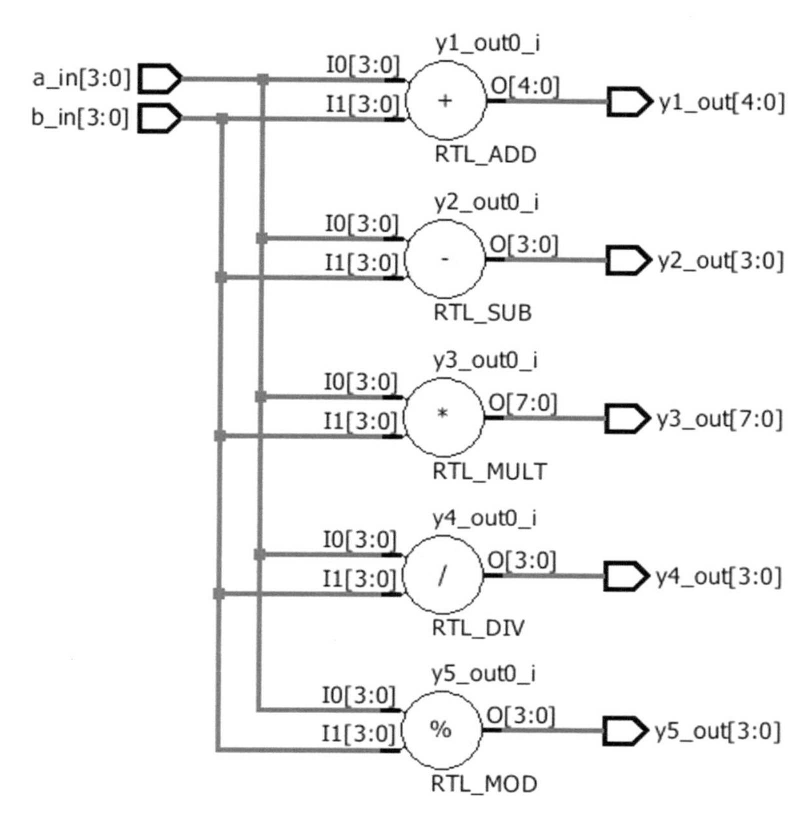

Fig. 2.4 RTL schematic of Example 4

2.5 Verilog Logical Operators

Verilog supports logical AND, OR, and negation operators to perform desired logical operation. Logical operators are used to return single-bit value at the end of the operation. Table 2.3 describes the functional use of logical operators.

The logic inferred is shown in Fig. 2.5, and as shown, it consists of the logic to perform logical operations.

Table 2.3 Verilog logical operators

Operator	Name	Functionality
&&	Logical AND	To perform logical AND on two binary operands
||	Logical OR	To perform logical OR on two binary operands
!	Logical negation	To perform logical negation for the given binary number

```
//////////////////////////////////////////////////////////////////////////
module logical_operators( input [2:0] a_in, b_in,c_in,d_in,e_in,f_in,
                          output reg y_out );
always@ (a_in, b_in, c_in,d_in,e_in,f_in)
begin
  if ( (a_in < b_in) && ((c_in ==d_in) || (e_in  > f_in)))
    y_out = 1;
  else
    y_out =0;
 end
endmodule
//////////////////////////////////////////////////////////////////////////
```

Example 5 Use of Verilog logical operators

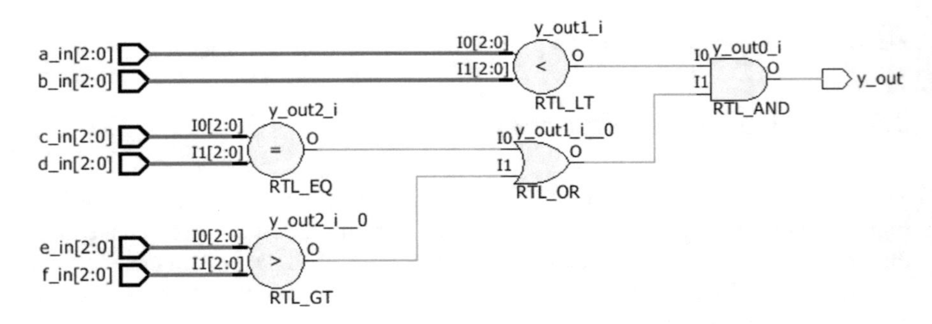

Fig. 2.5 RTL schematic of Example 5

2.6 Verilog Equality and Inequality Operators

Verilog equality operators are used to return true or false value after comparing two operands. Table 2.4 describes the functionality of the operators.

The logic inferred is shown in Fig. 2.6, and as shown, it consists of the logic to perform equality and inequality operations.

Table 2.4 Verilog equality and inequality operators

Operator	Name	Functionality
==	Case equality	To compare the two operands
!=	Case inequality	Used to find out inequality for the two operands
!	Logical negation	To perform logical negation for the given binary number

```
////////////////////////////////////////////////////////////////////////////////

module Equality_operator(input [7:0] a_in, b_in,

                        output reg  y1_out, y2_out,

                        output reg [7:0] y3_out );

always@ (a_in, b_in )

begin

// use of equality operator

  y1_out = (a_in == b_in);

  // use of inequality operator

  y2_out = (a_in != b_in);

// use of operator in if condition

  if ( a_in ==b_in)

  y3_out =a_in;

     else

  y3_out = b_in;

   end

endmodule

////////////////////////////////////////////////////////////////////////////////
```

Example 6 Verilog equality and inequality operators and use in RTL

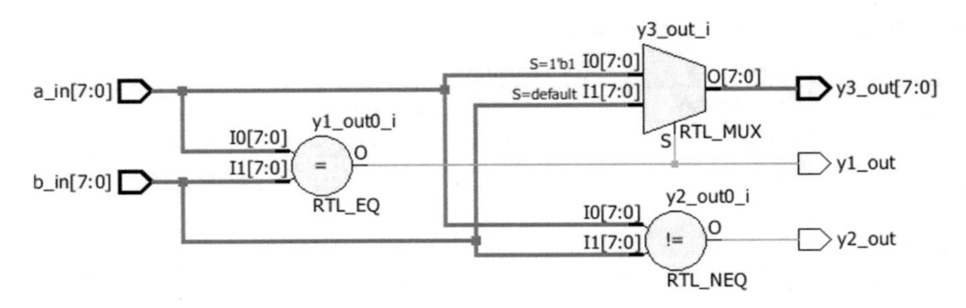

Fig. 2.6 RTL schematic of Example 6

2.7 Verilog Sign Operators

Verilog supports the operator positive '+' or '−' to assign sign to the operand. Table 2.5 describes the sign operands.

Table 2.5 Verilog sign operators

Operator	Name	Functionality
+	Unary sign plus	To assign positive sign to singular operand
−	Unary sign minus	To assign negative sign to singular operand

The logic inferred is shown in Fig. 2.7, and as shown, it consists of the logic to have the minus using the sign operator.

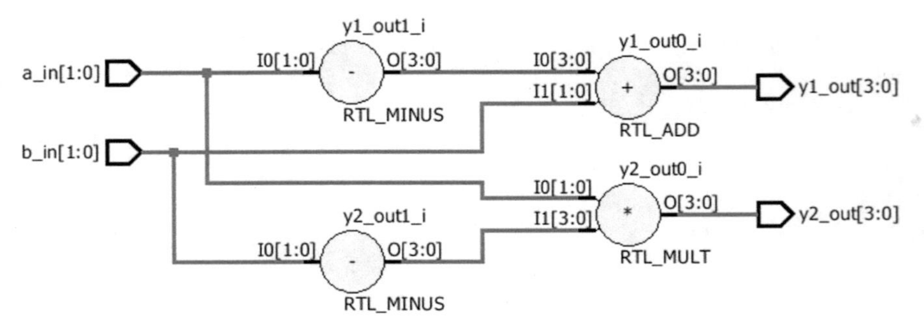

Fig. 2.7 RTL schematic of Example 7

```
/////////////////////////////////////////////////////////////////////////////////

module sign_operators (input [1:0] a_in, b_in,

                       output reg [3:0]  y1_out, y2_out

                       );

always@ (a_in, b_in )

begin

// use of sign operator

  y1_out = (-a_in) + b_in;

  // use of sign operator

  y2_out = a_in * (-b_in);

  end

endmodule

/////////////////////////////////////////////////////////////////////////////////
```

Example 7 Verilog sign operators and use in the RTL design

2.8 Verilog Bitwise Operators

Verilog supports the bitwise operations. Logical bitwise operators use two single- or multi-bit operands and return the multi-bit value. Verilog does not support NAND, NOR. Table 2.6 describes the functionality and use of bitwise operators.

Table 2.6 Verilog bitwise operators

Operator	Name	Functionality
&	Bitwise AND	To perform bitwise AND on two binary operands
|	Bitwise OR	To perform bitwise OR on two binary operands
^	Bitwise XOR	To perform bitwise XOR on two binary operands

The logic inferred is shown in Fig. 2.8, and as shown, it consists of the logic to perform bitwise operations such as NOT, AND, NAND, NOR, OR, XOR, and XNOR.

```
///////////////////////////////////////////////////////////////////////////////
module bit_wise_operators ( input [6:0] a_in,
                            input [5:0] b_in,
                            output reg [6:0] y_out );

always@ (a_in, b_in )
begin
// bit wise AND
 y_out[0] = a_in[0] & b_in[0];
 // bit wise NAND
 y_out[1] = !( a_in[1] & b_in[1]);
 // bit wise OR
  y_out[2] = a_in[2] | b_in[2];
  // bit wise NOR
  y_out[3] = !( a_in[3] | b_in[3]);
   // bit wise XOR
    y_out[4] = a_in[4] ^ b_in[4];
    // bit wise XNOR
   y_out[5] = ( a_in[5] ~^ b_in[5]);
      // bit wise NOT
   y_out[6] = ! a_in[6];
   end
endmodule
///////////////////////////////////////////////////////////////////////////////
```

Example 8 Verilog bitwise operators and use in the RTL design

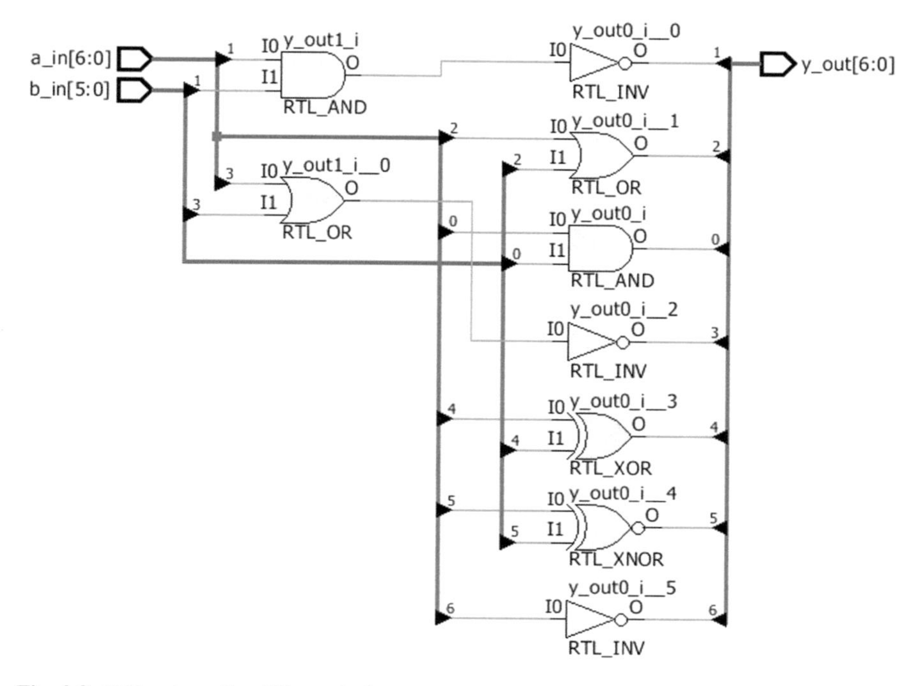

Fig. 2.8 RTL schematic of Example 8

2.9 Verilog Relational Operators

Verilog supports the relational operator to compare two binary numbers and returns true ('1') or false ('0') value after comparison of two operands. Table 2.7 describes the relational operators.

Table 2.7 Verilog relational operators

Operator	Name	Functionality
>	Greater than	To compare two numbers
>=	Greater than or equal to	To compare two numbers
<	Less than	To compare two numbers
<=	Less than or equal to	To compare two numbers

```
//////////////////////////////////////////////////////////////////////////////////

module Relational_operators (  input [7:0] a_in,
                          `      input [7:0] b_in,
                              output reg y1_out, y2_out, y3_out, y4_out);

always@ (a_in, b_in )

begin

// less than < operator

  y1_out = a_in < b_in;

    // less than equal to  <= operator

  y2_out = a_in <= b_in;

// greater than > operator

    y3_out = a_in > b_in;

     // greater than equal to >= operator

 if (a_in >= b_in)

    y4_out = 1;

   else

    y4_out =0;

  end

endmodule

//////////////////////////////////////////////////////////////////////////////////
```

Example 9 Verilog relational operators and use in the RTL design

The logic inferred is shown in Fig. 2.9, and as shown, it consists of the logic to perform less than, greater than, etc.

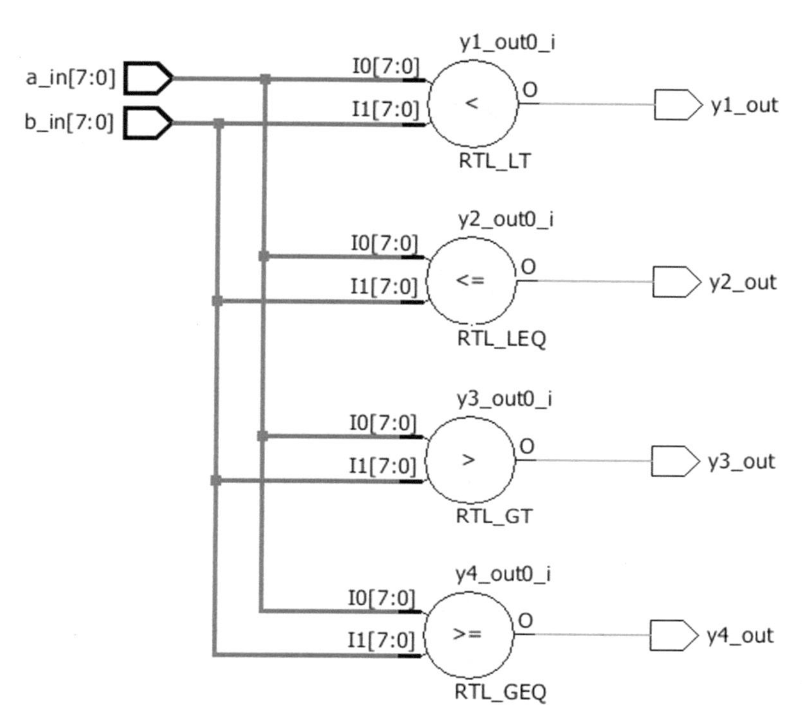

Fig. 2.9 RTL schematic of Example 9

2.10 Verilog Concatenation and Replication Operators

Verilog supports the concentration and replication for any binary string. Table 2.8 describes the functionality of concentration and replication operators.

Table 2.8 Verilog concentration and replication operators

Operator	Name	Functionality
{ }	Concatenation	To concatenate two binary strings
{m, { }}	Replication	To replicate the string m times

```
////////////////////////////////////////////////////////////////////////////////

module concatenation_operator ( input [2:0] a_in,

                               input [2:0] b_in,

                               output reg [15:0] y_out );

parameter c_in = 3'b010;

always@ (a_in, b_in )

begin

// use of concatenation{ } and replication n{} operator

  y_out = { a_in, b_in , {3{c_in}}, 3'b111};

 end

endmodule

////////////////////////////////////////////////////////////////////////////////
```

Example 10 Verilog concatenation and replication operators and use in the RTL design

The logic inferred is shown in Fig. 2.10, and as shown, it consists of the logic to perform string concatenation and replication.

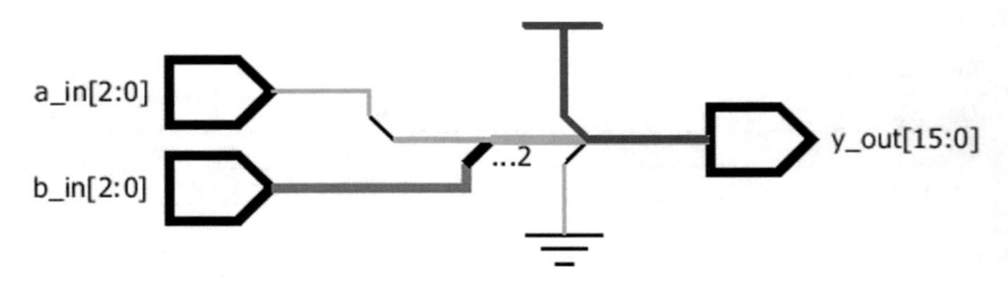

Fig. 2.10 RTL schematic of Example 10

```
////////////////////////////////////////////////////////////////////////////////
module reduction_operators (  input [3:0] a_in,
                              output reg [5:0] y_out);
always@ (a_in)
begin
// reduction AND
  y_out[0] = & a_in;
  // reduction NAND
  y_out[1] = ~& a_in;
// reduction OR
   y_out[2] = | a_in;
   // Reduction NOR
  y_out[3] = ~| a_in;
  // Reduction XOR
      y_out[4] = ^ a_in;
      // Reduction XNOR
     y_out[5] = ~^ a_in;
end
endmodule
////////////////////////////////////////////////////////////////////////////////
```

Example 11 Verilog reduction operators and use in the RTL design

2.11 Verilog Reduction Operators

Verilog supports the reduction operators and returns the single-bit value after bitwise reduction. Table 2.9 describes the reduction operators.

Table 2.9 Verilog reduction operators

Operator	Name	Functionality
&	Reduction AND	For performing the bitwise reduction
~&	Reduction NAND	For performing the bitwise reduction
|	Reduction OR	For performing the bitwise reduction
~|	Reduction NOR	For performing the bitwise reduction
^	Reduction XOR	For performing the bitwise reduction
~^ or ^~	Reduction XNOR	For performing the bitwise reduction

The logic inferred is shown in Fig. 2.11, and as shown, it consists of the logic to get the reduction output as single-bit result.

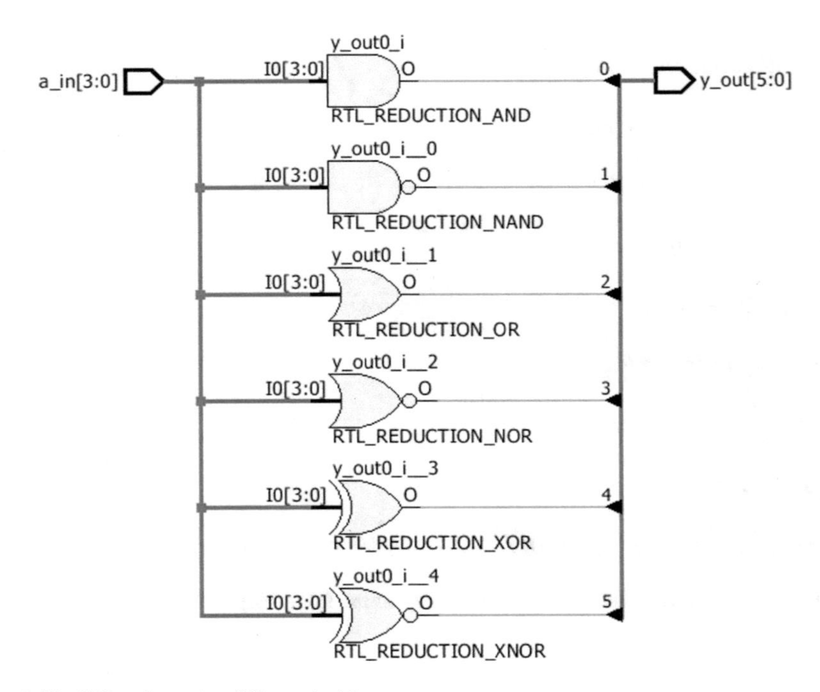

Fig. 2.11 RTL schematic of Example 11

2.12 Verilog Shift Operators

Verilog uses the shift operators and requires two operands. These operators are used to perform the shifting operation. Table 2.10 describes the functionality of shift operators.

Table 2.10 Verilog shift operators

Operator	Name	Functionality
<<	Shift left	To perform logical shift left
>>	Shift right	To perform logical shift right

```
/////////////////////////////////////////////////////////////////////////////

module shift_operators (          input [3:0] a_in,

                                  output reg [3:0] y1_out, y2_out);

parameter b_in = 2;

always@ (a_in)

begin

// use of left shift operator

  y1_out = a_in << b_in;

  // use of right shift operator

    y2_out = a_in >>  b_in;

end

endmodule

/////////////////////////////////////////////////////////////////////////////
```

Example 12 Verilog shift operators and use in the RTL design

The logic inferred is shown in Fig. 2.12, and as shown, it consists of the logic to perform left and right shift operations.

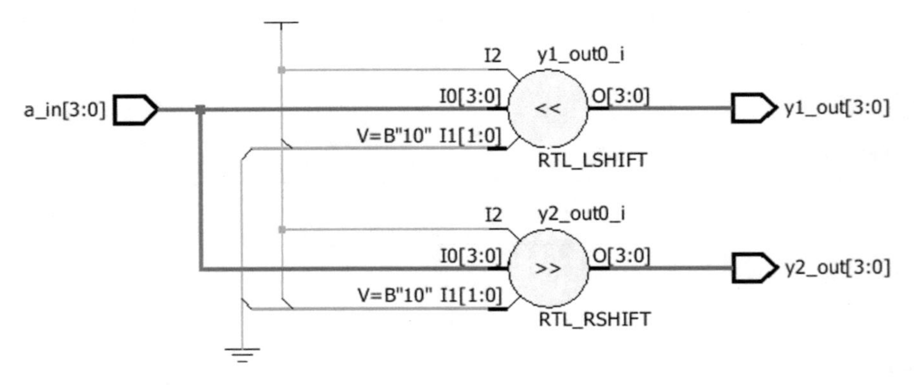

Fig. 2.12 RTL schematic of Example 12

2.13 Exercises

For better understanding of Verilog assign and always construct, complete the following exercises.

1. *The logic inferred by the following code is*

```
module comb_design( input a_in, b_in, output q_out);

always @ (a_in)

q_out = a_in ^^ b_in;

endmodule
```

   a. *XOR gate*
   b. *XNOR gate*
   c. *NOR gate*
   d. *Syntax error in the code*

2. The logic inferred by the following code is

```
module comb_design_1 ( input a_in, b_in, output reg q1, q2);

always @ (*)

begin

q1=  a_in | b_in;

q2= a_in & b_in;

end

endmodule
```

   a. XOR gate, AND gate
   b. OR gate, AND gate
   c. NOR gate, AND gate
   d. XNOR gate, AND gate

3. What is the logic inferred by the code?

```
module comb_design_2 ( input a_in, b_in, c_in, output y_out);

wire tmp;

always @ *

tmp = a_in & b_in;

assign y_out = tmp & c_in;
```

a. *Three input AND gate*
b. *Two input AND gate*
c. *Two AND gates connected in cascade*
d. *Syntax error in the RTL*

4. *The logic inferred by the following code is*

```
module comb_design_3 ( input a_in, b_in, output reg y_out);

always

 y_out = a_in || b_in;

endmodule
```

a. *XOR gate*
b. *OR gate*
c. *NOR gate*
d. *Syntax error*

5. *The logic inferred by the following code is*

```
module comb_design_4 ( input a_in, b_in, c_in, output y_out);

reg tmp;

always @ *

begin

 tmp = a_in ^ b_in;
```

```
y_out = tmp ^ c_in;

end

endmodule
```

a. *Addition(a,b,c)*
b. *Syntax error in the code*
c. *Two input XOR (a,c)*
d. *XNOR (a,b,c)*

2.14 Summary

The following are important points to conclude this chapter.

1. The continuous assignments are used to model the combinational logic.
2. The assign is a keyword used with the desired expression to code the combinational logic.
3. The continuous assignments are neither blocking nor non-blocking.
4. The always procedural block with sensitivity list is useful to code the combinational design.
5. The default net type of input and output is wire.
6. If assignments are used within the always procedural block, then the net type should be reg.
7. Verilog supports various operators and is useful to perform the arithmetic and logical operations.

Chapter 3
Verilog Constructs and Combinational Design-I

An efficient RTL design always uses minimum number of logic gates. This chapter discusses about the combinational Logic design using the synthesizable Verilog constructs. Also discusses about the practical and real-life scenarios, useful while implementing combinational designs.

Combinational logic is implemented by the logic gates, and in the combinational logic, output is function of present input. The goal of designer is to implement the logic using minimum number of logic gates or logic cells. Minimization techniques are K-map, Boolean algebra, Shannon's expansion theorems, and hyperplanes. The thought process of designer should be such that the design should have the better performance with lesser logic density. The area minimization techniques play important role in the design of combinational logic or Boolean functions. In the present scenario, designs are extraordinarily complex; the design functionality is described using the hardware description language Verilog. The subsequent section focuses on the use of Verilog RTL to describe the combinational design.

3.1 The Role of Constructs

As discussed in the Chap. 2, to model the combinational logic, we will use the ***assign*** and ***always*** @ (//sensitivity list) constructs. Within the always procedural block, we will use the ***if .... else*** construct, and most of the time it infers the multiplexer kind of logic.

The if-else is sequential construct, and the syntax is

V. Taraate, *Digital Logic Design Using Verilog*,
https://doi.org/10.1007/978-981-16-3199-3_3

```
if (condition)

//blocking assignment executed if condition is true

else

// blocking assignment is executed if condition specified in if () is false
```

Note: Use the ***begin….end*** for the multiple assignments within the *if* or ***else*** condition.

3.2 Logic Gates and Synthesizable RTL

This section discusses about the logic gates and the design using synthesizable Verilog constructs.

3.2.1 NOT or Invert Logic

NOT logic complements the input. NOT logic is also called as invert logic. Synthesizable design is shown in the Example 1. The truth table of NOT logic is shown in Table 3.1.

```
////////////////////////////////////////////////////////////////////////////////
module not_gate( input a_in, output reg y_out);

always@(a_in)

begin

  y_out = ~a_in;

end

endmodule

////////////////////////////////////////////////////////////////////////////////
```

Example 1 NOT logic synthesizable design

```
/////////////////////////////////////////////////////////////////////////////
module not_gate( input a_in, output y_out);

    assign y_out = ~a_in;

endmodule

/////////////////////////////////////////////////////////////////////////////
```

Example 2 NOT logic using continuous assignment

Table 3.1 Truth table of NOT gate

a_in	y_out
0	1
1	0

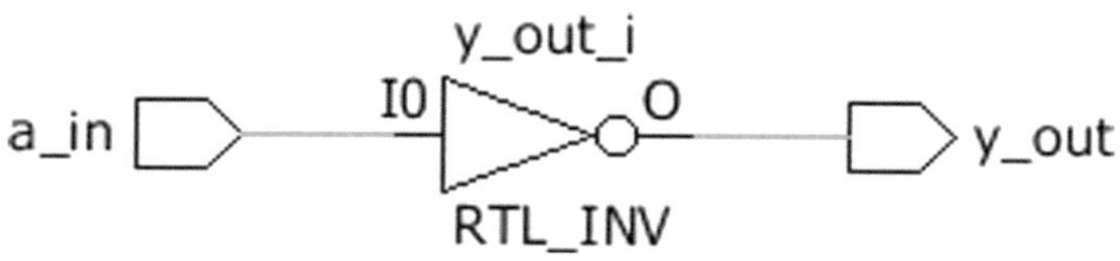

Fig. 3.1 RTL schematic of Example 2

The logic inferred is shown in Fig. 3.1, and input port of not logic gate is a_in and output as y_out.

3.2.2 *OR Logic*

OR logic generates output as logic 1 when one of the inputs is logic 1. Synthesizable design is shown in the Example 3. The truth table of OR logic is shown in Table 3.2.

Table 3.2 Truth table of two input OR gates

a_in	b_in	y_out
0	0	0
0	1	1
1	0	1
1	1	1

```
//////////////////////////////////////////////////////////////////////////////////

module or_gate(input a_in, b_in , output reg y_out);

always@(a_in, b_in)

begin

 if ( a_in==0 && b_in ==0)

   y_out = 0;

  else

   y_out = 1;

end

endmodule

/////////////////////////////////////////////////////////////////////
```

Example 3 OR logic synthesizable design

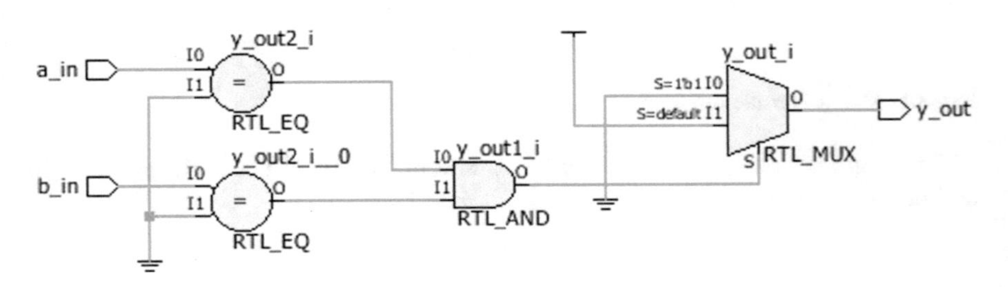

Fig. 3.2 RTL schematic of Example 3

RTL schematic of OR logic is shown in Fig. 3.2, input ports of OR logic gate are a_in, b_in, and output port is y_out. As shown in the schematic, the inferred logic uses 2:1 multiplexer and other operators to infer the logic. The synthesis result is EDA tool specific, and for ASIC and FPGA design, the synthesis result may differ!

For multiple input OR gate, you can use the bit-wise operator (|). RTL is coded as shown in the Example 4.

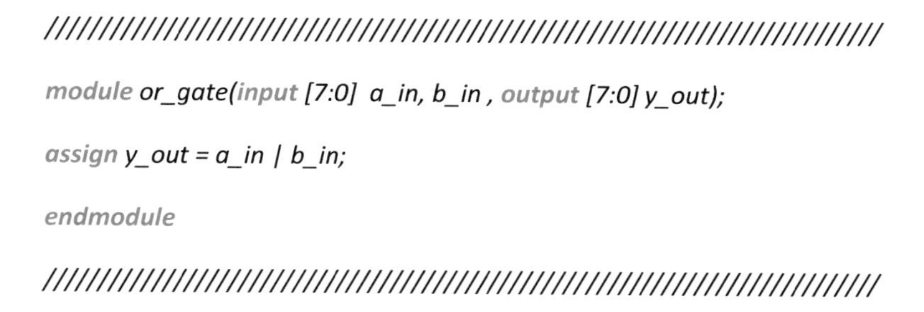

```
///////////////////////////////////////////////////////////////////////////////

module or_gate(input [7:0]  a_in, b_in , output [7:0] y_out);

assign y_out = a_in | b_in;

endmodule

///////////////////////////////////////////////////////////////////////////////
```

Example 4 Multi-input OR gate

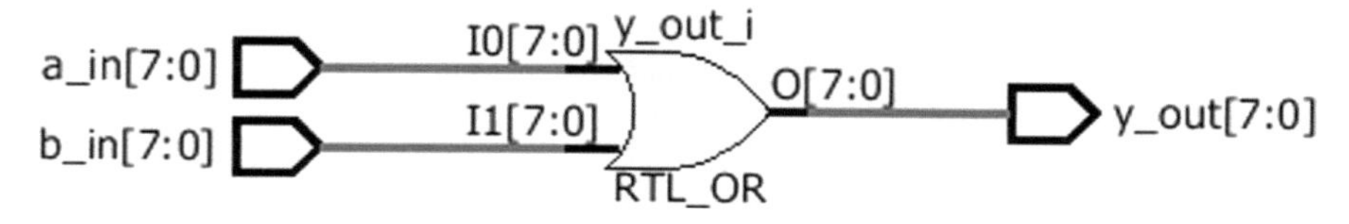

Fig. 3.3 Schematic of multi-input OR gate

The schematic of multiple input OR gate is shown in Fig. 3.3.

Note: While describing the design functionality, make sure that all the input ports are listed in the sensitivity list. Missing required signal from sensitivity list will create simulation and synthesis mismatch and will be discussed in Chap. 7.

3.2.3 NOR Logic

NOR logic is complement of the OR logic. Synthesizable design is shown in the Example 5. The truth table of NOR logic is shown in Table 3.3.

Table 3.3 Truth table of two input NOR gates

a_in	b_in	y_out
0	0	1
0	1	0
1	0	0
1	1	0

```
///////////////////////////////////////////////////////////////////////////////////

module nor_gate(input a_in, b_in , output reg y_out);

always@(a_in, b_in)

begin

 if ( a_in==0 && b_in ==0)

   y_out = 1;

  else

   y_out = 0;

end

endmodule

///////////////////////////////////////////////////////////////////////////////////
```

Example 5 NOR logic synthesizable design

The logic inferred is shown in Fig. 3.4, input ports of NOR logic gates are a_in, b_in, and output port is y_out. As shown, the inferred logic uses the equality operator and AND gate to implement the NOR gate (Fig. 3.5).

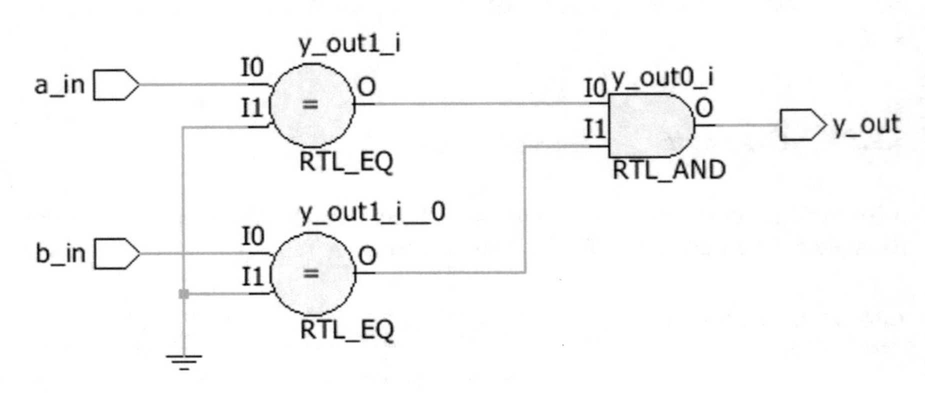

Fig. 3.4 RTL schematic of Example 5

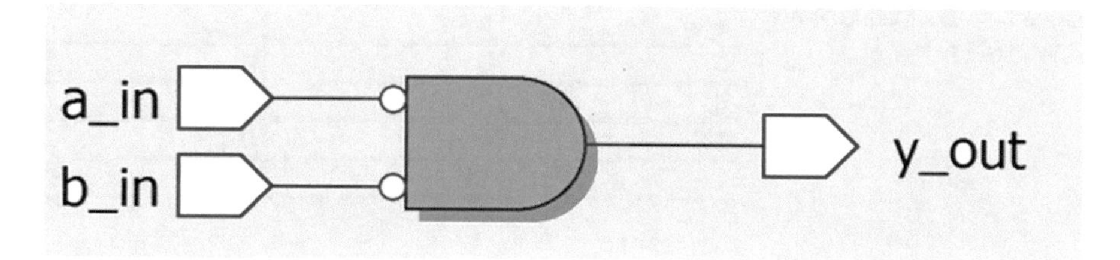

Fig. 3.5 Bubbled AND as NOR

Note: According to De Morgan's law, bubbled AND is equal to NOR.

For multiple input OR gate, you can use the bit-wise NOT (~) of bit-wise operator OR (|). RTL is coded as shown in the Example 6.

The schematic of multiple input NOR gate is shown in Fig. 3.6.

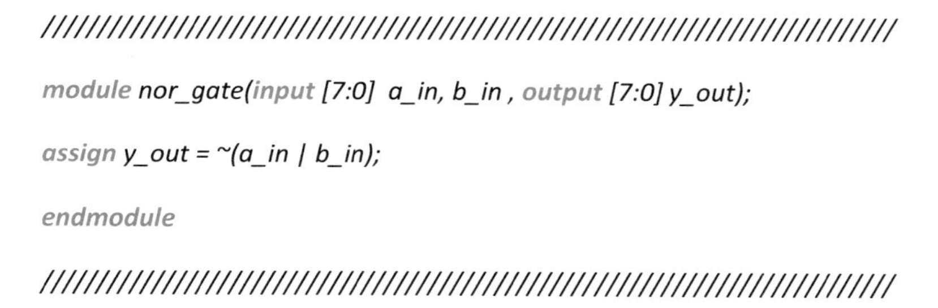

```
/////////////////////////////////////////////////////////////////////////////////////
module nor_gate(input [7:0]  a_in, b_in , output [7:0] y_out);
assign y_out = ~(a_in | b_in);
endmodule
/////////////////////////////////////////////////////////////////////////////////////
```

Example 6 Multi-input NOR

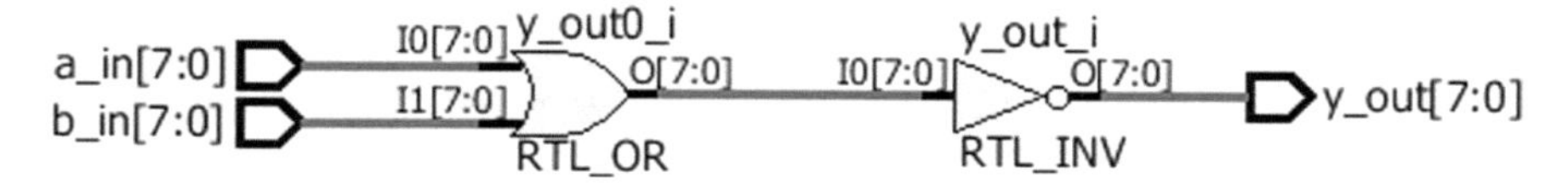

Fig. 3.6 Synthesis result of Example 6

Table 3.4 Truth table of two input AND gates

a_in	b_in	y_out
0	0	0
0	1	0
1	0	0
1	1	1

3.2.4 AND Logic

AND logic generates an output as logic 1 when both the inputs a_in, b_in are logic 1. Synthesizable design is shown in the Example 7. The truth table of AND logic is shown in Table 3.4.

```
////////////////////////////////////////////////////////////////////////////////

module and_gate(input a_in, b_in , output reg y_out);

always@(a_in, b_in)

begin

 if ( a_in==1 && b_in ==1)

    y_out = 1;

  else

    y_out = 0;

end

endmodule

////////////////////////////////////////////////////////////////////////////////
```

Example 7 AND logic synthesizable design

Note: And gate is visualized as series of two switches and used in programmable logic devices (PLD). Programmable AND plane can be created by using the AND gates having programmable inputs.

The logic inferred is shown in Fig. 3.7, and input port of AND logic gate is named as a_in, b_in and output as y_out.

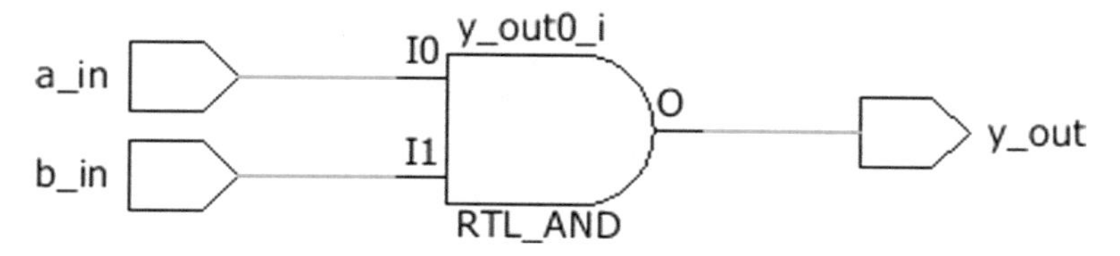

Fig. 3.7 RTL schematic of Example 7

The multiple input AND gate is coded using the assign construct and shown in the Example 8.

```
////////////////////////////////////////////////////////////////////////////////
module and_gate(input [7:0]  a_in, b_in , output [7:0] y_out);

assign y_out = (a_in & b_in);

endmodule

////////////////////////////////////////////////////////////////////////////////
```

Example 8 Multiple input AND gate

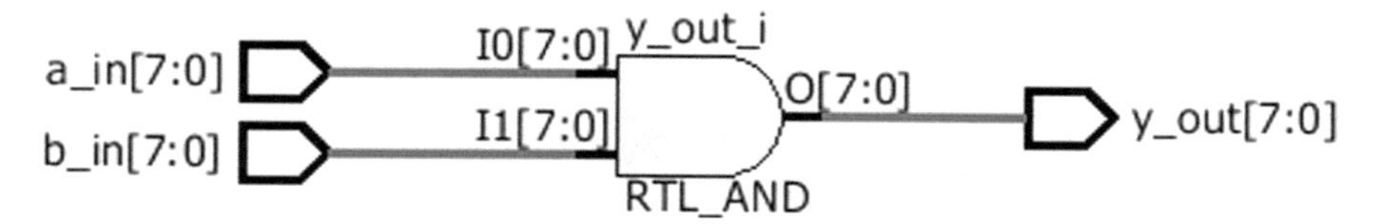

Fig. 3.8 Multi-input AND synthesis result

The logic inferred is shown in Fig. 3.8.

3.2.5 NAND Logic

NAND is complement of the AND logic. Synthesizable design is shown in the Example 9. The truth table of NAND logic is shown in Table 3.5.

Table 3.5 Truth table of two input NAND logic

a_in	b_in	y_out
0	0	1
0	1	1
1	0	1
1	1	0

```
//////////////////////////////////////////////////////////////////////////
module nand_gate(input a_in, b_in , output reg y_out);
always@(a_in, b_in)
begin
 if ( a_in==1 && b_in ==1)
   y_out = 0;
  else
   y_out = 1;
end
endmodule
//////////////////////////////////////////////////////////////////////////
```

Example 9 NAND logic synthesizable design

Note: NAND logic is a universal logic. By using NAND logic, all possible logic functions can be realized. NAND logic is used to implement the storage elements like latches or flip-flops and to realize combinational functions.

RTL schematic of NAND logic is shown in Fig. 3.9, and input ports of NAND logic gate are named as a_in, b_in and output as y_out. The logic inferred uses the multiplexer to generate NOT of AND.

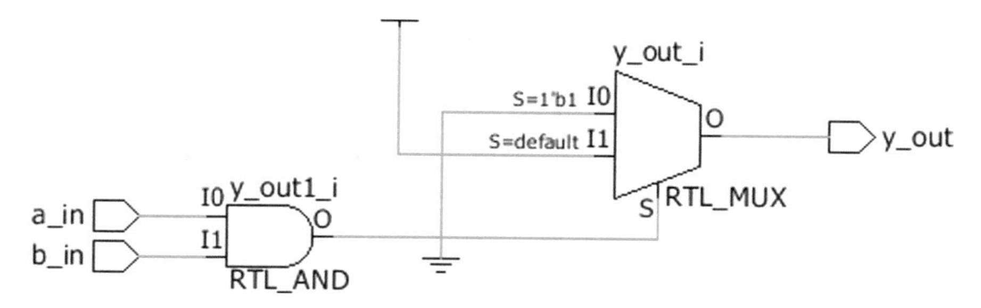

Fig. 3.9 RTL schematic of Example 9

The multiple input NAND gate is coded using the assign construct and by using the expression which has bit-wise NOT (~) and bit-wise AND (&) operator and is shown in the Example 10.

```
//////////////////////////////////////////////////////////////////////////////
module nand_gate(input [7:0]  a_in, b_in , output [7:0] y_out);
assign y_out = ~(a_in & b_in);
endmodule
//////////////////////////////////////////////////////////////////////////////
```

Example 10 Multiple input NAND gate

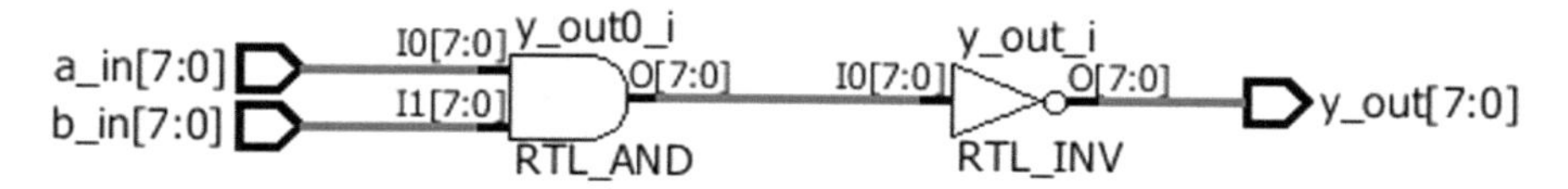

Fig. 3.10 Schematic of multiple input NAND

The logic inferred is shown in Fig. 3.10.

Note: According to De Morgan's law, bubbled OR is equal to NAND.

Table 3.6 Truth table of two input XOR gates

a_in	b_in	y_out
0	0	0
0	1	1
1	0	1
1	1	0

3.2.6 Two Input XOR Logic

Two input XOR is called as exclusive OR logic and generates output as logic 1 when both inputs are not equal. Synthesizable design is shown in Example 11. The truth table of XOR logic is shown in Table 3.6.

```
///////////////////////////////////////////////////////////////////////////////

module xor_gate(input a_in, b_in , output reg y_out);

always@(a_in, b_in)

begin

 if ( a_in != b_in )

   y_out = 1;

  else

   y_out = 0;

end

endmodule

///////////////////////////////////////////////////////////////////////////////
```

Example 11 XOR logic synthesizable design

Note: XOR gate can be implemented by using two input NAND gates. The number of two input NAND gates required to implement two input XOR gates is equal to 4. XOR gates are used to implement arithmetic operations like addition and subtraction.

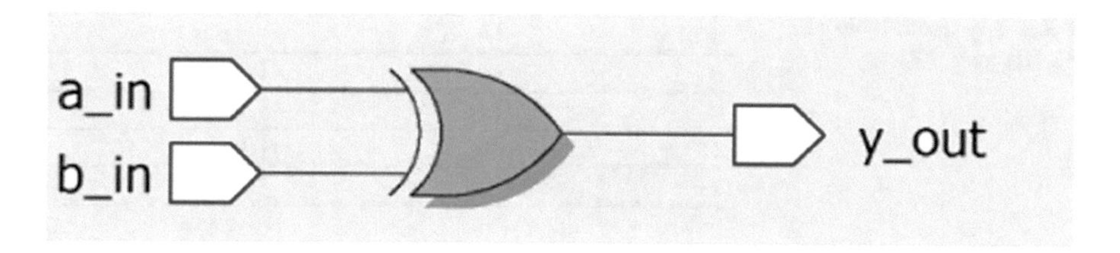

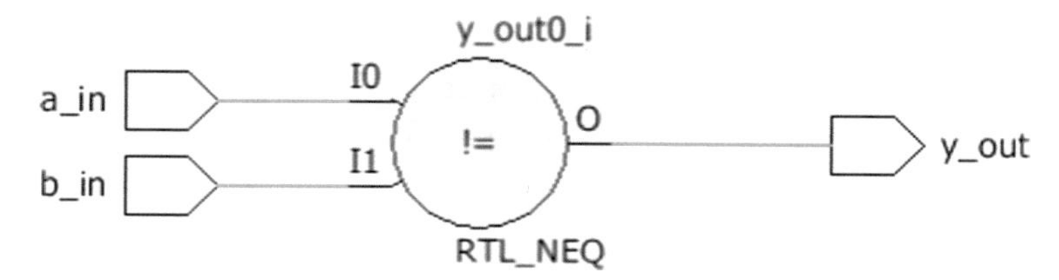

Fig. 3.11 RTL schematic of Example 11

RTL schematic of two input XOR logic is shown in Fig. 3.11; input ports of XOR logic gate are named as a_in, b_in and output as y_out.

If XOR gate is not available in the library, then XOR logic can be realized using AND-OR-invert or by using minimum number of NAND gates.

The RTL for multiple input XOR gate is coded using bit-wise operator and shown in the Example 12.

```
////////////////////////////////////////////////////////////////////////
module xor_gate ( input [7:0]  a_in, b_in , output [7:0] y_out );

assign  y_out = (a_in  ^  b_in);

endmodule

////////////////////////////////////////////////////////////////////////
```

Example 12 Multiple input XOR gate

The logic inferred is shown in Fig. 3.12.

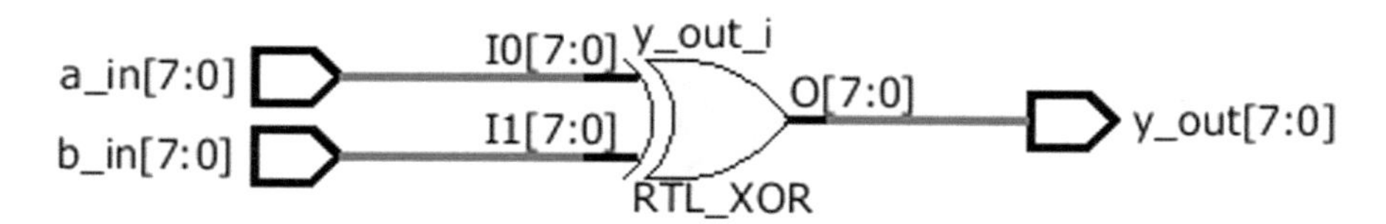

Fig. 3.12 Schematic of multiple input XOR gate

Table 3.7 Truth table of XNOR gate

a_in	b_in	y_out
0	0	1
0	1	0
1	0	0
1	1	1

3.2.7 Two Input XNOR Logic

Two input XNOR is called as exclusive NOR logic and generates output as logic 1 when two inputs are equal. XNOR is complement of XOR logic. Synthesizable RTL for XNOR is shown in the Example 13. The truth table of XNOR logic is shown in Table 3.7.

```
//////////////////////////////////////////////////////////////////////////////////
module xnor_gate(input a_in, b_in , output reg y_out);
always@(a_in, b_in)
begin
 if ( a_in == b_in )
   y_out = 1;
 else
   y_out = 0;
end
endmodule
//////////////////////////////////////////////////////////////////////////////////
```

Example 13 XNOR logic synthesizable design

RTL schematic of XNOR logic is shown in Fig. 3.13, input ports of XNOR logic gate are named as a_in, b_in, and output is named as y_out.

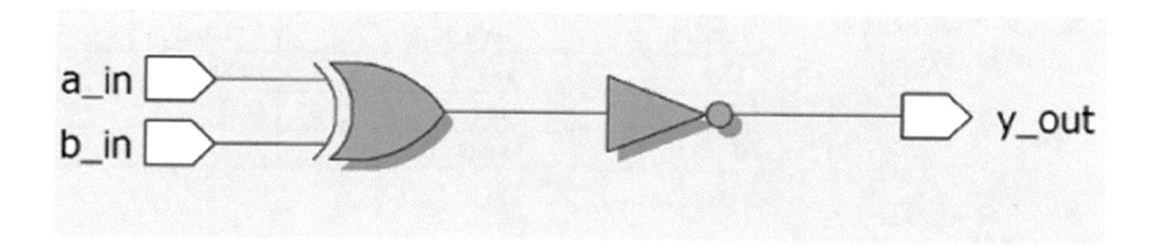

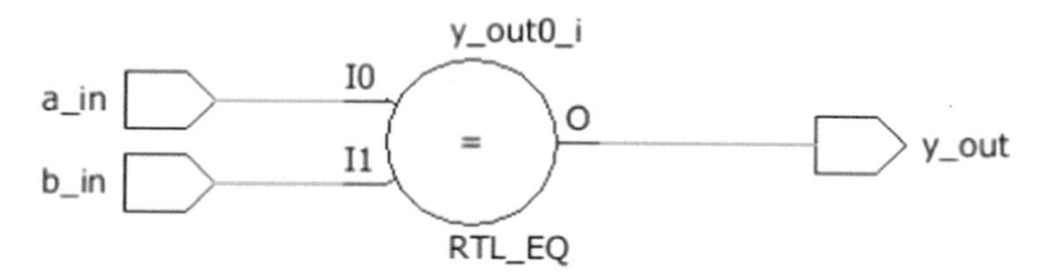

Fig. 3.13 RTL schematic of Example 13

If XNOR gate is not available in the library, then XNOR logic can be realized by using AND-OR-invert or by using minimum number of NAND or NOR gates.

The RTL for the multiple input XNOR is shown in the Example 14.

```
////////////////////////////////////////////////////////////////////////////////

module xnor_gate(input [7:0]  a_in, b_in , output [7:0] y_out);

assign y_out = (a_in  ~^  b_in);

endmodule

////////////////////////////////////////////////////////////////////////////////
```

Example 14 RTL of multiple input XNOR

The RTL schematic is shown in Fig. 3.14.

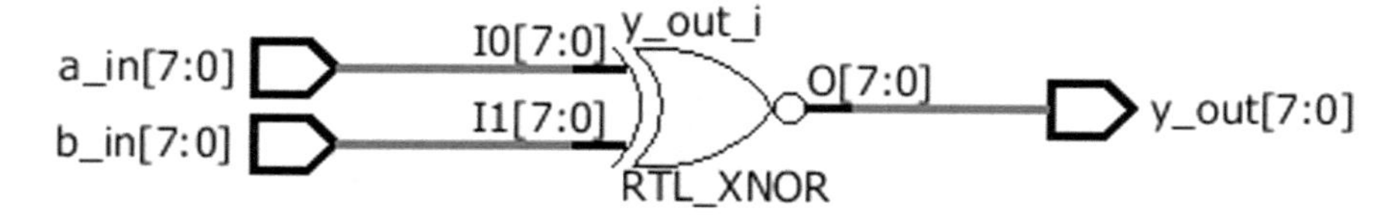

Fig. 3.14 Schematic of Example 14

Table 3.8 Truth table of tristate logic

Enable	data_in	data_out
1	0000	0000
1	1111	1111
0	xxxx	zzzz

3.3 Tristate Logic

Tristate has three logic states, namely logic 0, logic 1, and high impedance z. Synthesizable design is shown in the Example 15. The truth table of tristate buffer logic is shown in Table 3.8.

```
/////////////////////////////////////////////////////////////////////////////

module tri_state_logic ( input [3:0] data_in,
                         input enable,
                         output reg [3:0] data_out );

always@(data_in, enable)
begin
  if (enable)
    data_out = data_in;
  else
    data_out= 4'bZZZZ;
end
endmodule
/////////////////////////////////////////////////////////////////////////////
```

Example 15 Synthesizable Verilog code for tristate logic

> ***Note: Avoid use of tristate logic while developing the RTL. Tristate is difficult to test. Instead of tristate logic, it is recommended to use multiplexers to develop the logic with enable.***

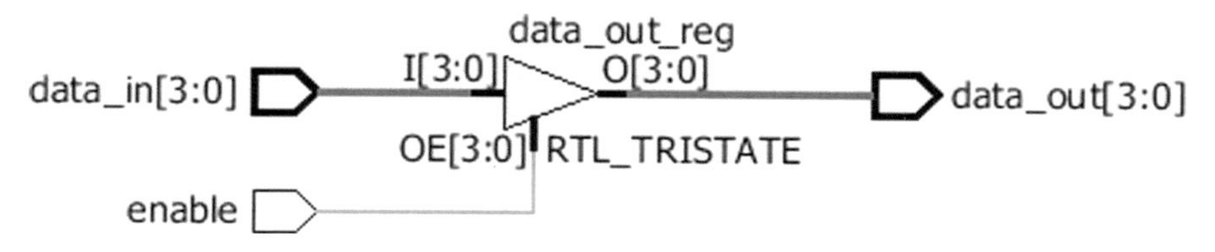

Fig. 3.15 RTL schematic of Example 15

RTL schematic of tristate logic is shown in Fig. 3.15, input port of tristate not logic is named as data_in, and enable input as enable and four-bit output port as data_out.

3.4 Arithmetic Circuits

Arithmetic operations like addition and subtraction are used frequently in the design of processor logic. Arithmetic and logical unit (ALU) of any processor is designed to perform the addition, subtraction, increment, and decrement operations. The arithmetic designs should be described by the synthesizable Verilog code to achieve the desired performance. This section discusses about RTL designs to perform arithmetic operations.

3.4.1 Adder

Adders are used to perform the binary addition of two binary numbers. Adders are used for signed or unsigned addition operations.

3.4.1.1 Half Adder

Consider half adder which has two inputs a_in, b_in and generates single bit outputs sum_out, carry_out. Where sum_out is adder result and carry_out is carry output. Table 3.9, is the truth Table for half adder and RTL is shown in the Example 16.

Table 3.9 Truth table of half adder

a_in	b_in	sum_out	carry_out
0	0	0	0
0	1	1	0
1	0	1	0
1	1	0	1

```
////////////////////////////////////////////////////////////////////////////////

module half_adder ( input a_in, b_in,

                    output sum_out, carry_out );

assign sum_out = a_in ^ b_in;

assign carry_out = a_in & b_in;

endmodule

////////////////////////////////////////////////////////////////////////////////
```

Example 16 Synthesizable code for half adder

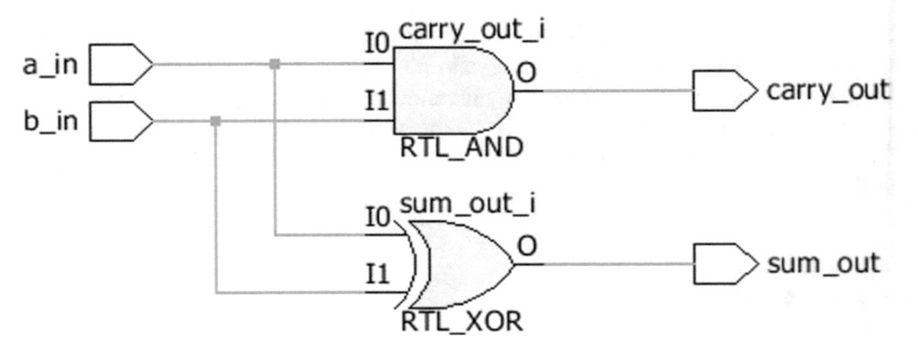

Fig. 3.16 RTL schematic of Example 16

> ***Note: Half adders are used as basic component to perform the addition. Full adder logic circuits are designed using the instantiation of half adders as components.***

RTL schematic of half adder is shown in Fig. 3.16, and as shown, the input ports of half adder are named as a_in, b_in and output as sum_out, carry_out.

3.4.1.2 Full Adder

Full adders are used to perform addition. Consider binary inputs as a_in, b_in, c_in and single bit binary outputs as sum_out, carry_out. Table 3.10 is the truth table for full adder, and RTL is described in the Example 17.

Table 3.10 Truth table of full adder

c_in	a_in	b_in	sum_out	carry_out
0	0	0	0	0
0	0	1	1	0
0	1	0	1	0
0	1	1	0	1
1	0	0	1	0
1	0	1	0	1
1	1	0	0	1
1	1	1	1	1

```
//////////////////////////////////////////////////////////////////////////////

module  full_adder ( input a_in, b_in, c_in,

                     output sum_out, carry_out );


assign  { carry_out, sum_out } = a_in + b_in + c_in;

endmodule

//////////////////////////////////////////////////////////////////////////////
```

Example 17 Synthesizable Verilog code for full adder

Note: Full adder consumes more area, so it is highly recommended to implement the adder logic using multiplexers.

RTL schematic of full adder is shown in Fig. 3.17, and input ports of full adder are named as a_in, b_in, c_in and output as sum_out, carry_out.

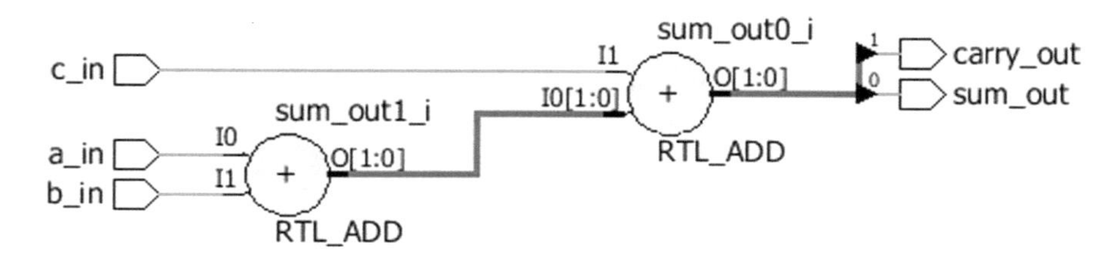

Fig. 3.17 RTL schematic of Example 17

Table 3.11 Truth table of half subtractor

a_in	b_in	diff_out	borrow_out
0	0	0	0
0	1	1	1
1	0	1	0
1	1	0	0

3.4.2 *Subtractor*

Subtractors are used to perform the binary subtraction of two binary numbers. This section describes about the half and full subtractors.

3.4.2.1 Half Subtractor

Consider the half subtractor which has inputs as a_in, b_in and generates single outputs diff_out, borrow_out. Where diff_out is difference output, and borrow_out is borrow output. Refer Table 3.11, and RTL is shown in the Example 18.

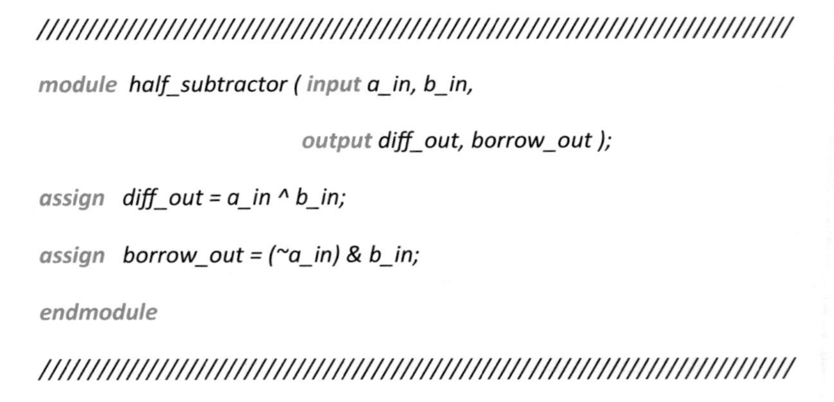

```
////////////////////////////////////////////////////////////////////////////

module half_subtractor ( input a_in, b_in,
                         output diff_out, borrow_out );

assign diff_out = a_in ^ b_in;

assign borrow_out = (~a_in) & b_in;

endmodule

////////////////////////////////////////////////////////////////////////////
```

Example 18 Synthesizable Verilog code for half subtractor

Note: Half subtractors are used as basic component to perform the binary subtractions. Full subtractor logic circuits are designed using the instantiation of half subtractors as components.

RTL schematic of half subtractor is shown in Fig. 3.18, and input ports of half adder are named as a_in, b_in and output as diff_out, borrow_out.

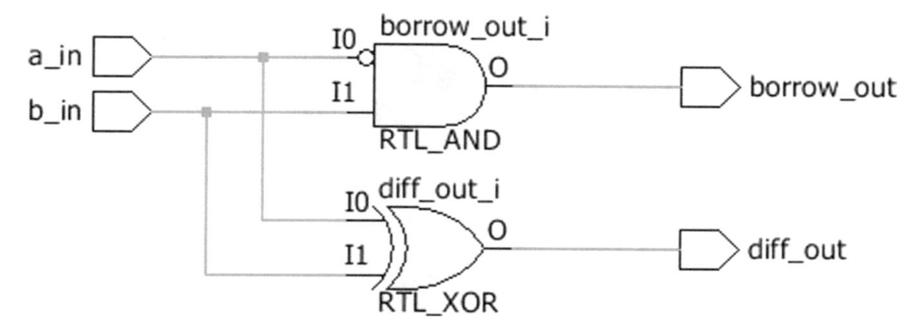

Fig. 3.18 RTL schematic of Example 18

3.4.2.2 Full Subtractor

Full subtractors are used to perform subtraction. Consider single bit inputs which are named as a_in, b_in, c_in and single bit binary outputs as diff_out, borrow_out. Refer Table 3.12, and RTL coded is shown in the Example 19.

Table 3.12 Truth table of full subtractor

c_in	a_in	b_in	diff_out	borrow_out
0	**0**	**0**	**0**	**0**
0	**0**	**1**	**1**	**1**
0	**1**	**0**	**1**	**1**
0	**1**	**1**	**0**	**1**
1	**0**	**0**	**1**	**0**
1	**0**	**1**	**0**	**0**
1	1	0	0	0
1	1	1	1	1

```
/////////////////////////////////////////////////////////////////////////////////

module full_subtractor ( input a_in, b_in, c_in,

                         output diff_out, borrow_out );


assign { borrow_out, diff_out } = a_in- b_in - c_in;

endmodule
```

Example 19 Synthesizable Verilog code for full subtractor

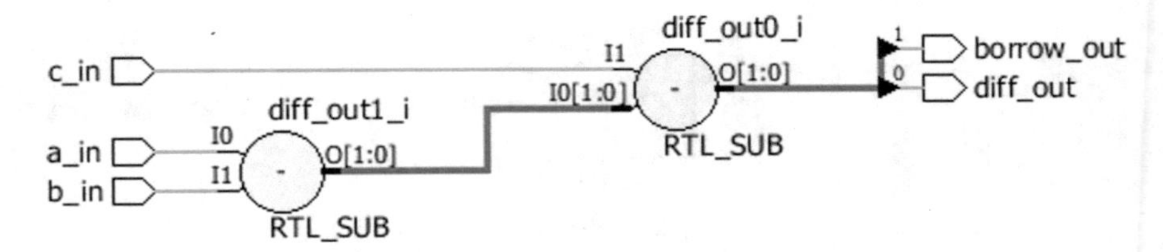

Fig. 3.19 RTL schematic of Example 19

Note: It is recommended to use the full adder to perform the subtraction operation. Subtraction is performed using 2's complement addition.

RTL schematic of full subtractor is shown in Fig. 3.19, and input ports of full subtractor are named as a_in, b_in, c_in and output as diff_out, borrow_out.

3.5 Exercises

The exercises are based on the understanding of ***assign*** and **always** procedural block to model the combinational design. Complete the exercises for better understanding and application of Verilog constructs.

1. *The logic inferred by the following code is*

```
module comb_design_logic ( input a, b, output y);

assign y = a ~^ b;

endmodule
```

   a. *XOR gate*
   b. *XNOR gate*
   c. *NOR gate*
   d. *Syntax error in the code*

2. *The logic inferred by the following code is*

```
module comb_design_logic_0 ( input a, b, output y);

assign y = a ~^ b;

assign y= a & b;
```

```
endmodule
```

a. *XOR gate*
b. *XNOR gate*
c. *NOR gate*
d. *Syntax error in the code*

3. *State true or false? The order of assign doesn't affect on the synthesis result.*

```
module comb_design_logic_1 ( input a, b, c, output y1);

wire y1;

assign y = a ~^ b;

assign y1= a & y;

endmodule
```

a. *True*
b. *false*

4. *The logic inferred by the following code is*

```
module comb_design_logic_2 ( input a, b, output reg y1, y2);

assign y1=  a ^ b;
```

```
always @ (a, b)

y2= a && b;

endmodule
```

a. XOR gate, AND gate
b. OR gate, AND gate
c. NOR gate, AND gate
d. XNOR gate, AND gate

5. *What is the logic inferred by the code?*

```
module design_logic2 ( input a, b, c, output reg y2);

reg y1;

always @ (a, b,c)

y1 = a & b;

y2 = y1 & c;

endmodule
```

a. Three input AND gate
b. Two input AND gate
c. Two AND gates connected in cascade
d. Syntax error in the RTL

3.6 Summary

As discussed already in this chapter, the following are important points need to be considered while implementing combinational logic RTL.

1. Use minimum area by sharing the arithmetic resources.
2. Use all the required signals in the sensitivity to avoid simulation and synthesis mismatch.
3. Avoid use of tristate logic and implement the logic required using multiplexers with proper enable circuit.
4. Verilog supports four states, and they are logic 0, logic 1, don't care x, and high impedance z.
5. Use minimum number of adders in design. Adders can be implemented using multiplexers.
6. NAND and NOR are universal logic gates and used to implement any combinational or sequential logic.
7. Bubbled AND is equal to NOR.
8. Bubbled OR is equal to NAND.

Chapter 4
Verilog Constructs and Combinational Design-II

An efficient RTL design always uses the synthesizable constructs, and the role of design engineer is to get the better design performance as per as area, speed and power is concern. The chapter discusses about RTL design for few of the arithmetic resources and the code converters.

As discussed in the Chap. 3 in the combinational design, an output is function of the present input only. If input changes, an output changes. Practically, the logic gates or combinational design can have the propagation delay. We have already discussed about the ***assign*** and ***always @ (// sensitivity list)*** constructs in the Chap. 3. This chapter discusses about the ***always*** @* and ***assign*** to code for the combinational designs such as multiple bit adders and subtractors and code converters.

4.1 Procedural Block always @*

Most of the time during combinational design we can have many number of inputs on which the always procedural block is sensitive to. As discussed in the previous chapter, the always procedural block with the sensitivity list is best suitable to model the combinational design. Consider the ***always*** procedural block which is sensitive to eight inputs. We can have

always @ (a_in, b_in, c_in, d_in, e_in, f_in, g_in, h_in)

If we miss one of input from the sensitivity list, then the issue is simulation, and synthesis mismatches. To avoid the simulation and synthesis mismatch, better strategy during the RTL design is to have always procedural block which can be sensitive to all the required inputs. We can use

always @*

where the character * is information provided to simulator to include all the inputs in the sensitivity list. This is efficient way of RTL coding.

V. Taraate, *Digital Logic Design Using Verilog*,
https://doi.org/10.1007/978-981-16-3199-3_4

4.2 Multi-bit Adders and Subtractors

Multi-bit adders and subtractors are used in the design of arithmetic units for the processors. The logic density depends upon the number of input bits of adder or subtractor.

4.2.1 *Four-Bit Full Adder*

Many practical designs use multi-bit adders and subtractors. It is industrial practice to use basic component as full adder to perform the addition operation. For example, if designer wishes to implement the 4-bit adder, then four full adders are required. As shown in the Example 1, addition is performed on two 4-bit binary numbers a_in, b_in. The result is 4-bit addition and will be available at output port sum_out. Carry input is c_in, and carry output is carry_out.

> ***Note: Four-bit addition uses four full adders. Depending on signed or unsigned addition requirements, the Verilog code can be modified***

Synthesis result of RTL coded for the 4-bit adder is shown in Fig. 4.1, and input ports of 4-bit adder are named as a_in, b_in, c_in and output port as sum_out, carry_out.

```
//////////////////////////////////////////////////////////////////////////////

module four_bit_adder ( input [3:0] a_in, b_in,

                        input c_in,

                        output [3:0] sum_out,

                        output carry_out);

assign { carry_out, sum_out } = a_in + b_in + c_in;

endmodule

//////////////////////////////////////////////////////////////////////////////
```

Example 1 Synthesizable Verilog code for 4-bit adder

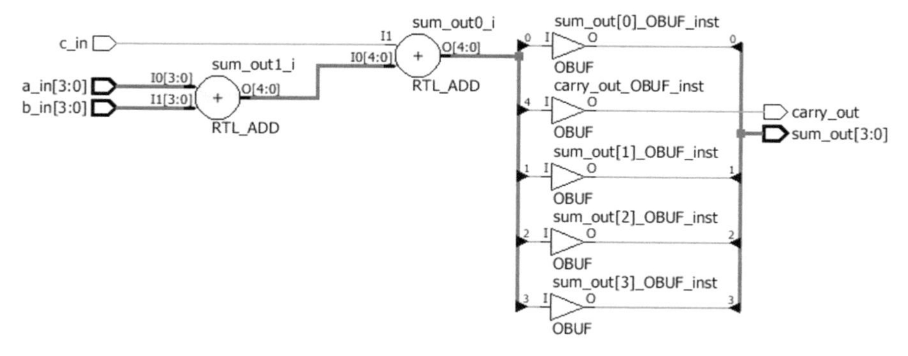

Fig. 4.1 RTL schematic of 4-bit adder

4.2.2 4-Bit Full Subtractor

RTL is coded for the 4-bit subtractor and shown in the Example 2, and subtraction is performed on two 4-bit binary numbers a_in, b_in. The result is 4-bit subtraction and is available at output port diff_out. Borrow input is c_in, and borrow output is borrow_out.

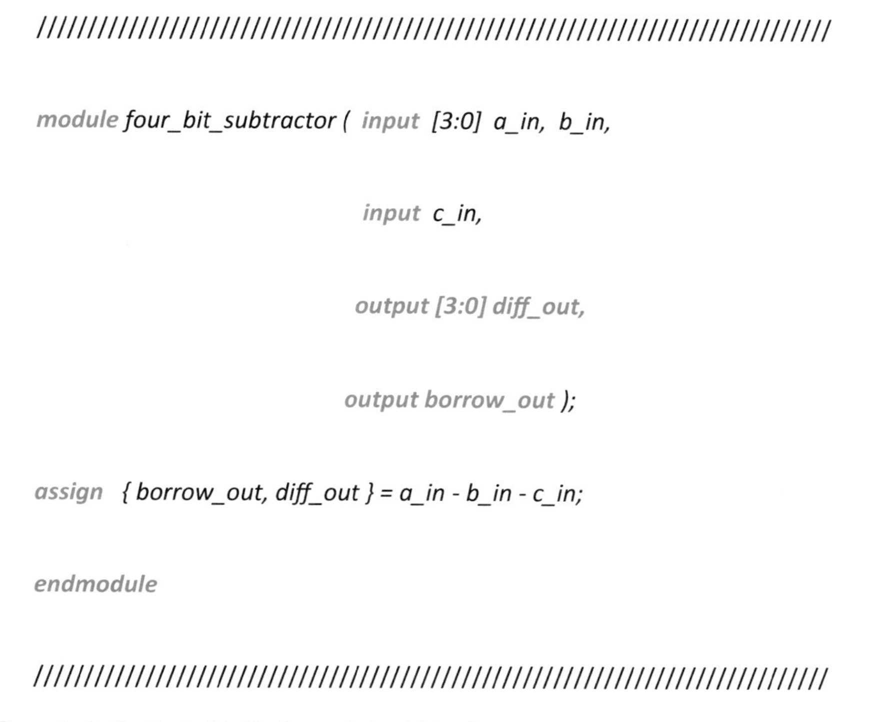
```
////////////////////////////////////////////////////////////////////////////////

module four_bit_subtractor ( input [3:0] a_in, b_in,

                             input c_in,

                             output [3:0] diff_out,

                             output borrow_out );

assign { borrow_out, diff_out } = a_in - b_in - c_in;

endmodule

////////////////////////////////////////////////////////////////////////////////
```

Example 2 Synthesizable Verilog code for 4-bit subtractor

Synthesis result of RTL coded for the 4-bit subtractor is shown in Fig. 4.2, and input ports of 4-bit subtractor are named as a_in, b_in, c_in and output port as diff_out, borrow_out.

4.2.3 4-Bit Adder and Subtractor

Design of addition and subtraction operation can be performed by using the adders only. Subtraction can be performed using 2's complement addition. For example, consider the scenario shown in Table 4.1.

The RTL is coded to implement the 4-bit adder and subtractor using synthesizable constructs. If control_in = 1, it performs addition; otherwise for the control_in = 0, it performs the subtraction.

Note: Here, the resource used is binary full adder to perform both the additions and subtractions. Subtraction operation is performed using adders only. Resource sharing and resource utilization are to be discussed in the Chap. 3.

Synthesized logic of 4-bit adder/subtractor is shown in Fig. 4.3, and input ports of 4-bit adder/subtractor are named as a_in, b_in,c_in. The control port is named as control_in, and output is named as result_out, carry_out.

4.3 Optimization of Resources

If we carefully observe the RTL design coded in the Example 3, then we can conclude that it uses a greater number of resources to perform the addition and subtraction. The logic performs both operations at a time, and at output selection, logic is used to select from one of the operations depending on the multiplexer select input (control_in). The design is inefficient and needs optimization. The following section discusses about few of the RTL tweaks to optimize for the design.

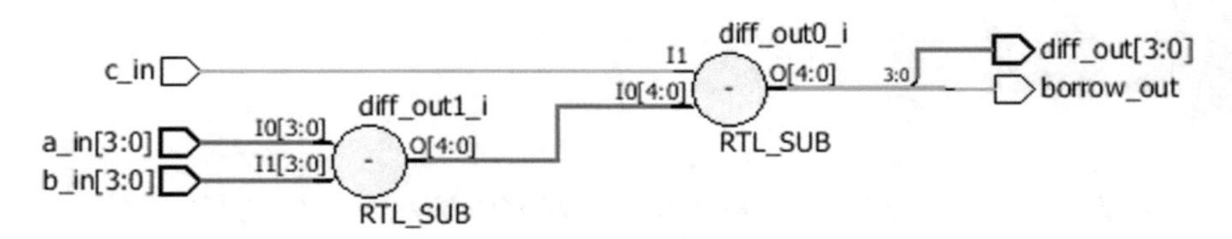

Fig. 4.2 RTL schematic of 4-bit subtractor

Table 4.1 Operational table for adder subtractor

Operation	Description	Expression
Addition	Unsigned addition of A, B	$A + B + 0$
Subtraction	Unsigned subtraction of A, B	$A - B = A + \sim B + 1$

```
////////////////////////////////////////////////////////////////////////////////

module four_bit_adder_subtractor (  input [3:0] a_in, b_in,
                                    input c_in,
                                    input control_in,
                                    output reg [3:0] result_out,
                                    output reg carry_out   );

always @ *
if ( control_in)
        { carry_out, result_out } = a_in + b_in + c_in;
else
        { carry_out, result_out } = a_in - b_in - c_in;
endmodule

////////////////////////////////////////////////////////////////////////////////
```

Example 3 Synthesizable Verilog code for 4-bit adder and subtractor

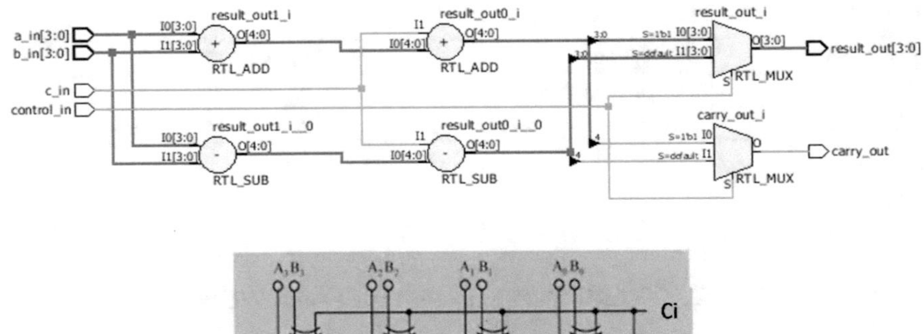

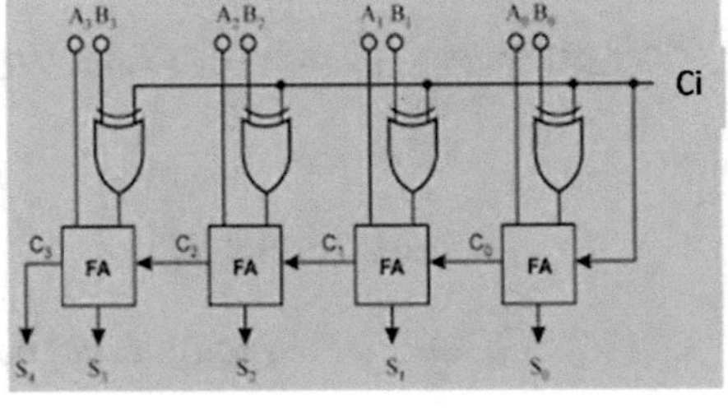

Fig. 4.3 Synthesis result of 4-bit adder/subtractor

4.3.1 *Optimization Using Only Adders*

To optimize for the resources, let us try to implement the subtraction using 2's complement addition. The strategy is described in Table 4.2.

The RTL for the operations shown in Table 4.2 is coded using ***if..else*** construct and shown in the Example 4.

Table 4.2 Operational table with optimization goal

Operation	control_in	Expression
Addition	0	A + B + 0
Subtraction	1	A − B = A + ∼B + 1

The strategy used to code the RTL is useful to optimize the logic, and the design uses only adders in the data path. But still there is issue as the chain of multiplexers is used at the output to select the result of one of the operations (Fig. 4.4).

```
////////////////////////////////////////////////////////////////////////////////////

module four_bit_adder_subtractor (  input [3:0] a_in, b_in,

                                    input control_in,

                                    output reg [3:0] result_out,

                                    output reg carry_out  );


always @ *

 if ( ~control_in)


  { carry_out, result_out } = a_in + b_in + control_in ;

 else

 { carry_out, result_out } = a_in +  (~b_in)  + control_in;

endmodule

////////////////////////////////////////////////////////////////////////////////////
```

Example 4 Synthesizable Verilog code for 4-bit adder and subtractor with optimization

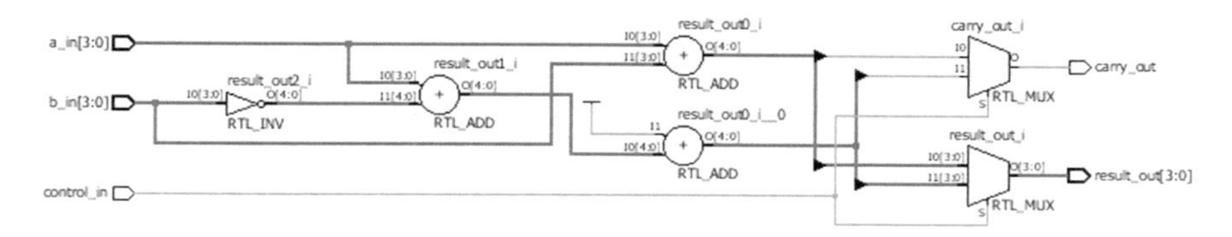

Fig. 4.4 RTL schematic after RTL tweak

4.3.2 *Optimization by Tweaking the Logic to Have Better Data and Control Path*

For the better data path and control path optimization, use the common resource as adder, and use the common expression to perform the addition and subtraction. The strategy is explained in Table 4.3.

Table 4.3 Use of the common expression in the RTL tweaks

Operation	control_in	Variable input	Common expression
Addition	0	b_in	A + control_in
Subtraction	1	~b_in	A + control_in

```
/////////////////////////////////////////////////////////////////////////////
module four_bit_adder_subtractor ( input [3:0] a_in, b_in,
                                   input control_in,
                                   output [3:0] result_out,
                                   output carry_out );
reg [4:0] temp_result;
assign { carry_out, result_out } = a_in + temp_result ;
always @ *
 if ( ~control_in)
  temp_result = b_in + control_in;
 else
  temp_result = (~ b_in) + control_in;
endmodule
/////////////////////////////////////////////////////////////////////////////
```

Example 5 Synthesizable Verilog code for 4-bit adder and subtractor with use of least adders

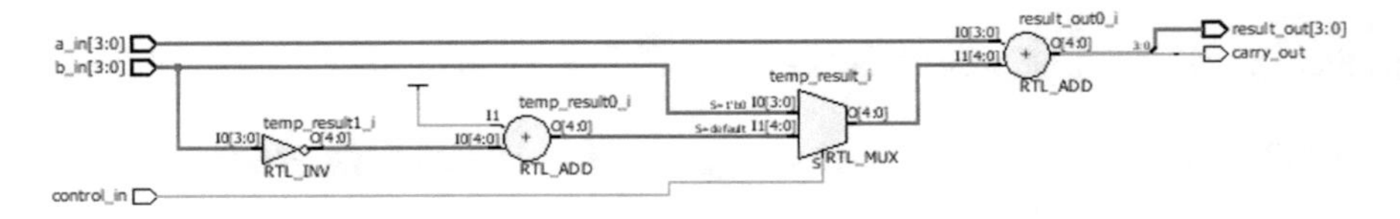

Fig. 4.5 RTL schematic after optimization

The RTL for the operations specified in Table 4.3 is coded using the synthesizable constructs and is shown in the Example 5.

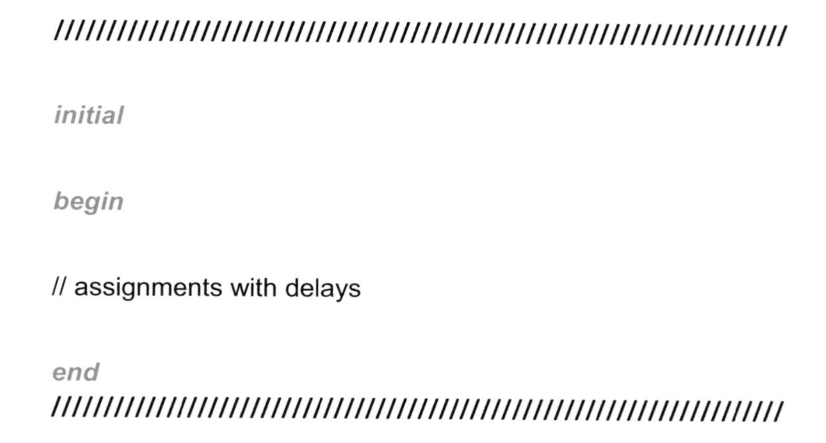

```
////////////////////////////////////////////////////////////////////

initial

begin

// assignments with delays

end
////////////////////////////////////////////////////////////////////
```

The synthesis result of Example 5 is shown in Fig. 4.5, and as shown, the inferred logic uses only two adders and multiplexer. At the output stage, it infers the adder and has the optimized data and control path. Now, the logic performs only one operation at a time.

4.4 Procedural Block initial

The procedural block ***initial*** is non-synthesizable construct and executes only once at 0 simulation time. It is mainly used for the initialization. The syntax is shown here.

The testbenches use ***initial*** procedural block, and the objective is to create the stimulus which can drive to the design under test. If I have blocking inter-delay assignments within the ***begin–end,*** then all these assignments will be executed sequentially. For more information about the inter- and intra-delay, refer Chap. 15.

4.5 Simulation Concepts: Basic Testbench

In most of the practical scenarios, we use the testbenches to check for the functional correctness of the design. The basic testbench we can visualize using the diagram which has driver and DUT.

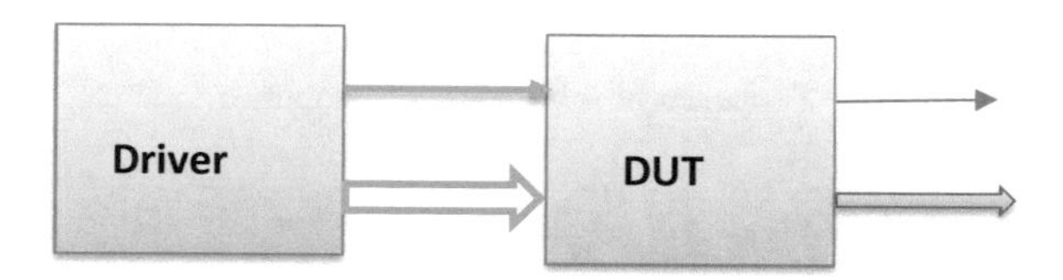

Fig. 4.6 Testbench basic architecture

```
//////////////////////////////////////////////////////////////////////
module test_add_sub;
reg [3:0] a_in, b_in;
reg control_in;
wire [3:0] result_out;
wire carry_out;
four_bit_subtractor UUT ( .a_in(a_in),
                          .b_in(b_in),
                          .control_in(control_in),
                          .result_out(result_out),
                          .carry_out(carry_out) );
initial
begin
a_in = 4'b0000;
b_in = 4'b0000;
control_in =0;
#10;
a_in = 4'b1000;
b_in = 4'b0010;
control_in =0;
#10;
```

Example 6 Testbench of 4-bit adder and subtractor

```
a_in = 4'b1000;
b_in = 4'b0110;
#10;
a_in = 4'b1000;
b_in = 4'b0111;
control_in =1;
#50;
a_in = 4'b1000;
b_in = 4'b1111;
control_in =1;
#10;
a_in = 4'b0111;
b_in = 4'b0111;
control_in =0;
#10;
a_in = 4'b0111;
b_in = 4'b0111;
control_in =1;
end
endmodule
////////////////////////////////////////////////////////////////////////
```

Example 6 (continued)

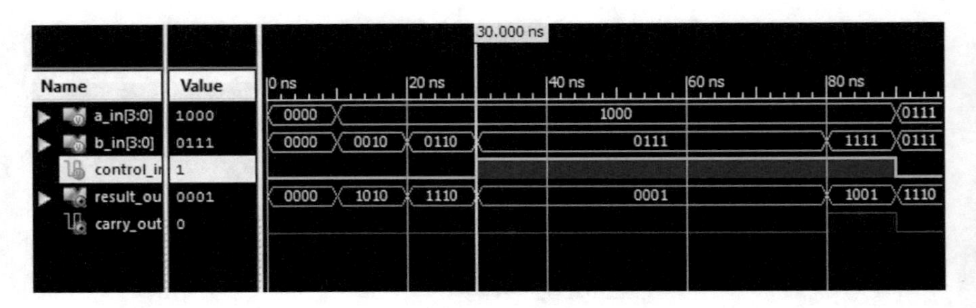

Fig. 4.7 Simulation result of 4-bit adder and subtractor

As shown in Fig. 4.6, the testbench has driver and DUT. The driver generates the stimulus of signals and is used to drive the DUT inputs. The goal is to check the output signals to understand and confirm about the functional correctness of the design.

The testbench for the Example 5 is coded using the non-synthesizable constructs and shown in the Example 6. The objective is to drive the signals a_in, b_in, control_in and to check for the output at result_out, carry_out.

As shown in the simulation waveform (Fig. 4.7) for time stamp t = 20 ns, control_in = 0, it performs addition of 1000, 0110 to get result_out as 1110 and carry_out = 0. At time stamp t = 30 ns, control_in = 1, and it performs the subtraction of 1000 and 0111 to get result_out = 0001, carry_out = 0. For other time stamp, you can observe the result_out, carry_out.

4.6 Comparators and Parity Detectors

In most of the practical scenarios, comparators are used to compare the equality of two binary numbers. Parity detectors are used to compute the even or odd parity for the given binary number. It becomes very essential for the design engineer to have the better understanding of this.

4.6.1 Binary Comparators

These are used to compare the two binary numbers. As discussed, earlier Verilog supports four states, and they are logic 0, logic 1, don't care x, and high impedance z. Verilog supports logical equality operator (==) and inequality operator (!=), and these are used to compare the two numbers. These operators are used in the synthesizable RTL design.

Table 4.4 Operational table for 1-bit comparator

a_in	*b_in*	*a_greater_b*	*a_equal_b*	*a_less_b*
0	0	0	1	0
0	1	0	0	1
1	0	1	0	0
1	1	0	1	0

```
////////////////////////////////////////////////////////////////////////////////
module binary_comparator ( input [3:0] a_in , b_in,
                                   output reg a_greter_b, a_equal_b, a_less
always @ *
begin
if (a_in ==b_in)
 begin
 a_greter_b = 0;
 a_equal_b = 1;
 a_less_b =0;
 end
else if (a_in > b_in)
 begin
 a_greter_b = 1;
 a_equal_b = 0;
 a_less_b =0;
 end
else
```

Example 7 Synthesizable Verilog code for 1-bit comparator

```
begin
 a_greter_b = 0;
 a_equal_b = 0;
 a_less_b =1;
 end
end
endmodule
////////////////////////////////////////////////////////////////////////////
```

Example 7 (continued)

For example, consider the operational Table 4.4. As shown in the table, for a_in = b_in a_equal_b = 1, a_in > b_in a_greater_b = 1, a_in < b_in a_less_b. But it should generate the parallel output to reflect the result.

> ***Note: Logical equality and inequality operators are used in the synthesizable RTL code and if any of the operands are 'x' or 'z' comparison is false.***

Synthesized equivalent block representation is shown Fig. 4.8. Due to use of nested if…else construct, it infers the priority logic with the tree of multiplexers at the output and is inefficient design.

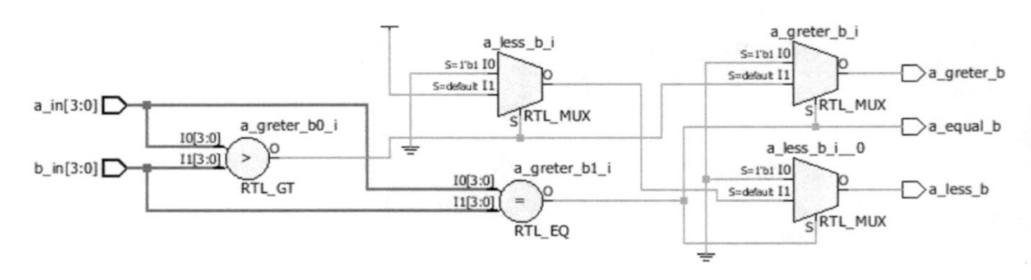

Fig. 4.8 Synthesis result of 1-bit comparator

4.6.2 Parity Detector

Parity detectors are used to detect the even or odd parity for the binary number string. Consider the design requirement to detect even or odd number of 1's, for even number of 1's the output is logic 0 and for odd number of 1's the output is logic 1. The RTL using Verilog can be coded as shown in the Example 8.

The operational table for the parity detector is shown below in Table 4.5. For odd number of 1s, the output is Logic 1, and for even number of 1s, output is assigned as logic 0.

> ***Note: Parity detectors are used in many of DSP applications and an integral module for encryption engines.***

Synthesized equivalent block representation is shown in Fig. 4.9.

Table 4.5 Operational table for parity detector

Condition	Description
Odd 1s	Assign output as logical 1
Even 1s	Assign output as logical zero

```
/////////////////////////////////////////////////////////////////////////////

module even_parity_detector ( input [7:0] data_in,

                                 output parity_out);

assign  parity_out = ^ data_in;

endmodule

/////////////////////////////////////////////////////////////////////////////
```

Example 8 Synthesizable Verilog code for parity detector

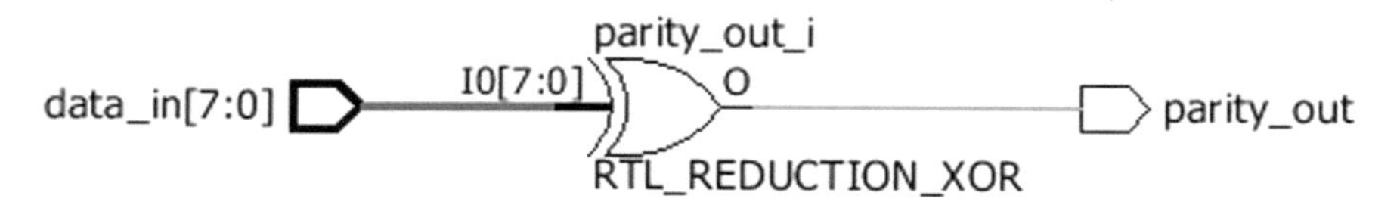

Fig. 4.9 Synthesized parity detector

4.7 Code Converters

This section deals with the commonly used code converters in the design. As name itself indicates, the code converters are used to convert the code from one number system representation to another number system. In the practical scenarios, binary to gray and gray to binary converters are used.

4.7.1 Binary to Gray Code Converter

Base of binary number system is 2, for any multi-bit binary number, one or more than one bit changes in two successive numbers. In gray code, only one bit changes at a time in two successive gray codes.

Refer Table 4.6.

The RTL is coded using the synthesizable constructs that is continuous assignments for 4-bit binary to gray code conversion described in Example 9.

Note: Gray codes are used in the multiple clock domain designs to transfer the control information from one of the clock domains to another clock domain.

The synthesis result is shown in Fig. 4.10.

Table 4.6 Binary and its equivalent gray code

4-bit binary	4-bit gray
0000	0000
0001	0001
0010	0011
0011	0010
0100	0110
0101	0111
0110	0101
0111	0100
1000	1100
1001	1101
1010	1111
1011	1110
1100	1010
1101	1011
1110	1001
1111	1000

```
////////////////////////////////////////////////////////////////////////
module binary_to_gray ( input [3:0] binary_in,

                        output [3:0] gray_out );


assign gray_out [3] = binary_in[3];

assign gray_out [2] = binary_in[3] ^ binary_in[2];

assign gray_out [1] = binary_in[2] ^ binary_in[1];

assign gray_out [0] = binary_in[1] ^ binary_in[0];

endmodule

////////////////////////////////////////////////////////////////////////////////
```

Example 9 Synthesizable Verilog code for 4-bit binary to gray code converter

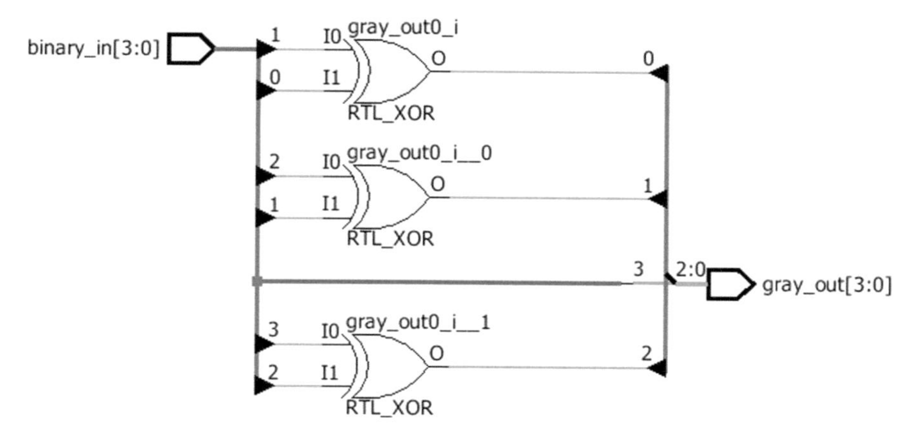

Fig. 4.10 RTL schematic of 4-bit binary to gray converter

Table 4.7 Gray code and its binary equivalent

4-bit gray	4-bit binary
0000	0000
0001	0001
0011	0010
0010	0011
0110	0100
0111	0101
0101	0110
0100	0111
1100	1000
1101	1001
1111	1010
1110	1011
1010	1100
1011	1101
1001	1110
1000	1111

4.7.2 Gray to Binary Code Converter

The truth table of the 4-bit gray to binary code converter is shown in Table 4.7.

```
////////////////////////////////////////////////////////////////////////////////

module binary_to_gray ( input [3:0] gray_in,

                        output [3:0] binary_out );

assign binary_out[3] = gray_in[3];
assign binary_out [2] = gray_in[3] ^ gray_in[2];

assign binary_out [1] = ( gray_in[3] ^ gray_in[2] ) ^ gray_in[1] ;

assign binary_out [0] = ( gray_in[3] ^ gray_in[2] ^ gray_in[1] ) ^
gray_in[0];

endmodule

////////////////////////////////////////////////////////////////////////////////
```

Example 10 Synthesizable Verilog code for 4-bit gray to binary code converter

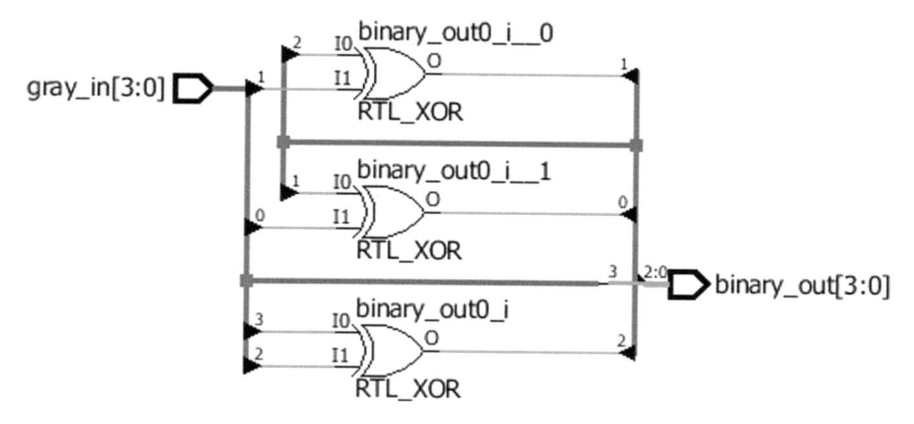

Fig. 4.11 Synthesis result of 4-bit gray to binary converter

The RTL description of 4-bit gray to binary code conversion is described in Example 10.

Note: Gray codes are used in the gray counter implementation and in the error correcting mechanism.

Synthesized representation of 4-bit gray to binary code converter is shown in Fig. 4.11.

Table 4.8 Operational table for combinational design

Condition	Output	Description
a_in = b_in	y_out = a_in ~^ b_in	If both inputs are at same logic level, then output should be XNOR (a_in, b_in)
a_in != b_in	y_out = a_in ^ b_in	If both inputs are at different logic level, then output should be XOR (a_in, b_in)

4.8 Let Us Think About the Design from Specifications

Let us design logic using the ***if…else*** construct for the functionality described in Table 4.8.

The RTL description to get the y_out as XOR (a_in, b_in) and XNOR (a_in, b_in) for a_in != b_in and a_in = b_in, respectively, is coded using the synthesizable constructs and shown in the Example 11.

```
//////////////////////////////////////////////////////////////////////////////////
module combo_design ( input a_in, b_in,
                      output reg y_out);

always@*
begin
 if (a_in == b_in)
  y_out = a_in ~^ b_in;
 else
  y_out = a_in ^ b_in;
end
endmodule
//////////////////////////////////////////////////////////////////////////////////
```

Example 11 Synthesizable Verilog code of combinational design

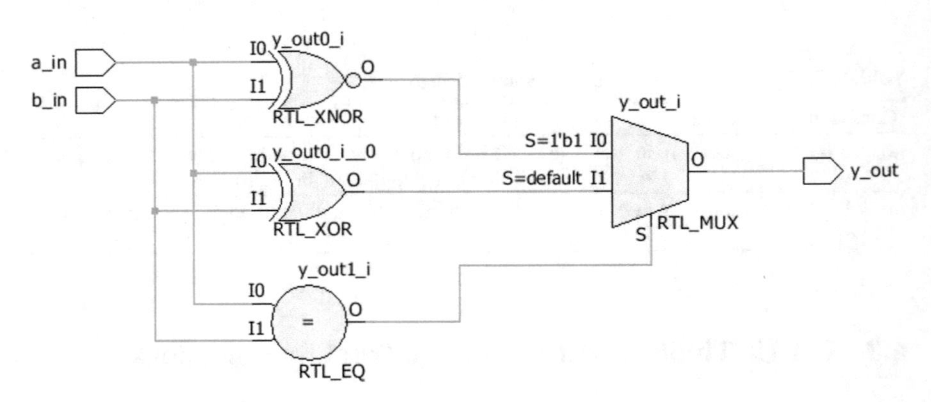

Fig. 4.12 RTL schematic of combinational design

The RTL schematic for Example 11 is shown in Fig. 4.12, and logic inferred has the 2:1 multiplexer with control input and the XOR and XNOR gate in the data path.

4.9 Exercises

The exercises are based on the understanding of ***assign*** and **always** procedural block to model the combinational design. Complete the exercises for better understanding and application of Verilog constructs.

1. *The logic inferred by the following code is*

```
module comb_design_logic ( input a, b, output reg y);

assign y = a ~^ b;

endmodule
```

   a. *XOR gate*
   b. *XNOR gate*
   c. *NOR gate*
   d. *Syntax error in the code*

2. *The logic inferred by the following code is*

```
module comb_design_logic_0 ( input a, b, output y);

always@*

begin

assign y = a ~^ b;

assign y= a & b;

end

endmodule
```

   a. *XOR gate, AND gate*
   b. *XNOR gate, AND gate*
   c. *NOR gate , AND gate*
   d. *Syntax error in the code*

3. *State true or false? The order of always doesn't affect on the synthesis result.*

```
module comb_design_logic_1 ( input a, b, c, output reg y1);

reg y1;

always@*

y = a ~^ b;

always@*

y1= a & y;

endmodule
```

   a. *True*
   b. *false*

4. *The logic inferred by the following code is*

```
module comb_design_logic_2 ( input a, b, output reg y1, y2);

always@ *

y1= a ^ b;
```

```
always @ (a, b)

y2= a && b;

endmodule
```

a. *XOR gate, AND gate*
b. *OR gate, AND gate*
c. *NOR gate, AND gate*
d. *XNOR gate, AND gate*

5. *What is simulation result for the code?*

```
module design_logic2 ();

reg y1;

initial

begin

y1=1'b0;

#10 y1= 1'b1;

#20 y1=1'b0;

end

endmodule
```

a. *Syntax error in the testbench*
b. *At t=0 y1=0, t=10ns y1=1, t=20 y1=0;*
c. *At t=0 y1=0, t=10ns y1=1, t=30 y1=0;*
d. *At t=10 y1=0, t=20ns y1=1, t=30 y1=0;*

4.10 Summary

As discussed already in this chapter, the following are important points need to be considered while implementing combinational logic RTL.

1. The ***if..else*** construct is used within the always procedural block.
2. The ***if..else*** construct infers the 2:1 multiplexer.
3. Use the adders to perform the subtraction.
4. Use the RTL tweaks for better data and control path optimization.
5. Multiple ***assign*** constructs execute concurrently.
6. The procedural block ***initial*** executes at 0 time and executes only once.
7. The testbench is used to check for the functional correctness of the design.
8. In two consecutive binary numbers, one or more than one bit changes.
9. In two consecutive gray numbers, only one bit changes.

Chapter 5
Multiplexers as Universal Logic

For an RTL design engineer, it is especially important to have better understanding of multiplexers. In most of the applications these are used as a functional blocks. This chapter discusses about the efficient RTL coding for multiplexers and parallel verses priority logic.

In most of the application, we experience the use of multiplexers. The multiplexer or MUX is used to implement the Boolean functions or any of the logic gates, and it is called as universal logic. The main application of multiplexer is to select from one of the inputs and hence called as the switch. The next few subsequent sessions are useful to understand about the RTL design for the multiplexers.

5.1 Multiplexers

Multiplexers are used to select one of the inputs from many. Multiplexers are also called as universal logic, and terminology used in the practical world is MUX. By using the suitable multiplexers, any of combinational logic function can be realized. Multiplexers are used as selection logic in ASIC- and FPGA-based designs. Multiplexer consumes lesser area as compared to adders, and most of the time MUX are used to implement arithmetic components such as adders and subtractors.

The block diagram of n:1 MUX is shown in Fig. 5.1, and it consists of n input lines, m select lines, and one output line. Input lines are denoted as i[0], i[1] … i [n − 1]; select lines by s[0], s[1], … s[m − 1], and output line by ‘y’.

As shown in Fig. 5.1, multiplexer has n input lines, m select lines, and single output line. Relation between the input lines and select lines is given by $n = 2^m$. For example, for 4:1 MUX input lines are four so $m = \log_2 n$, that is, select lines equal to two.

V. Taraate, *Digital Logic Design Using Verilog*,
https://doi.org/10.1007/978-981-16-3199-3_5

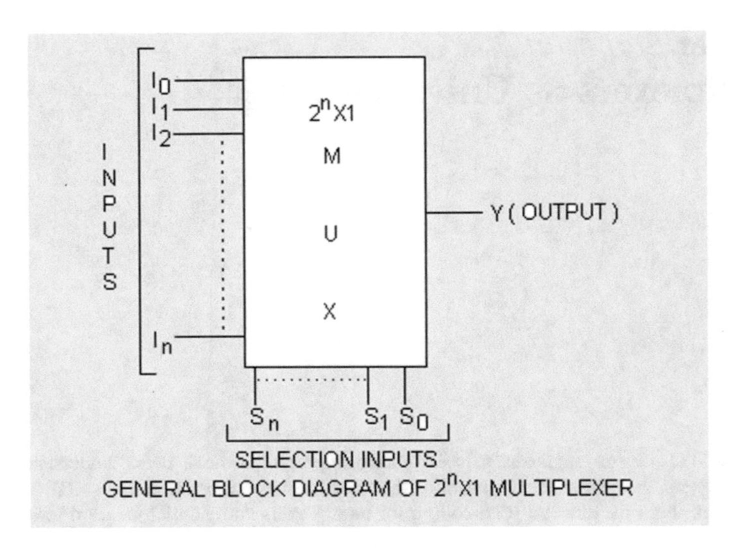

Fig. 5.1 Block diagram of n:1 MUX

5.2 Multiplexer as Universal logic

As discussed, earlier multiplexer is treated as universal logic as all combinational logic functions can be realized using MUX.

5.2.1 2:1 MUX

A 2:1 MUX has two input lines: one select line and one output line. When sel_in input is logic 0, output y_out is assigned as a_in and output is assigned as 'b_in' for sel_in equal to logic 1. Table 5.1 describes the truth table of 2:1 MUX and gate level design (Fig. 5.2).

> *Note* Conditional assignments are used to select from many inputs so infers the multiplexer.

The RTL schematic is shown in Fig. 5.3.

Table 5.1 Truth table of 2:1 MUX

sel_in	y_out
O	a_in
1	b_in

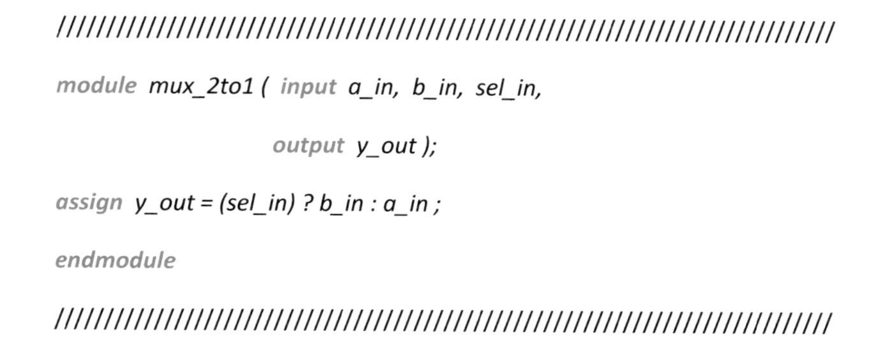

```
////////////////////////////////////////////////////////////////////////////////
module mux_2to1 ( input a_in, b_in, sel_in,
                  output y_out );
assign y_out = (sel_in) ? b_in : a_in ;
endmodule
////////////////////////////////////////////////////////////////////////////////
```

Example 1 Synthesizable design of 2:1 MUX

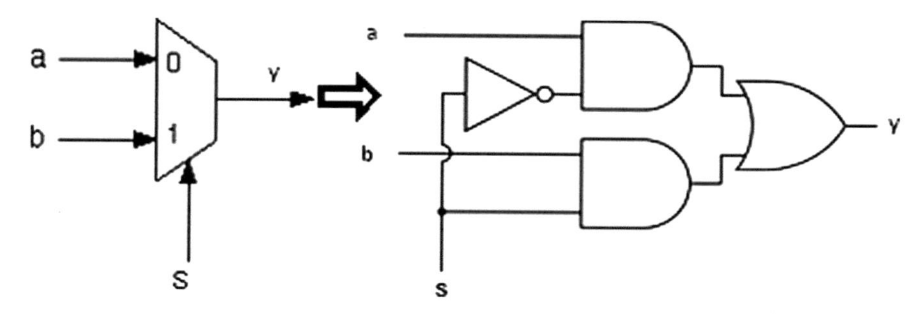

Fig. 5.2 Gate-level structure of 2:1 multiplexer

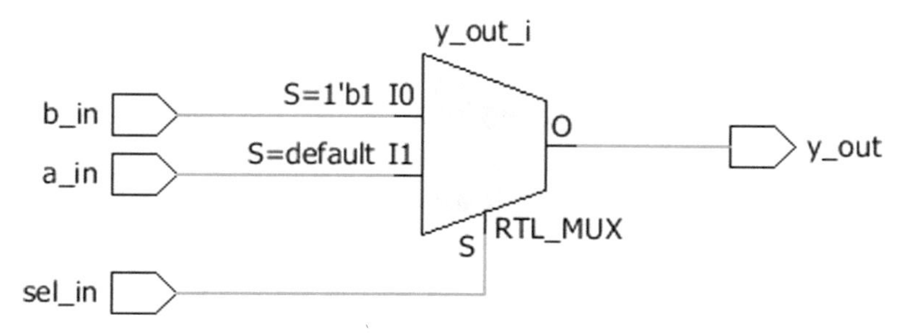

Fig. 5.3 RTL schematic of 2:1 multiplexer

A 2:1 multiplexer symbolic representation is used to describe the implementation of higher density multiplexers. Multiplexer is treated as universal logic. Using multiplexers, all possible combinational logic can be realized.

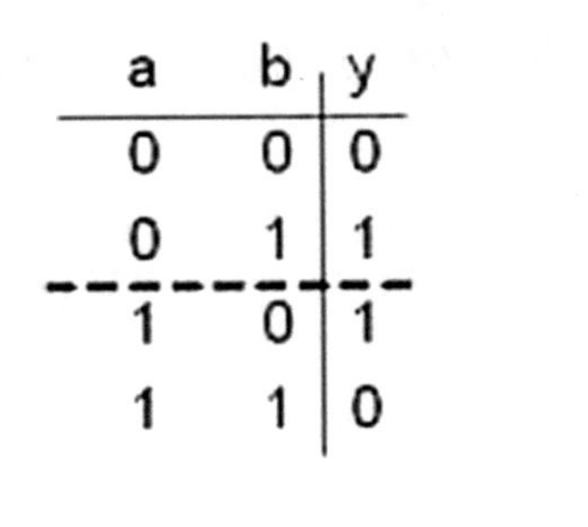

a	b	y
0	0	0
0	1	1
1	0	1
1	1	0

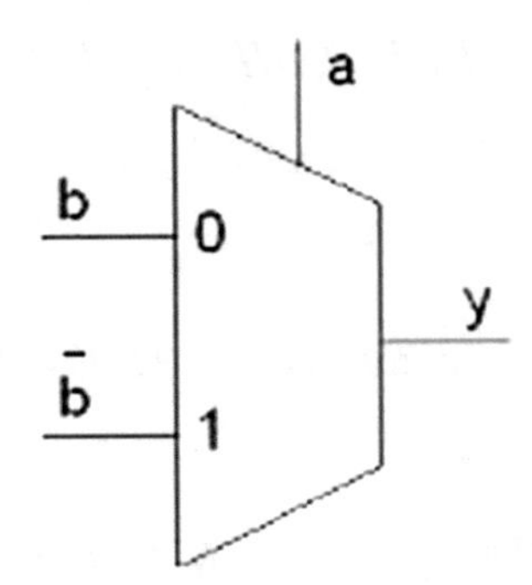

Fig. 5.4 Two-input XOR logic using 2:1 MUX

The reason for using MUX as universal logic is due to its easy to understand and simple structure. Figure 5.4 is useful to understand how 2:1 MUX is used to implement the two input XOR logic gates. Consider XOR logic gate has two inputs a, b and an output y. The implementation of two input XOR logic gates using 2:1 MUX is shown in Fig. 5.4.

> *Note* if-else generates priority logic, and case…endcase generates parallel logic. It is recommended to use case…endcase construct to code the MUX RTL. It is recommended to use if-else to code the RTL design to infer the priority logic.

5.3 The if...else Versus case Construct

Let us discuss the other ways to code the 2:1 MUX. There are different ways in which 2:1 MUX can be coded. It can be coded by using ***if-else*** or by using ***case… endcase.*** Example 2 describes synthesizable design using 'if-else', and Example 3 describes synthesizable design using 'case' statement.

```
//////////////////////////////////////////////////////////////////////////////
module mux_2to1( input a_in, b_in, sel_in,
            output reg y_out );
always@*
begin
if(sel_in)
  y_out = b_in;
else
  y_out = a_in;
end
endmodule
//////////////////////////////////////////////////////////////////////////////
```

Example 2 Synthesizable Verilog code for 2:1 MUX using if-else

```
//////////////////////////////////////////////////////////////////////////////
module mux_2to1 ( input a_in, b_in, sel_in,
                  output reg y_out );
always@*
begin
  case(sel_in)
  1'b0 : y_out = a_in;
  1'b1 : y_out = b_in;
  endcase
end
endmodule
//////////////////////////////////////////////////////////////////////////////
```

Example 3 Synthesizable Verilog code of 2:1 MUX using case

5.4 The 4:1 MUX Using if...else

Four is to one MUX has four input lines and single output lines. The 4:1 MUX has two select lines and been used to select one of the inputs at a time. The truth table of 4:1 MUX is shown in Table 5.2, and Example 4 is synthesizable design of 4:1 MUX (Fig. 5.6).

Table 5.2 Truth table of 4:1 MUX

sel_in[1]	sel_in[0]	y_out
0	0	d_in[0]
0	1	d_in[1]
1	0	d_in[2]
1	1	d_in[3]

```
/////////////////////////////////////////////////////////////////////////////////

module mux_4to1 ( input [3:0] d_in,

                  input [1:0] sel_in,

                 output reg y_out );

always @*

begin

  if (sel_in ==2'b00)

    y_out = d_in[0];

  else if (sel_in ==2'b01)

    y_out = d_in[1];

  else if (sel_in ==2'b10)

    y_out = d_in[2];

  else

    y_out = d_in[3];

end

endmodule

/////////////////////////////////////////////////////////////////////////////////
```

Example 4 Synthesizable Verilog design of 4:1 MUX

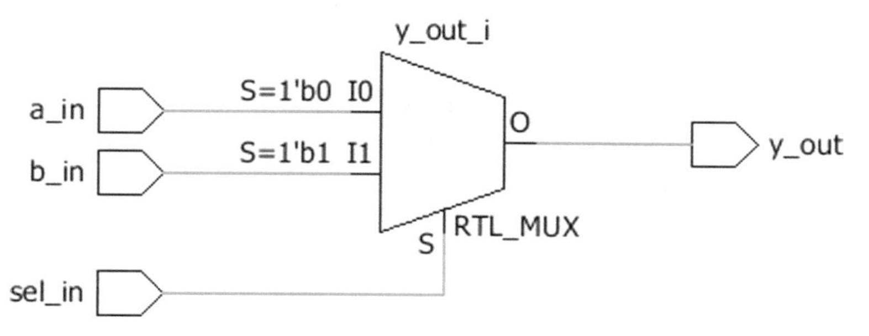

Fig. 5.5 RTL schematic of 2:1 multiplexer

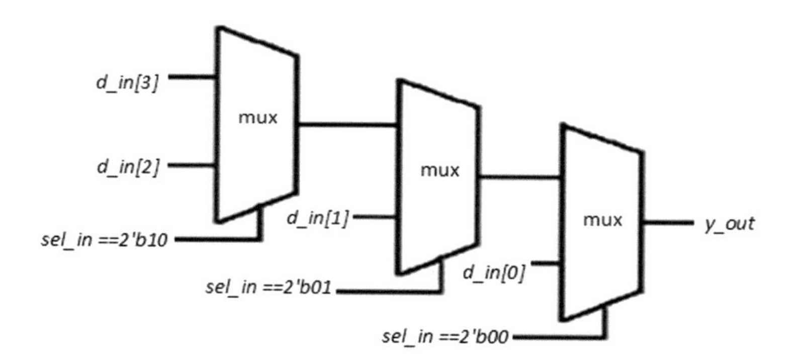

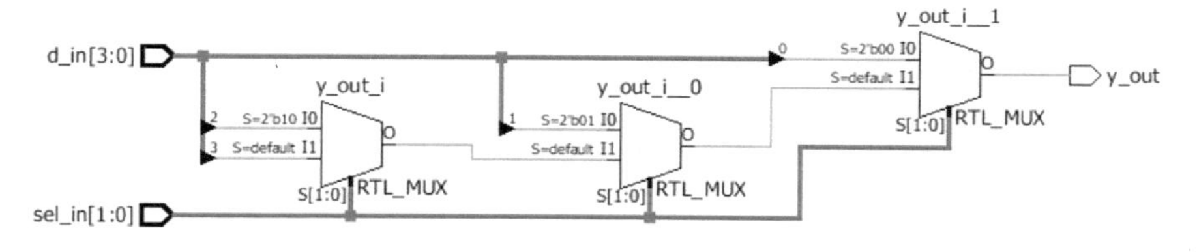

Fig. 5.6 Synthesis result of 4:1 MUX as priority logic

The logic inferred for the 4:1 MUX is shown in Fig. 5.6. As shown, input d_in [0] has highest priority as compared to other inputs, and an input d_in[3] has least priority.

5.5 The 4:1 MUX Using case Construct

The 4:1 MUX is described by using the ***case-endcase construct,*** and it is described in Example 5. The logic inferred is shown in Fig. 5.7. As shown, the ***case-endcase*** construct is used to infer the parallel logic.

```
////////////////////////////////////////////////////////////////////////////////
module mux_4to1 ( input [3:0] d_in,
                  input [1:0] sel_in,
                  output reg y_out);
always @*
begin
  case ( sel_in )
   2'b00 :  y_out = d_in[0];
   2'b01 :  y_out = d_in[1];
   2'b10 :  y_out = d_in[2];
   2'b11 :  y_out = d_in[3];
  endcase
end
endmodule
////////////////////////////////////////////////////////////////////////////////
```

Example 5 Synthesizable Verilog design of 4:1 MUX

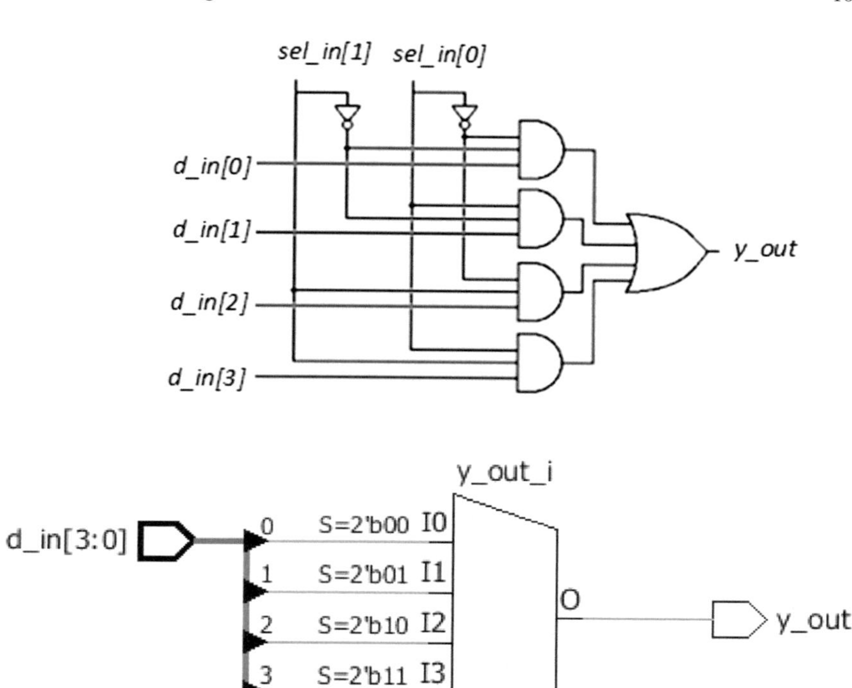

Fig. 5.7 Synthesis result of 4:1 MUX RTL design coded using case

5.6 The 4:1 Mux Using 2:1 MUX

The 4:1 MUX can be implemented by using 2:1 MUX, and the equivalent representation is shown in Fig. 5.8.

As shown in Fig. 5.8, 4:1 MUX is implemented by using three 2:1 multiplexers. The Verilog design is coded using the ***case…endcase*** construct and shown in Example 6 (Fig. 5.9).

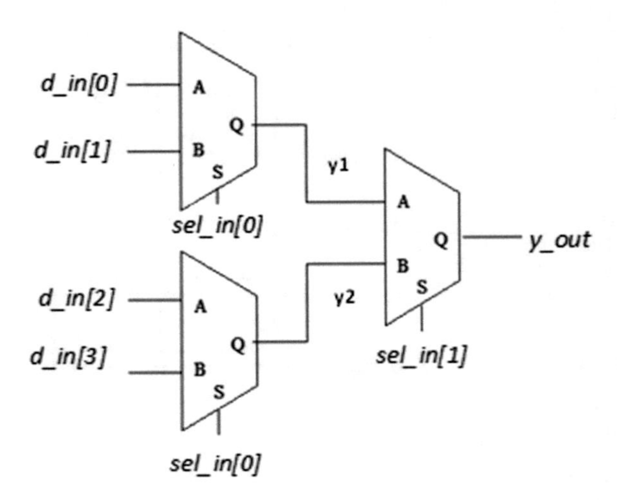

Fig. 5.8 4:1 MUX implementation using 2:1 MUX

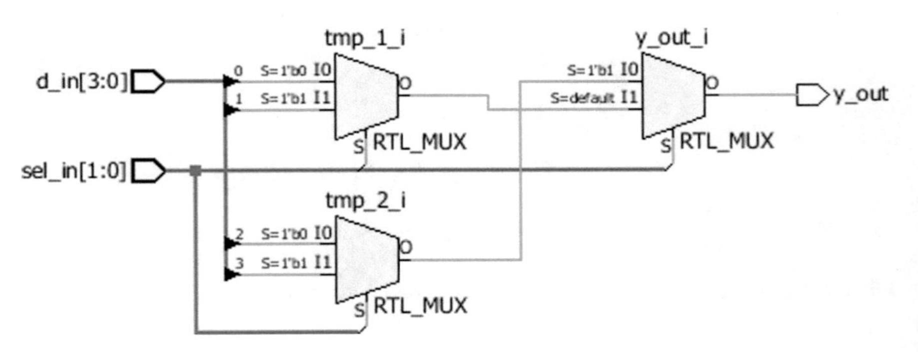

Fig. 5.9 RTL schematic of Example 6

```
////////////////////////////////////////////////////////////////////////////////
module mux_4to1 ( input [3:0] d_in,
                  input [1:0] sel_in,
                  output y_out );
reg tmp_1, tmp_2;
always @*
begin
  case ( sel_in[0] )
   1'b0 : begin
           tmp_1 = d_in[0];
           tmp_2 = d_in[2];
          end
   1'b1 : begin
           tmp_1 = d_in[1];
           tmp_2 = d_in[3];
          end
  endcase
end
assign y_out = (sel_in[1]) ? tmp_2 : tmp_1;
endmodule
////////////////////////////////////////////////////////////////////////////////
```

Example 6 Synthesizable Verilog design of 4:1 MUX

5.7 Let Us Design Combinational Logic Using Multiplexers

Now let us implement the single-bit adder and subtractor shown in Table 5.3 using the synthesizable constructs.

Due to use of the ***if…else*** construct the logic inferred uses the chain of multiplexers at the output. In Chap. 4, we have discussed the similar kind of example (Fig. 5.10).

Table 5.3 Operational table for the combinational design

Control input	Operation	Description
0	a_in + b_in	Perform addition of (a_in, b_in)
1	a_in − b_in	Perform subtraction of (a_in, b_in)

```
////////////////////////////////////////////////////////////////////////////////

module add_sub ( input a_in, b_in, control_in,

                    output reg result_out, carry_out );

always @ *

 if (control_in)

        { carry_out, result_out } = a_in + b_in;

 else

        { carry_out, result_out } = a_in + (~b_in) + 1;

endmodule

////////////////////////////////////////////////////////////////////////////////
```

Example 7 Use of if ... else to implement the design

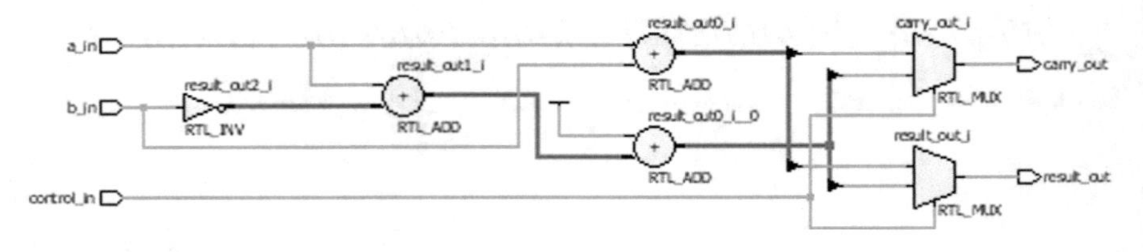

Fig. 5.10 RTL schematic of Example 7

5.8 Optimization Strategies Using RTL Tweaks

Using the RTL tweaks recommended in Chap. 4, the RTL design is coded by using the synthesizable constructs and shown in Example 8.

The synthesis result for Example 8 is shown in Fig. 5.11, and the logic inferred performs only one operation at a time and uses least resources as compared to Example 7.

```
//////////////////////////////////////////////////////////////////////////////

module add_sub ( input a_in, b_in, control_in,
                 output reg result_out, carry_out );

reg tmp_1;

always @*
        { carry_out, result_out } = a_in + tmp_1 + control_in ;

always @ *
if (control_in)
        tmp_1 = ~b_in;
else
        tmp_1 = b_in;

endmodule

//////////////////////////////////////////////////////////////////////////////
```

Example 8 RTL tweaks to improve the area for design

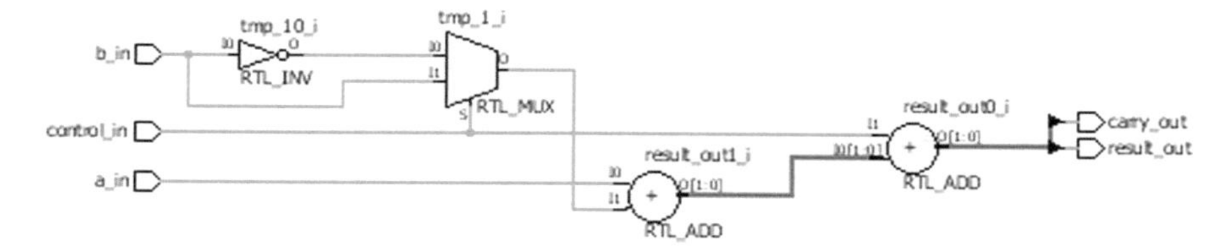

Fig. 5.11 RTL schematic of Example 8

5.9 Exercises

The exercises are based on the understanding of ***assign*** and **always** procedural block and use of ***if…else*** to model the combinational design. Complete the exercises for better understanding and application of Verilog constructs.

1. *The logic inferred by the following code is*

```
module comb_design_logic ( input a, b, sel, output y);

assign y = sel ? a : b ;

endmodule
```

   a. *Wire logic*
   b. *AND gate*
   c. *2:1 mux*
   d. *Syntax error in the code*

2. *The logic inferred by the following code is*

```
module comb_design_logic_0 ( input a, b,sel, output reg y);

always@*

begin

y = sel ? a : b;

end

endmodule
```

   a. *2:1 mux*
   b. *AND gate*

c. *NAND gate*
d. *Syntax error in the code*

3. *The logic inferred by the following code is*

```
module comb_design_logic ( input a, b, c,d, input [1:0] sel,
      output y);

wire y1,y2;

assign y1 = sel [0] ? a : b ;

assign y2 = sel [0] ? a : b ;

assign y = sel[1] ? y1 : y2;

endmodule
```

a. *4:1 mux*
b. *Single 2:1 mux*
c. *Two, 2:1 mux*
d. *Syntax error in the code*

4. *The logic inferred by the following code is*

```
module comb_design_logic_2 ( input a, b, output reg y1, y2);

always@ *
```

```
if (a==b)

y1=  a ^ b;

else

y2= a ~^ b;

endmodule
```

a. XOR gate, XNOR gate
b. OR gate, NOR gate
c. NOR gate, NAND gate
d. Multiplexer having one input as XOR and another as XNOR

5. The logic inferred by the following code is

```
module comb_design_logic ( input a, b, sel, output y);

always@*

assign y = sel ? a : b ;

endmodule
```

a. Wire logic
b. AND gate
c. 2:1 mux
d. Syntax error in the code

5.10 Summary

As discussed in this chapter, the combinational logic RTL using Verilog can be efficiently coded by using the Verilog constructs and the following are important points to summarize.

1. assign with the conditional expression is used to infer the 2:1 MUX.
2. MUX is treated as universal logic.
3. if-else is used to infer the 2:1 MUX, and nested if…else is used to infer the priority logic.
4. case-endcase is used to model the parallel logic and used within the procedural block.
5. default condition in the case-endcase is used to include the non-covered conditions.
6. Synthesis tool ignores the sensitivity list specified in the procedural block used to model combinational logic.
7. Using nested if-else, the priority designs are inferred and not recommended to code multiplexers.

Chapter 6
Decoders and Encoders

For an RTL design engineer, it is especially important to have better understanding of decoders, encoders. In most of the applications these are used as a functional blocks. This chapter discusses about the efficient RTL coding for decoders and encoders. The RTL design strategies for these combinational design elements are discussed using the synthesizable constructs.

In most of the system designs, we experience the use of the decoders and encoders. Most of the time in the system design, we use the decoder to select from one of the memory or IO devices. Even we experience the use pf the priority encoders in the design of level-sensitive interrupt controller. The chapter discusses the decoder and encoder RTL design using synthesizable Verilog constructs.

6.1 Decoders

Decoder has n select lines or input lines and m output lines and used to generate either active high or active low output. The relation between select lines and output lines is given by $m = 2^n$. Depending on the logic status on 'n' input lines at a time, one of the output line goes high or low. Figure 6.1 represents 3:8 decoder; as shown in the figure, X_2, X_1, X_0 are select inputs and Y_0 to Y_7 are active high output lines.

The truth table of 3–8 decoder is shown in Table 6.1. For the active high output decoder, at a time one of the output line is active high.

Note In the practical applications, decoders are used to select one of the memory or input–output device at a time. To enable the expansion of decoder, decoder can have either active high enable or active low enable

V. Taraate, *Digital Logic Design Using Verilog*,
https://doi.org/10.1007/978-981-16-3199-3_6

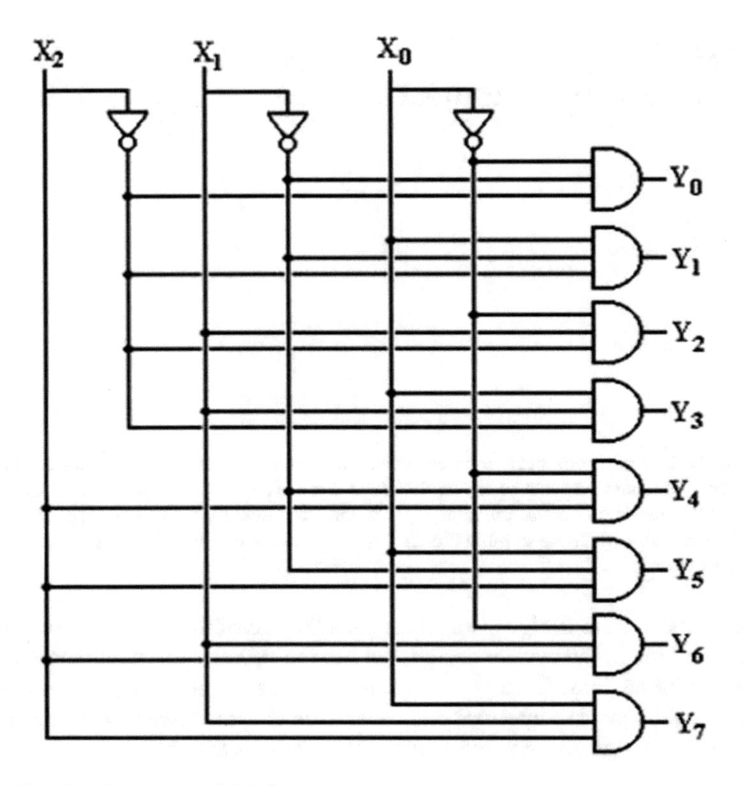

Fig. 6.1 Gate-level structure of 3:8 decoder

Table 6.1 Truth table of 3:8 decoder

X_2	X_1	X_0	Y_7	Y_6	Y_5	Y_4	Y_3	Y_2	Y_1	Y_0
0	0	0	0	0	0	0	0	0	0	1
0	0	1	0	0	0	0	0	0	1	0
0	1	0	0	0	0	0	0	1	0	0
0	1	1	0	0	0	0	1	0	0	0
1	0	0	0	0	0	1	0	0	0	0
1	0	1	0	0	1	0	0	0	0	0
1	1	0	0	1	0	0	0	0	0	0
1	1	1	1	0	0	0	0	0	0	0

Figure 6.2 is symbolical representation of 3:8 decoder having active high enable input en. The truth table shown holds good for the decoder which has active high enable en = 1. When en = 0, decoder is disabled and output Y = 8'b0000_0000. The synthesizable design is shown in Example 1.

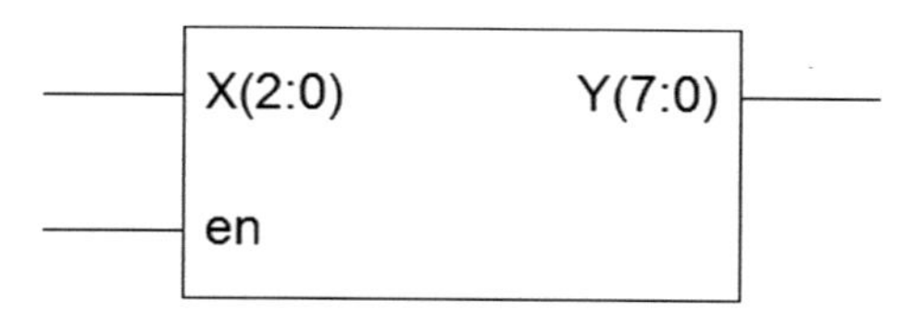

Fig. 6.2 Block-level representation of 3:8 decoder

```
//Verilog RTL for 3 to 8 decoder
module decoder_3to8 (X,en,Y) ;
input [2:0] X;
input en;
output [7:0] Y;
reg [7:0] Y;
// Functionality of design
always @ (X or en)
begin
        if (en)
        case(X)
        3'b000 : Y= 8'b0000_0001;
        3'b001 : Y= 8'b0000_0010;
        3'b010 : Y= 8'b0000_0100;
        3'b011 : Y= 8'b0000_1000;
        3'b100 : Y= 8'b0001_0000;
        3'b101 : Y= 8'b0010_0000;
        3'b110 : Y= 8'b0100_0000;
        3'b111 : Y= 8'b1000_0000;
        endcase
    else
        Y=8'b0000_0000;
end
endmodule
```

The decoder is enable for 'en=1' and generates one of the output as active high. The 'case' statement is used to describe the functionality of the decoder. Decoder generates parallel output.

In the RTL Verilog code 'if-else' is used to specify the select condition depending on the status of 'en'.

Example 1 Verilog RTL of 3:8 decoder (coding style Verilog-95)

6.1.1 1 Line to 2 Decoder Using case construct

The 1 line to 2 or (1:2) decoder has one select input Sel and two output lines Out_Y0 and Out_Y1. The truth table and logic equivalent are shown in Table 6.2 and Fig. 6.3, respectively.

Consider the 1:2 decoder which has input sel_in, active high enable input enable_in and output y_out[0], y_out[1]. The Verilog RTL is shown in Example 2, and the equivalent synthesis results in Fig. 6.4.

Table 6.2 Truth table for 1:2 decoder

Sel	Out_Y1	Out_Y0
0	0	1
1	1	0

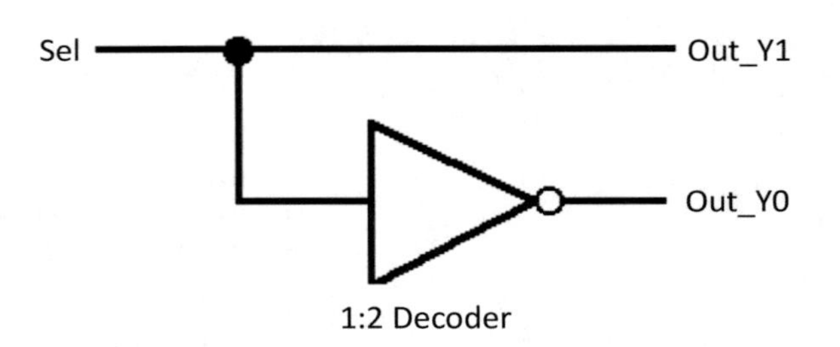

Fig. 6.3 1 line to 2-line decoder

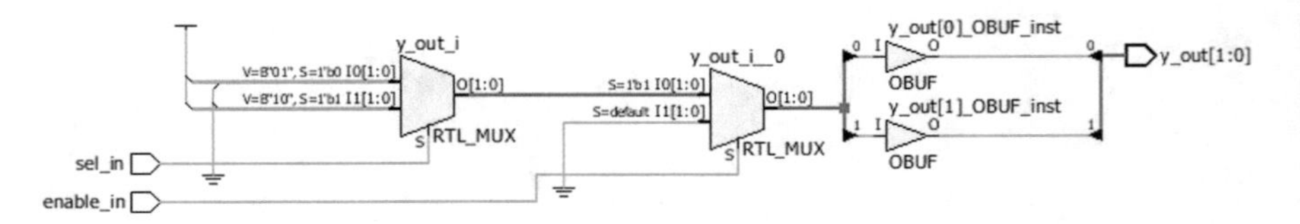

Fig. 6.4 RTL schematic of 1:2 decoder

```
//////////////////////////////////////////////////////////////////////////////

module decoder_1to2 ( input sel_in,

                      input enable_in,

                      output reg [1:0] y_out );

always @*

begin

 if(enable_in)

   case ( sel_in )

    1'b0 : y_out = 2'b01;

    1'b1 : y_out = 2'b10;

   endcase

 else

   y_out = 2'b00;

end

endmodule

//////////////////////////////////////////////////////////////////////////////
```

Example 2 Verilog RTL of 1:2 Decoder

6.1.2 1 Line to 2 Decoder Having Enable Using case

The 1 line to 2 or 1:2 decoder has one select input Sel, enable input En and two output lines Out_Y0 and Out_Y1. The relationship between the inputs and outputs is shown in Table 6.3.

The Verilog RTL is shown in the Example 3 and the equivalent synthesis result in Fig. 6.5.

Table 6.3 Truth table of 1:2 decoder having active high enable

En	Sel	Out_Y1	Out_Y0
1	0	0	1
1	1	1	0
0	X	0	0

```
//Verilog RTL for 1 Line to 2 Line decoder with active high enable input
module one_two_decoder_with_enable ( Sel, En, Out_Y1, Out_Y0);
input Sel;
input En;
output Out_Y1;
output Out_Y0;
reg Out_Y1;
reg Out_Y0;
always @ (Sel or En)
begin
    if (En)
            case (Sel)
                1'b0 : {Out_Y1, Out_Y0} = 2'b01;
                1'b1 : {Out_Y1, Out_Y0} = 2'b10;
                endcase
            else
             {Out_Y1, Out_Y0} = 2'b00;
end
endmodule
```

The decoder generates active high output 'Out_Y1, Out_Y0' depending on the select input 'sel'. For enable input 'En=1'

Sel=1 generates output as '10'.

Sel=0 generates output as '01'

For 'en=0' Output is '00'

Example 3 Verilog RTL of 1:2 decoder having enable input (Coding style Verilog-95)

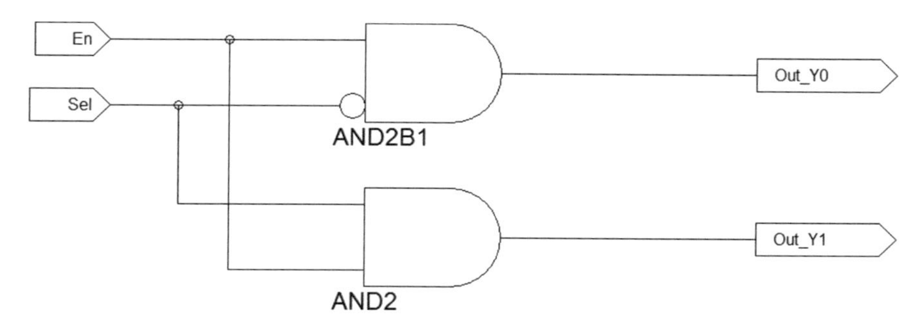

Fig. 6.5 Schematic of 1:2 decoder having active high enable

6.1.3 2 Line to 4 Decoder with Enable Using case

The 2 line to 4 or (2:4) decoder has two select inputs sel_in [1], sel_in [0], enable input enable_in and four output lines y_out[3], y_out[2], y_out[1], and y_out[0]. Table 6.4 gives information about the relationship between the select inputs and outputs.

The synthesizable Verilog design is coded and shown in Example 4, and the equivalent logic inferred is shown in Fig. 6.6.

Table 6.4 Truth table of 2:4 decoder

enable_in	Sel_in[1]	sel_in[0]	y_out[3]	y_out[2]	y_out[1]	y_out[0]
1	0	0	0	0	0	1
1	0	1	0	0	1	0
1	1	0	0	1	0	0
1	1	1	1	0	0	0
0	X	X	0	0	0	0

```
////////////////////////////////////////////////////////////////////////////////
module  decoder_2to4 (  input [1:0] sel_in,
                        input enable_in,
                       output  reg [3:0] y_out  );
always @*
begin
 if(enable_in)
   case ( sel_in )
    2'b00  :  y_out = 4'b0001;
    2'b01  :  y_out = 4'b0010;
    2'b10  :  y_out = 4'b0100;
    2'b11  :  y_out = 4'b1000;
    endcase
 else
    y_out = 2'b0000;
end
endmodule
////////////////////////////////////////////////////////////////////////////////
```

Example 4 Synthesizable Verilog design of 2:4 decoder

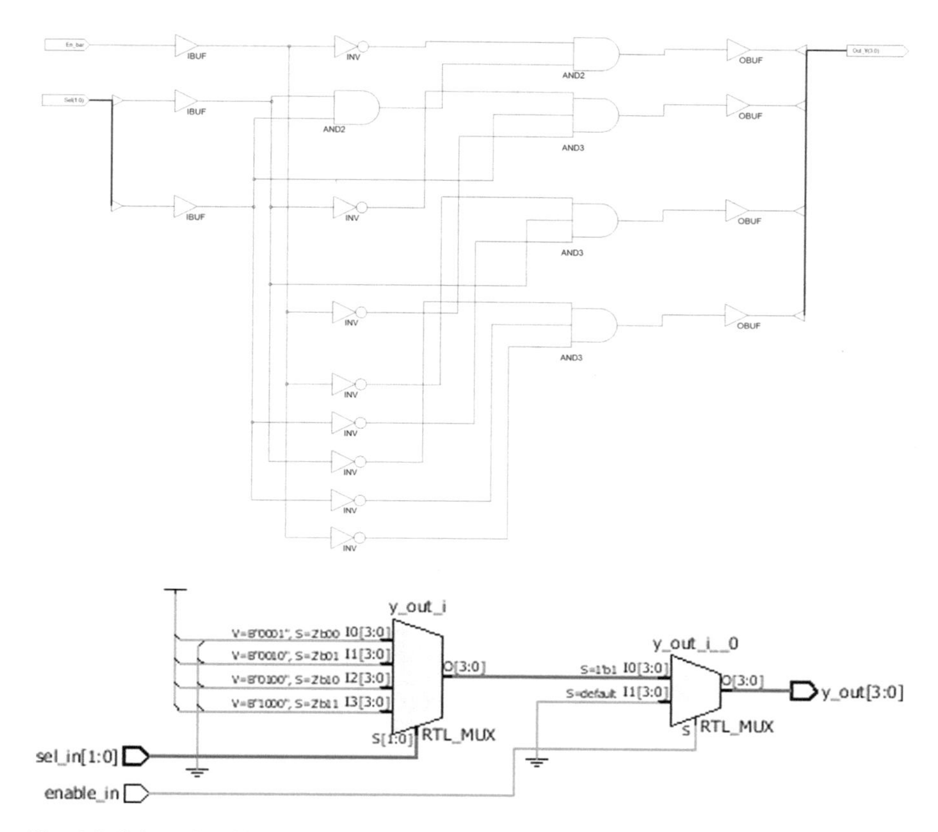

Fig. 6.6 Schematic of 2:4 decoder having active high enable input

6.1.4 2 Line to 4 Decoder with Active Low Enable Using case

The 2 line to 4 or (2:4) decoder has two select inputs Sel [1], Sel [0], active low enable input En_bar, and four active low output lines Out_Y[3], Out_Y[2], Out_Y [1], and Out_Y[0]. The truth table and equivalent representation is shown in Table 6.5.

Table 6.5 Truth table for 2:4 decoder having active low enable and active low output

En_bar	Sel[1]	Sel[0]	Out_Y[3]	Out_Y[2]	Out_Y[1]	Out_Y[0]
0	0	0	1	1	1	0
0	0	1	1	1	0	1
0	1	0	1	0	1	1
0	1	1	0	1	1	1
1	X	X	1	1	1	1

```
//Verilog RTL for 2 Line to 4 Line decoder with active low enable input and active low output lines
module Two_to_Four_decoder(Sel,En_bar, Out_Y);
input [1:0] Sel;
input En_bar;
output [3:0] Out_Y;
reg [3:0] Out_Y;
always @ (Sel or En_bar)
begin
    if (~En_bar)
            case (Sel)
                2'b00 : Out_Y = 4'b1110;
                2'b01 : Out_Y = 4'b1101;
                2'b10 : Out_Y = 4'b1011;
                2'b11 : Out_Y = 4'b0111;
                endcase
            else
             Out_Y = 4'b1111;
end
endmodule
```

The decoder generates active low outputs ' Out_Y[3]:Out_Y[0] ' depending on the select input 'sel[1:0]'. For enable input 'En_bar=0'

For 'En_bar=1' Output is '1111'

```
//Verilog RTL for 2 Line to 4 Line decoder with active low enable input and for active low output lines
module Two_to_Four_decoder(Sel,En_bar, Out_Y);
input [1:0] Sel;
input En_bar;
output [3:0] Out_Y;
assign Out_Y[3] = (~En_bar) && (~Sel[1]) && (~Sel[0]);
assign Out_Y[2] = (~En_bar) && (~Sel[1]) && (Sel[0]);
assign Out_Y[1] = (~En_bar) && (Sel[1]) && (~Sel[0]);
assign Out_Y[0] = (~En_bar) && (Sel[1]) && (Sel[0]);
endmodule
```

The decoder generates active low outputs ' Out_Y[3]:Out_Y[0] ' depending on the select input 'sel[1:0]'. For enable input 'En_bar=0'

For 'En_bar=1' Output is '1111'

Example 5 Synthesizable design of 2:4 decoder (coding style Verilog-95)

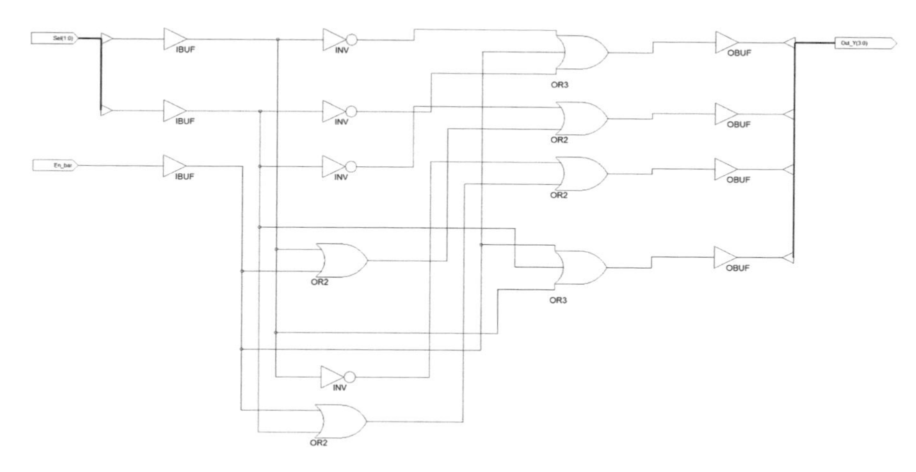

Fig. 6.7 Schematic of 2:4 decoder

The Synthesizable Verilog design using Verilog-95 coding style is shown in Example 5, and the equivalent logic inferred is shown in Fig. 6.7.

Both the RTL Verilog codes shown in Example 5 infer the same logic and shown in Fig. 6.7.

6.1.5 2 to 4 Decoder Using Continuous Assignments

Even the functionality of decoders can be described using the continuous assignment construct. The reason being multiple assign constructs executes in parallel and infers the parallel output logic. The 2:4 decoder description using continuous assign construct is shown in Example 6.

The synthesis result of Example 6 is shown in Fig. 6.8, as shown the logic inferred uses the 2 input AND gates to generate the four parallel output lines. At a time, one of the outputs is high depending on the status of select lines.

```
/////////////////////////////////////////////////////////////////////////////////
        module  decoder_2to4 (  input [1:0] sel_in,

                                input enable_in,

                                output  [3:0] y_out  );

  assign  y_out[0] = enable_in & (~sel_in[1])&(~sel_in[0]);

  assign  y_out[1] = enable_in & (~sel_in[1])&(sel_in[0]);

  assign  y_out[2] = enable_in & (sel_in[1])&(~sel_in[0]);

  assign  y_out[3] = enable_in & (sel_in[1])&(sel_in[0]);

  endmodule

/////////////////////////////////////////////////////////////////////////////////
```

Example 6 Synthesizable design of 2:4 decoder using continuous assignments

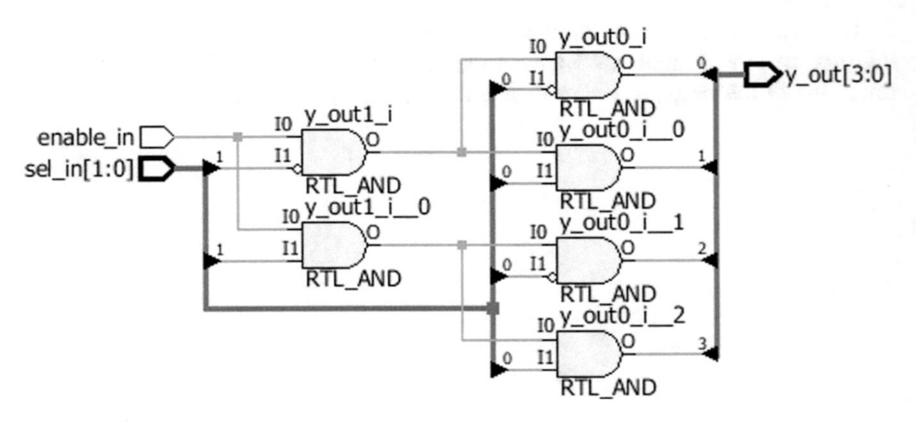

Fig. 6.8 Gate-level schematic of 2:4 decoder

6.1.6 Decoder Using Shift Operator

The better technique to implement the RTL design for 2:4 decoder is use of the shift operator. The RTL is coded to get one of the output as active high deepening on the status of select inputs and shown in Example 7.

```
////////////////////////////////////////////////////////////////////////////

module decoder_2to4 ( input [1:0] sel_in,

                      input enable_in,

                      output reg [3:0] y_out );

always @*

begin

 if(~enable_in)

y_out = 4'b0000;

else

 y_out = ( 4'b0001 << sel_in);

end

endmodule

////////////////////////////////////////////////////////////////////////////
```

Example 7 Synthesizable design of 2:4 decoder using shift operators

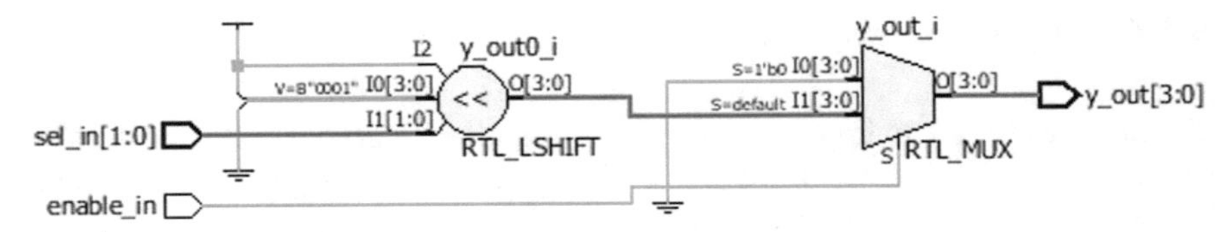

Fig. 6.9 2:4 decoder schematic using shift operator

The synthesis of Example 7 infers the logic which has the operator-specific hardware and the multiplexer chain at the output (Fig. 6.9).

6.1.7 Testbench of 2:4 Decoder

The testbench of decoder using the non-synthesizable constructs to generate the stimulus at the sel_in, enable_in is shown in Example 8.

The simulation result for the 2:4 decoder is shown in Fig. 6.10.

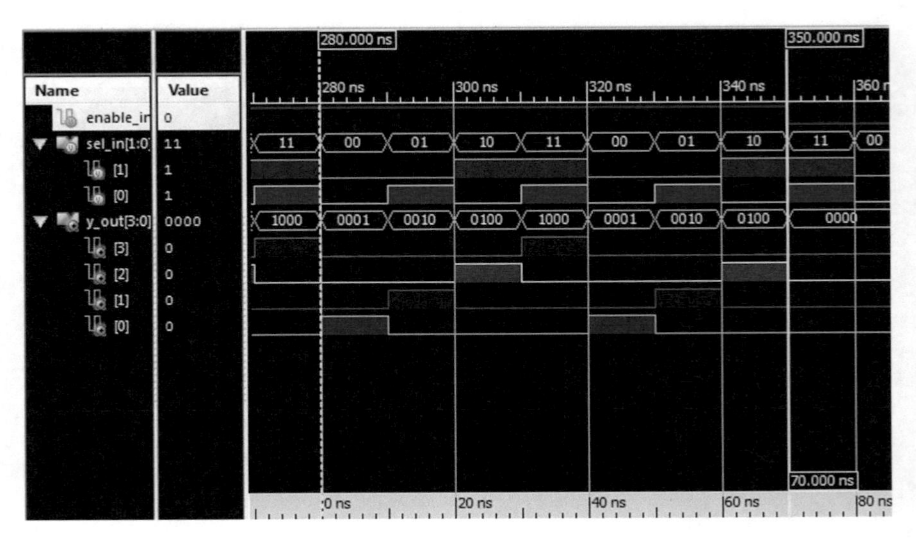

Fig. 6.10 Simulation waveform of 2:4 decoder

```
////////////////////////////////////////////////////////////////////////////////

module test_decoder;

// Inputs

reg [1:0] sel_in;

reg enable_in;

// Outputs

wire [3:0] y_out;

// Instantiate the Unit Under Test (UUT)

decoder_2to4 uut ( .sel_in(sel_in),

                   .enable_in(enable_in),

                   .y_out(y_out)

                 );

always #10 sel_in[0] = ~sel_in[0];

always #20 sel_in[1] = ~sel_in[1];

initial

begin

    // Initialize Inputs

    sel_in = 0;

    enable_in = 0;

    // Wait 100 ns and then force enable_in =1

    #100;

    enable_in =1 ;
```

Example 8 Testbench of 2:4 decoder using non-synthesizable constructs

```
            // Wait 250 ns and then force enable_in =0
            #250
            enable_in =0;
      end
endmodule
////////////////////////////////////////////////////////////////////////////
```

Example 8 (continued)

6.1.8 4 Line to 16 Decoder Using 2:4 Decoder

The 4 line to 16 or (4:16) decoder has four select inputs sel_in[3]: sel_in[0], active low enable input enable_in and designed by using four, 2:4 decoders. Each 2:4 decoder has four active low output lines y_out[3], y_out[2], y_out[1], and y-out[0]. The equivalent representation is shown in Fig. 6.11.

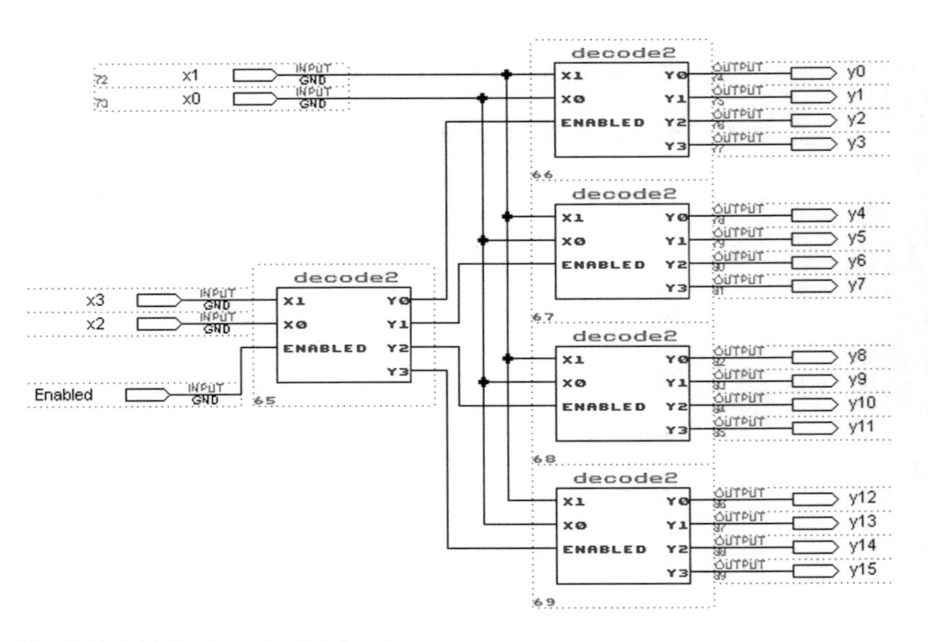

Fig. 6.11 4:16 decoder using 2:4 decoders

The logic inferred is shown in Fig. 6.12 and has 16, 2:1 multiplexers at the output. The FPGA synthesis uses most of the time more multiplexers, and ASIC synthesis result differs as compared with FPGA synthesis.

```
////////////////////////////////////////////////////////////////////////////////
module decoder_4to16 ( input [3:0] sel_in,
                       input enable_in,
                       output reg [15:0] y_out );
reg [3:0] tmp_enable;
always@*
if (~enable_in)
 tmp_enable = 4'b0000;
else
 tmp_enable = ( 4'b0001 << sel_in[3:2]);
always@*
if (~tmp_enable[0])
 y_out[3:0] = 4'b0000;
else
 y_out [3:0] = ( 4'b0001 << sel_in[1:0]);
always@*
if (~tmp_enable[1])
 y_out[7:4] = 4'b0000;
else
```

Example 9 Synthesizable design of 4:16 decoder using 2:4 decoders

```
 y_out [7:4] = ( 4'b0001 << sel_in[1:0]);

always@*
if (~tmp_enable[2])
 y_out[11:8] = 4'b0000;
else
 y_out [11:8] = ( 4'b0001 << sel_in[1:0]);
always@*
if (~tmp_enable[3])
 y_out[15:12] = 4'b0000;
else
 y_out [15:12] = ( 4'b0001 << sel_in[1:0]);
endmodule
//////////////////////////////////////////////////////////////////////////////
```

Example 9 (continued)

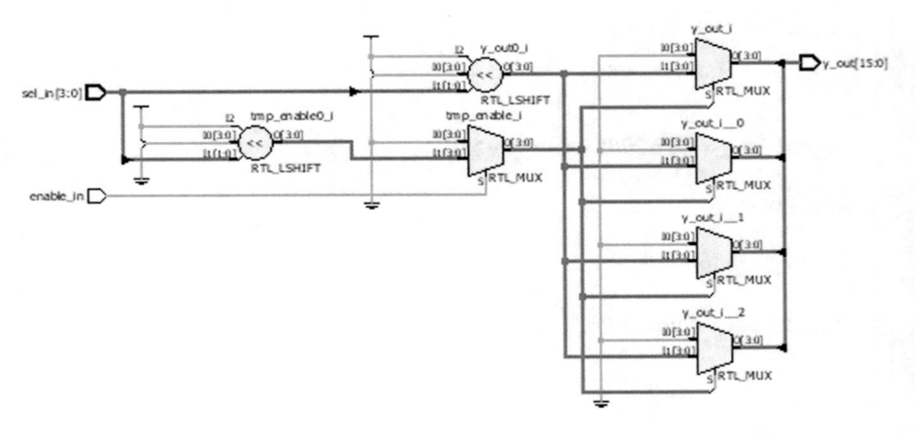

Fig. 6.12 Schematic of 4:16 decoder using 2:4 decoders

```
//////////////////////////////////////////////////////////////////////////////////

module test_decoder_4to16;

// Inputs

reg  [3:0]  sel_in;

reg  enable_in;

// Outputs

wire  [15:0]  y_out;

// Instantiate the Unit Under Test (UUT)

decoder_4to16  uut  (

                    .sel_in(sel_in),

                    .enable_in(enable_in),

                    .y_out(y_out)

                );

always  #5 sel_in[0] = ~sel_in[0];

always  #10 sel_in[1] = ~sel_in[1];

always  #20 sel_in[2] = ~sel_in[2];

always  #40 sel_in[3] = ~sel_in[3];

initial

begin

        // Initialize Inputs

        sel_in = 0;

        enable_in = 0;
```

Example 10 Testbench of 4:16 decoder using non-synthesizable constructs

```
// Wait 100 ns and then force enable_in =1

 #100;

enable_in =1 ;

// Wait 250 ns and then force enable_in =0

#250

enable_in =0;

    end

endmodule

////////////////////////////////////////////////////////////////////////////////
```

Example 10 (continued)

6.2 Testbench for 4:16 Decoder

The testbench for the 4:16 decoder is described using the non-synthesizable constructs and shown in Example 10.

The simulation result is shown in the waveform Fig. 6.13.

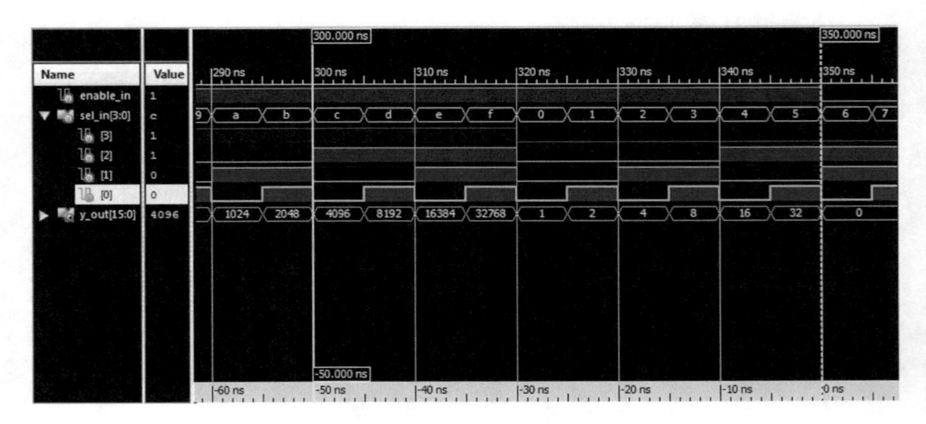

Fig. 6.13 Simulation result of 4:16 decoder

6.3 Encoders

Function of an encoder is exactly reverse of the decoder. Encoder has n input lines and m output lines, and the relation between input lines and output lines is given by $n = 2^m$ For example, consider 4:2 encoder. Number of input lines are n = 4, and output lines m = 2. The block diagram of 4:2 encoder is shown in Fig. 6.14 with the equivalent gate-level representation for 4:2 encoder, and the truth table is shown in Table 6.6.

The Verilog RTL description for 4:2 encoder is shown in Example 11. The Verilog RTL infers the similar logic as shown in Fig. 6.15.

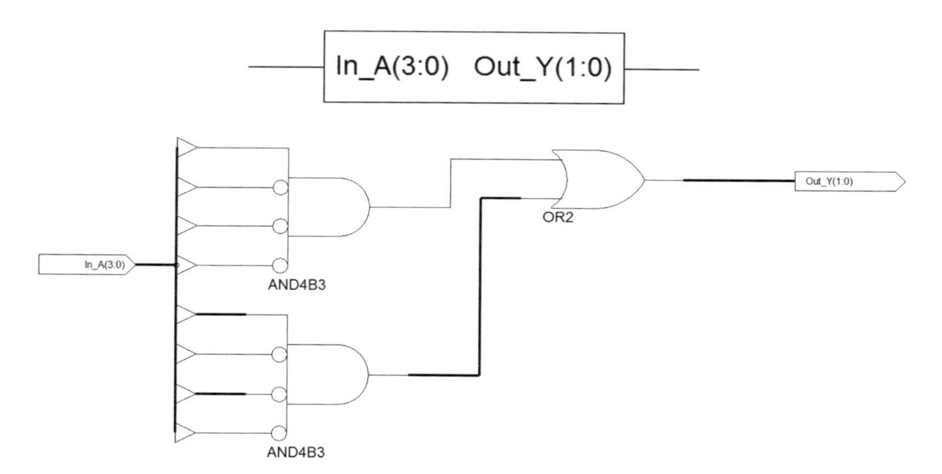

Fig. 6.14 4:2 encoder

Table 6.6 Truth table of 4:2 encoder

In[3]	In[2]	In[1]	In[0]	Out_Y[1]	Out_Y[0]
1	0	0	0	1	1
0	1	0	0	1	0
0	0	1	0	0	1
0	0	0	1	0	0

```
////////////////////////////////////////////////////////////////////////////////

module encoder_4to2 ( input [3:0] data_in,
                      output reg invalid_data,
                      output reg [1:0] y_out);

always @*

begin

   case ( data_in )

    4'b0001 : { invalid_data, y_out } = 3'b000;

    4'b0010 : { invalid_data, y_out } = 3'b001;

    4'b0100 : { invalid_data, y_out } = 3'b010;

    4'b1000 : { invalid_data, y_out } = 3'b011;

    default : { invalid_data, y_out } = 3'b100;

   endcase

end

endmodule

////////////////////////////////////////////////////////////////////////////////
```

Example 11 Synthesizable Verilog design of 4:2 encoder

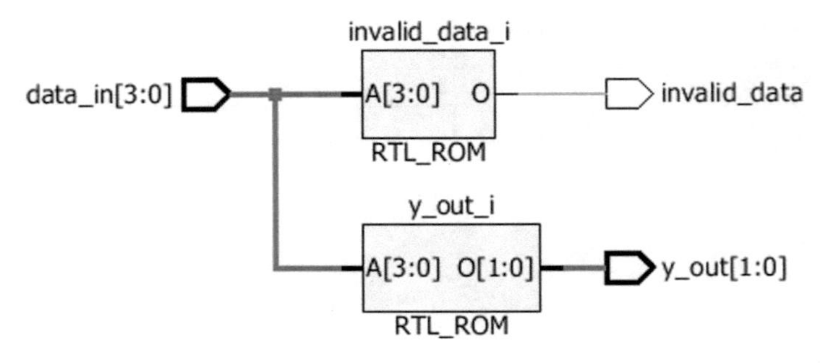

Fig. 6.15 RTL schematic of 4:2 encoder

6.3.1 Priority Encoders

Priority encoders are used in the practical applications and have n input lines and m output lines, and the relation between input lines and output lines is given by $n = 2^m$ For example, consider 4:2 priority encoder. Number of input lines are n = 4, and output lines m = 2. The block diagram of 4:2 priority encoder is shown in Fig. 6.16 with the equivalent gate-level representation for 4:2 priority encoder. The truth table is described in Table 6.7. The input In[3] has highest priority, and the input in[0] has lowest priority, where 'X' indicates the don't care.

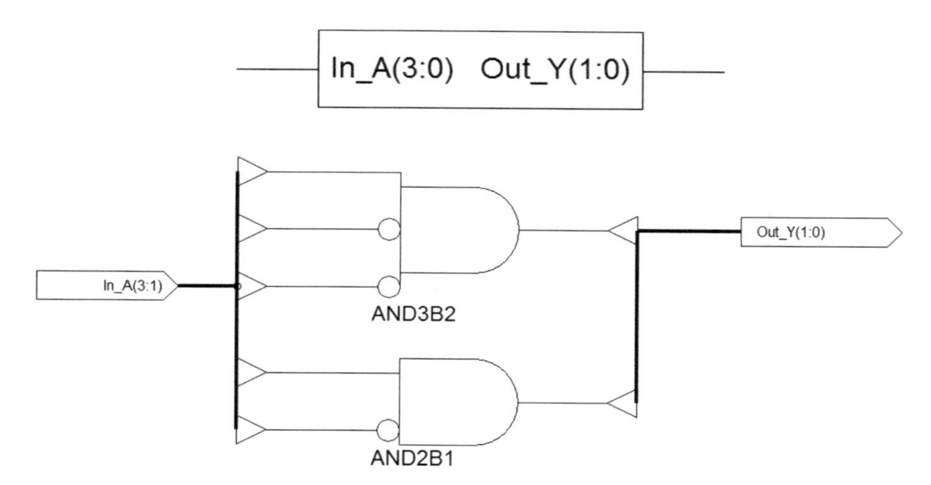

Fig. 6.16 4:2 Priority encoder

Table 6.7 Truth table of 4:2 priority encoder

In[3]	In[2]	In[1]	In[0]	Out_Y[1]	Out_Y[0]
1	X	X	X	1	1
0	1	X	X	1	0
0	0	1	X	0	1
0	0	0	1	0	0

```
/////////////////////////////////////////////////////////////////////////

module encoder_4to2 ( input [3:0] data_in,

                                     output reg invalid_data,

                                     output reg [1:0] y_out );

always @*

begin

   if( data_in[3])

    { invalid_data, y_out } = 3'b000;

   else if (data_in[2])

    { invalid_data, y_out } = 3'b001;

   else if (data_in[1])

    { invalid_data, y_out } = 3'b010;

   else if (data_in[0])

    { invalid_data, y_out } = 3'b011;

   else

    { invalid_data, y_out } = 3'b100;

end

endmodule

/////////////////////////////////////////////////////////////////////////
```

Example 12 Synthesizable Verilog design of 4:2 priority encoder

The Verilog RTL description for 4:2 priority encoder is coded using nested ***if… else*** construct and shown in Example 12. The Verilog RTL infers the logic as shown in Fig. 6.17.

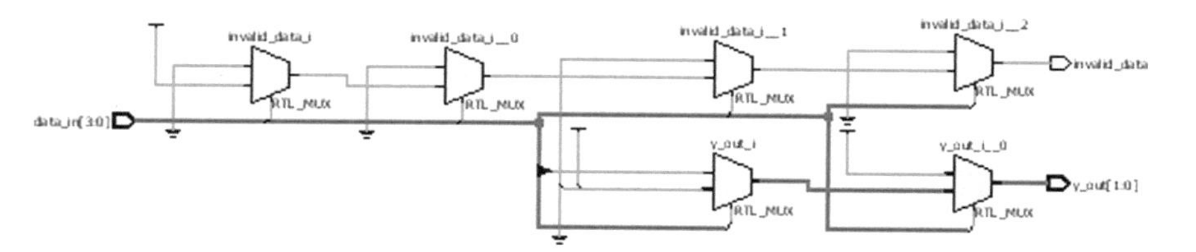

Fig. 6.17 RTL schematic of 4:2 priority encoders

Note: In the practical applications, encoders are used to design the control logic. As case..endcase generates the parallel logic and if-else generates the priority logic; case..endcase is used to code the behavior of encoder. if-else is used to code the behavior of priority encoder. Priority encoders are used to sense the level-sensitive interrupts.

6.4 Testbench of 4:2 Priority encoder

The testbench using the non-synthesizable constructs is shown in Example 13, and the simulation waveform is shown in Fig. 6.18.

```
//////////////////////////////////////////////////////////////////////////////////

module test_encoder_4to2;

// Inputs

reg [3:0] data_in;

// Outputs

wire invalid_data;

wire [1:0] y_out;

// Instantiate the Unit Under Test (UUT)

encoder_4to2 uut (

                    .data_in(data_in),

                    .invalid_data(invalid_data),

                    .y_out(y_out)

          );

always #5  data_in[0] = ~data_in[0];

always #10 data_in[1] = ~data_in[1];

always #20 data_in[2] = ~data_in[2];

always #40 data_in[3] = ~data_in[3];

initial

begin

     // Initialize Inputs

     data_in = 0;
```

Example 13 Testbench of 4:2 priority encoder

```
        // Wait 100 ns for global reset to finish

        #100;
    end

    endmodule

//////////////////////////////////////////////////////////////////////////////////
```

Example 13 (continued)

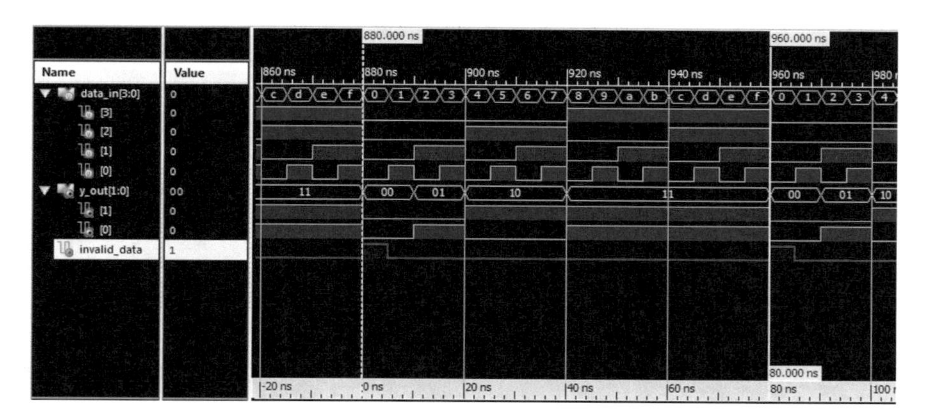

Fig. 6.18 Simulation waveform of 4:2 priority encoder

6.5 Exercises

Following are the exercises on the combinational design and verification. Complete the exercises for better understanding and application of the constructs.

1. *At t=200 ns the logic level at enable_in is equal to?*

```
initial
    begin
        // Initialize Inputs
        sel_in = 0;
        enable_in = 0;
        // Wait 100 ns and then force enable_in =1
        #100;

        enable_in =1 ;
        // Wait 250 ns and then force enable_in =0
        #250
        enable_in =0;
    end
```

a. 0
b. 1
c. X
d. Z

2. *At t=350 ns the logic level at enable_in is equal to?*

```
initial

    begin

        // Initialize Inputs

        sel_in = 0;

        enable_in = 0;

        // Wait 100 ns and then force enable_in =1

        #100;


        enable_in =1 ;

        // Wait 250 ns and then force enable_in =0

        #250

        enable_in =0;

    end
```

a. 0
b. 1
c. X
d. Z

3. The logic inferred by following code is ?

```
module  decoder_2to4 (  input [1:0] sel_in,

                        input enable_in,

                        output  [3:0] y_out  );

assign y_out = (enable_in) ? ( 4'b0001 << sel_in) : 4'b0000;

endmodule
```

a. 4:2 encoder
b. 2:4 decoder
c. 4:2 priority encoder
d. Syntax error in the code.

6.6 Summary

Following are important points to conclude this chapter.

1. Multiple continuous assignments using assign can be used to infer the decoder logic.
2. Decoders should be implemented using case construct as they have parallel outputs.
3. if-else generates priority logic and not recommended to use to model the decoders.
4. case-endcase is used to model the parallel logic that is decoders and encoders.
5. Decoders are used to select one of the memory or input–output device at a time.
6. Priority encoders are used in the design of interrupt control logic, and logic can be described by using nested if-else.

Chapter 7
Event Queue and Design Guidelines

It is essential to follow the coding and design guidelines while coding the RTL design. The design and coding guidelines will improve the performance of design, readability, and reusability. This chapter discusses about few important design and coding guidelines for the combinational logic design.

The design and coding guidelines are generally used to improve the design performance, readability, and the reusability. The combinational design where output is function of the present inputs should be coded in such a way that the design should have least propagation delay and the least area.

It is always recommended to follow certain coding guidelines while describing the design using Verilog. This chapter is more focused on the important design and coding guidelines used in the industries to code an efficient combinational logic.

7.1 Verilog Stratified Event Queue

Verilog supports the two kinds of the assignments in the procedural blocks. These assignments are named as blocking (=) and non-blocking (<=) assignments. It is always recommended to use the blocking assignments while describing the combinational logic design. The reason being quite simple to understand, but it is essential to understand the fundamental behind this!

To understand the blocking assignments, let us understand the concept of stratified event queue. According to IEEE 1364-2005 Verilog standard, the stratified event queue is classified into four major regions. These regions are named as: Active, Inactive, NBA, and Monitor.

But the major question is why to understand the stratified event queue? And what exactly the application of it? As the name itself indicates that the stratified event queue is used to hold the result after expression execution and useful to hold the results. Figure 7.1 shows according to the Verilog IEEE 1364-2005 standard.

V. Taraate, *Digital Logic Design Using Verilog*,
https://doi.org/10.1007/978-981-16-3199-3_7

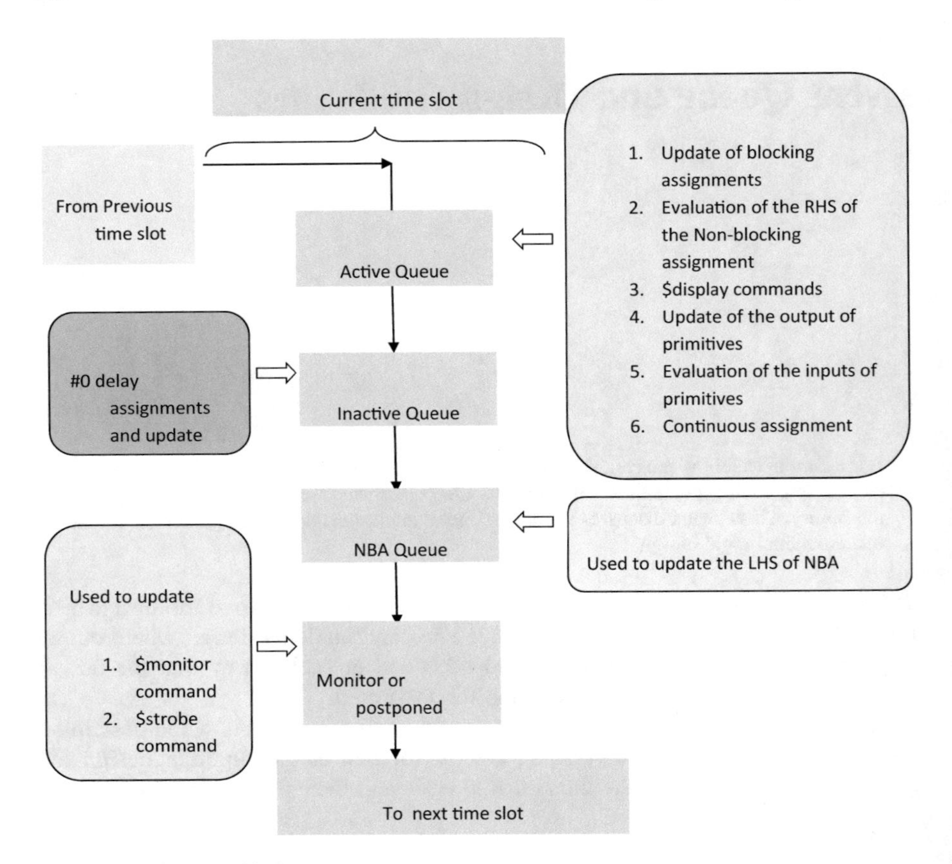

Fig. 7.1 Verilog stratified event queue

As shown in Fig. 7.1, the Verilog stratified event queue has four main regions and explained below.

i. **Active Queue**: Most of the Verilog events are scheduled in the active event queue. These events can be scheduled in any order and evaluated or updated in any order. The active queue is used to update the blocking assignments, continuous assignments, evaluation of RHS of the non-blocking assignments (LHS of NBA is not updated in the active queue), $display commands and to update the primitives.
ii. **Inactive Queue**: The #0delay assignments are updated in the inactive queue. Use of #0 delays in the Verilog is not good practice, and it unnecessarily complicates the event scheduling and ordering. Most of the times the designer uses the #0 delay assignments to fool the simulator to avoid the race around conditions.

iii. **NBA Queue**: The LHS of the non-blocking assignments updates in this queue.
iv. **Monitor Queue**: It is used to evaluate and update the $monitor and $strobe commands. The updates of all the variables are during the current simulation time.

7.2 Verilog Blocking Assignments

As discussed above, the blocking assignments execute sequentially inside procedural block. Blocking assignment blocks all the trailing statements in the procedural block while executing the current assignment. The execution of the blocking assignment is always considered as one-step process. In an active event queue, the RHS of blocking assignment is evaluated, and during the same time stamp, the LHS of blocking assignment is updated. Consider an Example 1 for the blocking assignments.

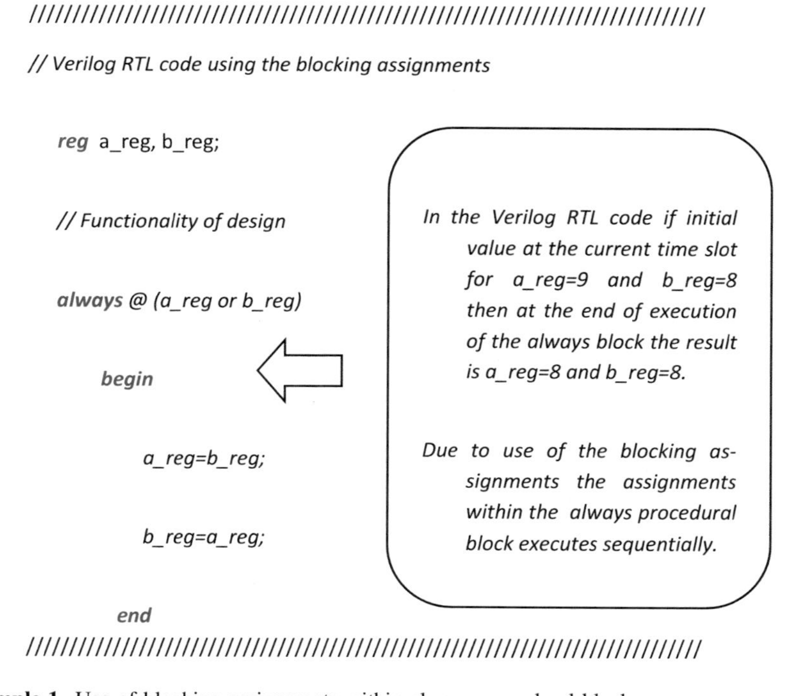

Example 1 Use of blocking assignments within always procedural block

Note: The major issue with the blocking assignments is while using the same variable on the RHS side in one procedural block and on LHS side in another procedural block. If both the procedural blocks are scheduled in the same simulation time or on the same clock edge it generates the race condition in the design. This will be discussed subsequently.

In the subsequent section, we will discuss the design and coding guidelines for combinational logic, and we will continue to use the blocking assignments to code for the combinational design.

7.3 Incomplete Sensitivity List

It is recommended to incorporate all the required signals and inputs in the sensitivity list while using ***always*** procedural block. Consider Example 2 to describe the functionality of two-input NAND logic.

In Example 2, the synthesis tool ignores the *sensitivity list* and infers the two input NAND gate as synthesizable output, but the simulator ignores the changes in the input b_in and generates the output waveform. This leads to simulation and synthesis mismatch. The testbench using the non-synthesizable constructs is coded to report the simulation and synthesis mismatch (Example 3). The simulation result is shown in Fig. 7.3.

Note: To avoid the simulation and synthesis mismatch, it is recommended to use the procedural block: always@(*). According to IEEE 1364-2001 standard, the '*' in the sensitivity list will include all the inputs and required signals.

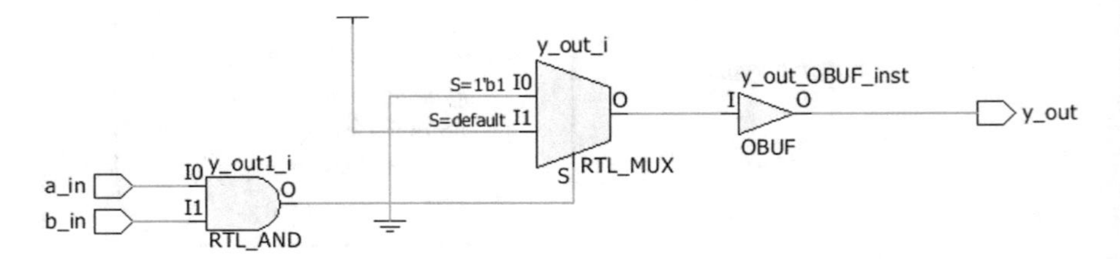

Fig. 7.2 RTL schematic of Example 2

```
///////////////////////////////////////////////////////////////////////////
// Verilog RTL code to understand the incomplete sensitivity list

module logic_design(

input a_in,

input b_in,

output reg y_out) ;

// Functionality of design

always @ (a_in)

  begin

    if (a_in==1'b1 && b_in==1'b1)

      y_out = 1'b0;

    else

      y_out =1'b1;

  end

endmodule
///////////////////////////////////////////////////////////////////////////
```

In the sensitivity list the a_in is specified but b_in input is missing and the simulator and synthesis Tool will flash the warning "Incomplete Sensitivity list"

This will cause the simulation and synthesis mismatch,.

Example 2 RTL design with missing one of the input from sensitivity list (Fig. 7.2)

```
/////////////////////////////////////////////////////////////////////////////////////

//testbench to find the simulation and synthesis mismatch

module test_logic;

// Inputs

reg a_in;

reg b_in;

// Outputs

wire y_out;

// Instantiate the Unit Under Test (UUT)

logic_design uut (

        .a_in(a_in),

        .b_in(b_in),

        .y_out(y_out)

);

always #25 a_in = ~a_in;

always #40 b_in = ~b_in;

initial

begin
                // Initialize Inputs

        a_in = 0;
        b_in = 0;

        // Wait 100 ns

        #100;

end

endmodule

/////////////////////////////////////////////////////////////////////////////////////
```

Example 3 Testbench to verify the functionality described in Example 2

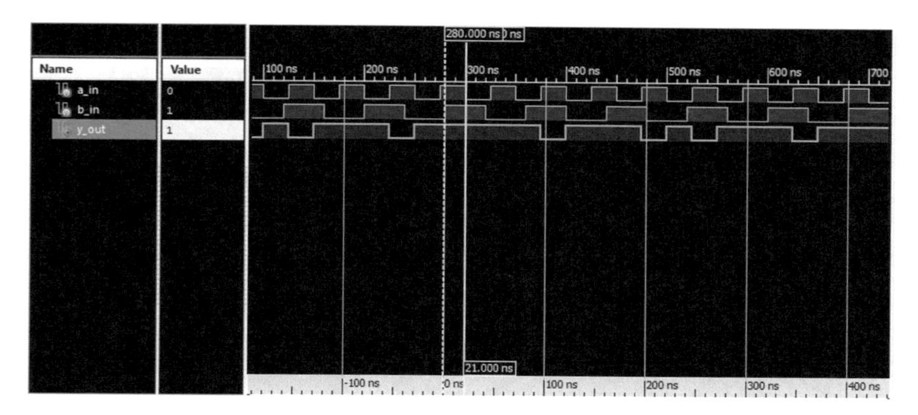

Fig. 7.3 Waveform to indicate the simulation and synthesis mismatch

As shown in Fig. 7.3, as b_in is missing from the sensitivity list, the y_out is 1 when a_in = 1 and b_in = 1. But NAND gate output is 0 when a_in = 1 and b_in = 1. Hence, simulation and synthesis mismatch.

7.4 Continuous Versus Procedural Assignments

Continuous assignments: Continuous assignments are used to assign the value to the net. These are used to code the combinational logic functionality. These assignments are updated in the active event queue, and the net values are updated upon evaluation of the right-hand side expression. The port or output is declared as ***wire*** while using the continuous assignment.

assign *y_out = sel_in ? a_in: b_in;*

Procedural assignments: Procedural assignments are used to assign value to the reg. These are used to code both the combinational and sequential logic. The output assigned to ***reg*** is hold until the next assignment is executed. These assignments are used in the procedural blocks ***always***, ***initial,*** and within the task and functions according to the requirements.

In the procedural block, if the blocking (=) assignments are used, then they are updated in the active event queue. All the non-blocking assignments (<=) are evaluated in the active event queue but updated in the non-blocking event queue.

```
//////////////////////////////////////////////////////////////////////////

always@(posedge clk) // Sequential design description

begin

    q_out<= data_in;

end

always@ *  // Combinational design description

begin

 y_out = sel_in ? a_in : b_in;

end
//////////////////////////////////////////////////////////////////////////
```

Example 4 Use of assignments in the RTL design

7.5 Combinational Loops in Design

The unintentional combinational loops in the design are extremely critical to debug and fix during the implementation phase and generate an oscillatory behavior. Example 5 is useful to understand what is exactly combinational loop?

Figure 7.4 describes the synthesizable output for the combinational loop.

> ***Note: It is recommended that the design should not have any combinational loop. To avoid the combinational loop, break the feedback by using the sequential elements.***

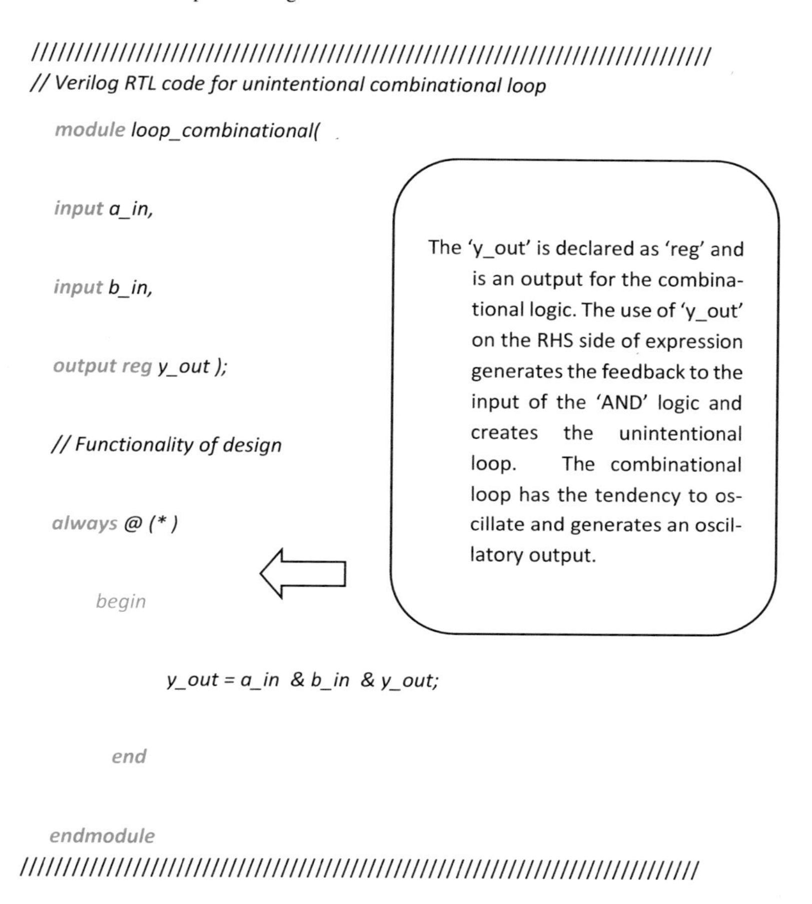

```
////////////////////////////////////////////////////////////////////////////////
// Verilog RTL code for unintentional combinational loop

module loop_combinational(

input a_in,

input b_in,

output reg y_out );

// Functionality of design

always @ (* )

    begin

        y_out = a_in  & b_in  & y_out;

    end

endmodule
////////////////////////////////////////////////////////////////////////////////
```

The 'y_out' is declared as 'reg' and is an output for the combinational logic. The use of 'y_out' on the RHS side of expression generates the feedback to the input of the 'AND' logic and creates the unintentional loop. The combinational loop has the tendency to oscillate and generates an oscillatory output.

Example 5 RTL to understand the combinational loop in the design

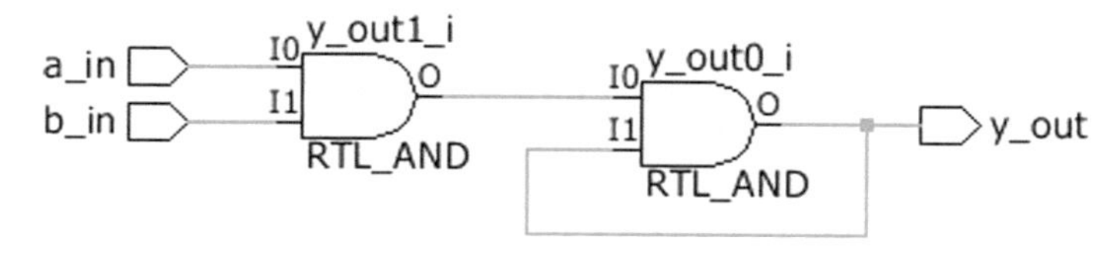

Fig. 7.4 Combinational loop in design

As discussed above, combinational loops in design are one of the dangerous and critical design errors. Combinational loop in the design occurs if the same signals are used or assigned in the multiple procedural blocks. If the same signal is present on the right-hand side of expression or on the left-hand side of expression, then the design has combinational loop.

Combinational loops exhibit the oscillatory behavior and during update phase, they can have race conditions. Consider the design scenario shown in Example 6.

In the above example, both ***always*** blocks execute concurrently and due to that while updating of the b, the b value is assigned to a. This is the race condition in the design. This design generates the oscillatory behavior due to events on a, b.

The oscillatory behavior can be understood from Fig. 7.5.

Combinational loops are not synthesizable, and the synthesis tool generates an error or warning for the combinational loop. Combinational looping can be potential hazard in the design and hence need to be avoided.

```
////////////////////////////////////////////////////////////////////////////

always@(a)

begin

    b=a;

end

always@(b)

begin

    a=b;

end
////////////////////////////////////////////////////////////////////////////
```

Example 6 Verilog RTL design having combinational loop

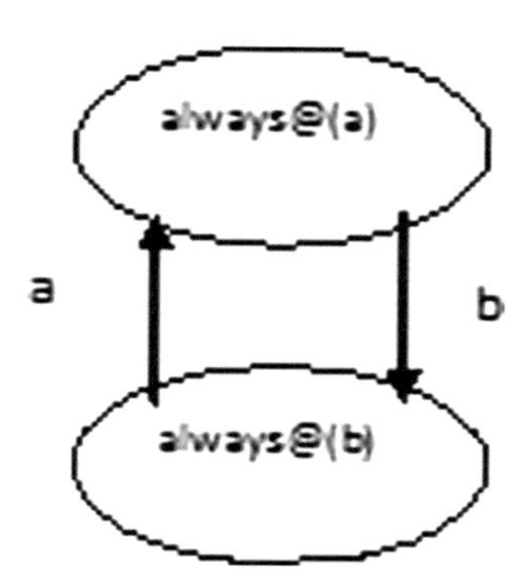

Fig. 7.5 Oscillatory behavior due to combinational loop

As shown in the above diagram, the ***always*** block ***always***@(a) is sensitive to the event on a and generates an output b. Eventually, changes on b input are used to trigger another ***always*** block ***always***@(b) and generate the output a. So, this goes on and exhibits the oscillatory behavior or the race around conditions in the design.

The solution to overcome this problem is to use of the register to avoid the dependency of signals to trigger multiple ***always*** block. Register can be inserted in the combinational loop to update the value.

To avoid combinational looping do the following. Use the ***non-blocking assignments,*** and use the register logic to break the combinational loops. The RTL tweak is shown in Example 7.

In Example 7, both ***always*** blocks are sensitive to positive edge of clock and assigns the value to b, a respectively. Although both the procedural blocks are executed concurrently, the non-blocking assignments are queued in the NBA queue and hence generates the structure as shown in Fig. 7.6.

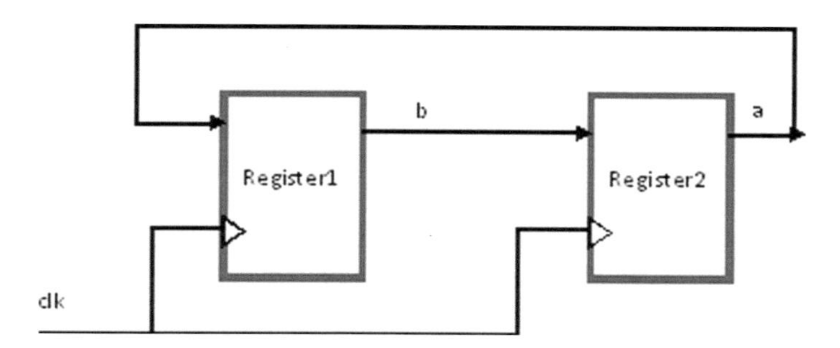

Fig. 7.6 Schematic indicates the use of register logic to avoid oscillatory behavior

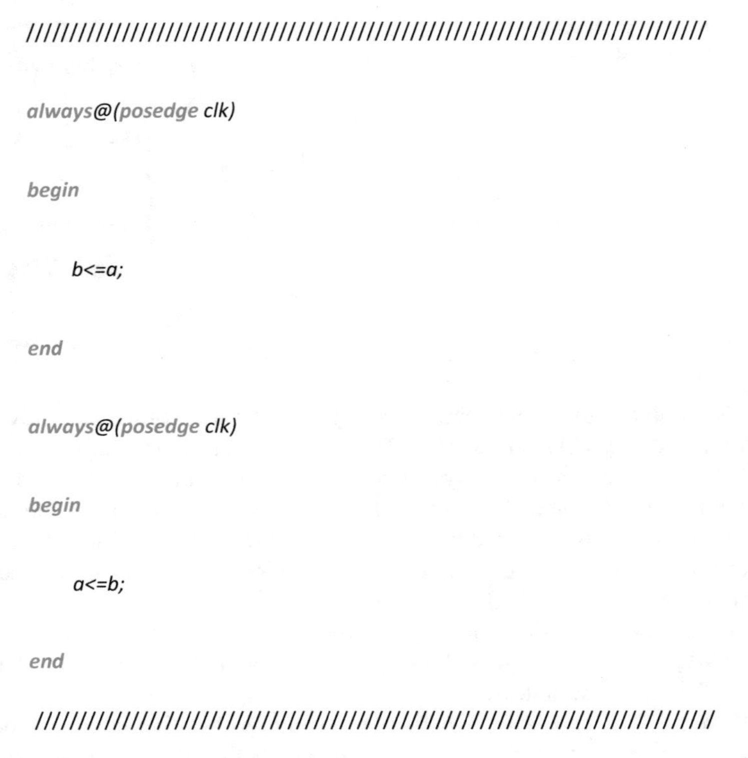

```
////////////////////////////////////////////////////////////////////////////////

always@(posedge clk)

begin

     b<=a;

end

always@(posedge clk)

begin

     a<=b;

end

////////////////////////////////////////////////////////////////////////////////
```

Example 7 RTL tweak to avoid combinational loop

7.6 Unintentional Latches in the Design

It is recommended that the design should not have unintentional latches as latch is transparent during active level and transfers the data to its output. The unintentional latches are not recommended in ASIC design as it causes the issues during the design testing or during DFT. Even during STA, the timing algorithm will be not able to understand whether to sample the data on positive edge of the clock or on negative edge of the clock. So, most of the time main intention of the designer is not reflected while using the latches, and STA for such paths is difficult. This will be discussed in subsequent chapters.

Consider the example coded and shown in Example 8.

In the above code as during the execution of ***else*** condition, the information about the assignment to b_in is not given so it infers the latch and holds the previous value of b_in. The logic representation is shown in the Fig. 7.7. The ***if-else*** construct infers multiplexer, and as assignment to b_in is missing within ***else***, it infers the positive level-sensitive latch which is controlled by enable input c_in.

As shown in Fig. 7.7, due to missing b_in assignment within the ***else*** clause, it generates the latch and holds the previous value assigned. Latches are inferred due

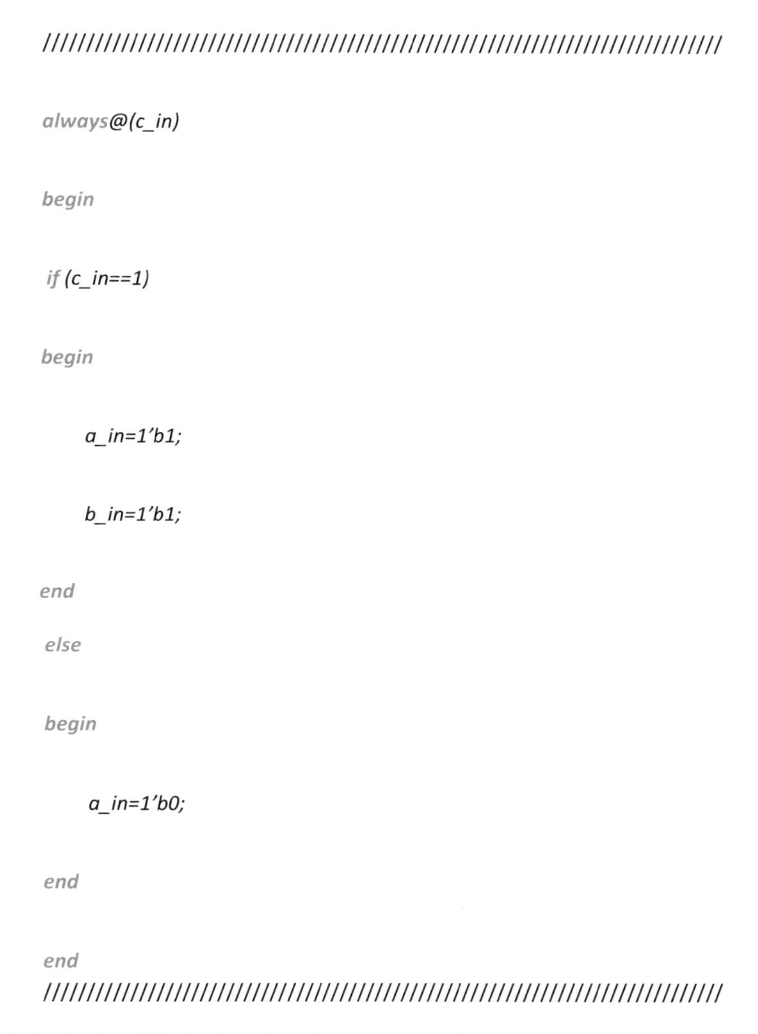

```
///////////////////////////////////////////////////////////////////////////////

always@(c_in)

begin

 if (c_in==1)

begin

      a_in=1'b1;

      b_in=1'b1;

end

 else

 begin

      a_in=1'b0;

 end

 end
///////////////////////////////////////////////////////////////////////////////
```

Example 8 Verilog RTL design with missing else condition

to the incomplete assignment within ***if-else*** or due to incomplete conditions covered while using the ***case*** construct. It is recommended that designer should take care of this while coding the RTL!

Consider one more example for better clarity to implement the AND gate (Example 9).

As shown in the RTL schematic (Fig. 7.8) as the else condition is missing, it infers the AND gate with the unintentional D latch. The intention of the RTL design engineer is to implement the AND gate, but due to missing else, it infers the latch-based design.

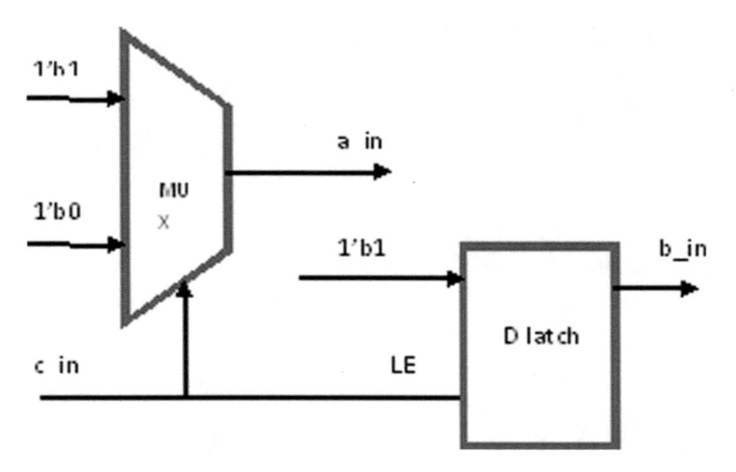

Fig. 7.7 Unintentional latches due to missing else condition

```
//////////////////////////////////////////////////////////////////////////////////
module  logic_design   ( input a_in, b_in,
                         output reg y_out );
// missing b_in from sensitivity list
always@(*)
begin
  if (a_in ==1 && b_in ==1)
     y_out = 1;
end
endmodule
//////////////////////////////////////////////////////////////////////////////////
```

Example 9 Verilog RTL design to understand the issue due to missing else

7.7 Use of Blocking Assignments

As discussed above, blocking assignments are denoted by (=) and used within the procedural block to describe the functionality of combinational logic design. Readers are requested not to get confused with the (=) assignment used while using

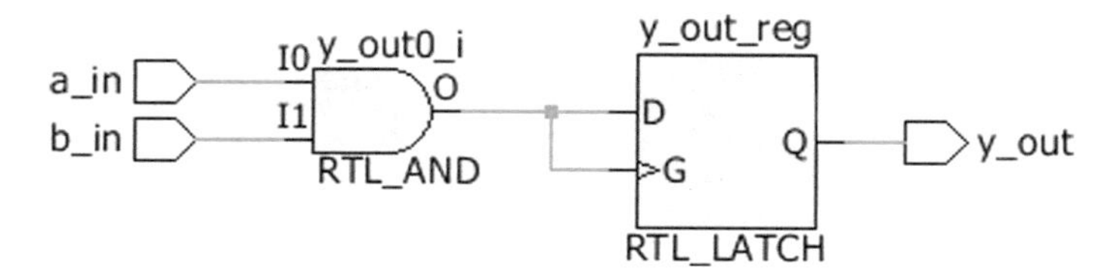

Fig. 7.8 Unintentional latches due to missing else

continuous assignment ***assign*** as they are neither blocking nor non-blocking'. Example 10 is shown below and uses the multiple ***assign*** constructs to describe the functionality of design.

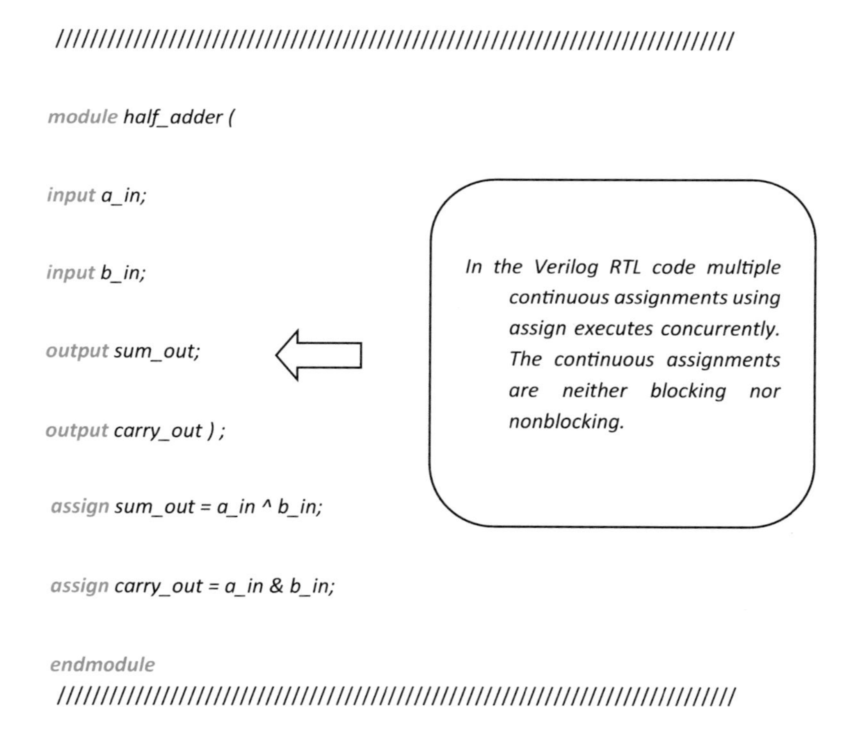

//
module half_adder (
input a_in;
input b_in;
output sum_out;
output carry_out) ;
assign sum_out = a_in ^ b_in;
assign carry_out = a_in & b_in;
endmodule
//
In the Verilog RTL code multiple continuous assignments using assign executes concurrently. The continuous assignments are neither blocking nor nonblocking.

Example 10 Concurrent execution due to multiple assign

Note: It is recommended to use the full adder to perform the subtraction operation. Subtrac-tion is performed using 2's complement addition. Multiple continuous assignments exe-cutes concurrently.

Consider the scenario of use of blocking assignment in the procedural block. If the order of the inter-dependent blocking assignments is not correct, then there is chance for the simulation and synthesis mismatch.

Example 11 is shown below, and in the example the issue in simulation and synthesis mismatch due to order of the blocking statements. Blocking assignment blocks the next immediate assignment execution unless and until current assignment is executed. Readers are encouraged to use only blocking assignments while modeling the combinational logic, but care should be taken while using these assignments to have the real intended results.

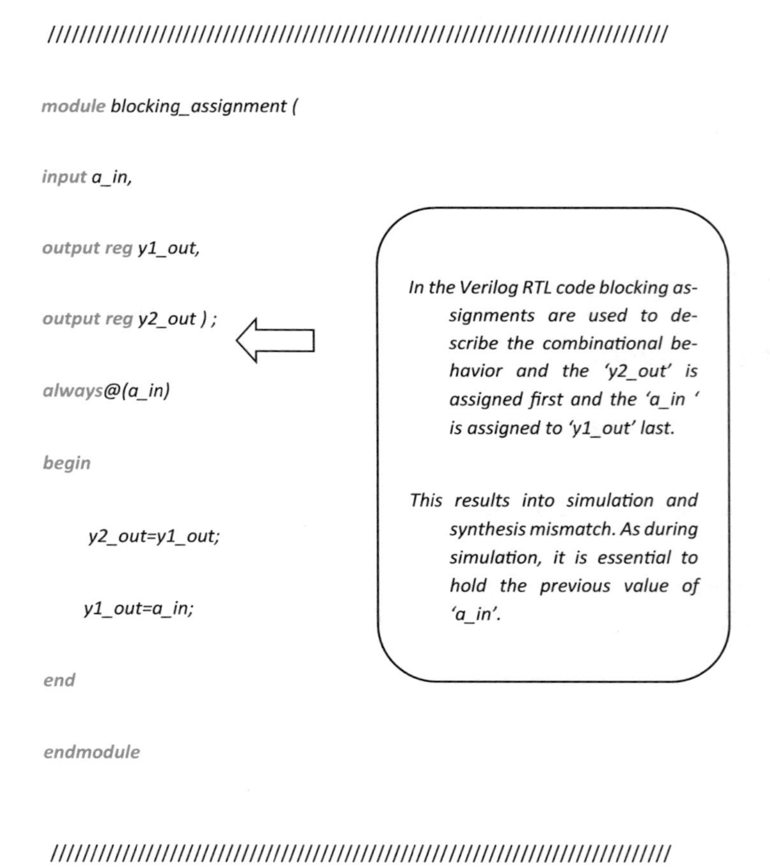

//
module blocking_assignment (
input a_in,
output reg y1_out,
output reg y2_out) ;
always@(a_in)
begin
y2_out=y1_out;
y1_out=a_in;
end
endmodule
In the Verilog RTL code blocking assignments are used to describe the combinational behavior and the 'y2_out' is assigned first and the 'a_in ' is assigned to 'y1_out' last.
This results into simulation and synthesis mismatch. As during simulation, it is essential to hold the previous value of 'a_in'.
//

Example 11 RTL having blocking assignments within always procedural block

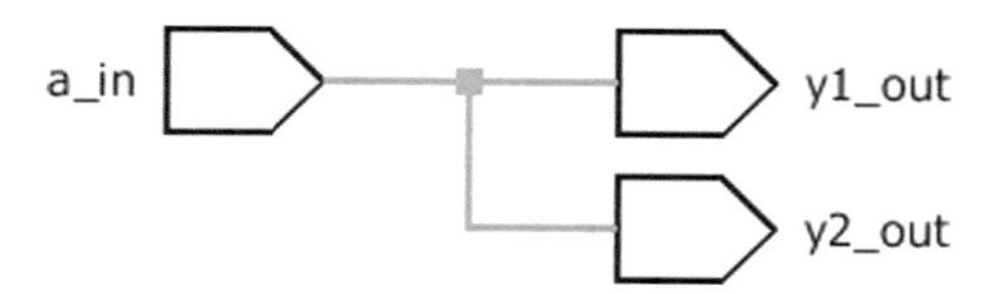

Fig. 7.9 Synthesis result due to use of blocking assignment

The synthesis result of Example 11 within always procedural block is shown in Fig. 7.9, and it generates two wires. But while simulating, the 'y2_out' is updated with the previous time stamp value 'a_in'. So, results in simulation and synthesis mismatch.

7.8 Use of if...else Versus case constructs

If all the ***case*** conditions covered while using the ***case-endcase*** construct, then it is full-case construct. For combinational design within ***case*** construct, all the blocking assignments should be included!

The synthesis result of Example 12 is shown in Fig. 7.10, and it infers parallel logic.

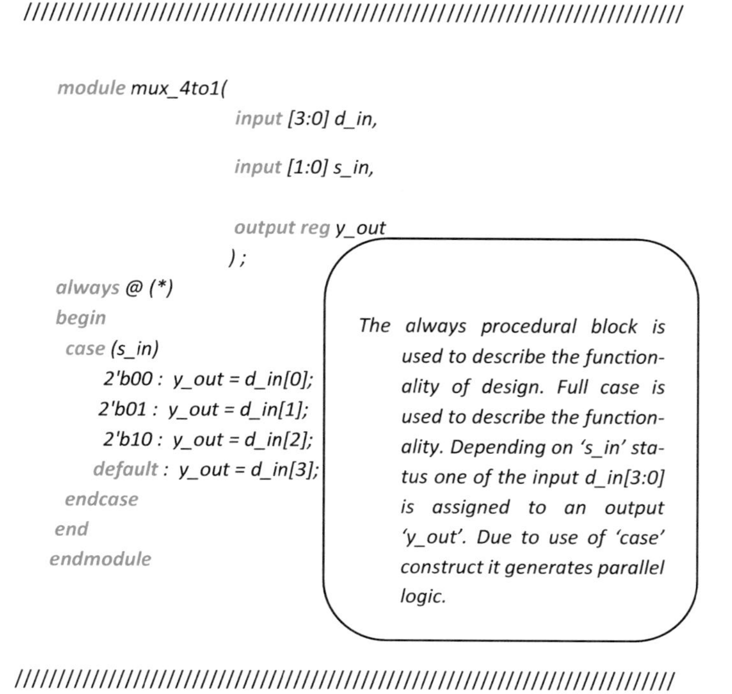

//
module mux_4to1(
input [3:0] d_in,
input [1:0] s_in,
output reg y_out
);
always @ (*)
begin
case (s_in)
2'b00 : y_out = d_in[0];
2'b01 : y_out = d_in[1];
2'b10 : y_out = d_in[2];
default : y_out = d_in[3];
endcase
end
endmodule
The always procedural block is used to describe the functionality of design. Full case is used to describe the functionality. Depending on 's_in' status one of the input d_in[3:0] is assigned to an output 'y_out'. Due to use of 'case' construct it generates parallel logic.
//

Example 12 RTL design using full case construct

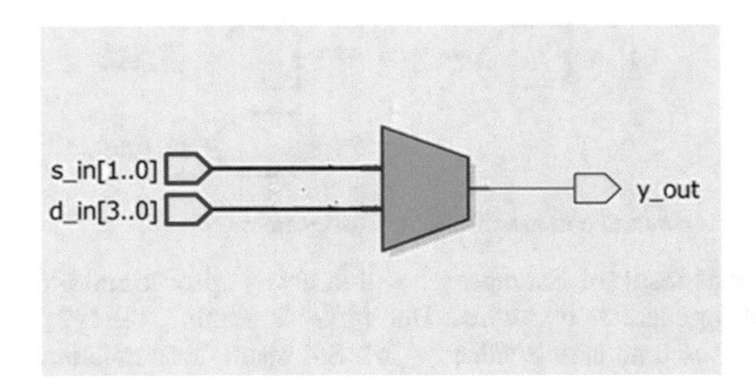

Fig. 7.10 Parallel logic inferred for 4:1 MUX due to use of 'case'

7.9 Nested Multiplexer or Priority Logic

If the functionality is described by using ***if-else*** construct, then the synthesis outcome results into priority logic. It is recommended to use ***if-else*** construct to describe the priority logic. Example 13 is RTL description of the functionality of 4:1 mux using nested ***if-else*** construct.

```
////////////////////////////////////////////////////////////////////////////////

module mux_4to1(
                    input[3:0]d_in,
                    output reg y_out,
                    input [1:0] s_in ) ;
 always @ (*)
 begin
  if( s_in == 2'b00)
      y_out = d_in[0];
     else if( s_in == 2'b01)
      y_out = d_in[1];
     else if( s_in == 2'b10)
      y_out = d_in[2];
     else
      y_out = d_in[3];
  end

 endmodule

////////////////////////////////////////////////////////////////////////////////
```

Nested if-else generates priority logic. The functionality using if-else is described for 4:1 MUX and in this d_in[3] has last priority and d_in[0] has the highest priority.

Example 13 Verilog RTL design to infer the priority logic

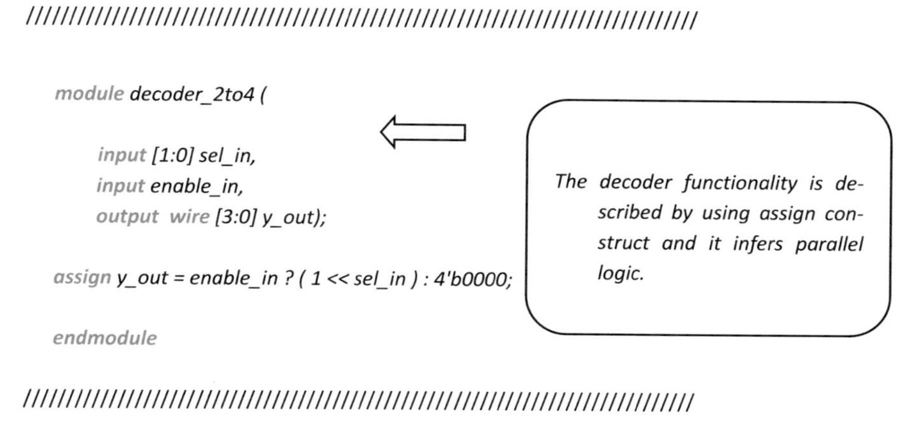

```
////////////////////////////////////////////////////////////////////////////

module decoder_2to4 (

    input [1:0] sel_in,
    input enable_in,
    output  wire [3:0] y_out);

assign y_out = enable_in ? ( 1 << sel_in ) : 4'b0000;

endmodule

////////////////////////////////////////////////////////////////////////////
```

The decoder functionality is described by using assign construct and it infers parallel logic.

Example 14 Parallel logic using continuous assignments

7.10 Parallel Logic or Decoding Logic

While describing the functionality of decoding logic, use the continuous assignment or ***case*** construct. Both will generate the parallel logic. As discussed in Chap. 6, decoder has parallel select inputs and should generate parallel outputs.

If the decoder is described using ***case-endcase*** construct. then it also infers the parallel logic. The logic inferred for decoder implementation using ***assign*** and ***case-endcase*** is shown in Fig. 7.11.

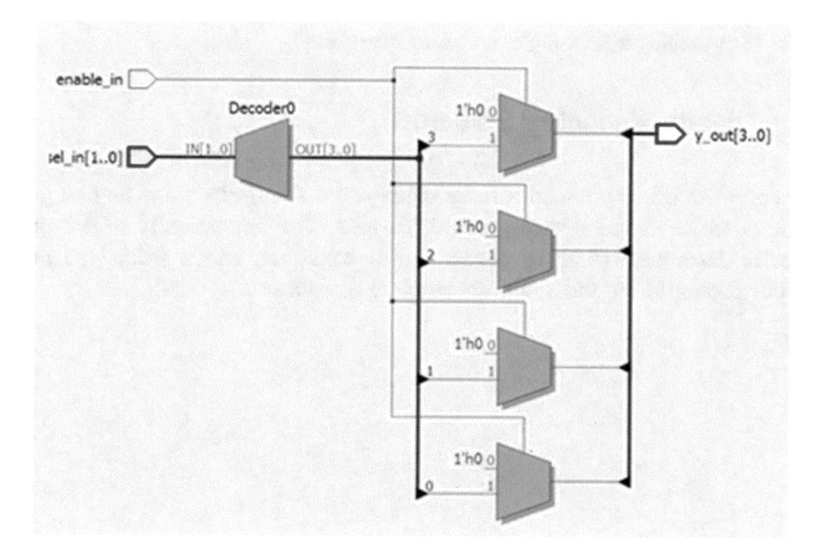

Fig. 7.11 Decoding logic due to use of 'case-endcase' construct

```
///////////////////////////////////////////////////////////////////////////////
//Decoder using 'case-endcase'


module decoder_2to4 (

                    input [1:0] sel_in,
                    input enable_in,
                    output  reg [3:0] y_out ) ;

always @ (*)
begin
 if (enable_in)
   case (sel_in)
       2'b00 : y_out = 4'b0001;
       2'b01 : y_out =4'b0010;
       2'b10 : y_out =4'b0100;
       default : y_out =4'b1000;
       endcase
 else
      y_out = 4'b0000;
end


endmodule

///////////////////////////////////////////////////////////////////////////////
```

In this always procedural block is used with the case construct to describe the functionality of the decoder.

For enable_in=1 the decoder generates the valid output depending on the status of sel_in.

Example 15 Decoding logic using case-endcase construct

7.11 Priority Encoding Structure

To describe the priority encoder functionality, use the ***if-else*** construct as priority definition can be included using nested ***if…else***. The functionality of 4:2 priority encoder is described by using nested ***if-else*** construct, and it infers the priority logic. For Example 16, the synthesis result is shown in Fig. 7.12.

//

```
module encoder_4to2 (
input [3:0] data_in,
output  reg [1:0] y_out ) ;
always @ (*)
begin
 if (data_in[3])
    y_out = 2'b11;
 else if (data_in[2])
    y_out = 2'b10;
 else if (data_in[1])
    y_out = 2'b01;
 else
   y_out = 2'b00;
end
endmodule
```

The RTL is described using always block. The data_in[3] has highest priority over all the reaming inputs.

Input datain[0] has the least priority.

//

Example 16 Priority encoder logic using 'if-else'

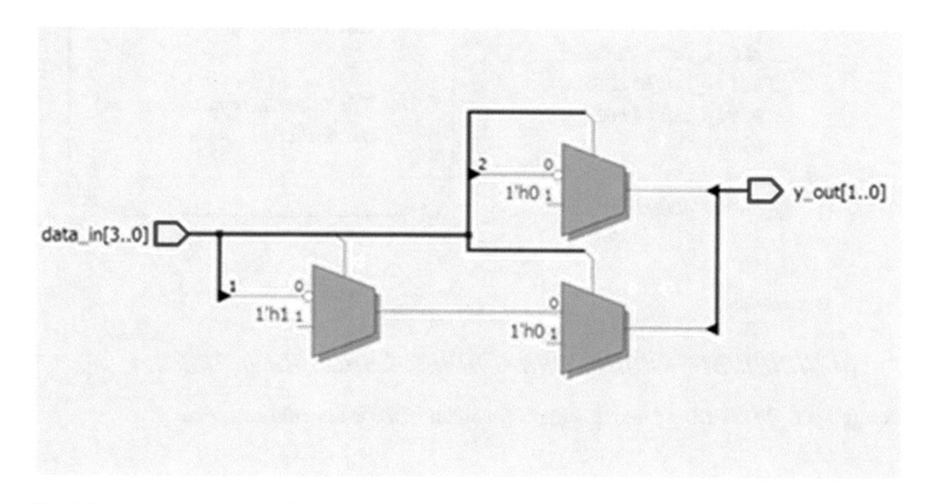

Fig. 7.12 Synthesis result of priority encoder using if-else

7.12 Missing default Condition in case construct

If all conditions are not covered while using the ***case-endcase*** expression, then it infers the logic having unintentional latches in the design. If all case conditions are not specified in the design functionality, then it is recommended to use ***default*** clause. If default is missing and all conditions are not covered, then synthesis tool flashes warning as missing case conditions and infers logic with unintentional latches.

The synthesis result of Example 17 is shown in Fig. 7.13.

```
/////////////////////////////////////////////////////////////////////////////

module decoder_2to4 (
                input [1:0] sel_in,
                input enable_in,
                output reg [3:0] y_out );

always @ (*)

begin
 if (enable_in)
   case (sel_in)
     2'b00 : y_out = 4'b0001;
     2'b01 : y_out =4'b0010;
     2'b10 : y_out =4'b0100;
     endcase
 else
     y_out = 4'b0000;
end

endmodule

/////////////////////////////////////////////////////////////////////////////
```

In this case expression, it is not full-case as 2'b11 condition is not specified. Even default clause is not used to describe the functionality for the missing case conditions.

This infers unintentional latches in the design

Example 17 RTL having missing default condition while using case construct

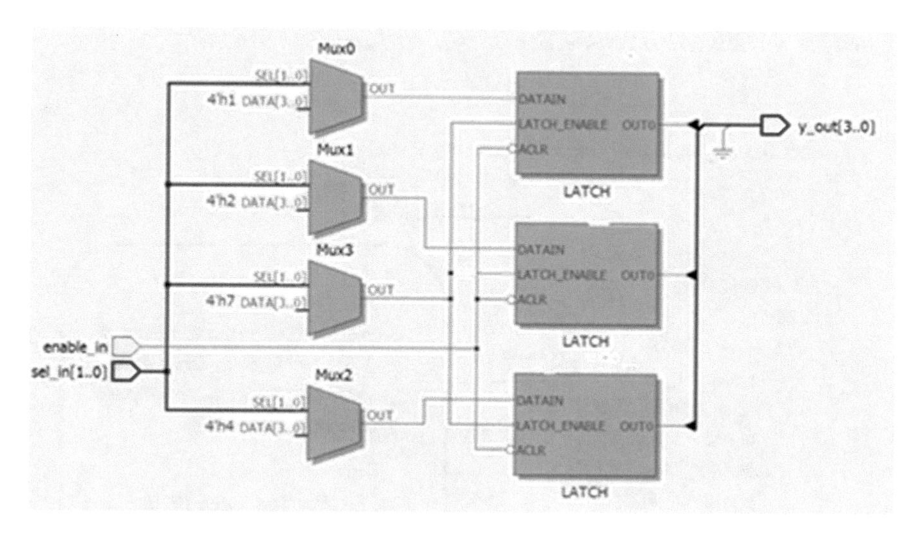

Fig. 7.13 Synthesis result of the decoder due to missing default condition in case construct

7.13 Nested if...else with Missing else Condition

As shown, the 4:1 mux functionality is described by using nested ***if-else*** but due to missing ***else*** condition it infers 4:1 mux with the unintentional latches. It is recommended to avoid the unintentional latches by incorporating the ***else*** condition in the RTL design.

For Example 18, the logic is generated having unintentional latches and shown in Fig. 7.14.

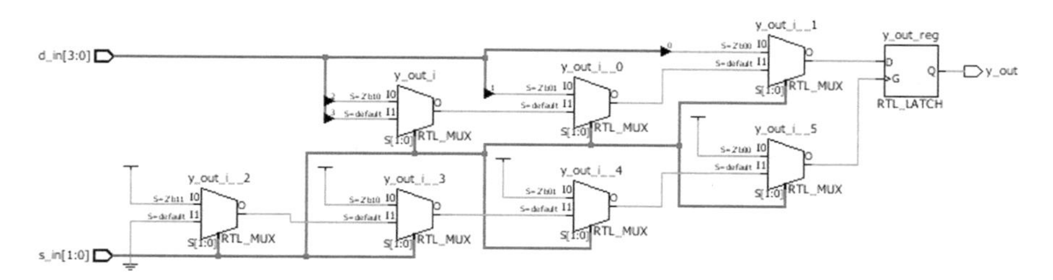

Fig. 7.14 Synthesis result of the 4:1 mux with missing else condition

```
////////////////////////////////////////////////////////////////////////////////

module mux_4to1_else_mising (output reg y_out,
                             input [3:0] d_in,
                             input [1:0] s_in );

always @ (*)
begin
 if( s_in == 2'b00)
       y_out = d_in[0];
     else if( s_in == 2'b01)
       y_out = d_in[1];
     else if( s_in == 2'b10)
       y_out = d_in[2];
     else if (s_in == 2'b11)
       y_out = d_in[3];
end
endmodule

////////////////////////////////////////////////////////////////////////////////
```

As else is missing in the nested if-else, the logic inferred generates combinational multiplexers with the latches.

The latches are inferred due to missing else while using the if-else construct.

Example 18 Verilog RTL design with missing else condition

7.14 Logical Equality Versus Case Equality

Logical equality (==) and logical inequality (!=) operators are used in the synthesizable designs, whereas case equality (===) and case inequality (!==) are not recommended in the synthesizable design.

7.14.1 Logical Equality and Logical Inequality Operators

1. Recommended to be used in the synthesizable design
2. If any one of the operand has either x or z value, then the result is unknown (x), and it results into logical comparison result as false.
3. The comparison outcome is non-deterministic if any one of the operand has x or z value.
4. Consider example of comparing a_in with b_in. In this if either of the operand has x or z value, then the *else* clause will be executed and infers the logic specified in the else condition.

```
////////////////////////////////////////////////////////////////////////////

always@(a_in, b_in)
begin
if (a_in==b_in)
        y_out= a_in ^b_in;
else
        y_out =a_in &b_in;
end

//For either of a_in, b_in has 'x' or 'z' value then the result is y_out=
a_in & b_in;

////////////////////////////////////////////////////////////////////////////
```

Example 19 RTL design using logical equality operators

7.14.2 Case Equality and Case Inequality Operators

1. Recommended to be used in the non-synthesizable design
2. If any one of the operand has either x or z value, then the result is known value, and it results either true or false.
3. The comparison outcome is deterministic if any one of the operand has x or z value.
4. Consider example of comparing a_in with b_in. In this if either of the operand as x or z value, then if a_in is equal to b_in the if clause will be executed and infers the logic specified in the if condition.

```
////////////////////////////////////////////////////////////////////////////

always@(a_in, b_in)
begin
if (a_in===b_in)
        y_out= a_in ^b_in;
else
        y_out =a_in &b_in;
end

//For either of a_in, b_in has 'x' or 'z' value then the result is y_out=
a_in ^ b_in;

////////////////////////////////////////////////////////////////////////////
```

Example 20 RTL code using case equality operators

7.15 Multiple Driver Assignments

If same net (wire) is driven by multiple expressions while using the continuous assignments, then the synthesis tool flashes an error 'Multiple Driver Assignment'. Similarly, if same 'reg' variable is driven by the different expressions within multiple procedural block, then it is multiple driver error. Exception for this is tri-state.

Consider the Example 21. In this example, net y_tmp is driven by two different continuous assignment expressions.

```
/////////////////////////////////////////////////////////////////////////////

wire y_tmp;

assign y_tmp = a_in ^ b_in;

assign y_tmp = a_in & b_in;

//in this example y_tmp is assigned by using 'xor' and 'and' at a time
    and hence multiple driver assignment error.

/////////////////////////////////////////////////////////////////////////////
```

Example 21 Verilog RTL with multiple driver assignment

7.16 Exercises

The exercises are based on the understanding of ***assign*** and **always** procedural block and use of ***if…else*** and ***case…endcase*** to model the combinational design. Complete the exercises for better understanding and application of Verilog constructs.

1. *The logic inferred by the following code is*

```
module comb_design_logic ( input a, b, sel, output reg y);

always@*

begin

    if (sel)
```

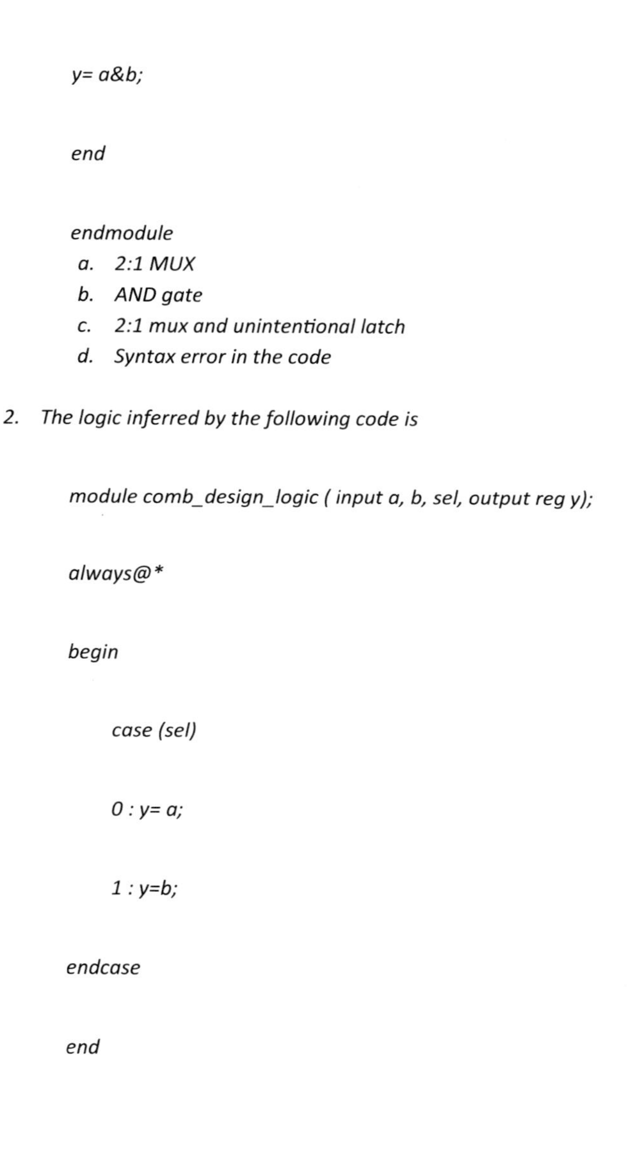

```
y= a&b;

end

endmodule
```

a. 2:1 MUX
b. AND gate
c. 2:1 mux and unintentional latch
d. Syntax error in the code

2. The logic inferred by the following code is

```
module comb_design_logic ( input a, b, sel, output reg y);

always@*

begin

    case (sel)

    0 : y= a;

    1 : y=b;

endcase

end
```

endmodule

a. *2:1 MUX*
b. *AND gate*
c. *2:1 mux and unintentional latch*
d. *Syntax error in the code*

3. *The logic inferred by the following code is*

```
module comb_design_logic ( input a, b, c, d , input [1:0] sel,
     output y);

assign y = sel [0] ? a : b ;

assign y = sel [1] ? c : d ;

endmodule
```

a. *Multiple driver error*
b. *Single 2:1 mux*
c. *Two, 2:1 mux*
d. *Syntax error in the code*

4. *The logic inferred by the following code is*

```
module comb_design_logic_2 ( input a, b, output reg y1, y2);

always@ (a)

if (a==b)
```

```
y1=  a ^ b;

else

y1= a ~^ b;

endmodule
```

a. XOR gate, XNOR gate
b. OR gate, NOR gate
c. NOR gate, NAND gate
d. Multiplexer having one input as XOR and another as XNOR

5. The following RTL has

```
module comb_design_logic_2 ( input a, b, output reg y1, y2);

always@ (a)

if (a==b)

y1=  a ^ b;

else

y1= a ~^ b;

endmodule
```

a. Multiple driver assignment error
b. Simulation synthesis mismatch issue
c. No simulation and synthesis mismatch
d. None of the above

7.17 Summary

As discussed in this chapter, following are important design guidelines.

1. Use blocking assignments for design of combinational logic.
2. All the blocking assignments are evaluated and updated in the active event queue.
3. Use case-endcase to infer parallel logic and use if-else to infer priority logic.
4. Cover all the case conditions or include default while using the case-endcase to avoid unintentional latches.
5. Use all the required inputs or signals in the sensitivity list while using always block. This is recommended to avoid simulation and synthesis mismatch.
6. Avoid use of multiple assignments to same net while using assign. This is recommended to avoid the multiple driver assignment error.
7. Avoid use of combinational looping as it exhibits the oscillatory behavior.
8. Cover all the case conditions and else conditions as missing case conditions or else conditions infers the unintentional latches in the design.
9. For decoders and multiplexers, code the RTL using case--endcase to infer parallel logic.
10. For priority encoders, use the nested if-else while coding the RTL to infer the priority logic.
11. To include all required inputs in the sensitivity list, use always@*.

Chapter 8
Basics of Sequential Design Using Verilog

The chapter is useful to understand about the RTL design for the latches and flip-flop. The concept of the synchronous and asynchronous reset is also discussed.

As discussed in previous chapters, in the combinational logic an output is a function of the present input. Examples of combinational design are multiplexers, decoders, adders, subtractors, code converters, encoders, and other priority encoders. In the sequential design, an output is a function of the present inputs and past outputs. The chapter is useful to understand about the RTL design techniques and strategies for the sequential circuit elements such as D latch and D flip-flop. The following few sections also discuss the synchronous and asynchronous reset strategies used in the sequential design.

8.1 Sequential Logic

Sequential logic is defined as the digital logic whose output is a function of present input and past output. So, the sequential logic holds the binary data. Sequential logic elements are latches and flip-flops and used to design the sequential logic for the given design functionality. For the RTL design engineer, it is essential to understand the efficient RTL design for clock-based logic circuits. The sequential logic is used to hold the larger amount of data in the complex designs. The logic is sensitive to the active edge of the clock. In the practical applications, it is always essential to describe the logic circuit which can be sensitive to either positive edge of clock or the negative edge of clock. It is always expected that the logic circuit should generate the finite output during the specified clock period. Figure 8.1 describes the basic sequential logic sensitive to positive edge of clock. The output from the logic is function of a present input and past output.

V. Taraate, *Digital Logic Design Using Verilog*,
https://doi.org/10.1007/978-981-16-3199-3_8

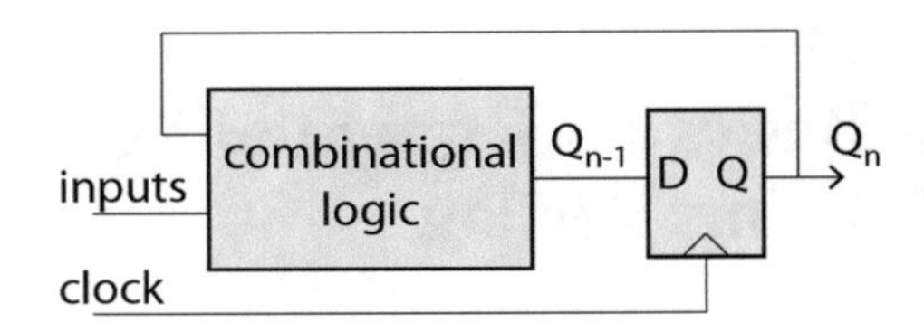

Fig. 8.1 Sequential design schematic

8.1.1 Positive-Level Sensitive D Latch

Latches are sensitive to the level. In the D latch, D stands for the data input. The latches are sensitive to either positive or negative level of clock or enable. Positive-level sensitive latch is shown in Fig. 8.2, and the truth table is described in Table 8.1. As shown in Table 8.1, for latch enable (E) is equal to positive-level (logic 1) output Q is equal to data input D; else output remains in the previous state (past output) and is indicated by Q_{n-1}. The timing sequence is shown in Fig. 8.3.

From the timing sequence, the output Q is equal to data input D during the time for which enable input E is equal to positive level. So, D latch acts as transparent

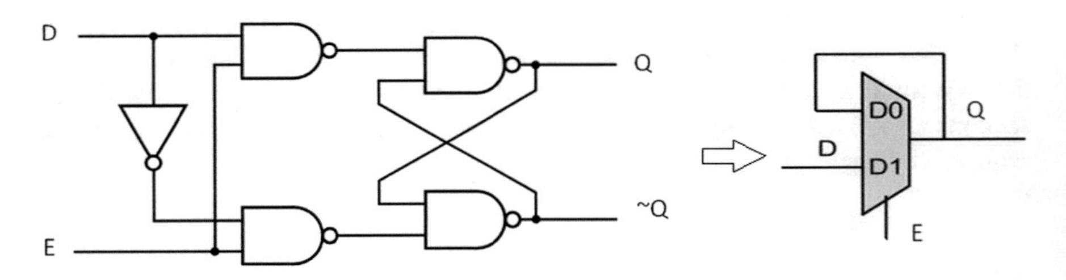

Fig. 8.2 Positive-level sensitive D latch

Table 8.1 Truth table of positive-level sensitive D latch

E	D	Q	~Q
1	0	0	1
1	1	1	0
0	X	Q_{n-1}	$\sim Q_{n-1}$

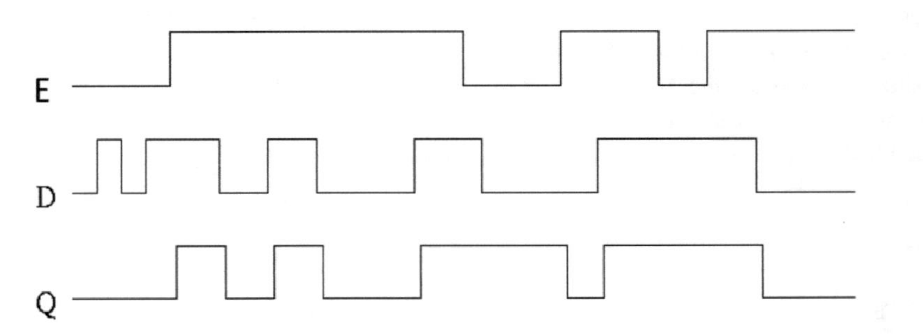

Fig. 8.3 Timing sequence of positive-level sensitive D latch

during this period. During negative level (logic 0) of enable E, D latch holds the previous value.

Now, the important point in the mind of you is how to code the RTL for positive-level sensitive D latch using Verilog? It is quite simple to visualize and to describe the RTL of latch! Example 1 describes the Verilog RTL of the positive-level sensitive D latch, and the synthesis outcome is shown in Fig. 8.4.

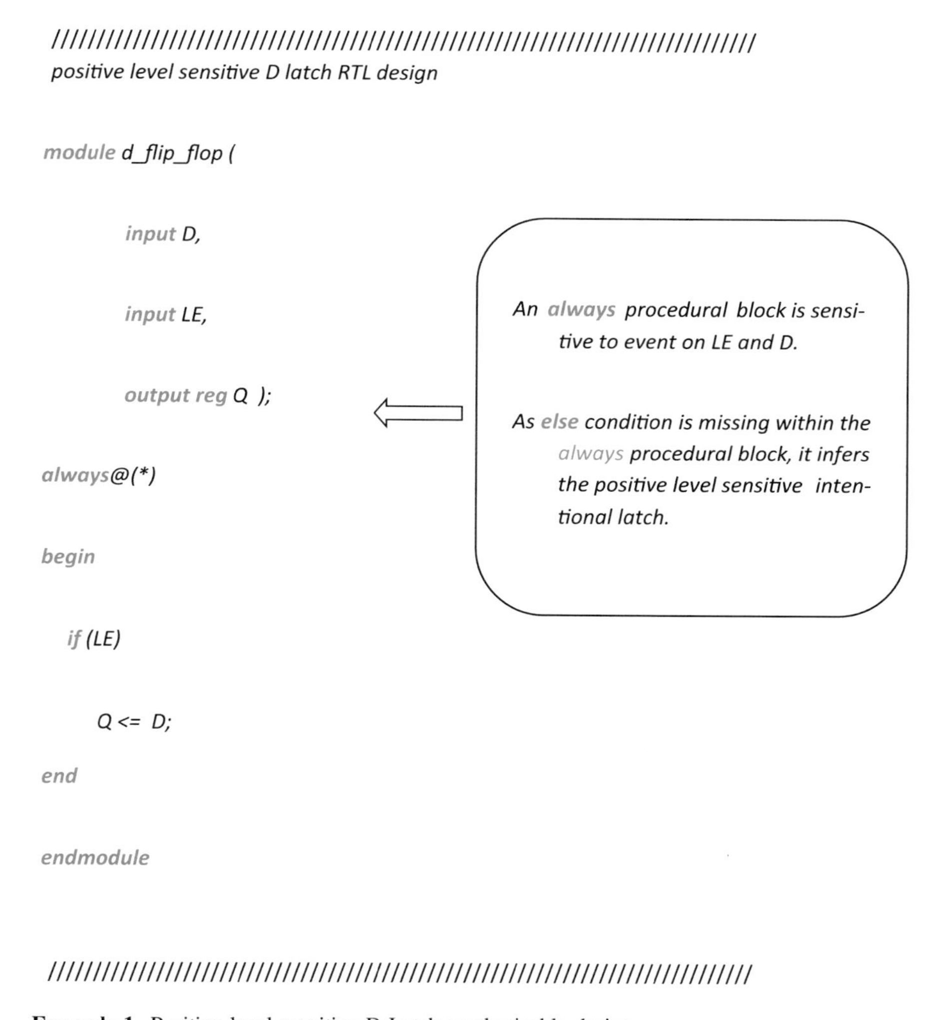

//
positive level sensitive D latch RTL design
module d_flip_flop (
input D,
input LE,
output reg Q);
always@(*)
begin
if (LE)
Q <= D;
end
endmodule
//
An always procedural block is sensitive to event on LE and D.
As else condition is missing within the always procedural block, it infers the positive level sensitive intentional latch.

Example 1 Positive level sensitive D-Latch synthesizable design

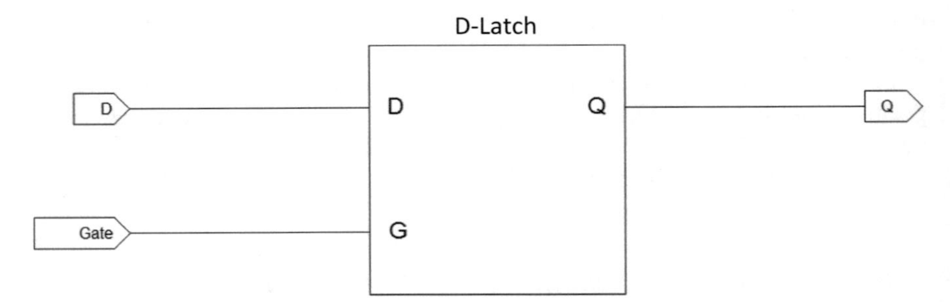

Fig. 8.4 Positive level sensitive D-Latch

8.1.2 Negative-Level Sensitive D Latch

The truth table of the negative-level sensitive D latch is shown in Table 8.2, and it has active low or negative-level sensitive latch enable input (LE_n), data input D, and output Q.

The equivalent logic gate-level representation is shown in Fig. 8.5. The latch acts as transparent during negative level of LE_n and holds the previous output during the positive level of LE_n. The timing sequence is shown in Fig. 8.6.

The Verilog RTL description is shown in Example 2, and the synthesis result is shown in Fig. 8.7.

Table 8.2 Truth table of negative-level sensitive D latch

LE_n	D	Q	~Q
0	0	0	1
0	1	1	0
1	X	Q_{n-1}	$\sim Q_{n-1}$

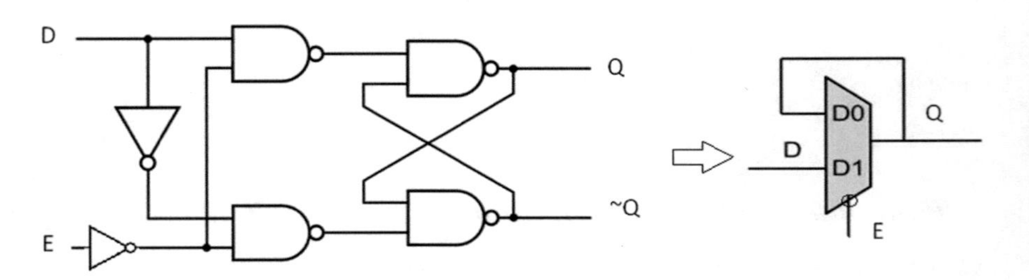

Fig. 8.5 Negative level sensitive D-Latch

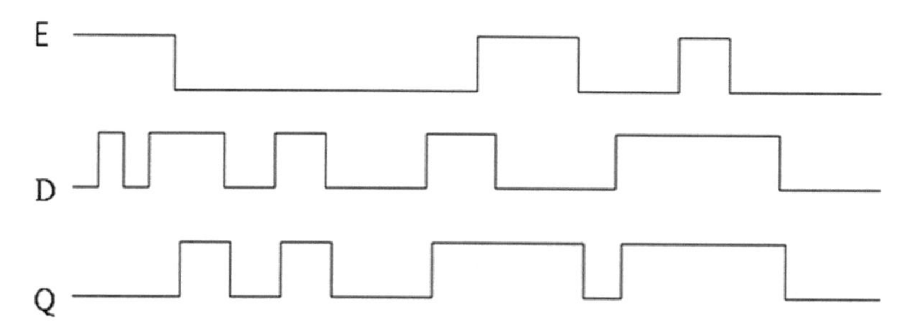

Fig. 8.6 Timing sequence of negative level sensitive latch

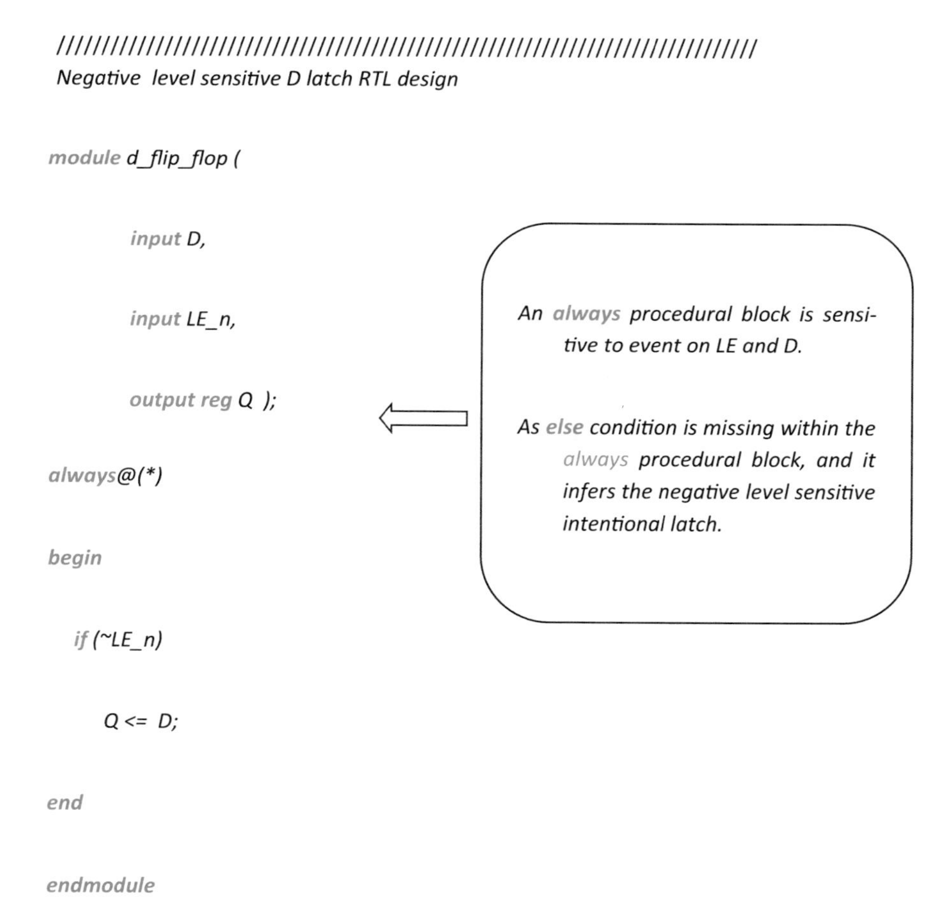

```
////////////////////////////////////////////////////////////////////////////
Negative  level sensitive D latch RTL design

module d_flip_flop (
        input D,
        input LE_n,
        output reg Q  );
always@(*)
begin
   if (~LE_n)
      Q <=  D;
end
endmodule

////////////////////////////////////////////////////////////////////////////
```

An always procedural block is sensitive to event on LE and D.
As else condition is missing within the always procedural block, and it infers the negative level sensitive intentional latch.

Example 2 Negative level sensitive D-Latch synthesizable design

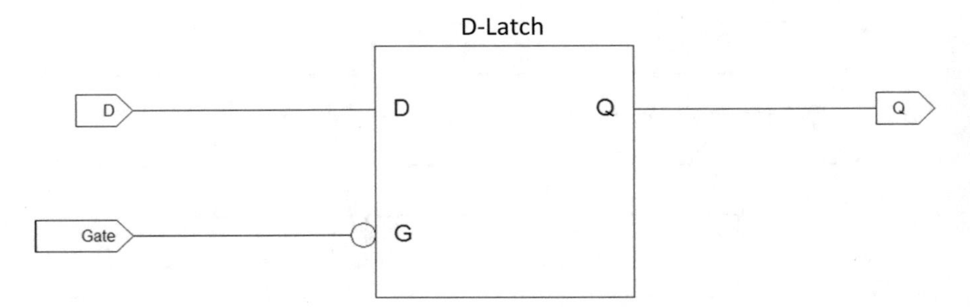

Fig. 8.7 Synthesis result of negative level sensitive latch

8.2 Flip-Flop

Flip-flop is an edge-triggered sequential logic element. It can be triggered either on positive edge of clock or on negative edge of clock. Flip-flop can be realized by using positive- and negative-level sensitive latches connected in cascade. Flip-flop is used as a memory storage element. Flip-flops are set–reset (SR), JK, D, and toggle. In an ASIC design, the D flip-flop is used as a memory storage element, where D stands for the data input. The subsequent section discusses the positive- and negative-edge triggered flip-flop.

8.2.1 Positive Edge-Triggered D Flip-Flop

Positive edge-triggered D flip-flop is sensitive to positive edge of clock. Practically, there is no any logic gate which can be sensitive to edge! Positive edge-sensitive flip-flop is realized by using negative-level sensitive latch followed by positive-level sensitive latch. The logic circuit of the positive edge-triggered D flip-flop is shown in Fig. 8.8.

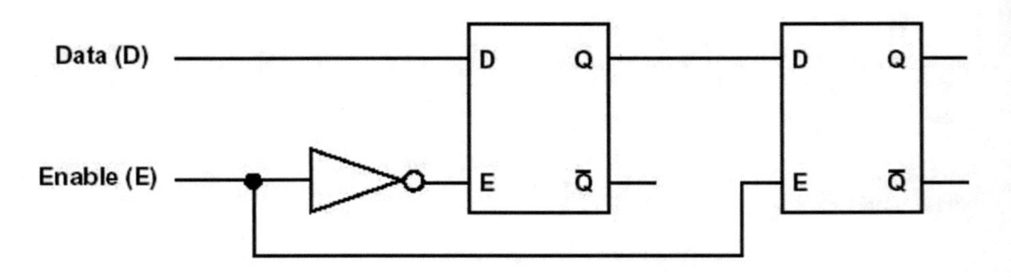

Fig. 8.8 Positive edge triggered D flip-flop

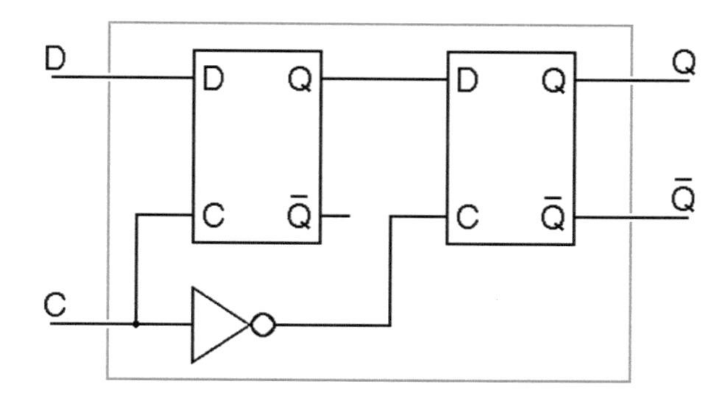

Fig. 8.9 Negative edge triggered D flip-flop

8.2.2 Negative Edge-Triggered D Flip-Flop

Negative edge-triggered D flip-flop is sensitive to the negative edge on clock. Negative edge-triggered flip-flop is realized by using positive-level sensitive latch followed by negative-level sensitive latch. The logic circuit of positive edge-triggered D flip-flop is shown in Fig. 8.9.

8.2.3 Synchronous and Asynchronous Reset

One of the important questions to address while coding the RTL for ASIC design is when to use asynchronous reset or synchronous reset? This always leads to confusion in the mind of design engineers. Synchronous reset signal is sampled on active clock edge and has some logic in the data path! whereas asynchronous reset signal is sampled irrespective of active clock edge, hence no reset-related logic in the data path. This section discusses Verilog RTL of D flip-flop using asynchronous and synchronous reset.

8.2.3.1 D Flip-Flop Having Asynchronous Reset

Asynchronous reset does not have reset-related logic in the data path and is used to initialize flip-flop irrespective of active clock edge, and hence, reset technique is named as asynchronous reset. This technique to initialize flip-flop is not recommended for internal reset signal generation as it is prone to glitches. Care should be taken by designer to synchronize this reset signal internally to avoid the glitches. The internally synchronized reset signal is applied to the sequential logic. The reset deassertion is the main problem in the asynchronous reset signals, and this problem can be overcome by using two-stage level synchronizer. Level synchronizer avoids the race around conditions during reset deassertion.

Verilog RTL is shown in Example 3 and uses active low asynchronous reset signal reset_n.

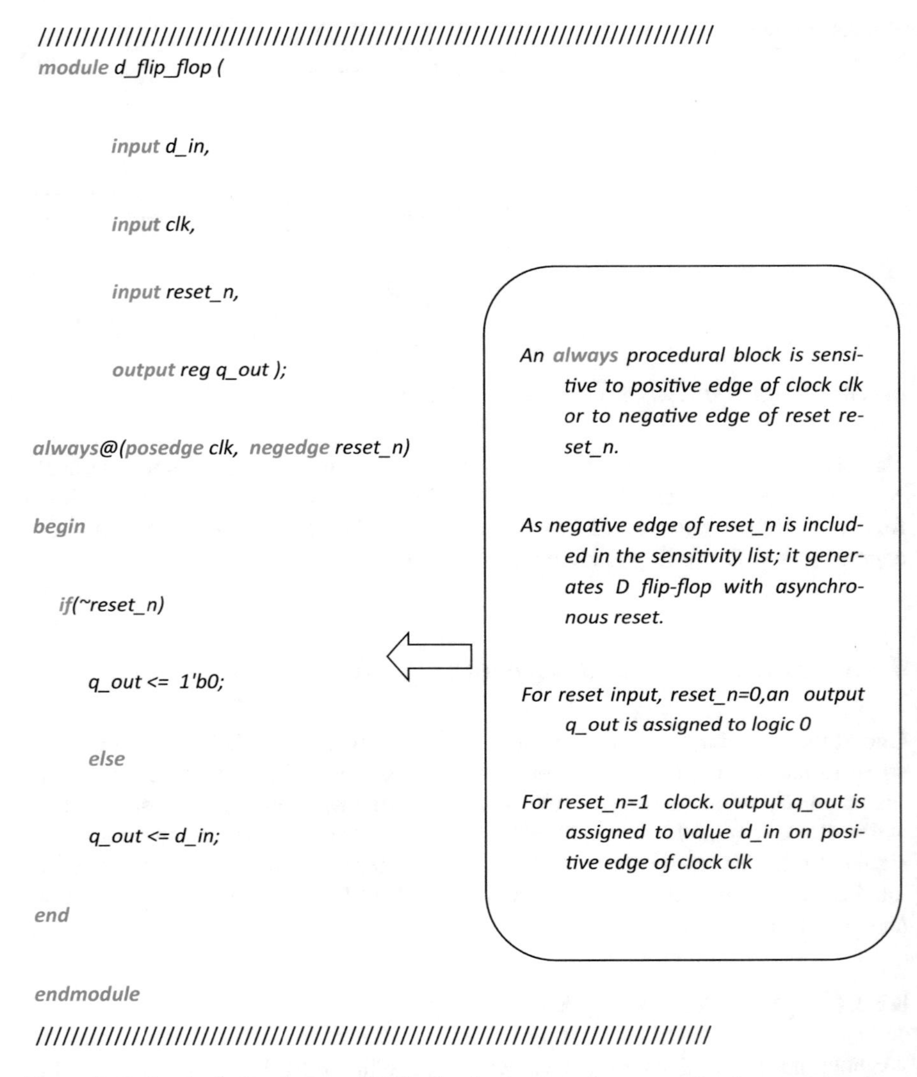

```
//////////////////////////////////////////////////////////////////////////////
module d_flip_flop (

        input d_in,

        input clk,

        input reset_n,

        output reg q_out );

always@(posedge clk,  negedge reset_n)

begin

   if(~reset_n)

      q_out <=  1'b0;

      else

      q_out <= d_in;

end

endmodule
//////////////////////////////////////////////////////////////////////////////
```

An always procedural block is sensitive to positive edge of clock clk or to negative edge of reset reset_n.
As negative edge of reset_n is included in the sensitivity list; it generates D flip-flop with asynchronous reset.
For reset input, reset_n=0,an output q_out is assigned to logic 0
For reset_n=1 clock. output q_out is assigned to value d_in on positive edge of clock clk

Example 3 Synthesizable design of D flip-flop having asynchronous active low reset input

The RTL schematic of D flip-flop having asynchronous reset reset_n is shown in Fig. 8.10. As shown, the inferred logic does not have any combinational element in the data path.

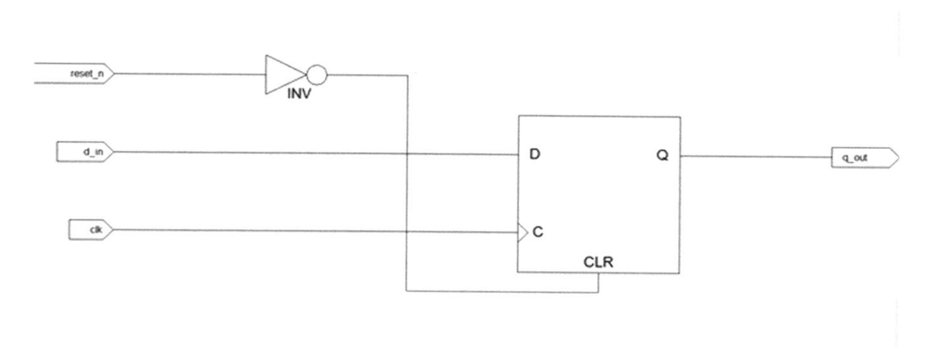

Fig. 8.10 RTL schematic of D flip-flop having asynchronous active low reset input

8.2.4 D Flip-Flop Having Synchronous Reset

In synchronous reset technique, the reset-related logic is included in the data path and reset is sampled on the active clock edge. The synchronous reset does not have issues of glitches or hazards, so this approach is best suited during the design. This mechanism does not require the additional synchronization circuit.

Verilog RTL is described in Example 4 and uses active low synchronous reset signal reset_n.

The synthesis result of positive edge-triggered D flip-flop having synchronous reset input is shown in Fig. 8.11.

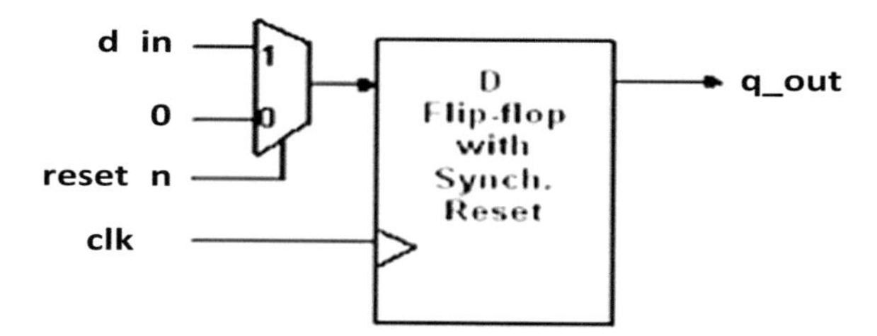

Fig. 8.11 Synthesis result of D flip-flop with synchronous reset

```
/////////////////////////////////////////////////////////////////////////////////
module d_flip_flop (

    input d_in,

    input clk,

    input reset_n,

    output reg q_out );

always@(posedge clk)

begin

  if(~reset_n)

    q_out <=  1'b0;

    else

    q_out <= d_in;

end

endmodule

/////////////////////////////////////////////////////////////////////////////////
```

An always procedural block is sensitive to positive edge of clock clk or to negative edge of reset reset_n.

As negative edge of reset_n is not included in the sensitivity list; it generates D flip-flop with synchronous reset.

Reset input is sampled on positive edge of clock. For reset reset_n=0, output q_out is assigned as logic '0'

During reset_n=1 output q_out iis assigned to value d_in on positive edge of clock clk

Example 4 Synthesizable design of D flip-flop having active low synchronous reset input

8.2.5 Flip-Flop Having Synchronous Load Enable and Asynchronous Reset

In most of the practical applications, multiple asynchronous or synchronous inputs are required. Consider an application where it requires to load the input data when enable input is active. And another requirement is to initialize register when reset signal is active and valid. If both asynchronous inputs arrive at a time, then the output is dependent on the priority assignment of these signals.

As shown in Example 5, two inputs are named as reset_n and load_en. The reset_n has highest priority, and load_en has the lowest priority. The priority is scheduled using nested ***if-else*** construct.

The synthesis result is shown in Fig. 8.12.

```
////////////////////////////////////////////////////////////////////////////
module d_flip_flop (

input d_in,

input load_en,

input clk,

input reset_n,

output reg q_out );

always@(posedge clk , negedge reset_n)

begin

   if(~reset_n)

        q_out <= 1'b0;

        else if (load_en)

        q_out <= 1'b1;

        else

        q_out <= d_in;

end

endmodule

////////////////////////////////////////////////////////////////////////////
```

An always procedural block is sensitive to positive edge of clock clk or to negative edge of reset reset_n.

As negative edge of reset_n is included in the sensitivity list; it generates D flip-flop with asynchronous reset.

For reset reset_n=0, output q_out is assigned as logic 0 and has high priority.

Another input load_en acts as an enable input and used to set the output q_out of flip-flop to logic 1 and has second priority after reset.

During reset_n=1 output q_out is assigned to value d_in on positive edge of clock clk.

Example 5 Synthesizable design of D flip-flop having synchronous load and asynchronous reset

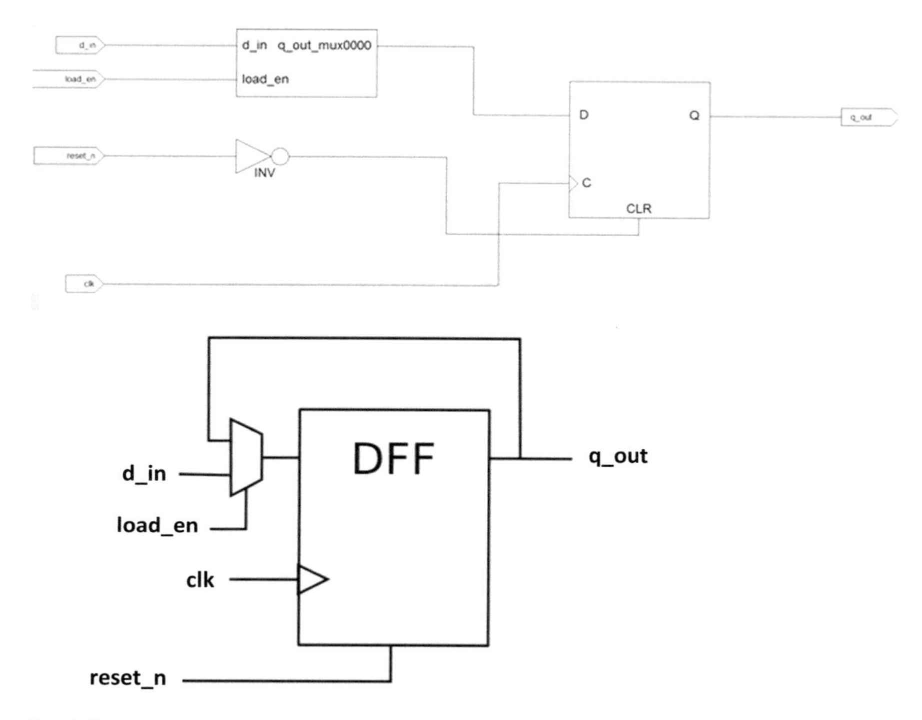

Fig. 8.12 Synthesized output of D flip-flop having asynchronous reset

8.2.6 Flip-Flop with Synchronous Load and Synchronous Reset

If multiple signals or inputs are sampled on the active edge of clock, then they are called as synchronous inputs. Consider the Verilog RTL shown in Example 6, and inputs reset_n and load_en are synchronous inputs and sampled on the positive edge of the clock. Synchronous input reset_n has highest priority, and load_en has the lowest priority.

The synthesis result is shown in Fig. 8.13 and having reset_n and load_en as synchronous inputs.

```
module d_flip_flop (

input d_in,

input load_en,

input clk,

input reset_n,

output reg q_out );

always@(posedge clk)

begin

   if(~reset_n)

      q_out <= 1'b0;

      else if (load_en)

      q_out <= 1'b1;

      else

      q_out <= d_in;

end

endmodule
```

An always procedural block is sensitive to positive edge of clock clk or to negative edge of reset reset_n.

As negative edge of reset_n is not included in the sensitivity list; it generates D flip-flop with synchronous reset.

For reset reset_n=0, output q_out is assigned as logic 0 and has high priority.

Another input load_en acts as an enable input and used to set the output q_out of flip-flop to logic 1 and has second priority after reset.

For reset_n=1, output q_out is assigned to value d_in on positive edge of clock clk.

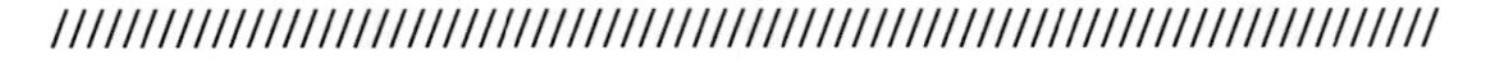

Example 6 Synthesizable design of D flip-flop with synchronous load_en and synchronous reset_n

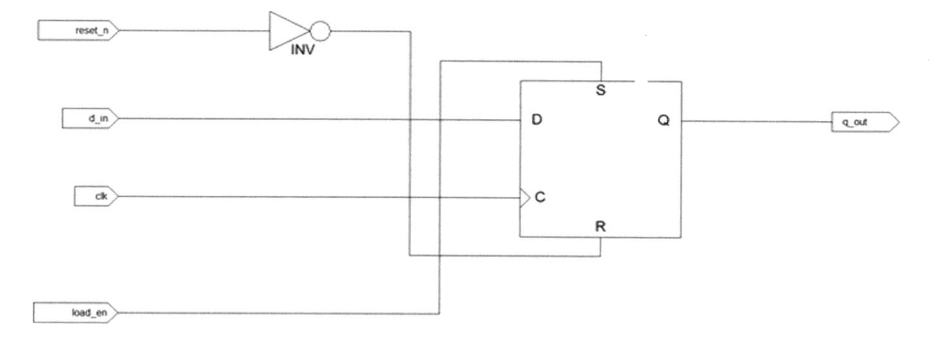

Fig. 8.13 Synthesized logic with synchronous reset_n and synchronous load

8.3 Exercises

The exercises are based on the understanding of procedural always block used to model the latches and flip-flops. Complete the exercises for better understanding and application of Verilog constructs.

1. The logic inferred by the following code is

```
module design_logic ( input a, b, clk, output reg y);

wire tmp;

assign tmp = a & b;

always @ (*)

begin

if (~ clk)

 y = y1;

end
```

```
endmodule
```

a. Positive edge triggered D flip-flop having input a&b
b. Positive level sensitive D latch having input a&b
c. Negative edge triggered D flip-flop having input a&b
d. Negative level sensitive D latch having input a&b

2. The logic inferred by the following code is

```
module design_logic ( input d, clk, reset_n, output reg y);

always @ (posedge clk, negedge reset_n)

begin

if (~reset_n)

   y <= 1'b0;

else

 y <= d;

end

endmodule
```

a. Positive edge triggered D flip-flop having asynchronous active low reset
b. Positive level sensitive D latch having input d
c. Negative edge triggered D flip-flop having synchronous active low reset
d. Negative level sensitive D latch having input d

3. The logic inferred by the following code is

```
module design_logic7 ( input d, clk, output reg y);

always @ (posedge clk)

begin

y <= d;

end

endmodule
```

   a. Positive edge triggered D flip-flop.
   b. Two-bit shift register
   c. positive edge triggered D flip-flop having input logic 1.
   d. positive edge triggered D flip-flop having input logic 0.

4. The logic inferred by the following code is

```
module design_logic ( input d, clk, output reg y);

always @ (clk)

begin

y <= d;

end

endmodule
```

   a. Positive level sensitive latch.
   b. Just a wire and clk not connected
   c. Negative level sensitive latch
   d. Syntax error in code

5. The logic inferred by the following code is

```
module design_logic ( input d, clk, reset_n, output reg y);

reg y1;

always @ (negedge clk )

if (~reset_n)

y<= 1'b0;

else

begin

y <= d;

end

endmodule
```

a. Negative edge triggered D flip-flop.
b. Two-bit shift register having synchronous reset.
c. Two-bit shift register having asynchronous reset.
d. Syntax error in the RTL

8.4 Summary

The following are key points to summarize the sequential logic design.

1. Latches are level sensitive and not recommended in the ASIC designs.
2. Flip-flops are edge triggered and are recommended in the ASIC designs.
3. Flip-flops are described by using procedural block always and sensitive to either ***posedge*** clk or ***negedge*** clk.
4. The reset can be of asynchronous reset type or synchronous reset type.
5. Using the asynchronous reset, the reset is sampled irrespective of the active clock edge and used to initialize the sequential logic.
6. Using the synchronous reset technique, the reset is sampled on the active edge of the clock and used to initialize the sequential logic.
7. If the asynchronous reset is used in the ASIC or FPGA design, then they should be synchronized using the level synchronizer.

Chapter 9
Synchronous Counter Design Using Synthesizable Constructs

The RTL design of various synchronous counters using the synthesizable constructs is discussed in this chapter. The chapter discusses about the RTL design, simulation, and synthesis concepts.

In the synchronous design, the clock of all the flip-flops is driven from common clock source, and the maximum operating frequency of the design is dependent on the register-to-register path which has more delay. The chapter discusses the various binary and gray counters and their RTL design, simulation, and synthesis using efficient Verilog constructs.

9.1 Synchronous Counters

If all the sequential design elements (flip-flops) are driven by the same source clock source, then the design is said to be synchronous. The advantage of synchronous design is that they are faster as compared to asynchronous designs. STA is quite easy for the synchronous logic, and even the performance improvement is possible by using the pipelining. Most of the ASIC implementation uses the synchronous logic. This section discusses the synchronous counter design.

Four-bit binary counter is used to count from 0000 to 1111, and the four-bit BCD counter is used to count from the 0000 to 1001. Figure 9.1 is the representation of the counter where every register logic is divide by two counter.

As shown in Fig. 9.1, the counter has four output lines: QA, QB, QC, and QD, where QA is LSB and QD is MSB. The output at QA toggles on every rising edge of the clock and hence divided by two. Output at QB toggles for every two clock cycles, and hence, it is divisible by four, and at QC output toggles for every four clock cycles, and hence, the output is divided by eight. Similarly, the output at QD toggles for every eight cycle, and hence, output at QD is divided by sixteen. In the practical applications, counters are used as clock divider network. Even counters are used in the frequency synthesizers to generate variable frequency outputs.

V. Taraate, *Digital Logic Design Using Verilog*,
https://doi.org/10.1007/978-981-16-3199-3_9

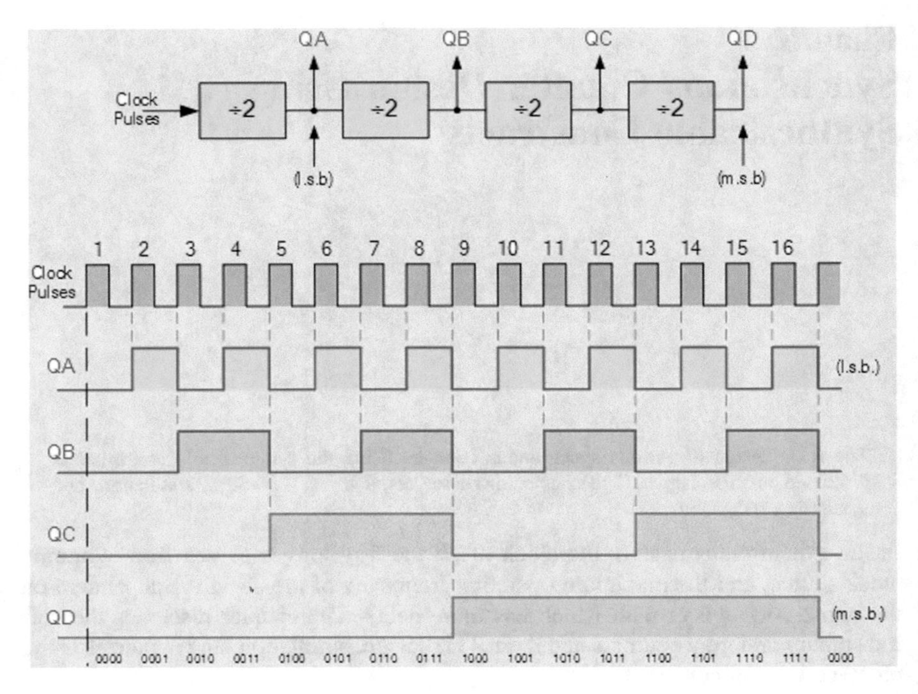

Fig. 9.1 Four-bit binary counter

9.1.1 Three-Bit Up Counter

Counters are used to generate the pre-defined and required count sequence on the active edge of clock. Most of the time in ASIC design, it is essential to have an efficient RTL code for the counter by using the synthesizable constructs. Three-bit up counter is described by using Verilog synthesizable constructs. Counter counts from 000 to 111 and works on the positive edge of the clock and wraps around to 000 on the next positive edge of the count. The counter described in Fig. 9.2 is presettable counter, and it has the synchronous active high 'load_en' input to sample the three-bit input as required. The data input is three bit and declared as data_in.

Counter has active low asynchronous reset_n input, and when it is active low the q_out is 000. During normal operation, reset_n is active high.

The synthesis result is shown in Fig. 9.2 and has three-bit data input lines data_in, active high load_en, and active low reset input reset_n. Output is denoted as q_out and postive edge-triggered clock by clk.

```
//////////////////////////////////////////////////////////////////////////////

module up_counter_3bit (

input [2:0] data_in,

input load_en,

input clk,

input reset_n,

output reg [2:0] q_out ) ;

always@(posedge clk , negedge reset_n)

begin

   if(~reset_n)

      q_out <= 3'b000;

      else if (load_en)

      q_out <= data_in;

      else

      q_out <= q_out +1'b1;

end

endmodule

//////////////////////////////////////////////////////////////////////////////
```

Procedural always block is sensitive to positive edge of clock clk, or negative edge of reset reset_n.

For reset reset_n=0 it assigns 3-bit output q_out=000. Here reset is asynchronous input and has highest priority over any other input.

For load signal load_en=1 the input data_in is assigned to output 'q_out'

For reset_n=1 and load_en=0 the else clause is executed and increments the counter output q_out by one on every positive edge of clock.

Example 1 Verilog RTL of three-bit up counter

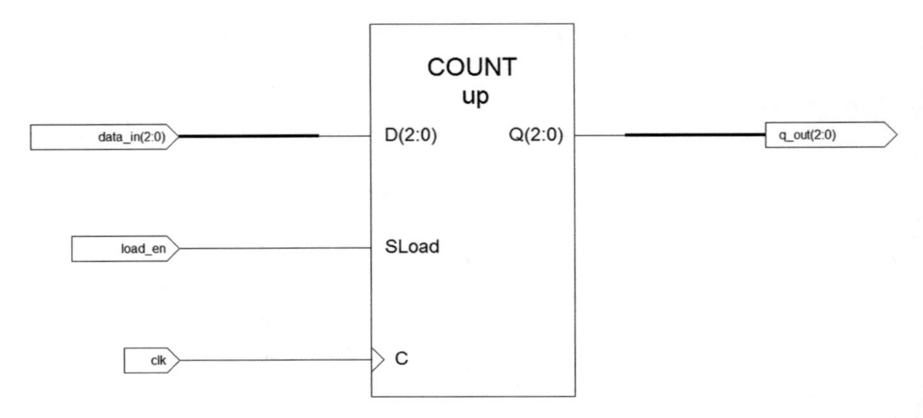

Fig. 9.2 Synthesized three bit up counter top level diagram

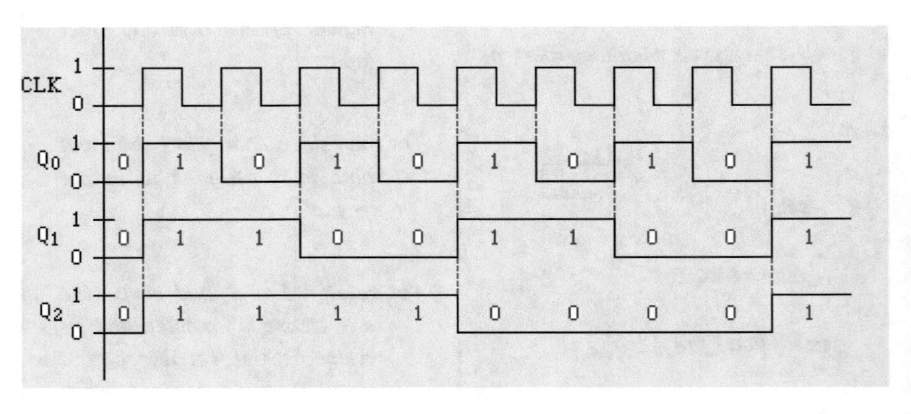

Fig. 9.3 Timing sequence for three-bit binary down counter

9.1.2 Three-Bit Down Counter

Three-bit down counter is coded by using synthesizable Verilog constructs. Counter counts from 111 to 000 and sensitive to the positive edge of the clock and wraps around to 111 on the next positive edge of the count after reaching to count value 000. The timing sequence of the three-bit down counter is shown in Fig. 9.3.

The counter shown in Example 2 is presettable counter, and it has the synchronous active high load_en input to sample the three-bit desired presettable value. The data input is three bit and declared as data_in.

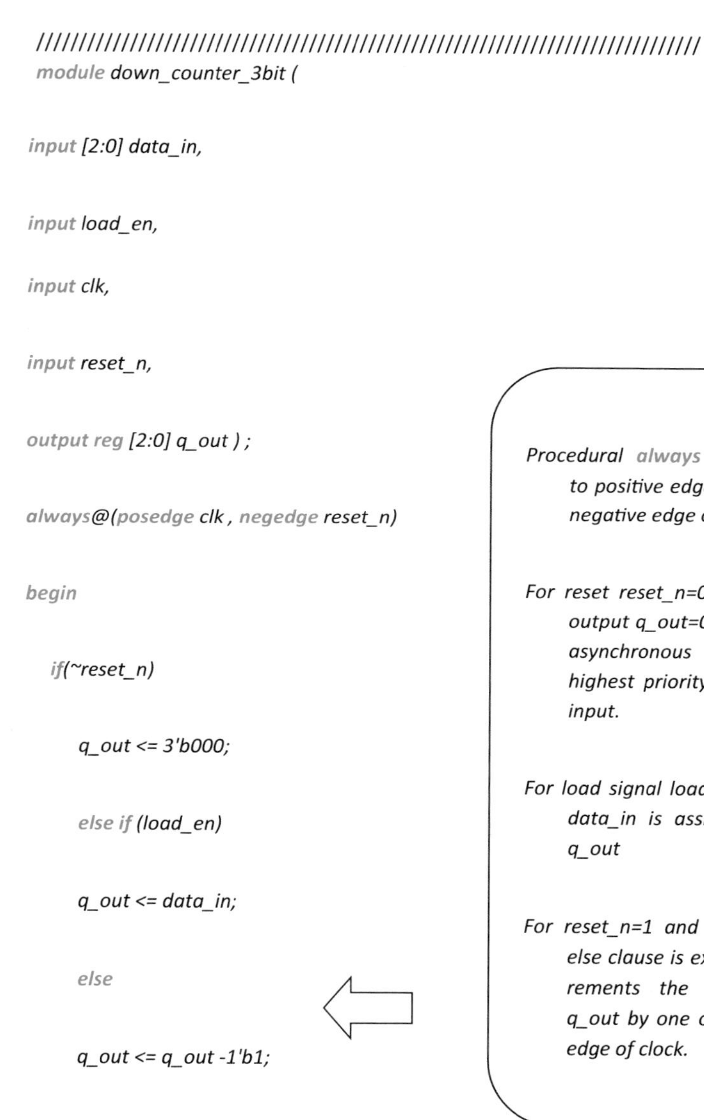

```
////////////////////////////////////////////////////////////////////////////
module down_counter_3bit (

input [2:0] data_in,

input load_en,

input clk,

input reset_n,

output reg [2:0] q_out ) ;

always@(posedge clk , negedge reset_n)

begin

   if(~reset_n)

      q_out <= 3'b000;

      else if (load_en)

      q_out <= data_in;

      else

      q_out <= q_out -1'b1;

end

endmodule

////////////////////////////////////////////////////////////////////////////
```

Procedural always block is sensitive to positive edge of clock clk , or negative edge of reset reset_n.

For reset reset_n=0 it assigns 3-bit output q_out=000. Here reset is asynchronous input and has highest priority over any other input.

For load signal load_en=1 the input data_in is assigned to output q_out

For reset_n=1 and load_en=0 the else clause is executed and decrements the counter output q_out by one on every positive edge of clock.

Example 2 Verilog RTL of three-bit down counter

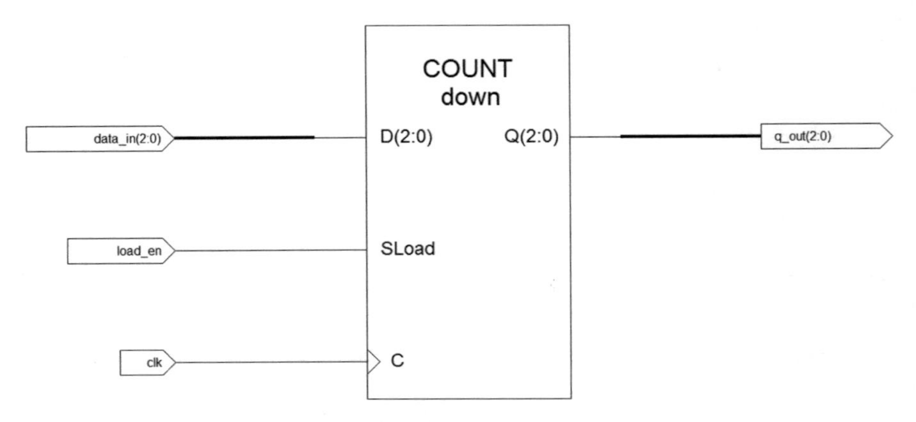

Fig. 9.4 Synthesized three bit down counter top level diagram

Counter has active low asynchronous reset_n input, and when it is active low, the output lines q_out is assigned to 000. During normal operation, reset_n is active high.

The RTL schematic is shown in Fig. 9.4 and has three-bit data input lines data_in, active high load_en, and active low reset input reset_n. Output is denoted as q_out. The counter output decrements by one on every postive edge of the clock by clk.

9.1.3 Three-Bit Up–Down Counter

Three-bit up–down counter is coded by using synthesizable Verilog constructs. Down counter decrements from 111 to 000 and sensitive to the positive edge of the clock and wraps to 111 on the next positive edge of the clock after reaching to count value 000. Up counter increments from 000 to 111 and sensitive to the positive edge of the clock and wraps to 000 on the next positive edge of the clock after reaching to count value 111.

For up_down is equal to logic 1, the counter acts as up counter, and for up_down is equal to logic 0, counter acts as down counter.

The counter is coded using the synthesizable constructs and shown in Example 3. It is presettable counter, and it has the synchronous active high load_en input to sample the three-bit desired presettable input. The data input is three bit and declared as data_in. The up or down counting operation is selected by the input up_down, for up_down = 1, counter acts as up counter, and for up_down = 0, counter acts as down counter.

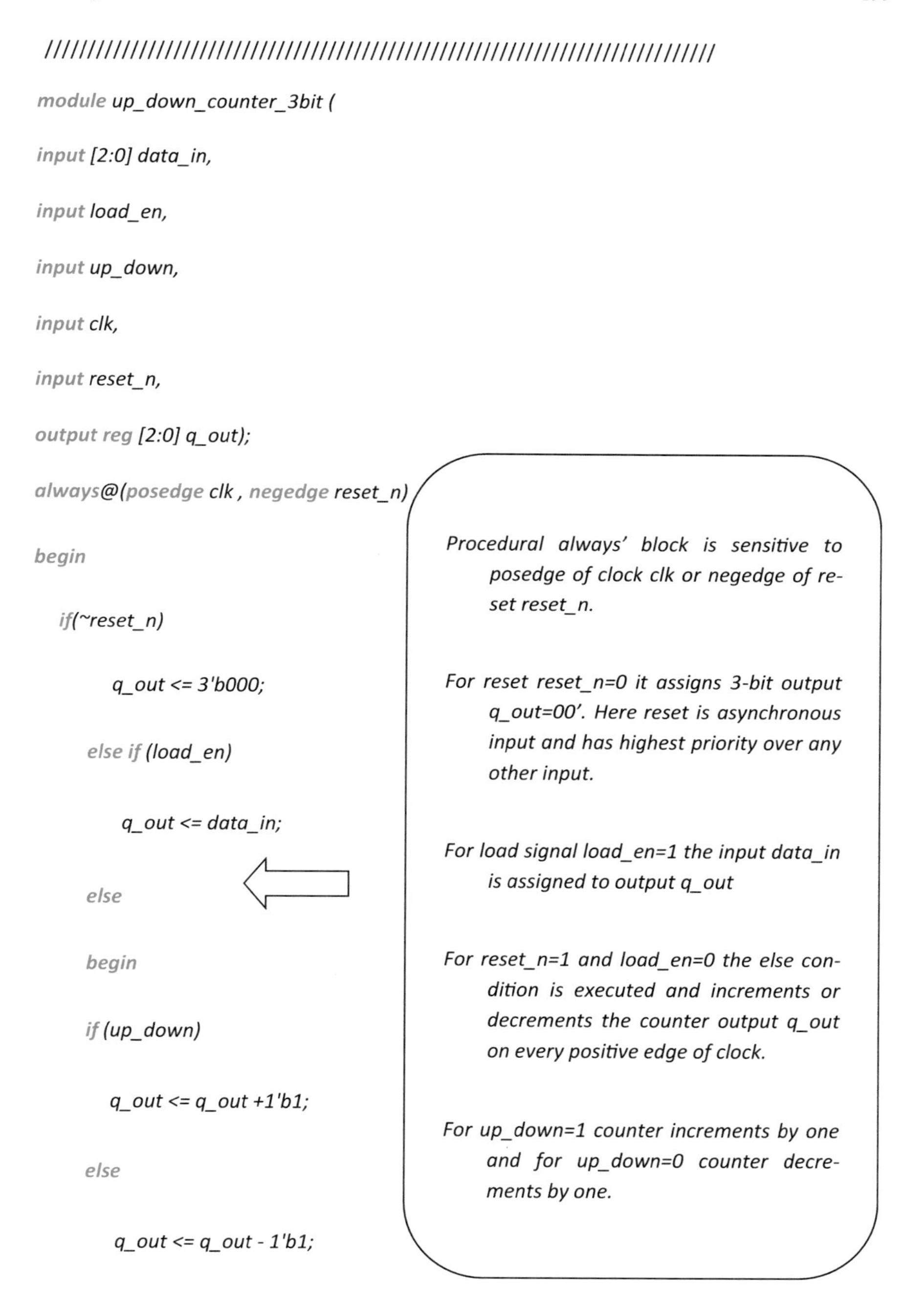
//
module up_down_counter_3bit (
input [2:0] data_in,
input load_en,
input up_down,
input clk,
input reset_n,
output reg [2:0] q_out);
always@(posedge clk , negedge reset_n)
begin
if(~reset_n)
q_out <= 3'b000;
else if (load_en)
q_out <= data_in;
else
begin
if (up_down)
q_out <= q_out +1'b1;
else
q_out <= q_out - 1'b1;
Procedural always' block is sensitive to posedge of clock clk or negedge of reset reset_n.
For reset reset_n=0 it assigns 3-bit output q_out=00'. Here reset is asynchronous input and has highest priority over any other input.
For load signal load_en=1 the input data_in is assigned to output q_out
For reset_n=1 and load_en=0 the else condition is executed and increments or decrements the counter output q_out on every positive edge of clock.
For up_down=1 counter increments by one and for up_down=0 counter decrements by one.

Example 3 Verilog RTL for three bit up-down counter

```
        end

    end

    endmodule

    ////////////////////////////////////////////////////////////////////////////
```

Example 3 (continued)

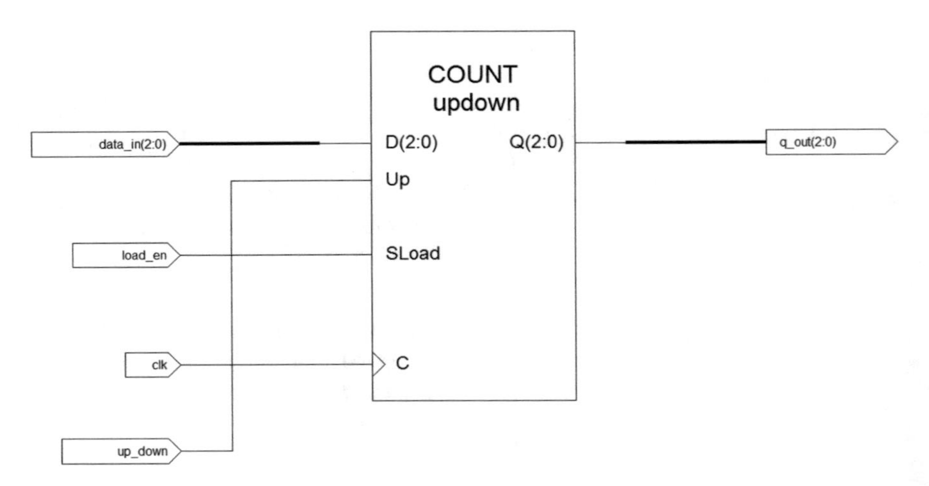

Fig. 9.5 Synthesized three bit up-down counter top level module

Counter has active low asynchronous reset_n input, and when it is active low, the output q_out is 000. During normal operation, reset_n is active high.

The synthesis outcome is shown in Fig. 9.5 and has three-bit data input lines data_in, active high load_en, and active low reset input reset_n. Three-bit output is denoted as q_out, and counter has postive edge-triggered clock clk and select line up_down to select for the up or down counting operation.

9.2 Gray Counters

Gray counters are used in the multiple clock domain designs as only one bit changes in two successive gray codes. Gray codes are used in the synchronizers. Gray counter is coded and shown in Example 4, and in this only one bit is changing on the active clock edge with reference to the previous output of the counter. In this active low, asynchronous reset input is reset_n. When reset_n = 0, the output of counter gray_out is assigned to 00.

```
////////////////////////////////////////////////////////////////////////////////

module gray_counte_2_bit(

                  input clk, reset_n,

                  output [1:0] gray_out

                  );

 reg [1:0] binary_out;

always @ (posedge clk, negedge reset_n)

begin

  if ( ~reset_n)

         binary_out <= 2'b00;

  else

        binary_out <= binary_out + 1;

  end

  assign gray_out [1] = binary_out[1];

  assign gray_out[0] = ^ binary_out;

   endmodule

////////////////////////////////////////////////////////////////////////////////
```

Two-bit gray counter functionality is described by using procedural always block which is sensitive to positive edge of clock clk.

The counter has asynchronous reset reset_n and when it is active low the counter output gray_out is assigned to logic 0.

Example 4 Two-bit gray counter

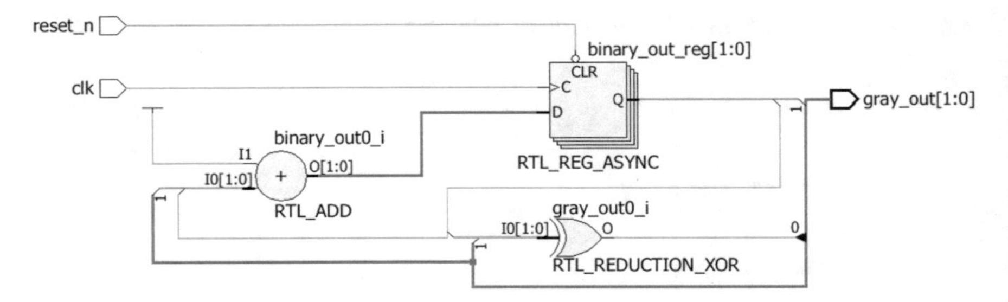

Fig. 9.6 Two-bit gray counter

Fig. 9.7 Waveform of gray counter

The counter described in Example 4 is presetscounter and has active low asynchronous reset reset_n input, and when it is active low, the output line gray_out is 00. During normal operation, reset_n is active high.

The synthesis result of Example 4 is shown in Fig. 9.6 and uses the binary counter with the XOR gate to generate the gray output.

As shown in the simulation waveform (Fig. 9.7), the gray_out sequence is 00,01,11,10,00…. And only one bit changes in the two consecutive codes.

9.2.1 *Gray and Binary Counter*

In most of the practical applications, binary and gray counters need to be used. Gray counter output can be generated from the binary counter output by using the combinational logic. Refer Sect. 9.2 for the binary and gray output.

Parameterized binary and gray counter is coded and shown in Example 5, and the Verilog RTL is described to generate four-bit binary and gray output. For 'reset_n = 0', binary and gray counter output is assigned to 0000. Four-bit gray code output is denoted as gray_out.

Simulation result for the four-bit binary and gray counter is shown in the timing sequence Fig. 9.9, and for every postive edge of clock, counter output increments by one (Fig. 9.8).

```
/////////////////////////////////////////////////////////////////////////////
module binary_gray_counter #(parameter data_size = 4)
    (input clk, reset_n, full, increment,
    output reg [data_size-1:0] gray_out);
    wire [data_size-1:0] gray_next, binary_next;
    reg [data_size-1:0] binary_data;

assign gray_next = (binary_next >> 1) ^ binary_next;

always@(posedge clk , negedge reset_n)
if (~reset_n)
{binary_data, gray_out} <= 4'b0000;
else
{binary_data, gray_out} <= {binary_next, gray_next};

assign binary_next = !full ? binary_data + increment : binary_data;
endmodule
/////////////////////////////////////////////////////////////////////////////
```

Four-bit parameterized gray and binary counter is described and having asynchronous reset reset_n and triggered on positive edge of clock clk.

For reset reset_n=0 the counter output is assigned as zero. Binary counter output is binary_data and gray counter output is gray_out.

The next value of binary counter and gray counter is evaluated by using continuous assignment constructs.

Example 5 Verilog RTL of Parameterized Binary and Gray counter

The testbench to check for the functional correctness of the design is shown in Example 6.

The simulation result is shown in Fig. 9.9 and generates the gray_out. If you compare the two successive codes, then only one-bit change you can notice!

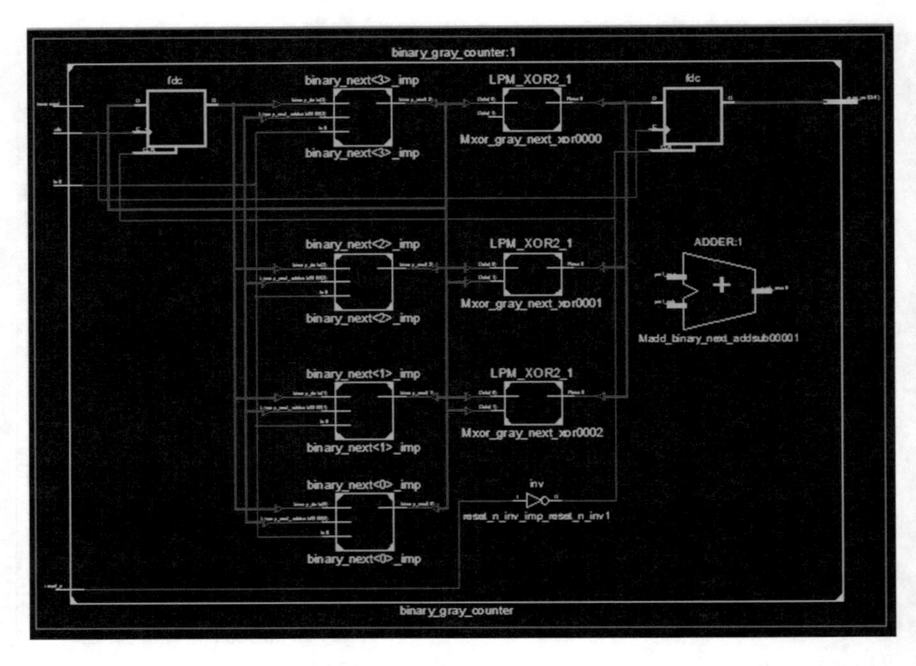

Fig. 9.8 Synthesis result of Example 5

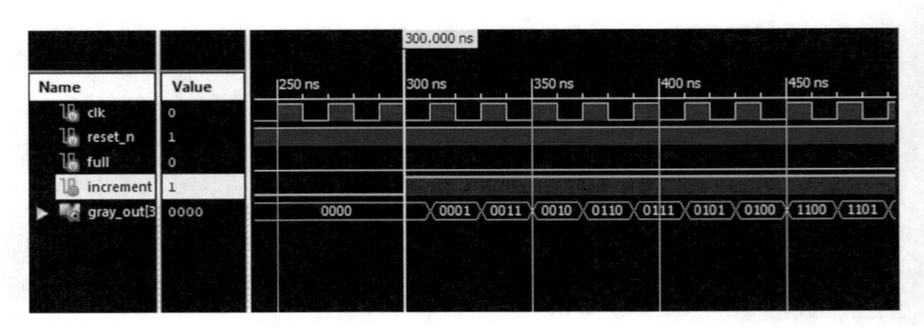

Fig. 9.9 Timing sequence for four-bit gray counter

```
////////////////////////////////////////////////////////////////////////////////
module test_binary_gray;

        // Inputs

        reg clk;

        reg reset_n;

        reg full;

        reg increment;

        // Outputs

        wire [3:0] gray_out;

        always #10 clk= ~clk;

        always #300 increment = ~increment;

        always #500 full =~full;

        // Instantiate the Unit Under Test (UUT)

        binary_gray_counter uut (

            .clk(clk),

            .reset_n(reset_n),
```

Example 6 Testbench for Example 5

```
        .full(full),
        .increment(increment),
        .gray_out(gray_out)
    );
    initial begin
        // Initialize Inputs
            clk = 0;
            reset_n = 0;
            full = 0;
            increment = 0;
        // Wait 10 ns for global reset to finish
            #10;
            reset_n=1;
      end
  endmodule
////////////////////////////////////////////////////////////////////////////////
```

Example 6 (continued)

9.2.2 *Ring Counters*

Ring counters are used in the practical applications to include the pre-defined delay. These counters are synchronous in nature and used in the practical applications like traffic light controllers and timers to introduce the certain amount of pre-defined delay. The internal logic structure using the D flip-flops for four-bit ring counter is shown in Fig. 9.10, as shown the output of LSB flip-flop is fed back to the MSB flip-flop input, and the counter shifts the data on every active edge of clock signal.

The Verilog RTL for the four-bit ring counter is coded and shown in Example 7 Verilog RTL for four-bit ring counter, and the counter has set_n input to set the initial output value of 1000 and works on the positive edge of clock signal.

The synthesis schematic is shown in Fig. 9.11.

The simulation result is shown in the waveform (Fig. 9.12), and as shown, the count_out is 1000, 0100, 0010, 0001, 1000.... It is like a ring.

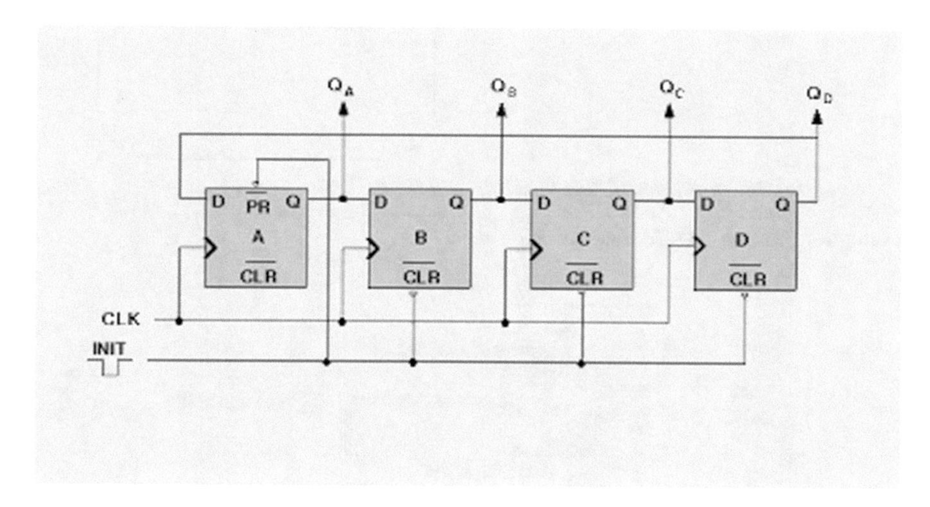

Fig. 9.10 Ring counter internal structure

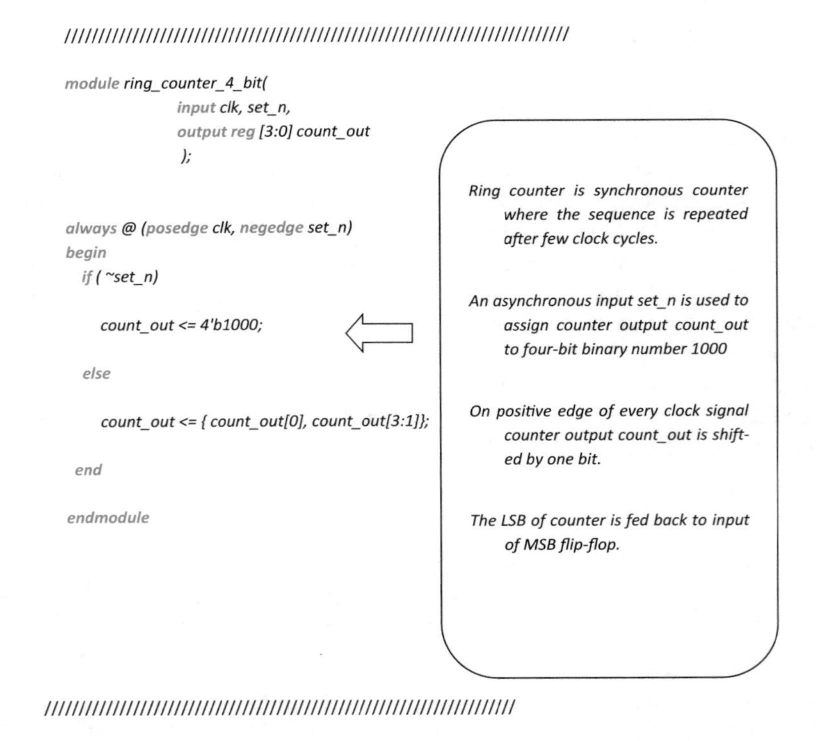

```
////////////////////////////////////////////////////////////////////////

module ring_counter_4_bit(
                input clk, set_n,
                output reg [3:0] count_out
                );

always @ (posedge clk, negedge set_n)
begin
  if ( ~set_n)

     count_out <= 4'b1000;

  else

     count_out <= { count_out[0], count_out[3:1]};

 end

endmodule

////////////////////////////////////////////////////////////////////
```

Ring counter is synchronous counter where the sequence is repeated after few clock cycles.
An asynchronous input set_n is used to assign counter output count_out to four-bit binary number 1000
On positive edge of every clock signal counter output count_out is shifted by one bit.
The LSB of counter is fed back to input of MSB flip-flop.

Example 7 Verilog RTL for four-bit ring counter

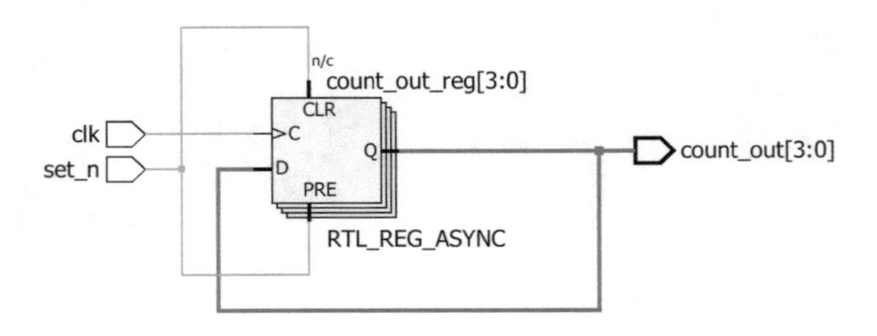
n/c
count_out_reg[3:0]
CLR
clk
set_n
C
D
Q
PRE
RTL_REG_ASYNC
count_out[3:0]

Fig. 9.11 Synthesized Logic for four-bit ring counter

Fig. 9.12 Simulation result of 4-bit ring counter

9.2.3 Johnson Counters

The Johnson counter is also called as twisted ring counter and designed by using the shift register. The internal structure for three-bit Johnson counter is shown in Fig. 9.13.

The Verilog RTL for four-bit Johnson counter is shown in Example 8.

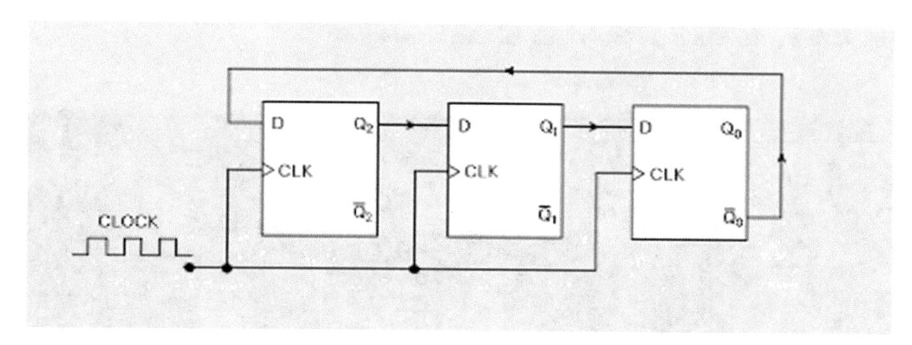

Fig. 9.13 Three-bit Johnson counter

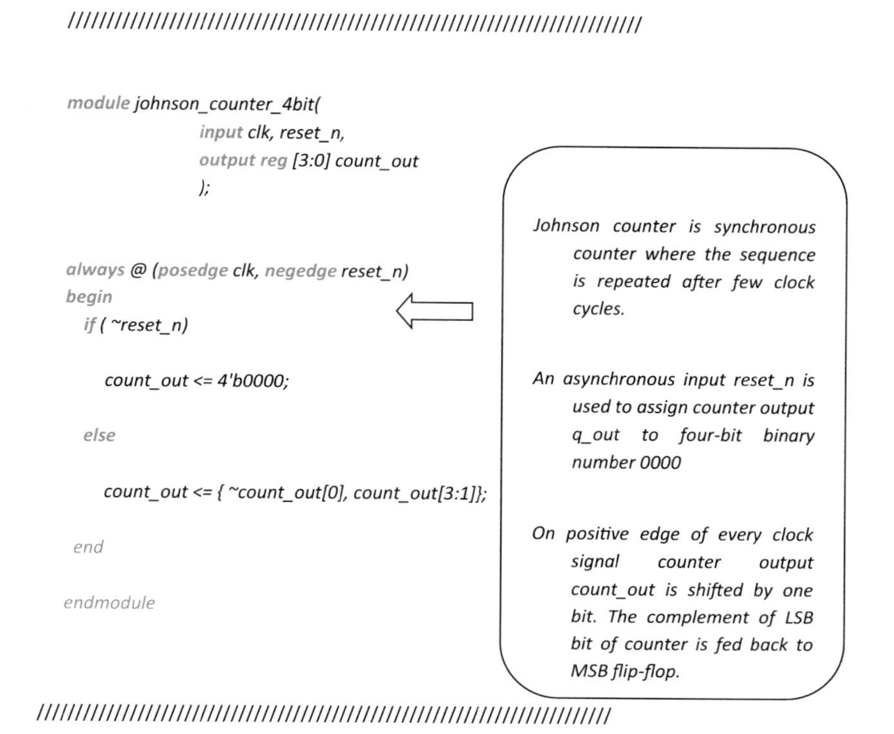

```
////////////////////////////////////////////////////////////////////////

module johnson_counter_4bit(
                input clk, reset_n,
                output reg [3:0] count_out
                );

always @ (posedge clk, negedge reset_n)
begin
  if ( ~reset_n)

    count_out <= 4'b0000;

  else

    count_out <= { ~count_out[0], count_out[3:1]};

 end

endmodule
////////////////////////////////////////////////////////////////////////
```

Johnson counter is synchronous counter where the sequence is repeated after few clock cycles.

An asynchronous input reset_n is used to assign counter output q_out to four-bit binary number 0000

On positive edge of every clock signal counter output count_out is shifted by one bit. The complement of LSB bit of counter is fed back to MSB flip-flop.

Example 8 Verilog RTL for four-bit Johnson counter

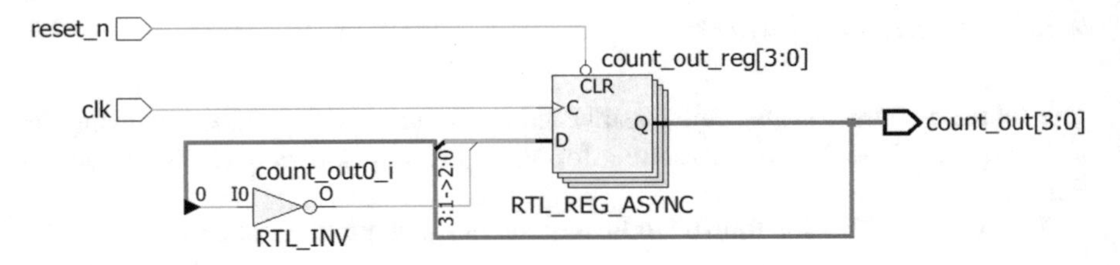

Fig. 9.14 Synthesized logic for four-bit Johnson counter

Fig. 9.15 Simulation result of 4-bit Johnson counter

The synthesis logic is shown in Fig. 9.14.

The simulation result for the 4-bit Johnson counter is shown in Fig. 9.15, and as shown, the output sequence at count_out is 1000, 1100, 1110, 1111, 0111, 0011, 0001, 0000, 1000…. This type of counter is also called as twisted ring counter.

9.3 BCD Up–Down Counter

The BCD counter or MOD-10 counter counts from 0000 to 1001. If it is up counter, then the sequence is from 0000 to 1001, and if it is down counter, then output sequence is from 1001 to 0000. The RTL using synthesizable Verilog constructs for the BCS up–down counter is coded, and it is shown in Example 9.

The synthesis result of Example 9 is shown in Fig. 9.16.

As shown in the simulation waveform, the counter counts from 0 to 9 in the up counter mode, and in the down counter mode, it counts from 9 to 0 (Fig. 9.17).

```
////////////////////////////////////////////////////////////////////////////////

module bcd_up_down_counter (

                    input clk, reset_n,

                    input up_down ,

                    output reg [3:0] count

                     );

always @ (posedge clk, negedge reset_n)

begin

   if ( ~reset_n)

       count <= 4'b0000;

   else

      if ( up_down)

         if ( count == 4'b1001)

           count <= 4'b0000;

         else

         count <= count + 1;
```

Example 9 RTL design of 4-bit BCD up-down counter

```
    else
      if ( count == 4'b0000)
            count <= 4'b1001;
        else
            count <= count - 1;
   end
   endmodule
   ////////////////////////////////////////////////////////////////////////////////
```

Example 9 (continued)

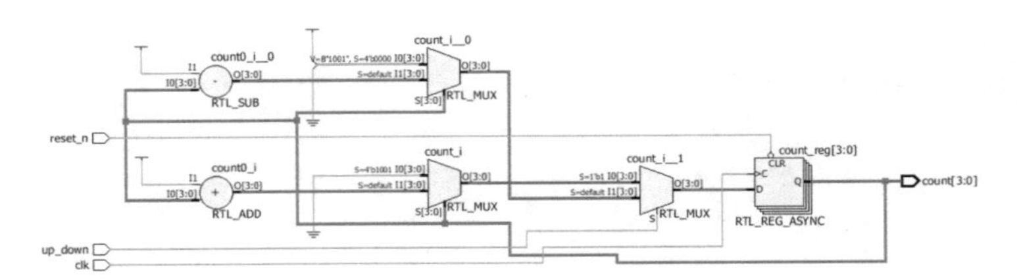

Fig. 9.16 Synthesis result of Example 9

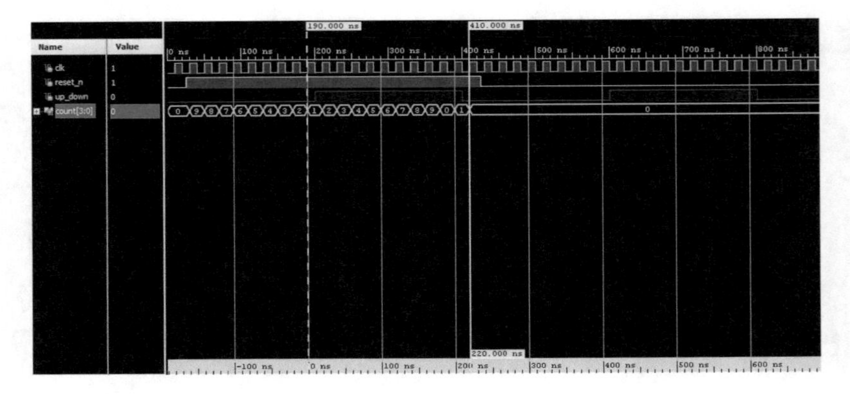

Fig. 9.17 Simulation waveform of Example 9

9.4 Exercises

The exercises are based on the understanding of procedural always block used to design the counters. Complete the exercises for better understanding and application of Verilog constructs.

2. The logic inferred by the following code is

```
module design_logic ( input  clk, reset_n, output reg [1:0] y);

always @ (posedge clk)

begin

if (~ reset_n)

  y <= 2'b00;

else if (y ==00)

 y <= 2'b10;

else

y<= y-1;

end

endmodule
```

a. MOD-4 binary up counter
b. MOD-3 binary up counter
c. MOD-2 binary up counter
d. Mod-3 binary down counter

1. The logic inferred by the following code is

```
module design_logic ( input  clk, reset_n, output reg [1:0] y);

always @ (posedge clk)

begin

if (~ reset_n)

  y <= 2'b00;

else if (y ==10)

 y <= 2'b00;

else

y<= y+1;

end

endmodule
```

   a. MOD-4 binary up counter
   b. MOD-3 binary up counter
   c. MOD-2 binary up counter
   d. Mod-3 binary down counter

3. The logic inferred by the following code is

```
module design_logic7 ( input d, clk, output reg y);

reg tmp;

always @ (posedge clk)

begin

y <= tmp;

tmp <= d;

end

endmodule
```

   a. Positive edge triggered D flip-flop.
   b. Two-bit shift register
   c. positive edge triggered D flip-flop having input logic 1.
   d. positive edge triggered D flip-flop having input logic 0.

9.5 Summary

The following are key points to summarize the chapter:

1. Binary counters can be designed by using synchronous counter design concept or asynchronous design techniques.
2. Gray counters can be designed by using the binary counters with the additional combinational logic.
3. Synchronous counters are recommended in the ASIC design as timing analysis will be easy, and they are not prone to the glitches.
4. Asynchronous counters are prone to glitches or spikes and hence not recommended in the ASIC designs.

5. Special counters like ring and Johnson can be designed by using the shift registers.
6. Johnson counter is also called as twisted ring counter.
7. Binary counters are used to count from 0000 to 1111.
8. BCD counters are used to count from 0000 to 1001.

Chapter 10
RTL Design of Registers and Memories

The chapter is useful to understand the RTL design techniques and strategies useful to code the RTL for registers, shift-registers and memories.

As discussed in previous chapters, the sequential design important element is flip-flop, and in the sequential design output is function of the present inputs and past outputs. Most of the time in the ASIC or FPGA design, we need to include the storage that is block of registers. In such scenario, the chapter is useful to understand about the RTL design techniques and strategies while coding the RTL for the registers and memories. The following few section discusses the RTL design of registers and memories.

10.1 Parallel Input and Parallel Output (PIPO) Register

In most of the processor design applications, the data needs to be transferred in parallel. Consider the four-bit data bus communicating with the external peripheral. If both processor and peripheral operate on the parallel data, then it is essential to transfer the data using parallel input parallel output logic.

In such scenarios, PIPO registers are used. The logic diagram of PIPO four-bit register is shown in Fig. 10.1. Four parallel input lines are named as P_A, P_B, P_C, and P_D, and four-bit parallel output lines are named as Q_A, Q_B, Q_c, and QD. The PIPO register is sensitive to the positive edge of clock.

The Verilog RTL is described in Example 1.

The synthesis result of the four-bit PIPO register is shown in Fig. 10.2.

V. Taraate, *Digital Logic Design Using Verilog*,
https://doi.org/10.1007/978-981-16-3199-3_10

```
////////////////////////////////////////////////////////////////////////////

module parallelin_parallelout(

                              input clk,

                              input reset_n,

                              input [3:0] d_in,

                              output  reg [3:0] q_out );

always @ ( posedge clk ,  negedge reset_n)

begin

  if (~reset_n)

     begin

        q_out <= 4'b0000;

     end

     else

       q_out <= d_in;

end

endmodule

////////////////////////////////////////////////////////////////////////////
```

For reset_n=0 the output of PIPO register is initialized to 0000.

During normal operation PIPO register has reset_n=1.

On positive edge of the clock input the data input d_in is assigned to an output q_out.

As data input is four-bit wide the logic inferred is four bit PIPO register.

Example 1 Verilog RTL for 4-bit PIPO register

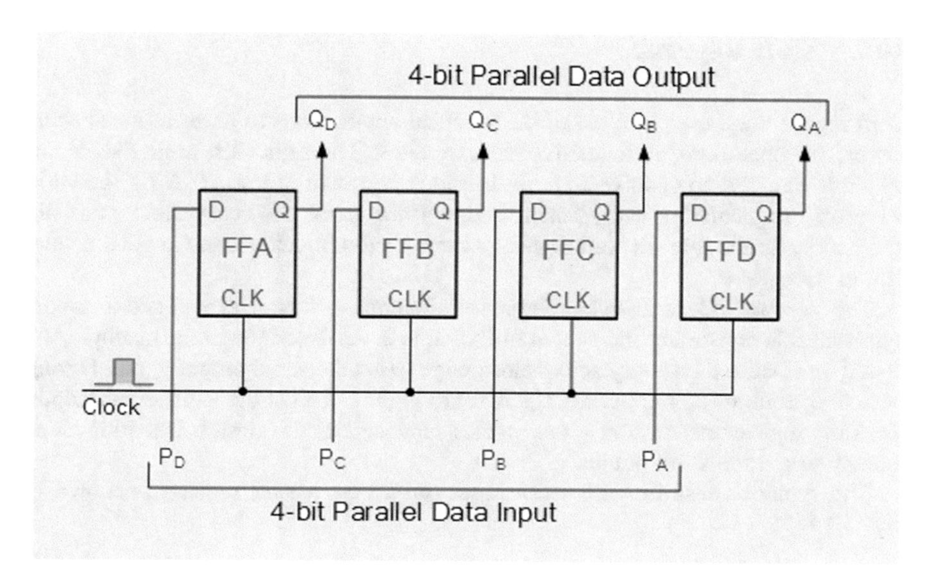

Fig. 10.1 Four-bit PIPO register

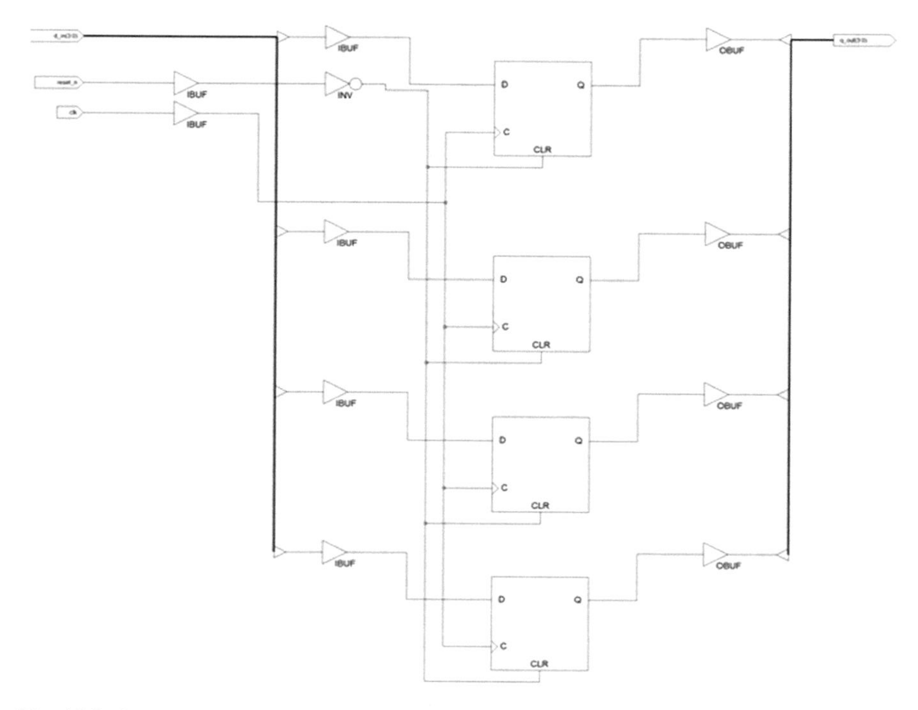

Fig. 10.2 Synthesized logic for four-bit PIPO register

10.2 Shift Register

Shift registers are used in most of the practical applications to perform the shifting or rotation operations on the active edge of clock. The right shift timing sequence which is sensitive to positive edge of the clock is shown in Fig. 10.3. As shown in the timing sequence for every positive edge of the clock, the data shifts by one bit, and hence, for the four-bit shift register it requires four clock latency to get the valid output data.

The Verilog RTL for the serial input serial output shift register is coded using the synthesizable constructs and shown in Example 2. As described in the example, the data d_in is shifted on every active clock edge to get the serial output q_out. During normal operation reset, input reset_n is set to logic 1. To get the valid serial output, the shift register needs four clock cycles, and hence, the design has four clock latency to get the valid output.

The synthesis result of the serial input serial output shift register is shown in Fig. 10.4.

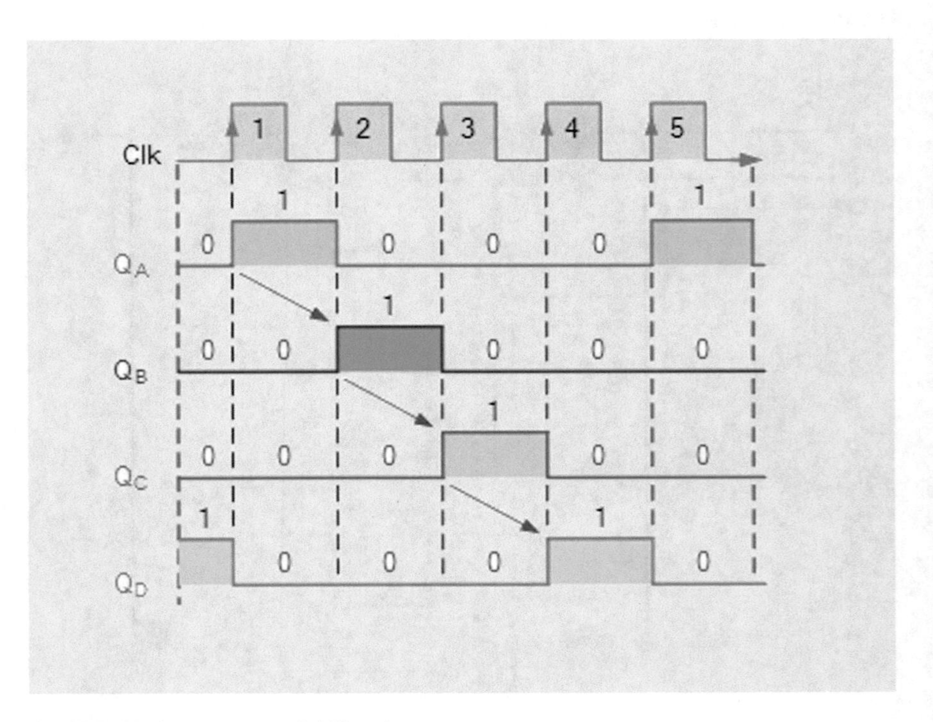

Fig. 10.3 Timing sequence of shift register

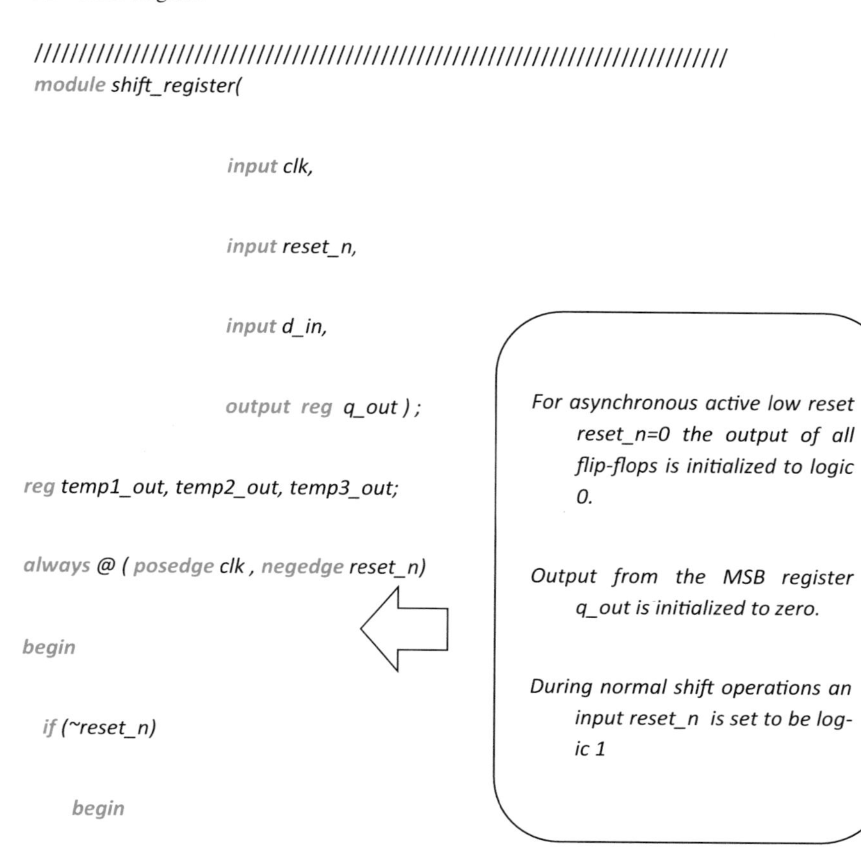

```
////////////////////////////////////////////////////////////////////////////////////
module shift_register(

                    input clk,

                    input reset_n,

                    input d_in,

                    output  reg  q_out ) ;

reg temp1_out, temp2_out, temp3_out;

always @ ( posedge clk , negedge reset_n)

begin

   if (~reset_n)

      begin

         {q_out , temp3_out, temp2_out, temp1_out}<= 4'b0000;

      end

      else
```

For asynchronous active low reset reset_n=0 the output of all flip-flops is initialized to logic 0.
Output from the MSB register q_out is initialized to zero.
During normal shift operations an input reset_n is set to be logic 1

Example 2 Serial input serial output shift register synthesizable design

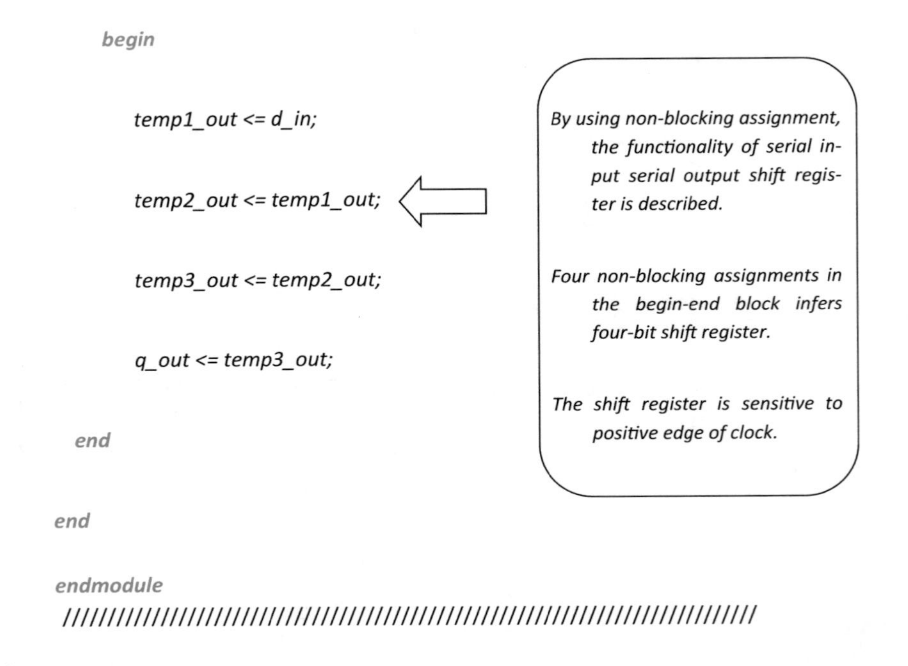

```
begin
    temp1_out <= d_in;
    temp2_out <= temp1_out;
    temp3_out <= temp2_out;
    q_out <= temp3_out;
  end
end
endmodule
////////////////////////////////////////////////////////////////////////////////
```

By using non-blocking assignment, the functionality of serial input serial output shift register is described.
Four non-blocking assignments in the begin-end block infers four-bit shift register.
The shift register is sensitive to positive edge of clock.

Example 2 (continued)

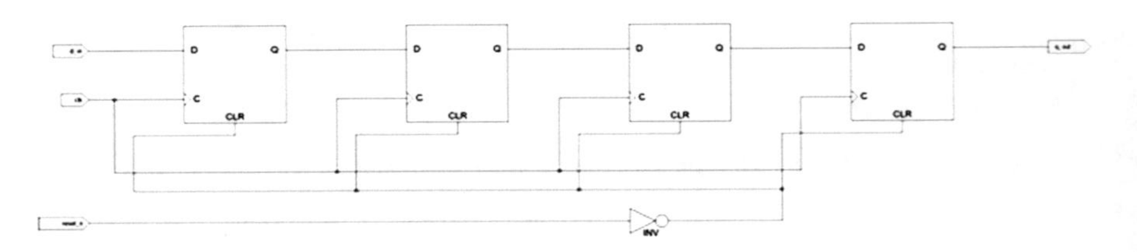

Fig. 10.4 Synthesized Logic for four-bit shift register

10.3 Right and Left Shift Operation

Most of the practical application demands the use of right or left shift of the data. Consider the RTL while implementing the protocol, in few serial protocols the requirement is to shift the data string to the right side or to the left side by one bit or by multiple bits. In such scenario, the bidirectional (right/left) shift registers can be used.

The Verilog RTL is shown in Example 2 for bidirectional shift register, and the direction of data is controlled by right_left input. For right_left = 1, the data is shifted toward right side, and for the right_left = 0, the data is shifted toward left side.

```
////////////////////////////////////////////////////////////////////////////////

module right_left_shift_register(

                                    input right_left,

                                    input clk,

                                    input reset_n,

                                    input d_in,

                                    output reg  [3:0] q_out );

always @ ( posedge clk , negedge reset_n)

begin

  if (~reset_n)

      begin

          q_out <= 4'b0000;

      end

      else

      begin

      if (right_left)
```

Example 3 Verilog RTL for the right/left shift register

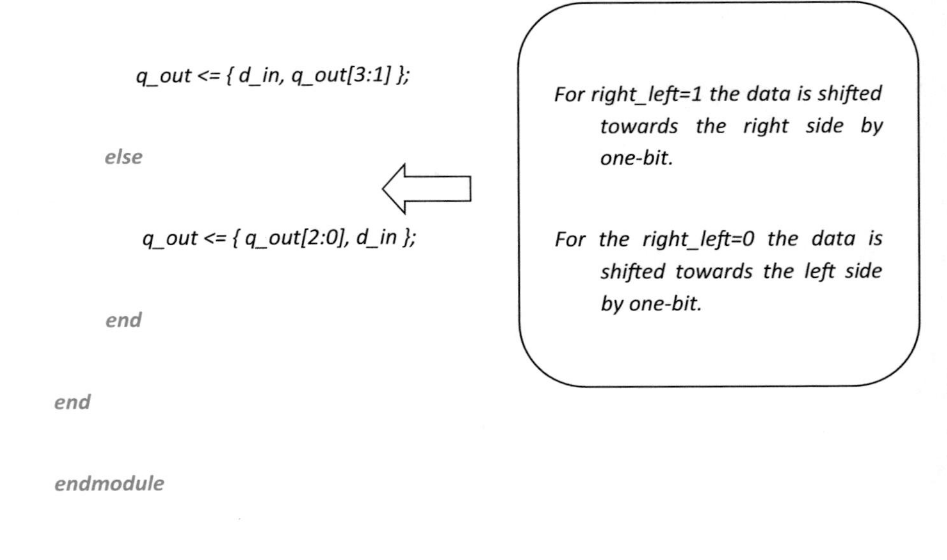

```
            q_out <= { d_in, q_out[3:1] };
        else
            q_out <= { q_out[2:0], d_in };
        end
end
endmodule
```

For right_left=1 the data is shifted towards the right side by one-bit.
For the right_left=0 the data is shifted towards the left side by one-bit.

//

Example 3 (continued)

The synthesis result is shown in Fig. 10.5, and the direction of data transfer is controlled by right_left input. The synthesized logic consists of four flip-flops with additional combinational logic to control, and the data flows either toward right or left side.

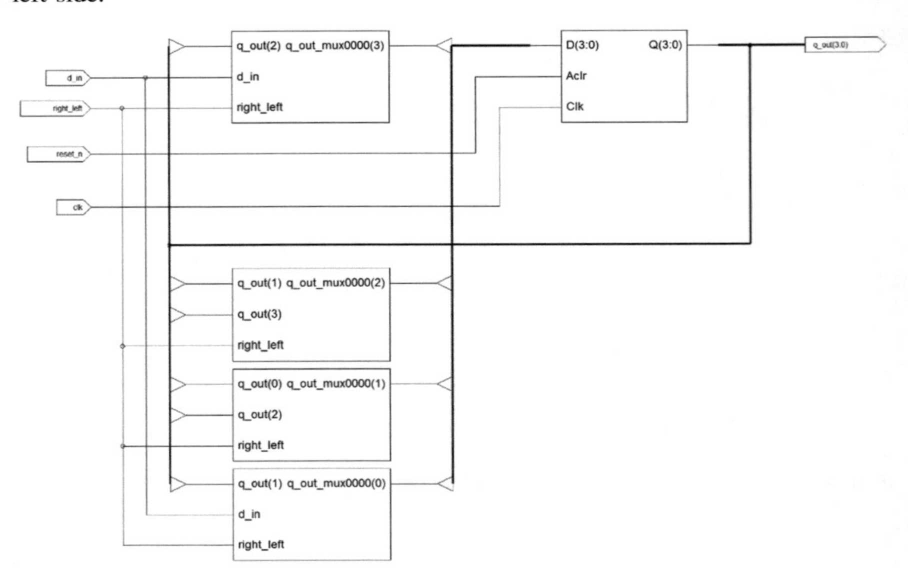

Fig. 10.5 Synthesized logic for bidirectional shift register

10.4 Timing and Performance Evaluation

The timing is particularly an important parameter for ASIC or FPGA designs. Meeting timing for the sequential circuits is very crucial for the complex designs. The timing analysis and frequency calculations will be discussed in Chap. 20.

For the better design performance, it is recommended to use the register inputs and register outputs. In the practical designs, the RTL using Verilog should be coded using efficient Verilog constructs and should have the registered inputs and registered outputs. The reason for the same is to have clean timing paths (reg to reg paths) to get the better timing.

The Verilog RTL which has the registered output is shown in Example 4. It is assumed that another module drives the input signals 'a', 'b', 'c,' and 'select'. All these inputs are registered inputs. This is useful to have clean register path and easy timing analysis.

```
/////////////////////////////////////////////////////////////////////////////////////

module mux_registered_output (
                    input [7:0] a, b, c, d,
                    input [1:0] select,
                    input clock,
                    output reg [7:0] y ) ;


always @ (posedge clock)
// Use of non-blocking assignments
  case (select)
    0: y <= a;
    1: y <= b;
    2: y <= c;
    3: y <= d;
    default y <= 8'b0;
  endcase

endmodule

/////////////////////////////////////////////////////////////////////////////////////
```

The case construct is used within the procedural always block.

The procedural always block is sensitive to the positive edge of clock.

Example 4 Verilog RTL of 4:1 mux having the registered output

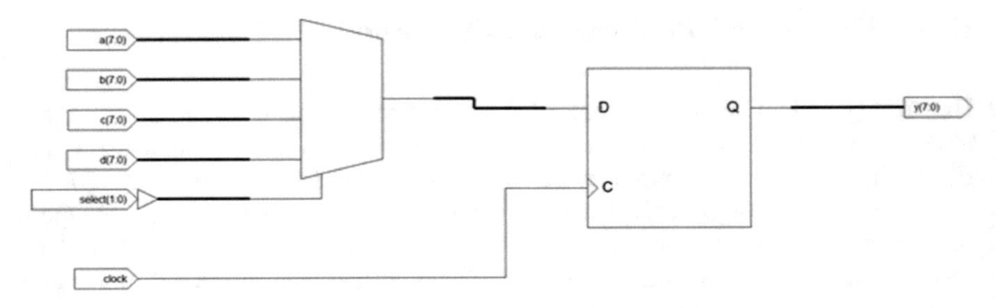

Fig. 10.6 Synthesized logic for the registered output logic

The synthesis result is shown in Fig. 10.6 and infers the 8-bit parallel input parallel output register at the output of the 4:1 mux. Each input of 4:1 mux is 8-bit wide. The logic is sensitive to the positive edge of clock.

10.5 Asynchronous Counter Design

In the asynchronous counters, the clock signal of all the flip-slops is not driven by the common clock source. If the output of LSB flip-flop is used as a clock input to the subsequent flip-flop, then the design is said to be asynchronous. The issue with the asynchronous design is the addition of clock to q delay of flip-flop due to the cascading of the number of flip-flop stages. Asynchronous counters are not recommended in the ASIC or FPGA design due to the issue of glitches or spikes, and even the timing analysis for such kind of design is extraordinarily complex.

10.5.1 Ripple Counters

The ripple counter is an asynchronous counter and shown in Fig. 10.7. As shown in the logic diagram, all the toggle flip-flops are positive edge triggered, and the LSB flip-flop receives the clock from the master clock source. The output of LSB flip-flop is used as clock input to the next subsequent stage flip-flop.

The Verilog RTL for the four-bit ripple up counter is shown in Example 5.

The synthesis result is shown in Fig. 10.8.

The simulation of the 4-bit ripple counter is shown in the waveform (Fig. 10.9). As shown, when toggle_in and reset_n both are at logic level 1, the counter counts from 0 to F. The output format is hexadecimal.

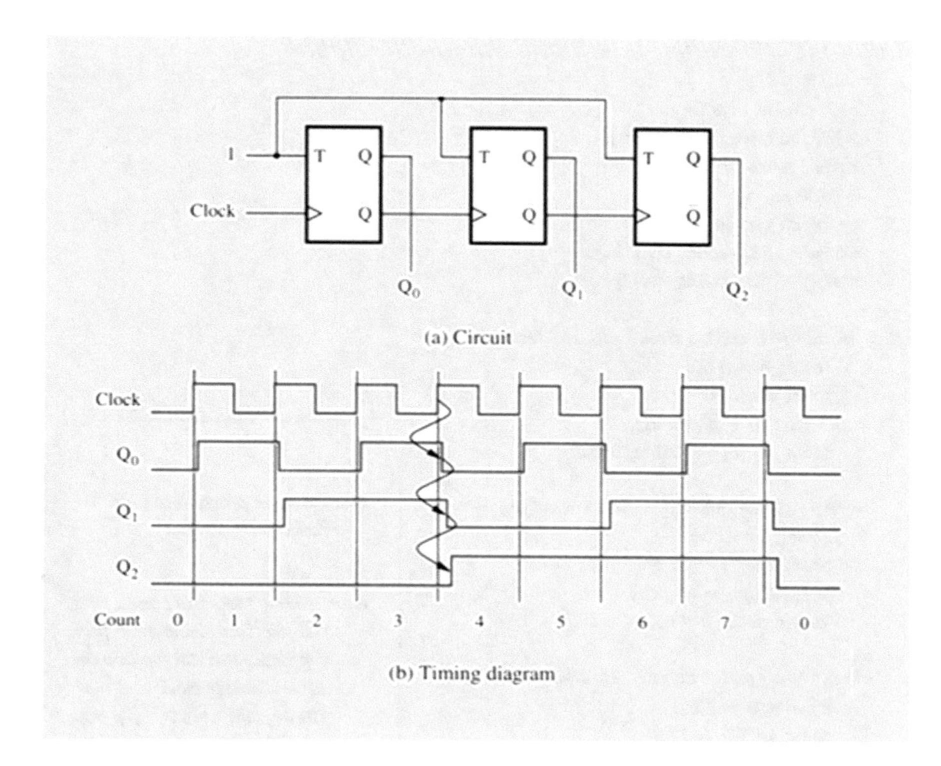

Fig. 10.7 Logic diagram of three-bit ripple counter

```
////////////////////////////////////////////////////////////////////////////////

module ripple_counter (
  input clock, toggle_in, reset_n,
  output reg [3:0] count_out );
  wire c0, c1, c2;
  assign c0 = count_out[0];
  assign    c1 = count_out[1];
  assign    c2 = count_out[2];

  always @ (posedge clock , negedge reset_n )
   if (reset_n == 1'b0)
     count_out[0] <= 1'b0;
   else if (toggle_in == 1'b1)
     count_out[0] <= ~count_out[0];

  always @ (negedge c0, negedge reset_n )
   if (reset_n == 1'b0)
     count_out[1] <= 1'b0;
   else if (toggle_in == 1'b1)
     count_out[1] <= ~count_out[1];

 always @ (negedge c1, negedge reset_n )
    if (reset_n == 1'b0)
    count_out[2] <= 1'b0;
   else if (toggle_in == 1'b1)
    count_out[2] <= ~count_out[2];

 always @ (negedge c2, negedge reset_n )
   if (reset_n == 1'b0)
    count_out[3] <= 1'b0;
   else if (toggle_in == 1'b1)
    count_out[3] <= ~count_out[3];
 endmodule

////////////////////////////////////////////////////////////////////////////////
```

Every always procedural block infers toggle flip-flop

Four always blocks are used and the inferred logic has four flip-flops. The LSB flip-flop receives master clock and next subsequent stage flip-flop uses the previous stage flip-flop output as a clock.

Example 5 Verilog RTL of four-bit ripple up counter

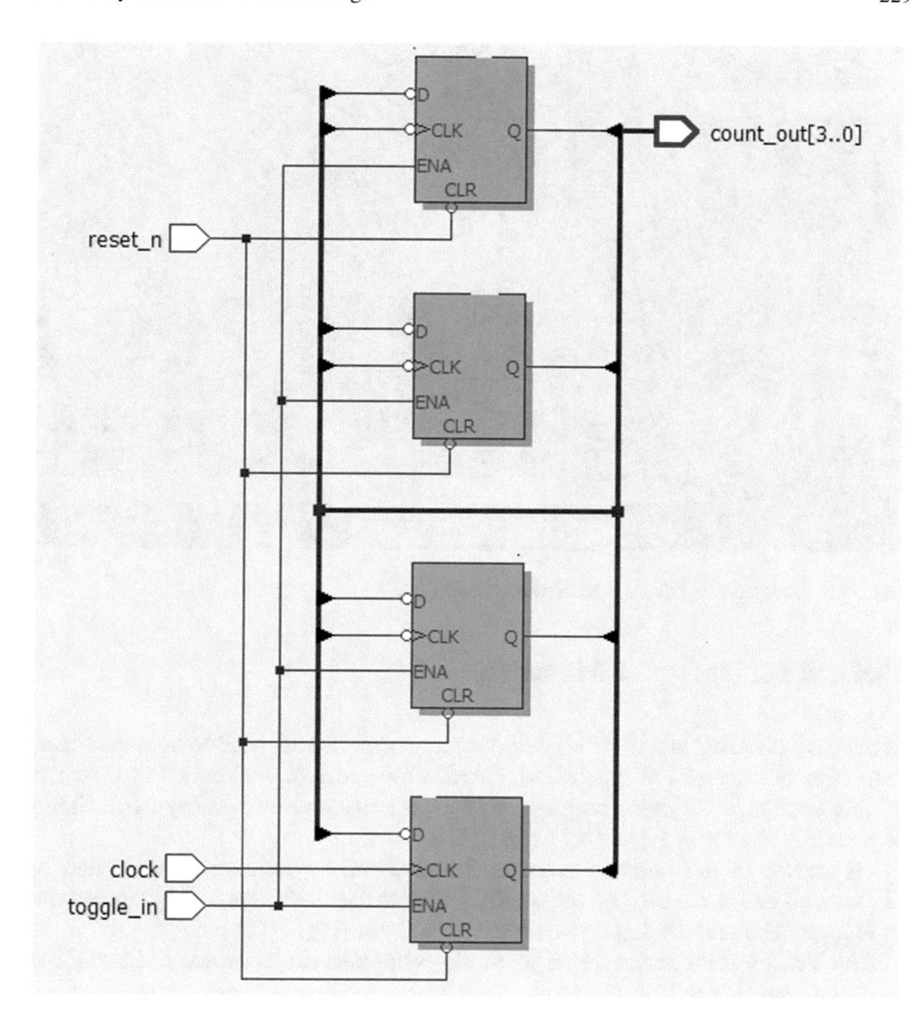

Fig. 10.8 Synthesized logic of four bit ripple up counter

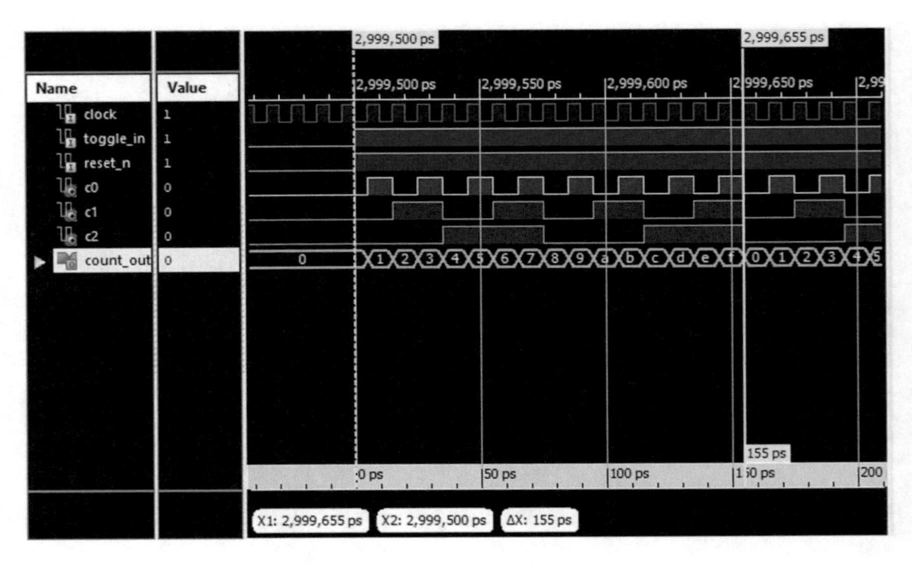

Fig. 10.9 Simulation result of 4-bit ripple counter

10.6 RTL Design of Memories

In most of the ASIC and SOC0-based designs, memories are used to hold the binary data. Memories can be of type ROM, RAM, single port, or dual port. The objective of this section is to describe basic single port read–write memory. The timing sequence is shown in Figs. 10.11 and 10.13.

As shown in the timing sequence, the read–write operation is controlled by rd_wr, and data is sampled on the positive edge of the clock signal clk provided that cs is high. The address input is denoted as address (Fig. 10.11).

The Verilog RTL of the single port read–write memory is shown in Example 6.

The synthesis result is shown in Fig. 10.12.

The simulation result is shown in Fig. 10.13. As shown, the memory location (01) holds the data AA, and location (02) holds data BB.

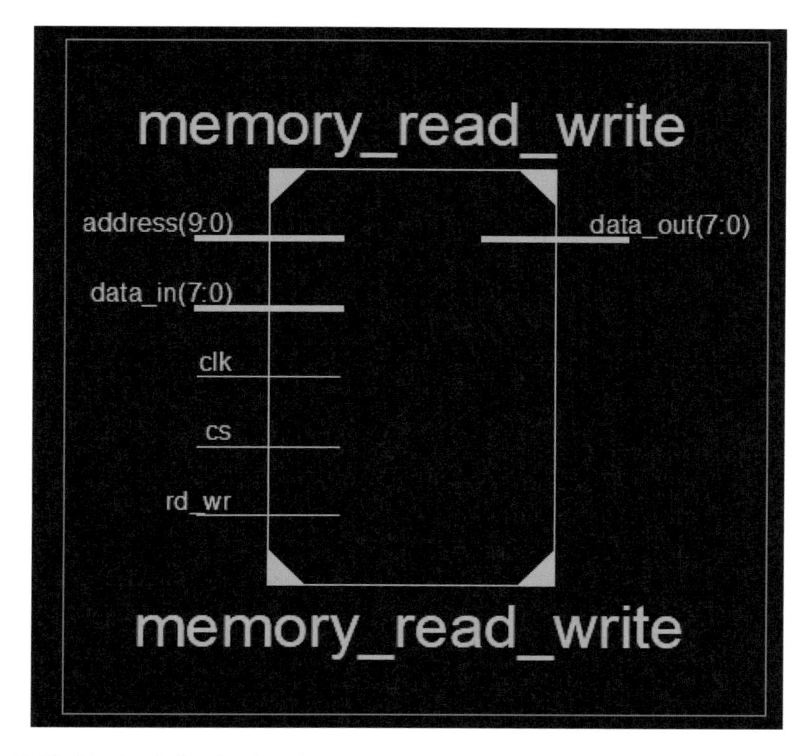

Fig. 10.10 Top-level signals of read–write memory

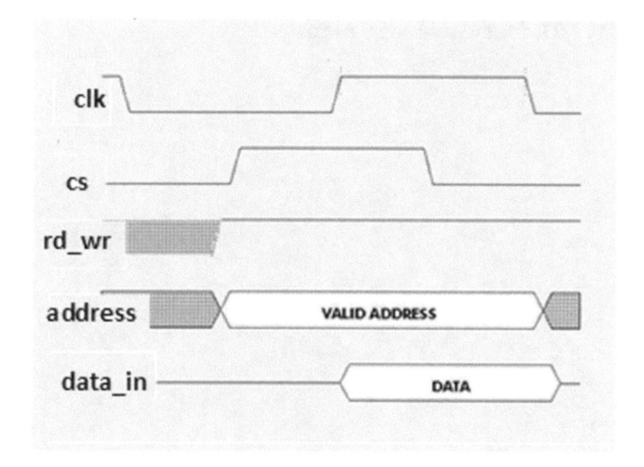

Fig. 10.11 Timing sequence for the memory

```
////////////////////////////////////////////////////////////////////////////////

module memory_read_write (

                input clk,
                input [7:0] data_in,
                input [9:0] address,
                input rd_wr,
                input cs,
                output reg  [7:0] data_out ) ;

reg [7:0] memory [0:1023];

always@ (posedge clk)
    if ( cs==1 && rd_wr==1)
    memory [address] <= data_in;

always@ (posedge clk)
    if ( cs==1 && rd_wr==0)
    data_out <= memory [address];

endmodule

////////////////////////////////////////////////////////////////////////////////
```

One always block is used to perform the write operation when rd_wr=1.

Another always block is used to perform the read operation when rd_wr=0

Example 6 Verilog RTL for the read write memory

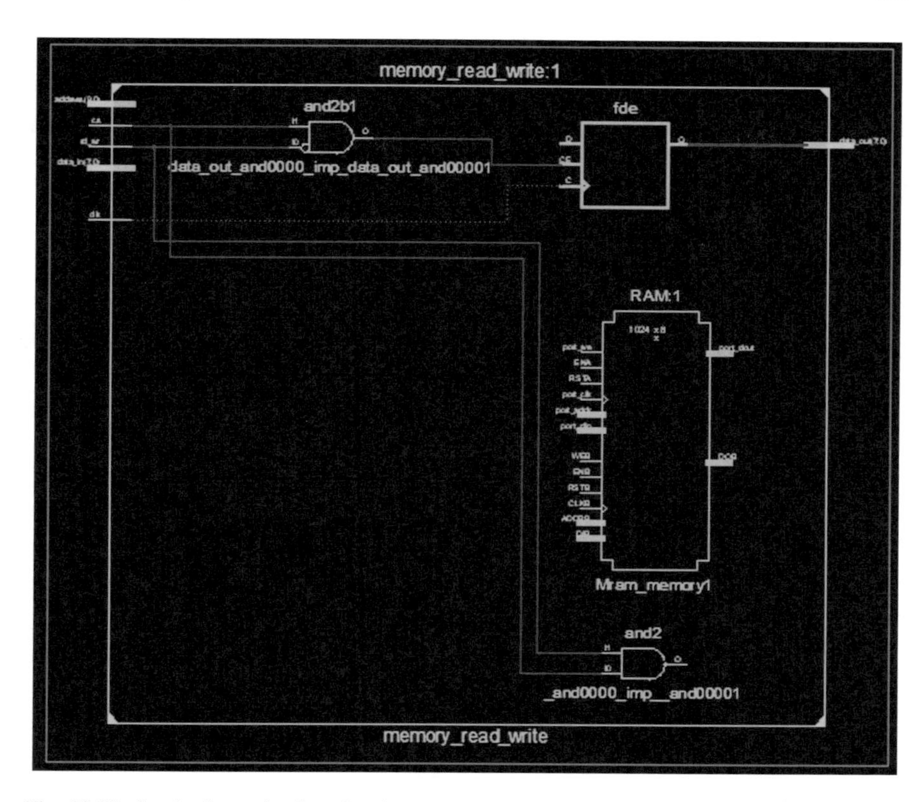

Fig. 10.12 Synthesis result of read–write memory

Fig. 10.13 Simulation waveform of memory

10.7 Parameterized Read–Write Memory

The parameterized memory buffer used in the FIFO design and having 8 location depth is coded using the synthesizable constructs and shown in Example 7. The top-level pin-out of the memory is shown in Fig. 10.14.

```
//////////////////////////////////////////////////////////////////////////////

module FIFO_memory #(parameter data_size = 8, parameter ad-
    dress_size = 3) // let us define the data and address size to get 8 loca-
    tion X 8 bit memory

input [data_size-1:0] write_data,
(

input [address_size-1:0] write_address, read_address,

input write_clk_en, write_full, write_clk,

output [data_size-1:0] read_data);

localparam FIFO_depth = 1<<address_size;

reg [data_size-1:0] mem [0:FIFO_depth-1];

assign read_data = mem[read_address];

always @(posedge write_clk)

    if (write_clk_en && !write_full) mem[write_address] <= write_data;

endmodule

//////////////////////////////////////////////////////////////////////////////
```

Example 7 RTL design of the memory used in the FIFO of depth 8

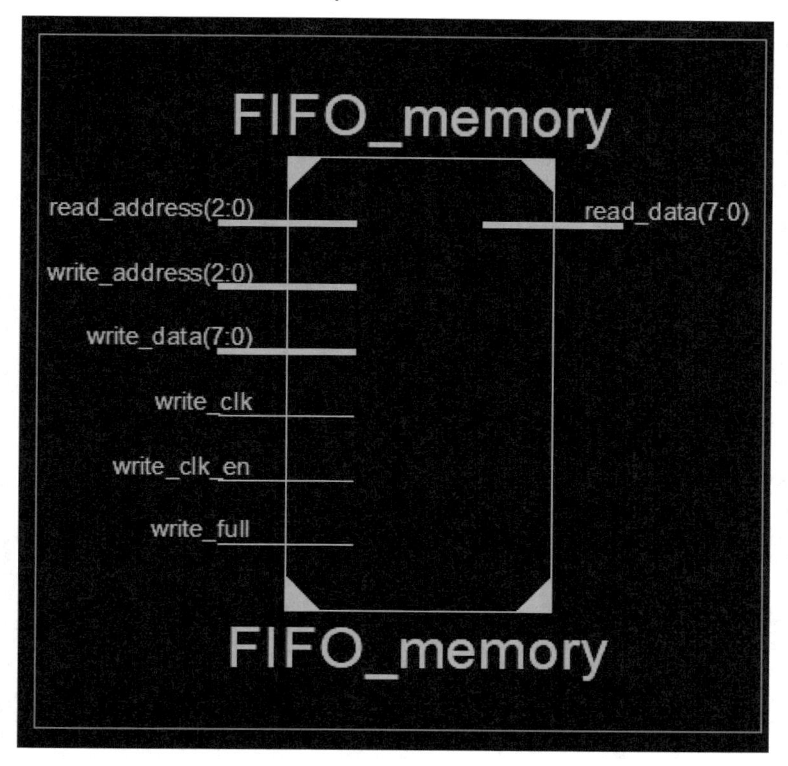

Fig. 10.14 FIFO memory buffer

The RTL uses ***parameter*** to input the size of the data and address.

The hashtag using character # is used to declare the parameters as parameter data_size = 8 and parameter address_size = 3.

The synthesis result of the 8 location wide FIFO is shown in Fig. 10.15. As shown during FPGA synthesis, it infers the BRAM. Refer Chaps. 16 and 17 for more information about FPGA design flow and FPGA synthesis.

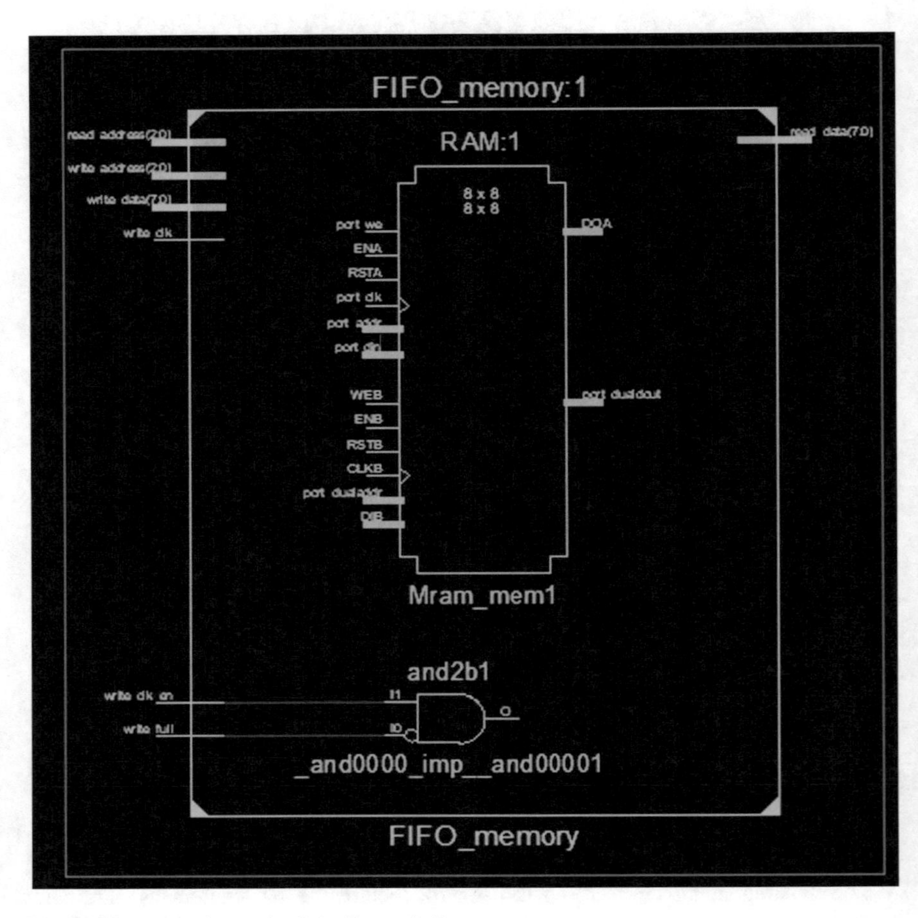

Fig. 10.15 Synthesis result of the Example 7

10.8 Exercises

The exercises are based on the understanding of procedural ***always*** block used to design the counters. Complete the exercises for better understanding and application of Verilog constructs.

1. The logic inferred by the following code is

```
module design_logic ( input  clk, reset_n, input[1:0] d_in, out-
      put reg [1:0] y);

always @ (posedge clk)

begin

if (~ reset_n)

 y <= 2'b00;

else

y<=d_in;

end

endmodule
```

   a. 2-bit right shift register
   b. 2-bit left shift register
   c. 2-bit PIPO register
   d. 2-bit counter

2. The logic inferred by the following code is

```
module design_logic ( input  d_in, clk, reset_n, output reg y);

reg tmp1, tmp2;

always @ (posedge clk)

begin

if (~ reset_n)

 {y,tmp2,tmp1} <= 3'b00;

else

begin

 y <= tmp2;

tmp2 <= tmp1;

tmp1 <= d_in;

end

endmodule
```

a. Two-bit shift register
b. Three-bit shift register
c. PIPO register
d. Binary up-counter

3. The logic inferred by the following code is

```
module design_logic ( input  d_in, clk, reset_n, output reg y);

reg tmp1, tmp2;

always @ (posedge clk)

begin

if (~ reset_n)

 {y,tmp2,tmp1} <= 3'b00;

else

begin

 y <= tmp2;

tmp2 <= tmp1;

tmp1 <= d_in;

end

endmodule
```

   a. Two-bit shift register
   b. Three-bit shift register
   c. PIPO register
   d. Binary up-counter

10.9 Summary

The following are important points to conclude the chapter:

1. Shift registers are used in most of the practical applications to perform the shifting or rotation operations on the active edge of clock.
2. For the better design performance, it is recommended to use the register inputs and register outputs.
3. In the asynchronous counters, the clock signal of all the flip-flops is not driven by the common clock source.
4. The issue with the asynchronous design is the addition of clock to q delay of flip-flop due to the cascading of the number of flip-flop stages.
5. The memories can be described by using the synthesizable Verilog constructs to perform the read and write operation.

Chapter 11
Sequential Circuit Design Guidelines

The RTL design coded without use of sequential design guidelines can result into inefficient performance. This chapter discusses about the sequential design guidelines which need to be followed while coding an efficient RTL using synthesizable Verilog constructs. Use of non-blocking assignments is recommended while coding the sequential design.

As discussed in previous chapters the sequential design such as counters, shift registers are coded using the synthesizable Verilog constructs using non-blocking assignments. For the complex RTL design, the design team needs to follow few guidelines and the important guidelines are documented and discussed in this chapter. The following few section discusses the blocking versus non-blocking assignments, guidelines for reset, clock.

11.1 What Happens If Blocking Assignments Are Used to Code Sequential Logic?

As discussed in Chap. 4, blocking assignments are recommended while coding the combinational logic. But what happens if blocking assignments are used while coding the sequential logic? This is one of the most important questions needs to be addressed as it is an important for the subsequent discussions!

If blocking assignments are used for coding the behavior of sequential logic, then it is observed that the synthesis does not result in the intended sequential logic.

This section discusses few design scenarios by using blocking assignments to code the sequential designs.

V. Taraate, *Digital Logic Design Using Verilog*,
https://doi.org/10.1007/978-981-16-3199-3_11

11.1.1 Blocking Assignments and Multiple always Blocks

The RTL using blocking assignments is coded in Example 1 blocking assignments which are used in the multiple *always* block. Procedural block ***always*** is sensitive to the positive edge of clock, and synthesis tool infers the sequential logic. As discussed already all the blocking assignments are evaluated and updated in active queue. Readers are requested to refer Chap. 6 to understand more about the stratified event queuing.

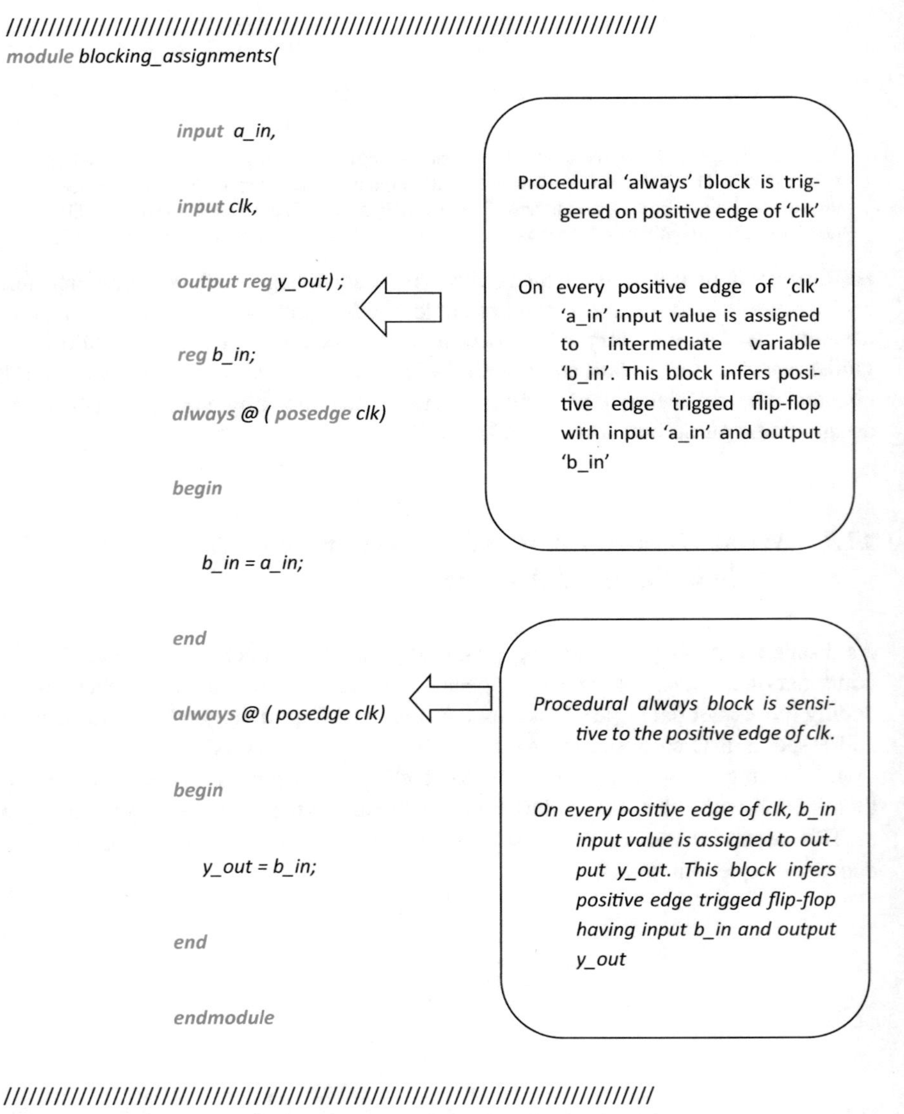

//
module blocking_assignments(
input a_in,
input clk,
output reg y_out) ;
reg b_in;
always @ (posedge clk)
begin
b_in = a_in;
end
always @ (posedge clk)
begin
y_out = b_in;
end
endmodule
//
Procedural 'always' block is triggered on positive edge of 'clk'
On every positive edge of 'clk' 'a_in' input value is assigned to intermediate variable 'b_in'. This block infers positive edge trigged flip-flop with input 'a_in' and output 'b_in'
Procedural always block is sensitive to the positive edge of clk.
On every positive edge of clk, b_in input value is assigned to output y_out. This block infers positive edge trigged flip-flop having input b_in and output y_out

Example 1 Blocking assignments in multiple always blocks

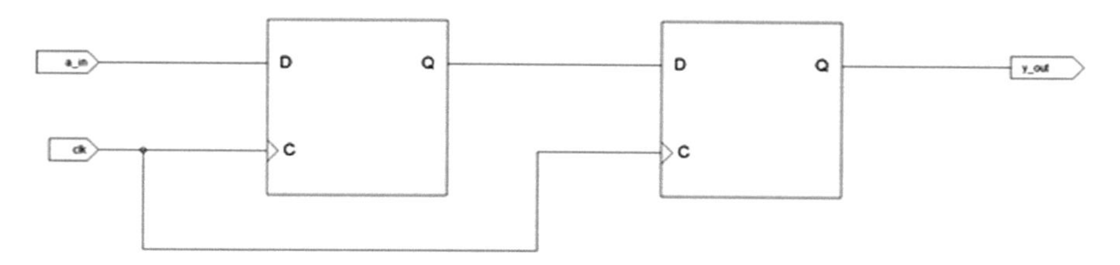

Fig. 11.1 Synthesized logic for blocking assignments in multiple always block

As described in Example 1, both ***always*** procedural blocks execute in parallel and generate the output as two-bit serial-in serial-out shift register. First ***always*** block generates an output b_in. The output generated from first ***always*** block is used as an input by another ***always*** block. Hence, synthesis tool understands this as cascading of the flip-flops and infers logic as two-bit serial-input serial-output shift register.

Synthesis result of Example 1 is shown in Fig. 11.1 and has input a_in, clk, and an output y_out.

11.1.2 Multiple Blocking Assignments Used in the Single always Block

If blocking assignments are used to describe the sequential logic and multiple assignments are used in the same ***always*** procedural block, then the desired intended result may or may not match with the synthesis result. The reason being in the blocking assignment all the trailing statements (next immediate) is blocked unless and until the present assignment is executed. This results in truncation of the logic and may infer the unintended synthesis result.

Consider the design scenario shown in Example 2, and intention is to create the three-bit serial-input and serial-output shift register but after synthesis of Example 2 it infers the logic having single flip-flop.

Synthesis logic schematic is shown in Fig. 11.2 and has inputs a, clk, and an output y. The intended functionality is serial-input serial-output shift register but the above example infers the single flip-flop due to the use of blocking assignments. So it is recommended to use the non-blocking assignments while coding the RTL for the sequential logic.

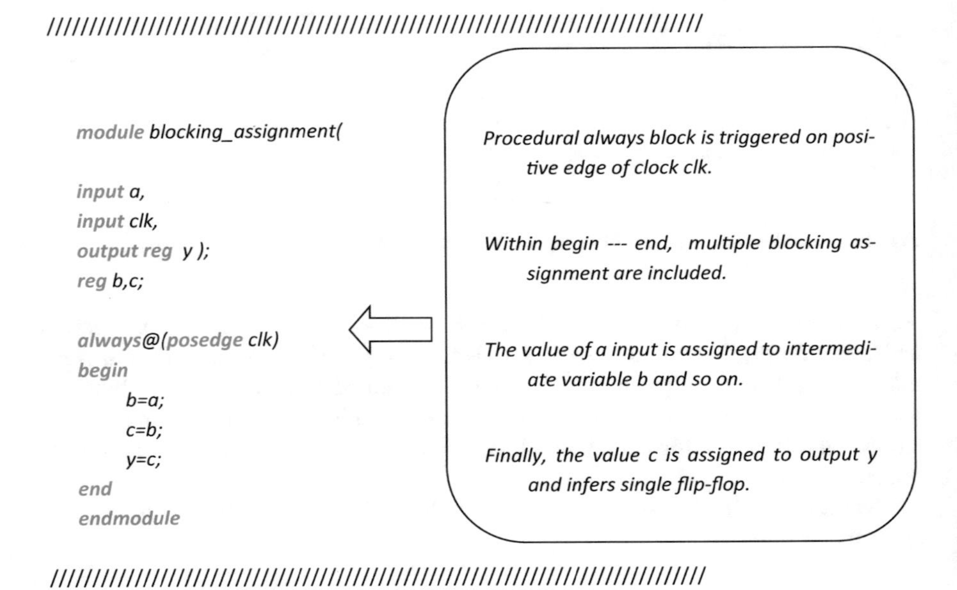
//
module blocking_assignment(
input a,
input clk,
output reg y);
reg b,c;
always@(posedge clk)
begin
b=a;
c=b;
y=c;
end
endmodule
Procedural always block is triggered on positive edge of clock clk.
Within begin --- end, multiple blocking assignment are included.
The value of a input is assigned to intermediate variable b and so on.
Finally, the value c is assigned to output y and infers single flip-flop.
//

Example 2 Blocking assignments in same always block

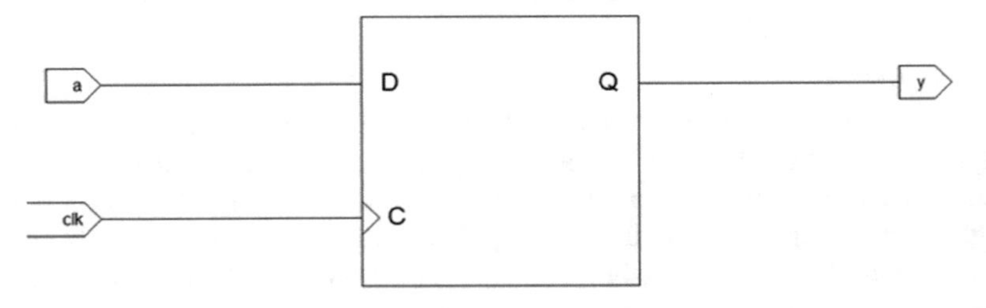
a
D
Q
y
clk
C

Fig. 11.2 Synthesized logic for the blocking assignments in same always block

11.1.3 Example Blocking Assignment

Consider the design scenario shown in Example 3, and intention is to create the three-bit serial-input and serial-output shift register and due to the reorder of the blocking assignment used within the ***begin…end*** it infers the three-bit serial-input serial-output shift register.

Synthesis result is shown in Fig. 11.3 and has inputs a, clk, and an output y. The desired functionality is serial-input serial-output shift register, and it infers the serial-input serial-output shift register. So the important point to remember is that order of the blocking assignments within the procedural ***always*** block is an important factor and decides the synthesis result.

```
//////////////////////////////////////////////////////////////////////////////

module blocking_assignment(

        input a,

        input clk,

        output reg y );

reg b,c;

always@(posedge clk)

begin

    y=c;

    c=b;

    b=a;

end

endmodule

//////////////////////////////////////////////////////////////////////////////
```

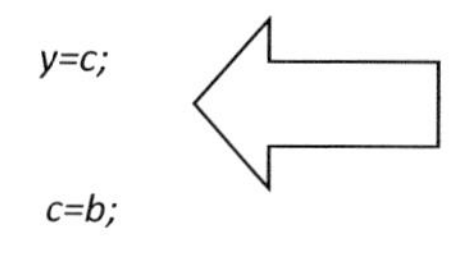

Procedural always block is triggered on positive edge of clock clk.

Within begin --- end, multiple blocking assignment are included.

An intermediate variable value c is assigned to output y and infers D flip-flop.

The value of intermediate variable b is assigned to intermediate variable c and infers D flip-flop.

The value of an input a is assigned to intermediate variable b and generates D flip-flop.

Example 3 Blocking assignments in the same always block (ordering)

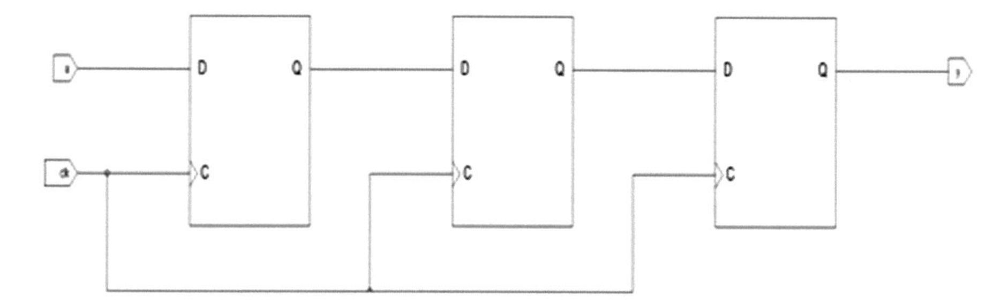

Fig. 11.3 Synthesizable logic after reordering of blocking assignments

11.2 Non-blocking Assignments

As discussed in Chap. 7, non-blocking assignments are evaluated in the active event queue and updated in the NBA queue. Non-blocking assignments are used to describe the sequential logic. These assignments are used in the procedural block ***always*** to get the desired synthesis results. All the non-blocking assignments execute in parallel within the ***always*** procedural block.

As coded in Example 4, non-blocking assignments are used within the multiple ***always*** procedural block. Procedural block ***always*** is sensitive to the positive edge of clock, and synthesis tool infers the sequential logic.

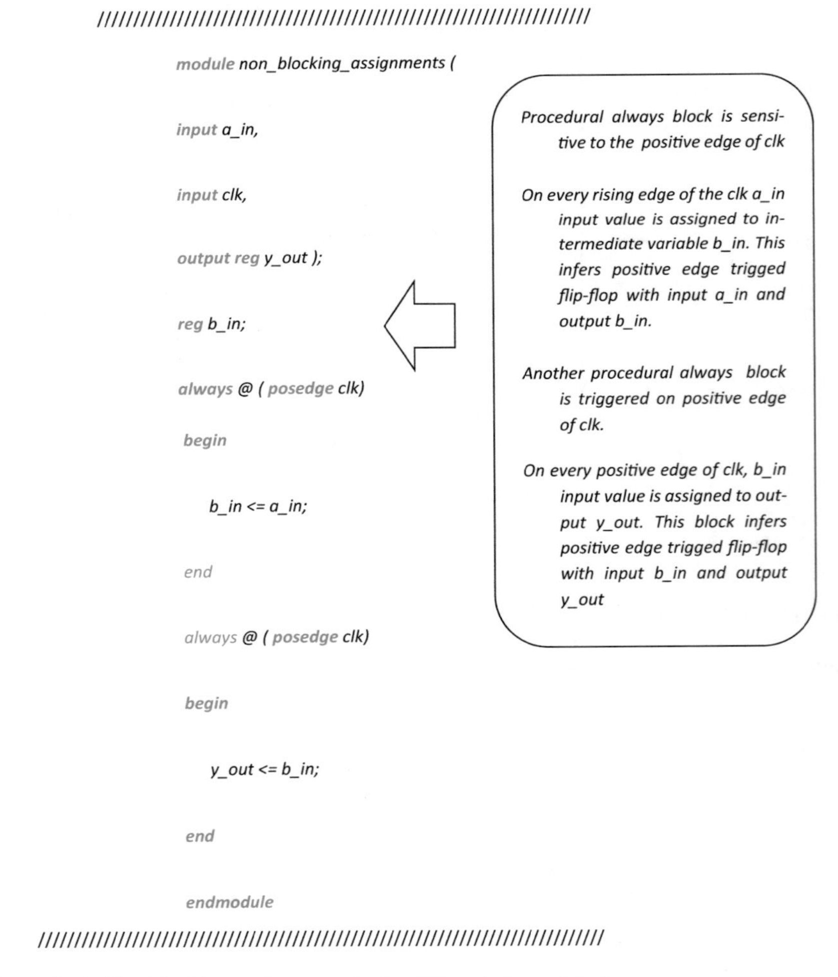

//
module non_blocking_assignments (
input a_in,
input clk,
output reg y_out);
reg b_in;
always @ (posedge clk)
begin
b_in <= a_in;
end
always @ (posedge clk)
begin
y_out <= b_in;
end
endmodule
//
Procedural always block is sensitive to the positive edge of clk
On every rising edge of the clk a_in input value is assigned to intermediate variable b_in. This infers positive edge trigged flip-flop with input a_in and output b_in.
Another procedural always block is triggered on positive edge of clk.
On every positive edge of clk, b_in input value is assigned to output y_out. This block infers positive edge trigged flip-flop with input b_in and output y_out

Example 4 Non-blocking assignments in the different always blocks

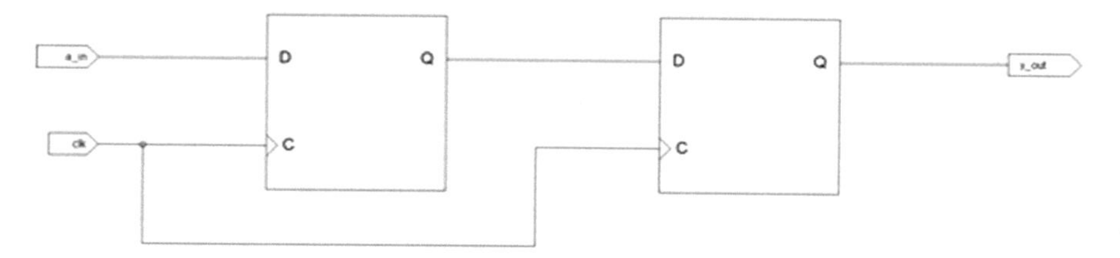

Fig. 11.4 Synthesized logic for non-blocking assignments in the different always blocks

11.2.1 Example Non-blocking Assignments

The synthesis result is shown in Fig. 11.4.

11.2.2 Example Non-blocking Assignment

If non-blocking assignments are used to code the sequential logic and multiple assignments are used within the ***always*** procedural block, then the desired intended logic is always inferred by synthesis tool. The reason being in the non-blocking assignment all the assignments within the ***begin-end*** is executed concurrently. This results in the sequential logic.

Consider the design scenario shown in Example 5, and intention is to create the three-bit serial-input and serial-output shift register by using non-blocking assignments.

Synthesis result is shown in Fig. 11.5 and has inputs a, clk, and an output y. The intended functionality is serial-input serial-output shift register, and it infers the serial-input serial-output shift register.

11.2.3 Example Using Non-blocking Assignments

Consider the design scenario shown in Example 6, and intention is to create the three-bit serial-input and serial-output shift register.

The order of non-blocking assignments used in Example 5 is reordered within ***begin…end*** and shown in Example 6.

Synthesis result is shown in Fig. 11.5 and has inputs a, clk, and an output y. The desired functionality is serial-input serial-output shift register, and it infers the serial-input serial-output shift register. So the important point to remember is that order of the non-blocking assignments within the procedural ***always*** block is not affecting the synthesis result for these kinds of designs!

```
////////////////////////////////////////////////////////////////////////////////

module non_blocking_assignment (
input a,
input clk,
output reg y );
reg b, c;
always@(posedge clk)
begin
    y<=c;
    c<=b;
    b<=a;
end
endmodule

////////////////////////////////////////////////////////////////////////////////
```

Procedural always block is triggered on positive edge of clock clk.

Within begin --- end, multiple non-blocking assignment are included.

An intermediate variable value c is assigned to output y and infers D flip-flop.

The value of intermediate variable b is assigned to intermediate variable c and infers D flip-flop.

The value of an input a is assigned to intermediate variable b and generates D flip-flop.

It infers serial input serial output shift register.

Example 5 Non-blocking assignment in the same always block

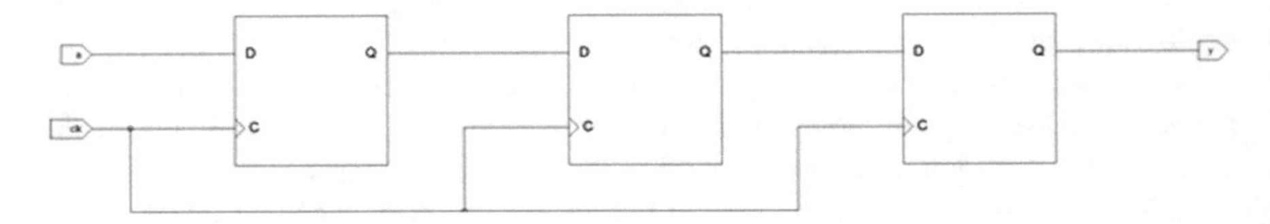

Fig. 11.5 Synthesized logic for non-blocking assignments in the same always block

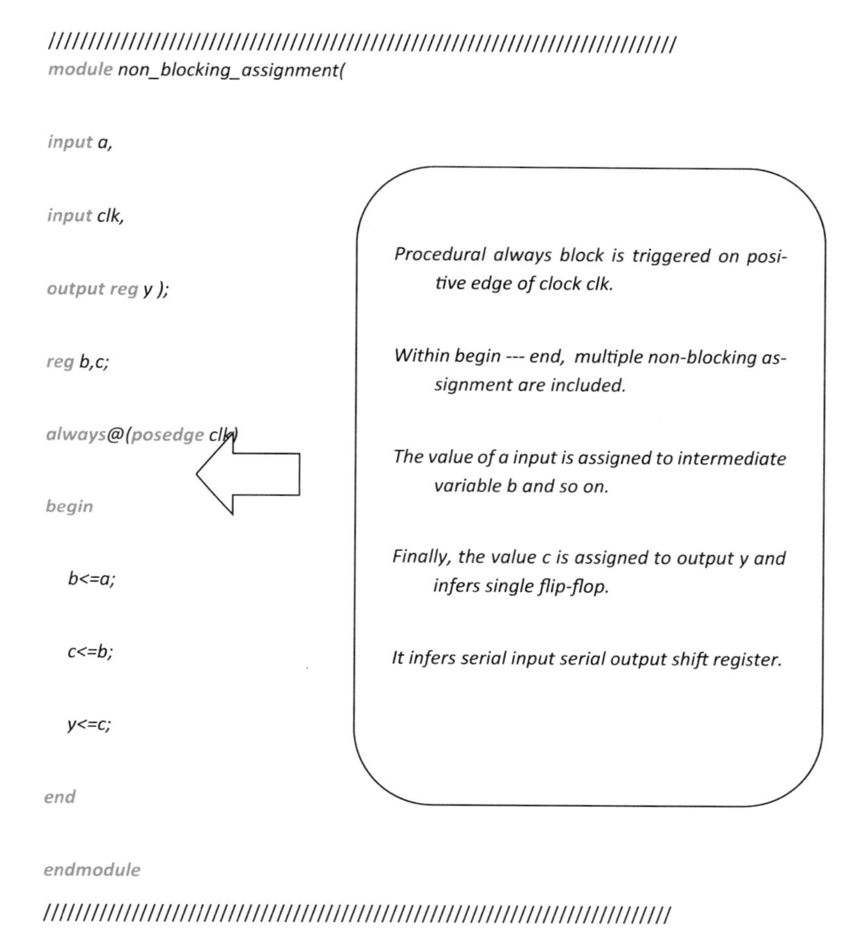

//
module non_blocking_assignment(
input a,
input clk,
output reg y);
reg b,c;
always@(posedge clk)
begin
b<=a;
c<=b;
y<=c;
end
endmodule
//
Procedural always block is triggered on positive edge of clock clk.
Within begin --- end, multiple non-blocking assignment are included.
The value of a input is assigned to intermediate variable b and so on.
Finally, the value c is assigned to output y and infers single flip-flop.
It infers serial input serial output shift register.

Example 6 Non-blocking assignment with order change in the same always block

11.3 Latch Versus Flip-Flop

In the practical sequential designs, latches and flip-flops are used as important elements to design the required intended design functionality. Latch is level-sensitive, and flip-flop is edge-triggered. Most of the ASIC designs use flip-flops as sequential element.

11.3.1 D Flip-Flop

As discussed, earlier flip-flop is edge-triggered and the area for the flip-flop is more as compared to latch and even for flip-flop additional power control logic is required as power consumption due to free-running clock is higher. Flip-flop does not have the cycle stealing or time borrowing concept. The operation needs to be completed in one clock cycle. For flip-flop-based design, the setup and hold time should be met and overall operating frequency of design depends upon the critical path in the design.

The D flip-flop RTL is coded and shown in Example 7 and uses the non-blocking assignment. Input D is assigned to output Q on positive edge of the clock.

The synthesized logic for the positive edge-triggered D flip-flop is shown in Fig. 11.6.

```
/////////////////////////////////////////////////////////////////////////////////
module D_flip_flop(
input D,
input CLK,
output reg Q );
always@(posedge CLK)
begin
   Q<=D;
end
endmodule
/////////////////////////////////////////////////////////////////////////////////
```

On the positive edge of clock signal CLK the input D is assigned to output Q.

It infers D flip-flop which is sensitive to the positive edge of clock.

Example 7 D flip-flop using non-blocking assignment

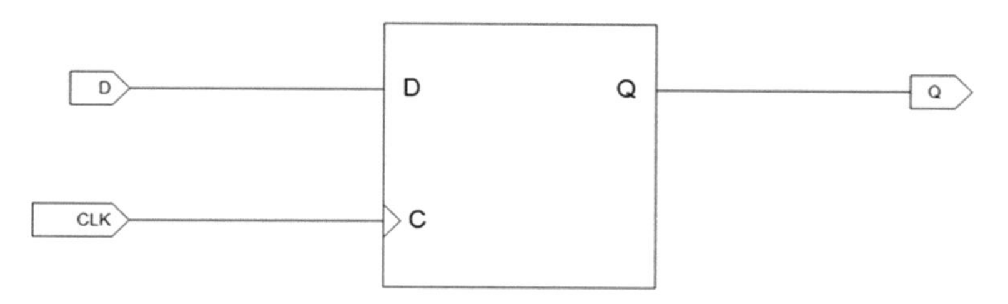

Fig. 11.6 Synthesized D flip-flop

11.3.2 Latch

As discussed, earlier latch is level-sensitive, and the area of the latch is less as compare to the flip-flop and even for the latch additional power control logic is not required as power consumption is lesser due to low switching at latch enable input. Latch has the cycle stealing or time borrowing concept and useful in the pipelining. It is not necessary that the operation needs not to be completed in one clock cycle. For latch-based design, the overall operating frequency of design does not depend upon the slowest path in the design. Timing analysis and time budgeting are more difficult for latch-based designs.

The D Latch RTL is coded using the synthesizable constructs and shown in Example 8, and RTL uses the non-blocking assignments. Input D is assigned to output Q on positive level of latch enable input.

The synthesized logic for positive level-sensitive latch is shown in Fig. 11.7.

11.4 Use of Synchronous Versus Asynchronous Reset

Most of the time the design engineer gets confused while using reset input! When to use an asynchronous reset and when to use synchronous reset is one of the important challenge to the design team! So, for a ASIC design engineer it is required to have good understanding of the reset and rest issues as well as reset trees. This section discusses the synchronous and asynchronous reset description using Verilog synthesizable constructs.

Fig. 11.7 Synthesized D latch

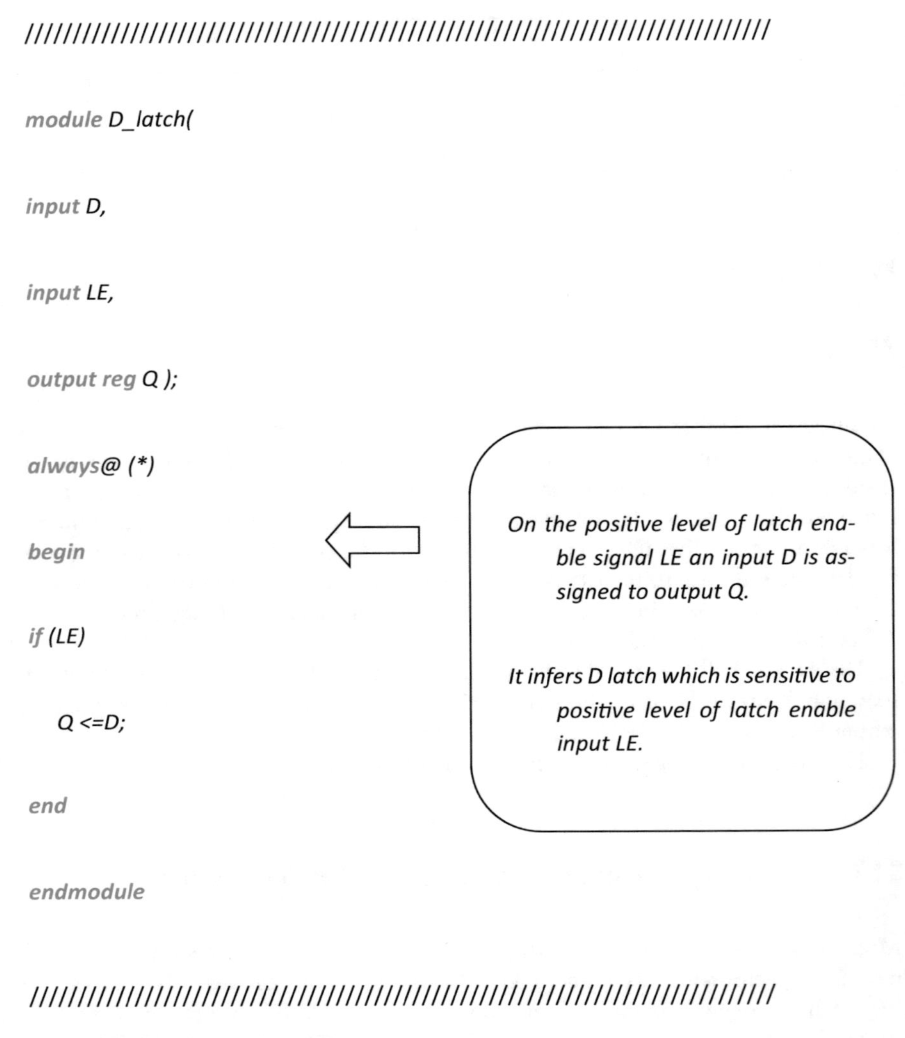

```
////////////////////////////////////////////////////////////////////////////////

module D_latch(

input D,

input LE,

output reg Q );

always@ (*)

begin

if (LE)

   Q <=D;

end

endmodule

////////////////////////////////////////////////////////////////////////////////
```

On the positive level of latch enable signal LE an input D is assigned to output Q.
It infers D latch which is sensitive to positive level of latch enable input LE.

Example 8 Positive level-sensitive D latch

11.4.1 D Flip-Flop Having Asynchronous Reset

As discussed in Chap. 8, an asynchronous reset is an issue in ASIC design as sampling of the reset is independent of active edge of the clock. The reset signal is used to initialize the sequential logic at any instance of time irrespective of active edge of the clock. Reset logic is not part of data path, and even internally generated resets or asynchronous resets are not recommended in the ASIC design as they are prone to glitches. Even reset recovery is an issue, and if asynchronous reset inputs are used, then it is recommended that an asynchronous reset input can be synchronized using two-stage level synchronizer.

As shown in Example 9 D flip-flop having an asynchronous reset is coded using Verilog constructs.

The synthesized logic of the D flip-flop having asynchronous reset is shown in Fig. 11.8.

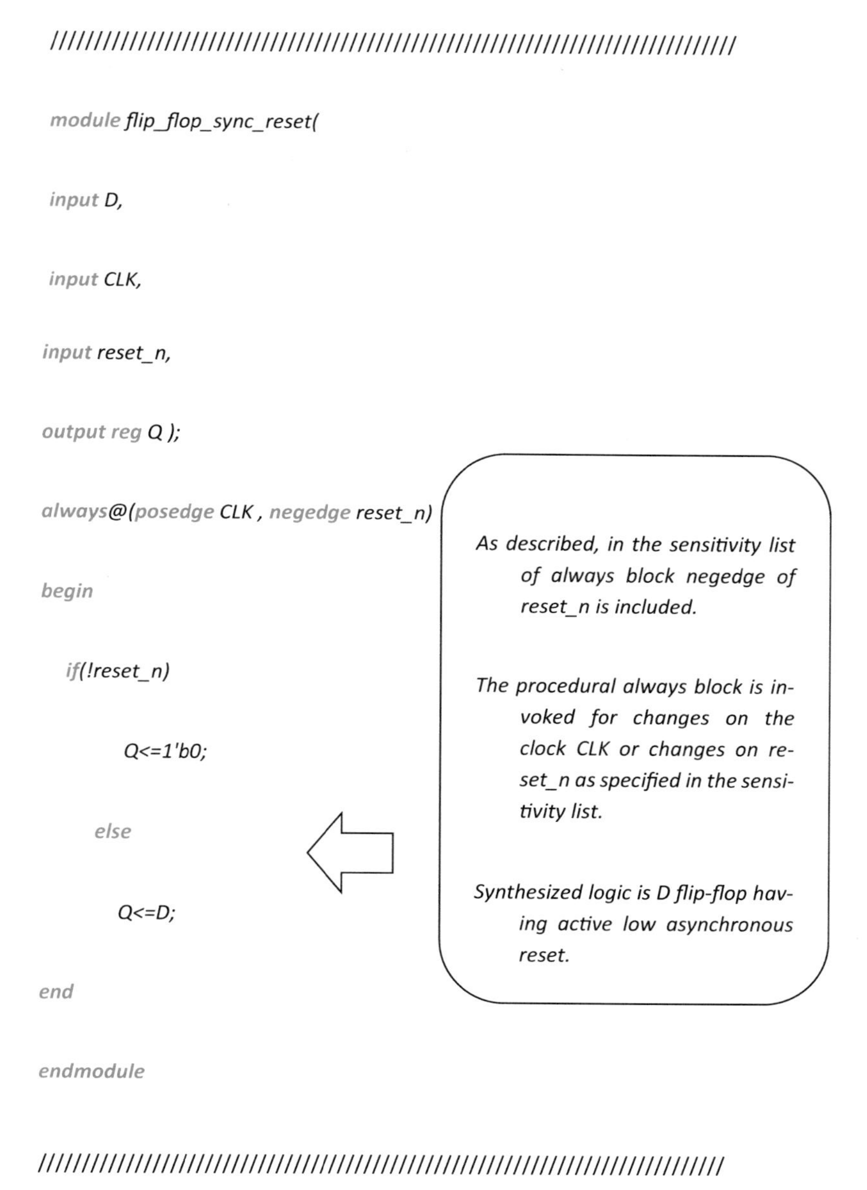

```
////////////////////////////////////////////////////////////////////////////

module flip_flop_sync_reset(
  input D,
  input CLK,
  input reset_n,
  output reg Q );
always@(posedge CLK , negedge reset_n)
begin
  if(!reset_n)
      Q<=1'b0;
    else
      Q<=D;
end
endmodule

////////////////////////////////////////////////////////////////////////////
```

As described, in the sensitivity list of always block negedge of reset_n is included.
The procedural always block is invoked for changes on the clock CLK or changes on reset_n as specified in the sensitivity list.
Synthesized logic is D flip-flop having active low asynchronous reset.

Example 9 Verilog RTL for D flip-flop with asynchronous reset

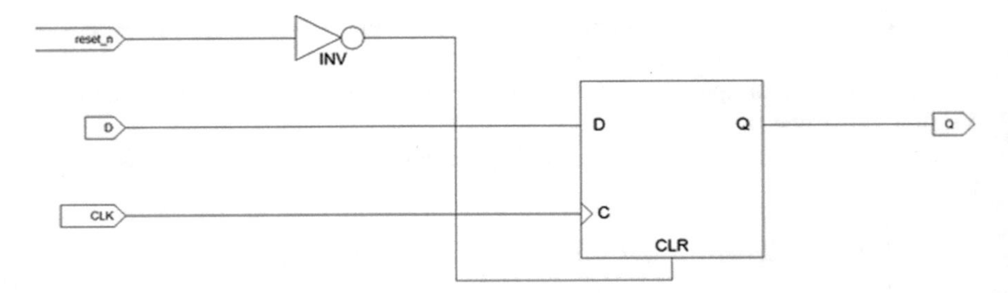

Fig. 11.8 Synthesized D flip-flop with asynchronous reset

11.4.2 Synchronous Reset D Flip-Flop

As discussed in Chap. 8, a synchronous reset is a better strategy in the ASIC design as sampling of the reset is dependent on the active edge of the clock. The reset signal is used to initialize the sequential logic on the rising edge of the clock. Reset logic is part of data path and not prone to glitches. Even reset recovery is not an issue and if synchronous reset inputs are used, then the design does not need the use of level synchronizer!

As shown in Example 10 D flip-flop having a synchronous reset is coded by using Verilog synthesizable constructs.

The synthesized logic is shown in Fig. 11.9 where reset logic is included in the data path.

```
////////////////////////////////////////////////////////////////////////////////

module flip_flop_sync_reset(

input D,

input CLK,

input reset_n,

output reg Q );

always@(posedge CLK)

begin

   if(!reset_n)

        Q<=1'b0;

      else

        Q<=D;

end

endmodule

////////////////////////////////////////////////////////////////////////////////
```

As described, in the sensitivity list of always block negedge of reset_n is not included.

The procedural always block is invoked for changes on the clock CLK as specified in the sensitivity list.

Synthesized logic is D flip-flop having active low synchronous reset.

Example 10 Verilog RTL for D flip-flop with synchronous reset

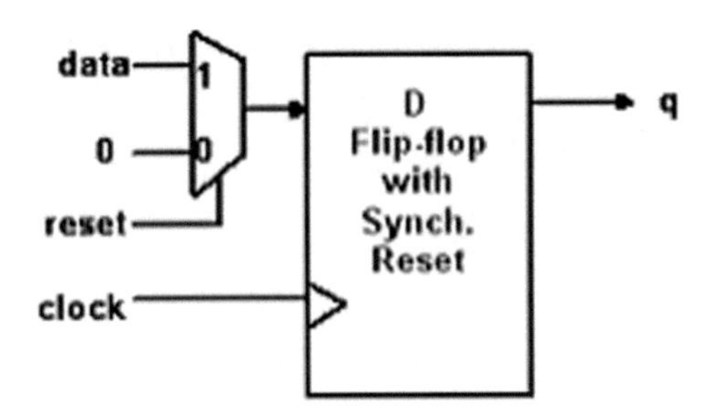

Fig. 11.9 Synthesized D flip-flop with synchronous reset

11.5 Use of if...else Versus case constructs

While coding the RTL of sequential designs use the ***if-else*** construct to code the priority logic functionality. To assign, the priority signals use the ***if-else*** construct. Use the ***case*** construct to code the parallel logic. Please refer Chap. 7 for the detail information about the use of ***if-else*** and ***case*** construct.

11.6 Internally Generated Clocks

Internally generated clock signals use system or master clock as an input and generates an output as internally generated clock signal. But internally generated clock signals need to be avoided as it causes the functional and timing issues in the design. The functional and timing problems are due to the combinational logic, and it introduces the propagation delays. The internal generated clock signals can generate the glitch or spike in the output. This can trigger the sequential logic multiple times or can generate undesired output. Even due to the violation of setup or hold time, these types of designs have the timing violations.

It is always recommended to generate the internal clocks by using register at the output. But still due to the propagation delay of the flip-flop the overall cumulative delay or skew can generate the glitches or spikes in the design.

As shown in Example 11, Verilog RTL is coded to generate the internal clocks. The internal clock signal is used by some other procedural block.

The synthesis schematic is shown in Fig. 11.10, and first flip-flop clock is driven by clk and the second flip-flop clock is driven by internal generated clock signal int_clk.

```
//////////////////////////////////////////////////////////////////////////////

module internal_clock (
input in_1, in_2, clk,
output reg out_1 );
reg  int_clk;
always @ (posedge clk)
 begin
      int_clk<= in_1;
end
always @ (posedge int_clk)
begin
      out_1 <= in_2;
end
endmodule

//////////////////////////////////////////////////////////////////////////////
```

The internal generated clock signal int_clk is assigned in the first procedural always block.

Second procedural always block is sensitive on the positive edge of int_clk and used to generates an output out_1.

Example 11 Verilog RTL for internally generated clock

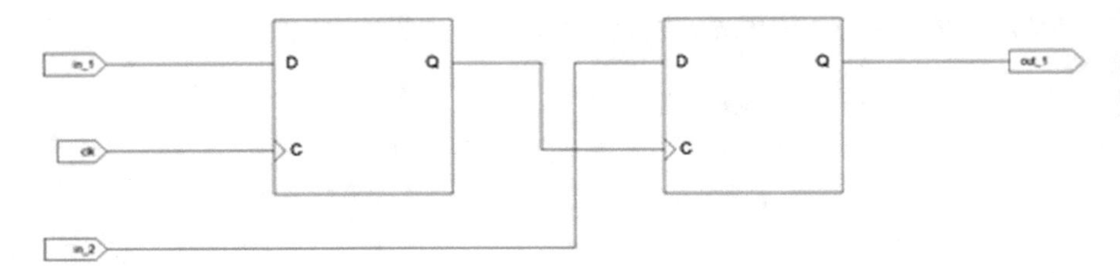

Fig. 11.10 Synthesized internally generated clock logic

11.7 Guidelines for Modeling Synchronous Designs

Following are important guidelines to be followed while coding the synchronous designs

1. To describe the functionality of synchronous designs, use the non-blocking assignments.
2. Do not use the latch-based designs as latches are transparent for half clock cycle.
3. Use the pipelined stages to improve the design performance and discussed in Chap. 13.
4. Use the synchronous reset signals as they are not prone to glitches or spikes.
5. If asynchronous signals are used, then use the dual-stage synchronizers to synchronize the internally generated resets.
6. Use clock gating cells for low-power design and discussed in Chap. 24.

11.8 Multiple Clocks in the Same module

Multiple clock sources are used in the multiple clock domain designs. These clock signals can be generated by using different clock sources and can be used in the ASIC design to invoke the multiple ***always*** procedural blocks. The data transfer from one clock domain to another clock domain needs additional synchronizers in the data path and control path, and these are discussed in Chap. 22.

Verilog RTL is coded using the synthesizable constructs and shown in Example 12. It uses two clock signals, clk1 and clk2. Two procedural blocks are used to describe the functionality.

Synthesis schematic is shown in Fig. 11.11, and generates an output f1_out, f2_out. The clock signal clk1 is used as clock source to the upper register. Upper register is triggered on positive edge of clock clk1. The lower register is triggered on negative edge of clk2.

```
////////////////////////////////////////////////////////////////////////////////

module multi_clock_gen (
input clk1,
input clk2,
input  a_in,
input  b_in,
input  c_in,
output reg f1_out,
output reg f2_out );
always @ (posedge clk1)
      f1_out <= a_in & b_in;
always @ (negedge clk2)
      f2_out <= b_in ^ c_in;
endmodule

////////////////////////////////////////////////////////////////////////////////
```

Multiple clocks can be defined in the same module and used to trigger two different procedural always blocks.

Multiple clocks are used in multiple clock domain designs.

Example 12 Verilog RTL for multiple clock definitions

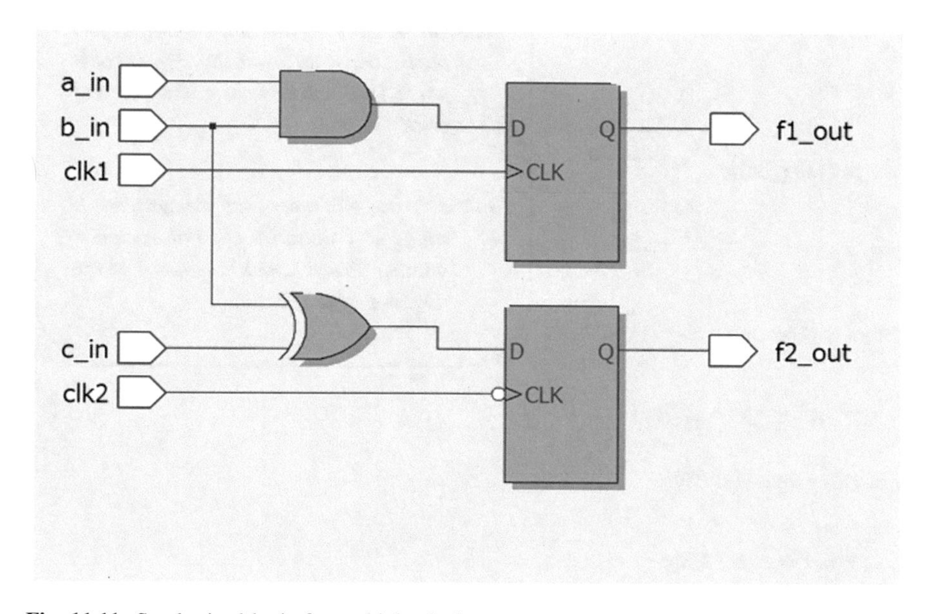

Fig. 11.11 Synthesized logic for multiple clock

11.9 Multi-phase Clocks in the Design

The signals used to trigger multiple procedural blocks and generated from the same clock source but having the difference in the arrival time are called as multi-phase clock signals. For example, if one of the procedural block is sensitive to the positive edge of clock and another procedural block is sensitive to the negative edge of clock, then there is phase difference of 180°; hence, these clocks are treated as phase-shifted signals.

Verilog RTL is shown in Example 13, and one of the procedural block is sensitive to the positive edge of clock and another is sensitive to the negative edge of clock. It is also recommended not to have mix edge-triggering in the design!

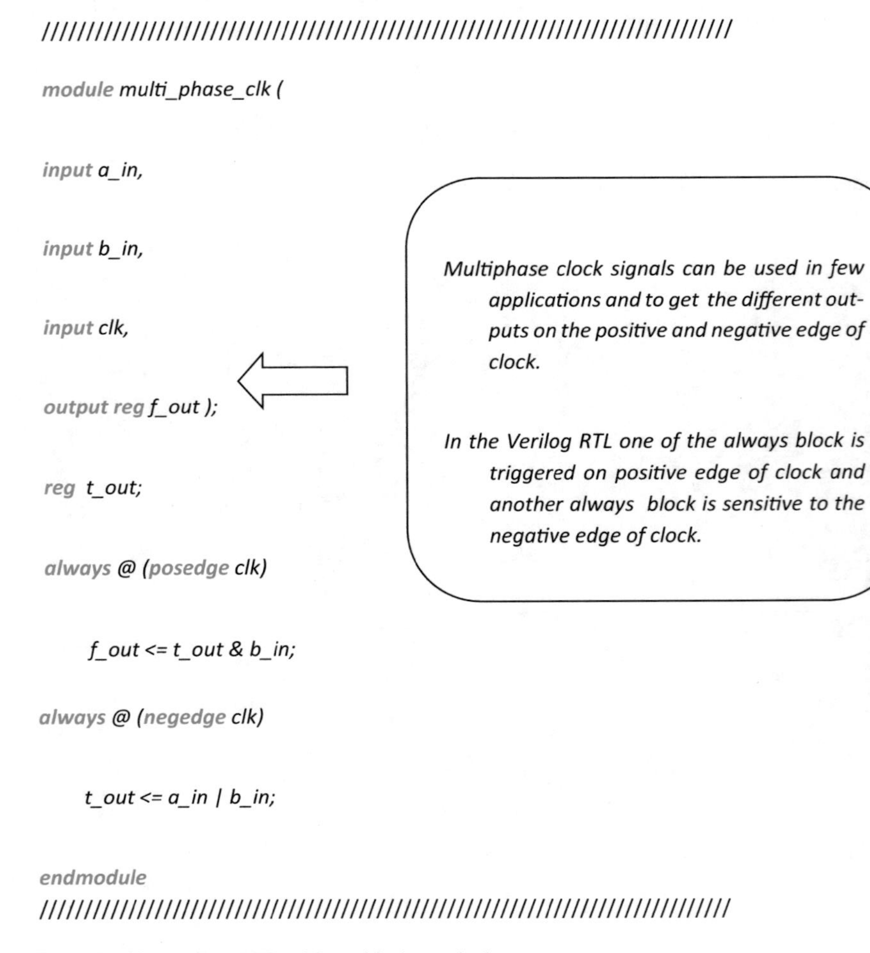

```
//////////////////////////////////////////////////////////////////////////////
module multi_phase_clk (
input a_in,
input b_in,
input clk,
output reg f_out );
reg  t_out;
always @ (posedge clk)
     f_out <= t_out & b_in;
always @ (negedge clk)
     t_out <= a_in | b_in;
endmodule
//////////////////////////////////////////////////////////////////////////////
```

Multiphase clock signals can be used in few applications and to get the different outputs on the positive and negative edge of clock.
In the Verilog RTL one of the always block is triggered on positive edge of clock and another always block is sensitive to the negative edge of clock.

Example 13 Verilog RTL with multi-phase clock

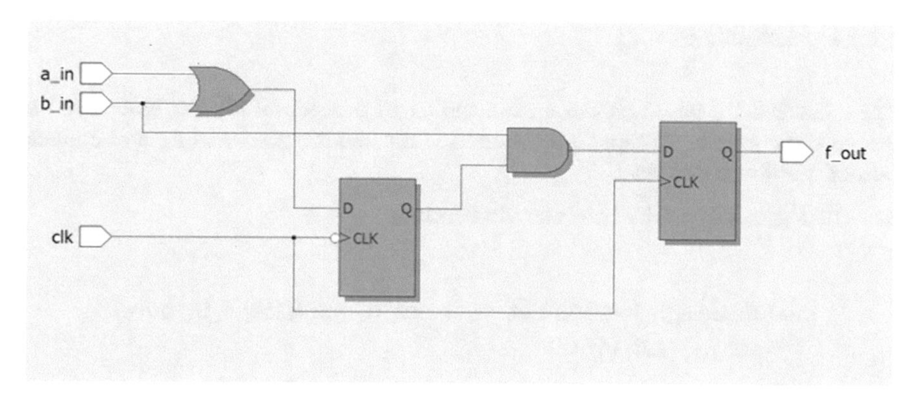

Fig. 11.12 Synthesized logic for the multi-phase clock

The synthesis schematic is shown in Fig. 11.12 where two different flip-flops are sensitive to the different edges of clocks.

11.10 Guidelines for Modeling Asynchronous Designs

Following are important guidelines that need to be followed while modeling the asynchronous design

1. If asynchronous reset signals are used, then use the level synchronizer to synchronize the internally generated reset signals.
2. Avoid the use of driving the flip-flop output to the asynchronous reset of the subsequent flip-flop as this can have the race conditions.
3. Avoid the use of asynchronous pulse generator as it creates the issue in the design during the timing closure and even during the place and route.
4. If power consumption is the goal, then only use the efficient ripple counter but there is an issue due to the performance degrade while using the ripple counters.

11.11 Exercises

The exercises are based on the understanding of procedural ***always*** block used to design the counters. Complete the exercises for better understanding and application of Verilog constructs.

1. The logic inferred by the following code is

```
module design_logic ( input  clk, reset_n, input[1:0] d_in, out-
      put reg [1:0] y);

always @ (posedge clk, negedge reset_n)

begin

if (~ reset_n)

 y <= 2'b00;

else

y<=d_in;

end

endmodule
```

   a. 2-bit shift register having asynchronous reset
   b. 2-bit shift register having synchronous reset
   c. 2-bit PIPO register having asynchronous reset
   d. 2-bit counter having synchronous reset

2. The logic inferred by the following code is

```
module design_logic ( input  d_in, clk, reset_n, output reg y);

reg tmp1;

always @ (posedge clk, negedge reset_n)

begin

if (~ reset_n)

  {y,tmp1} <= 2'b00;

else

begin

tmp1 <= d_in;

y <= tmp1;

end

endmodule
```

   a. Two-bit shift register having asynchronous reset
   b. Two-bit shift register having synchronous reset
   c. Binary down counter
   d. Binary up-counter

3. The logic inferred by the following code is

```
module design_logic ( input  d_in, clk, reset_n, output reg y);

reg tmp1;

always @ (posedge clk, negedge reset_n)

begin

if (~ reset_n)

  {y,tmp1} = 2'b00;

else

begin

tmp1 <= d_in;

y <= tmp1;

end

endmodule
```

   a. Two-bit shift register having asynchronous reset
   b. Two-bit shift register having synchronous reset
   c. Binary counter
   d. Synatx error and RTL should be tweaked

4. The logic inferred by the following code is

```
module design_logic ( input  d_in, clk, reset_n, output reg y);
reg tmp1, tmp2;

always @ (clk)

begin

if (~ reset_n)

 {y,tmp2,tmp1} <= 3'b00;

else

begin

 y <= tmp2;

tmp2 <= tmp1;

tmp1 <= d_in;

end

endmodule
```

a. Two-bit shift register
b. Three-bit shift register
c. PIPO register
d. Binary up-counter

5. The logic inferred by the following code is

```
module design_logic ( input  d_in, clk, reset_n, output reg y);

reg tmp1;

always @ (posedge clk)

begin

if (~ reset_n)

  {y,tmp1} = 2'b00;

else

begin

y = tmp1;

tmp1 = d_in;

 end

 endmodule
```

   a. Two-bit shift register having asynchronous reset
   b. Two-bit shift register having synchronous reset
   c. Binary down counter
   d. Binary up-counter

11.12 Summary

To summarize this chapter, the following are important highlights

1. Do not use blocking assignments while describing the sequential designs.
2. It is recommended to use the non-blocking assignments while coding the RTL for sequential design.
3. Use the flip-flop-based logic instead of latch-based logic.
4. Do not mix the blocking and non-blocking assignments within the single always procedural block.
5. Use the synchronous resets in the design and if asynchronous reset is used, then use the reset synchronizers.
6. Avoid the use of asynchronous pulse generator as it creates the issue in the design during the timing closure and even during the place and route.
7. If power consumption is the goal, then only use the efficient ripple counter but there is an issue due to the performance degrade while using the ripple counters.
8. Do not use the latch-based designs as latches are transparent for half clock cycle.
9. Use the pipelined stages to improve the design performance.
10. Use the synchronous reset signals as they are not prone to glitches or spikes.
11. If asynchronous signals are used, then use the dual-stage synchronizers to synchronize the internally generated resets.
12. Use clock gating cells for low-power design.

Chapter 12
RTL Design Strategies for Complex Designs

The complex designs can be efficiently implemented by using the synthesizable Verilog constructs. Now days design complexity has increased, and the design requirements are lower power, high speed and minimum area. This chapter discusses the use of synthesizable Verilog constructs to implement the complex designs for the desired functionality.

As discussed in the previous chapters, Verilog is efficient and useful to code the functionality of the design. The concurrent and sequential constructs discussed in the previous chapters can be used to infer the design with the desired performance. In the ASIC designs, the design functionality is complex and needs to be coded by using the synthesizable Verilog constructs to infer the intended design with better design performance. Most of the ASIC and SOCs use the processors, buses, arbiters, and protocols (predefined set of rules or transactions). An efficient Verilog coding is especially important aspect while coding the functionality of above blocks. In such scenario, RTL design team should use the synthesizable constructs with combinational and sequential design guidelines during the design phase.

The subsequent section discusses the few of the complex designs and practical scenario while coding the processor computational logic, and basics of synthesis for the tasks and functions. The subsequent sections are also useful to understand about the performance improvement and the registered input and registered output concept.

12.1 ALU Design

Arithmetic logic unit (ALU) is used in most of the processors to perform the arithmetic and logic operations. Processor performs one of the operation at a time depending on the operational code (op-code). For 8-bit processors, the ALU is used to perform the operations on two eight-bit operands. Operand is the data on which operation needs to be performed. Similarly, for the 16-bit processors, the ALU is used to perform the operations on two 16-bit numbers.

V. Taraate, *Digital Logic Design Using Verilog*,
https://doi.org/10.1007/978-981-16-3199-3_12

As shown in Fig. 12.1, a ALU architecture is shown to perform the operation on two four-bit numbers A (A3 is MSB and A0 is LSB), B (B3 is MSB and B0 is LSB), and carry input C0. The ALU generates an output F (F3 is MSB, and F0 is LSB) and an output carry C_{OUT3}. In the practical design scenario, one-bit ALU is designed to perform operation on the single bit of data. The operation is performed depending on the opcode specified by using the lines S1, S0. As shown in the figure, ALU is designed to perform the four operations. Table 12.1 is useful to understand various operations and the execution of the instruction depending on the status of select lines S1, S0. In this example, opcode is 2-bit and is indicated by input lines S1, S0.

12.1.1 Logic Unit Design

As stated in the previous section, in the practical design scenario, it is recommended to code the functionality of design using an efficient Verilog constructs. So, at the micro-architecture level, the design is partitioned into multiple functional blocks. The partitioning of design gives the better design understanding and visibility to design team. Consider a scenario to implement the design functionality of an 8-bit ALU, the design is petitioned as separate logic unit and arithmetic unit. Separate arithmetic and logical unit functionality can be coded by using synthesizable Verilog constructs with goal to have the better readability and better synthesis result.

Figure 12.2 is shown and used to implement the logic operations, and these logic operations are documented in the functional table. The logic unit is performing either AND, OR, XOR or complement operation. The requirement is that logic unit should perform only one operation at a time. Table 12.2 shown describes the different operations. The complement operation is performed by using adder having one input A_0 and another input logic 1.

The issue with this type of design is due to use of parallel inputs and multiplexing logic. The data path has the logic gates, and output of these gates is used by the multiplexer to generate an output for one of the logic operations. The control path is the control lines of multiplexers S1, S'. As shown in Fig. 12.2, the logic unit performs all the operations at a time and result F_0 which is the result for one of the operations. But this technique is inefficient as it requires more area and power, and it does not have the efficient data and control path. If S1, S0 are late arriving signals and if this block is used in the register-to-register path then there may be possibility of the timing violations. Another important aspect is the concept of resource sharing which is not used in this design.

Another issue is all the operations are performed at a time. For example, if S1 = 1 and S0 = 0, then the logic should perform the XOR operation. But as all the logic gates are the part of data path, the design performs all the operations. Unnecessary data waits at the multiplexer inputs, and result is dependent on control inputs S1, S0.

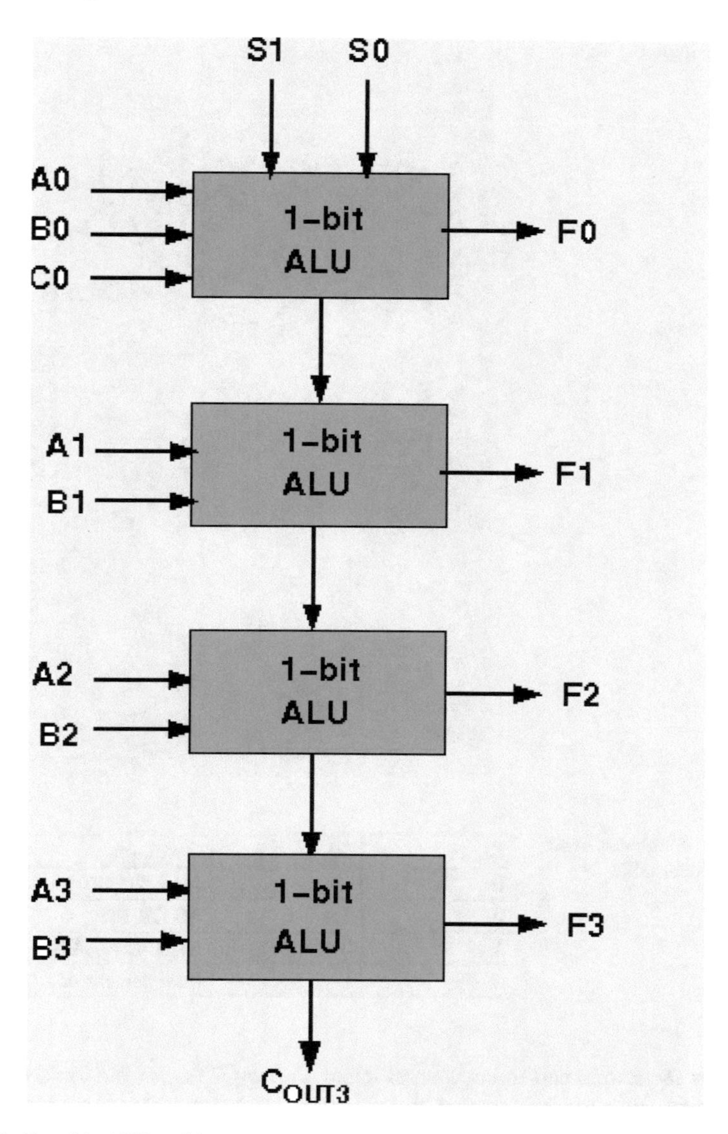

Fig. 12.1 Four-bit ALU architecture

Table 12.1 Four-bit ALU operational table

S1	S0	Operation
0	0	Addition of A, B without carry
0	1	Subtraction of A, B without borrow
1	0	XOR of A, B
1	1	Complement of A

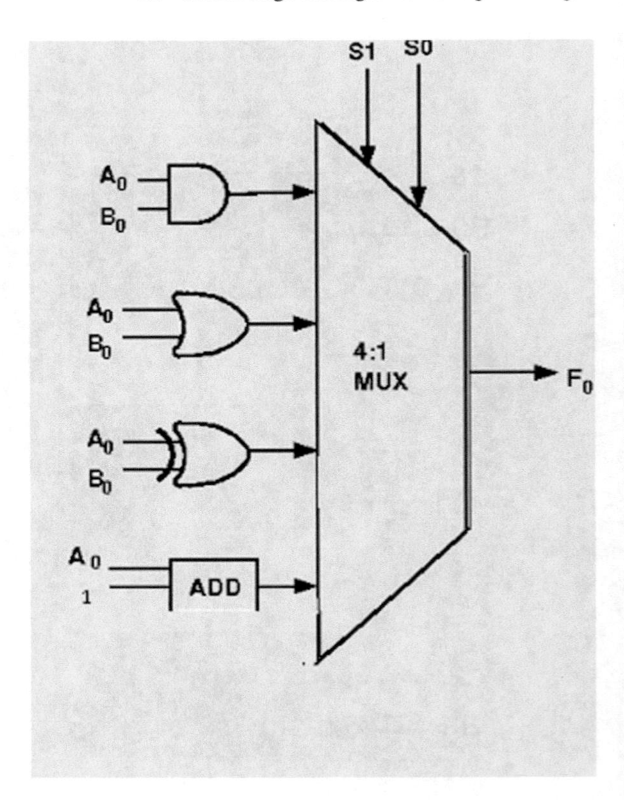

Fig. 12.2 Single-bit logic unit

Table 12.2 Single-bit logic unit operational table

S1	S0	Operation
0	0	A0 AND B0
0	1	A0 OR B0
1	0	XOR of A0, B0
1	1	Complement of A0

So, it is recommended to code an efficient Verilog RTL for the logic unit using the ***case*** construct but by sharing the common resources. The following section describes the Verilog RTL for the logic unit to infer the parallel logic and the logic having the registered inputs and outputs.

12.1.1.1 Logic Unit to Infer Parallel Logic

Example 1 shows RTL design to perform the operations on two 8-bit binary inputs a_in, b_in. The RTL is coded to meet the functionality documented in Table 12.3. The Verilog RTL infers the parallel logic with multiplexed encoding.

```
////////////////////////////////////////////////////////////////////////////

module logic_unit_8bit (

input [7:0] a_in,

input [7:0] b_in,

input [1:0] op_code,

output reg [7:0] result_out );

always@ ( *)

begin

case ( op_code)

2'b00 : result_out = a_in | b_in;

2'b01 : result_out = a_in ^ b_in;

2'b10 : result_out = a_in & b_in;

2'b11 : result_out = ~ a_in;

default : result_out = 8'b0000_0000;

endcase

end

endmodule

////////////////////////////////////////////////////////////////////////////
```

The procedural block is level sensitive to changes on a_in, b_in and op_code.

case construct is used to infer the parallel logic to perform the desired logic operation.

Please refer the functional table for the logic operations.

default condition is specified and the result during default condition is equal to 0000_00000.

Example 1 Verilog RTL for 8-bit logic unit

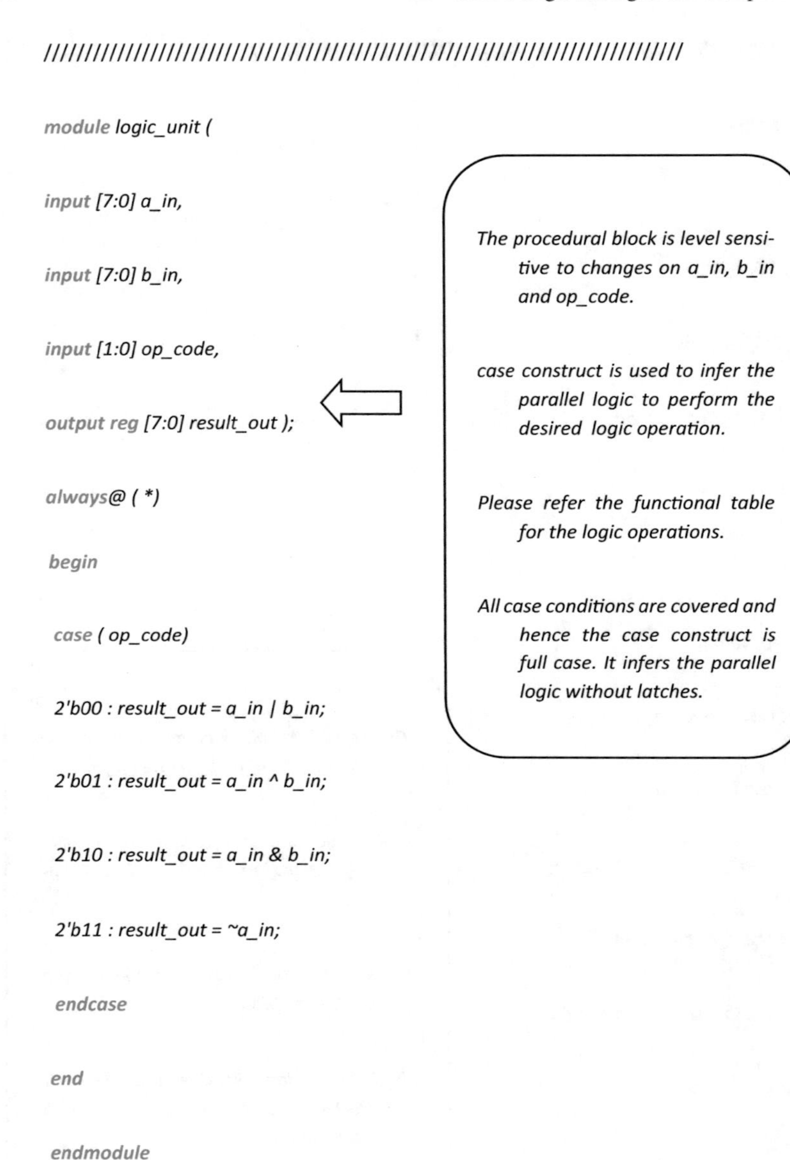

```
////////////////////////////////////////////////////////////////////////////////

module logic_unit (

input [7:0] a_in,

input [7:0] b_in,

input [1:0] op_code,

output reg [7:0] result_out );

always@ ( *)

 begin

  case ( op_code)

  2'b00 : result_out = a_in | b_in;

  2'b01 : result_out = a_in ^ b_in;

  2'b10 : result_out = a_in & b_in;

  2'b11 : result_out = ~a_in;

  endcase

 end

 endmodule

////////////////////////////////////////////////////////////////////////////////
```

The procedural block is level sensitive to changes on a_in, b_in and op_code.
case construct is used to infer the parallel logic to perform the desired logic operation.
Please refer the functional table for the logic operations.
All case conditions are covered and hence the case construct is full case. It infers the parallel logic without latches.

Example 2 Verilog RTL for 8-bit ALU using full-case construct

Table 12.3 Operational table for 8-bit ALU

op_code[1]	op_code[o]	Logic operation
0	0	a_in OR b_in
0	1	a_in XOR b_in
1	0	a_in AND b_in
1	1	Complement of a_in

As shown in Example 1, the functionality is coded by using a procedural ***always*** block using the ***case*** construct. All the ***case*** conditions are covered, and during ***default*** condition, the logic unit generates output result_out as 8'b0000_0000.

The functionality of the logic unit can be modeled using the full-case construct. As shown in Example 2, the functionality is described by using a procedural 'always' block with the full 'case' construct. All the case conditions are described using the full-case construct.

Synthesis result for the RTL which uses the full ***case*** construct is shown in Fig. 12.3. As shown in the figure, it infers the logic gates with multiplexer logic at the output. In the practical scenario, it is recommended to use the adders as common resources to implement both the logic and arithmetic unit.

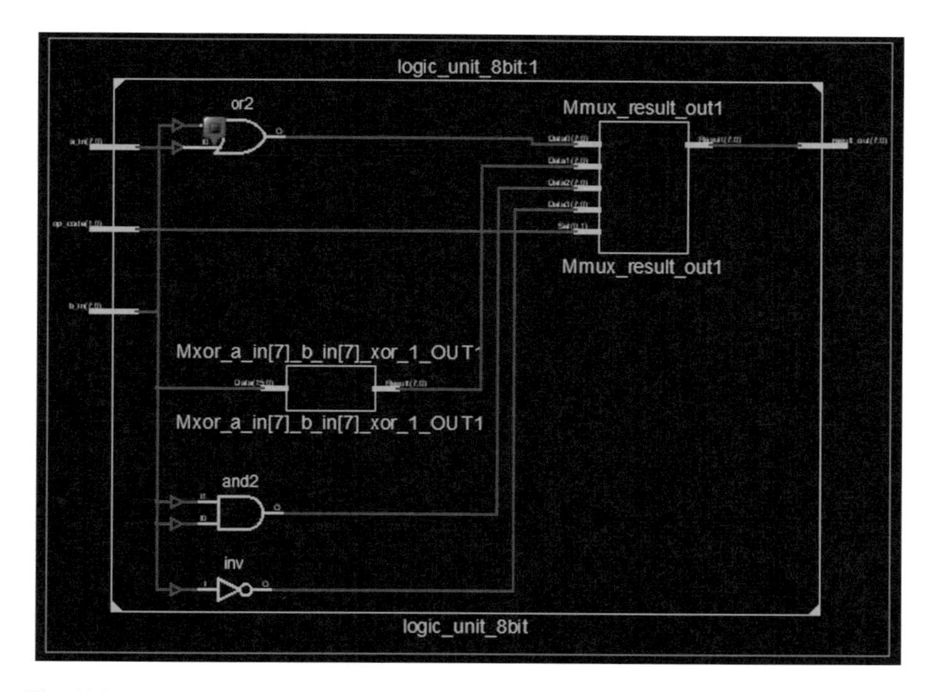

Fig. 12.3 Synthesis result of 8-bit logic unit

```
///////////////////////////////////////////////////////////////////////////
module logic_unit_registered_io # (parameter data_size=8,
parameter opcode_size=2)
(

input [data_size-1:0] a_in,
input [data_size-1:0] b_in,
input [opcode_size-1:0] op_code,
input clk,
input reset_n,
output reg [data_size-1:0] result_out ) ;
reg [data_size-1:0] reg_a_in;
reg [data_size-1:0] reg_b_in;
reg [opcode_size-1:0]  reg_op_code;
reg [data_size-1:0] reg_result_out;

always @ ( posedge clk or negedge reset_n)
begin
  if ( ~reset_n)
      { reg_a_in, reg_b_in, reg_op_code} <= 8'b0;
      else
      { reg_a_in, reg_b_in, reg_op_code} <= { a_in, b_in, op_code};

end

always@ (reg_a_in, reg_b_in, reg_op_code)
begin
 case ( reg_op_code)
 2'b00 : reg_result_out = reg_a_in | reg_b_in;
 2'b01 : reg_result_out = reg_a_in ^ reg_b_in;
 2'b10 : reg_result_out = reg_a_in & reg_b_in;
 default : reg_result_out = ~reg_a_in;
 endcase
end

always @ (posedge clk , negedge reset_n)
begin
  if (~reset_n)
      result_out <= 8'b0000_0000;
      else
      result_out <= reg_result_out;
end
endmodule
///////////////////////////////////////////////////////////////////////////
```

For the better and clean timing analysis it is recommended to use the registered inputs and registered outputs.

The procedural block is sensitive to positive edge of the clock and used to sample the data inputs a_in, b_in, opcode.

The procedural block is level sensitive to changes on registered inputs reg_a_in, reg_b_in and reg_op_code.

case construct is used to infer the parallel logic to perform the desired operation.

Please refer the functional table for the logic operations.

2'b11 or default condition results into output which is equal to complement of registered input reg_a_in.

Example 3 Verilog RTL of 8-bit logic unit having registered inputs and outputs

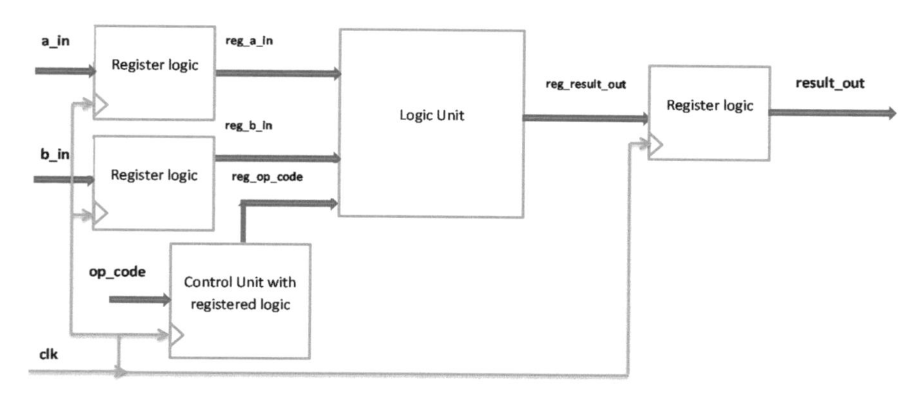

Fig. 12.4 Synthesis of logic unit having registered inputs and outputs

12.1.1.2 Logic Unit Having Registered Inputs and Outputs

For the better performance, it is recommended to use registered inputs and registered outputs. If all the inputs are registered that is sampled on the active edge of clock and if all the outputs are registered on the active edge of clock, then design can result into better performance. The registered inputs and registered outputs can give the clean data path, and even the output is free from glitches or hazards. For the performance improvement the pipelining can be used to improve the data arrival time in the reg-to-reg path. Please refer Chap. 20 for the detailed information about the timing analysis.

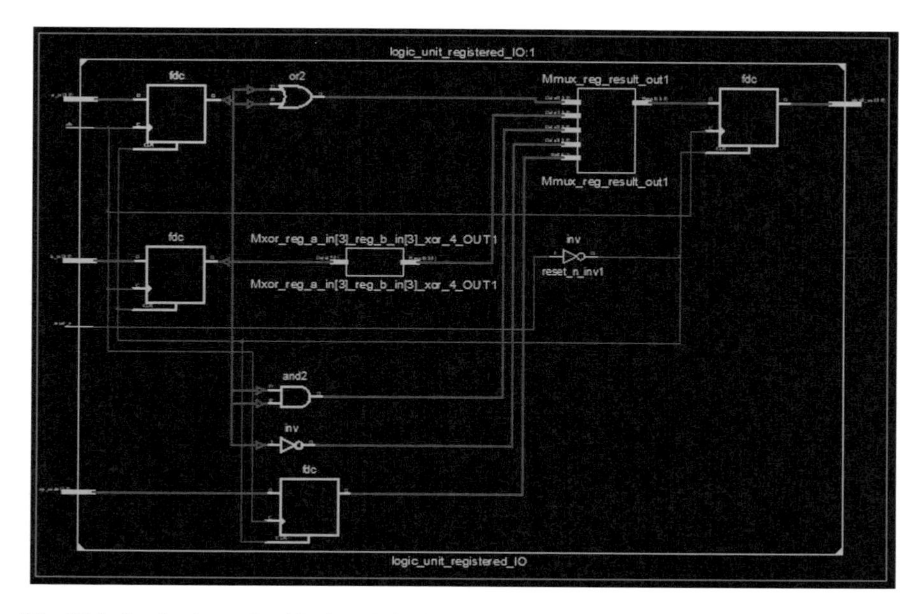

Fig. 12.5 Synthesis result of logic unit having registered IO

Example 3 uses the registered inputs and register outputs. The inputs are sampled on the positive edge of clock clk, and result is launched on the positive edge of clk. During reset condition reset_n = 0, the output of logic unit is initialized to 0.

Example 3 infers the logic with all the inputs and outputs registered on positive edge of clock. Readers are requested to assume that every register (flip-flop) has an asynchronous reset input reset_n. The synthesized logic is shown in Figs. 12.4 and 12.5.

12.1.2 Arithmetic Unit

Arithmetic unit is used to perform the arithmetic operations such as addition, subtraction, increment and decrement. The operations are performed by using two operands. The functional Table 12.4 gives information about the different operations need to be performed. Arithmetic unit should be coded in such a way that it performs only one operation at time. Figure 12.6 shows the resources required with input and output signals.

The top-level inputs and outputs of the arithmetic unit are shown in Fig. 12.7.

The parameterized 8-bit arithmetic unit is coded using synthesizable constructs and shown in Example 4.

The synthesis result of 1-bit arithmetic unit is shown in Fig. 12.8. Logic uses the full adder as component to perform the addition and subtraction. Subtraction is performed using 2's complement addition. The synthesized logic also consists of the multiplexer 4:1 to pass the required operand at one of the inputs of full adder depending on the status of the opcode.

The logic inferred for the 8-bit arithmetic unit is shown in Fig. 12.9 and consists of arithmetic resources and multiplexers. The logic can be optimized using RTL tweaks, and few techniques are discussed in Chap. 13.

Table 12.4 Operational table of the arithmetic unit

op_code_in [2]	op_code_in [1]	op_code_in [o]	Operation
0	0	0	Transfer a_in
0	0	1	a_in ADD b_in
0	1	0	a_in ADD b_in with carry input cin_in
0	1	1	a_in SUB b_in
1	0	0	a_in SUB b_in with borrow input cin_in
1	0	1	Increment a_in
1	1	0	Decrement b_in
1	1	1	No operation performed

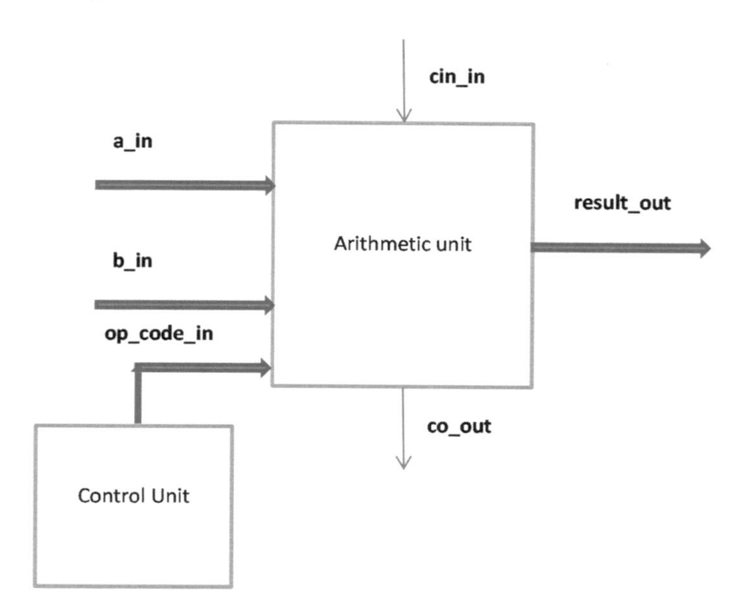

Fig. 12.6 Block diagram of arithmetic unit

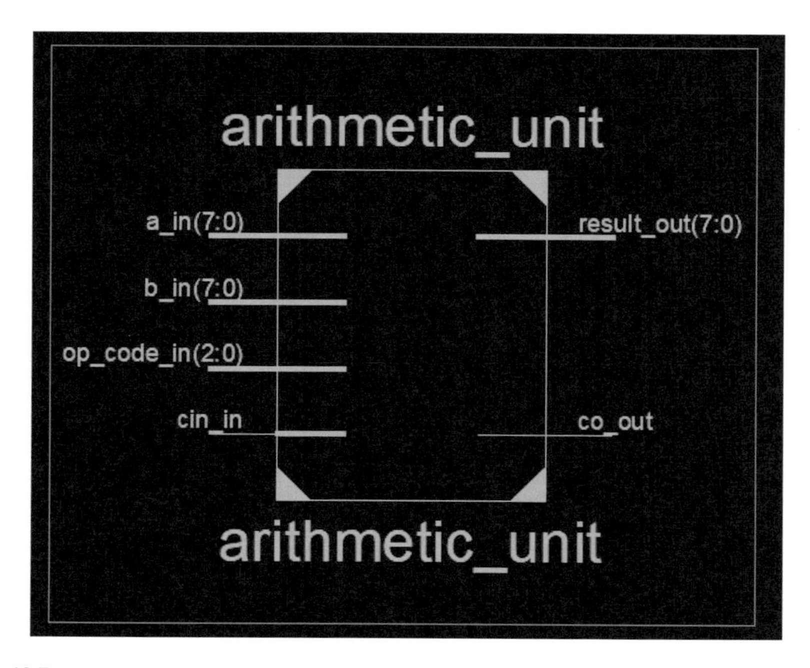

Fig. 12.7 Top-level signal description of arithmetic unit

```
////////////////////////////////////////////////////////////////////////
module arithmetic_unit # (parameter data_size=8, opcode_size=3)
(

input [data_size-1:0] a_in,
input [data_size-1:0] b_in,
input cin_in,
input [opcode_size-1:0] op_code_in,
output reg [data_size-1:0] result_out,

output reg co_out );

always @ ( *)
begin

  case ( op_code_in)
    3'b000 : {co_out,result_out} = {1'b0,a_in} ;
    3'b001 : {co_out,result_out} = a_in + b_in ;
    3'b010 : {co_out,result_out} = a_in + b_in + cin_in ;
    3'b011 : {co_out,result_out} = a_in - b_in ;
    3'b100 : {co_out,result_out} = a_in - b_in - cin_in;
    3'b101 : {co_out,result_out} = a_in + 1'b1 ;
    3'b110 : {co_out,result_out} = a_in - 1'b1 ;
    default : {co_out,result_out} = 9'b0_0000_0000 ;
    endcase
end

endmodule
////////////////////////////////////////////////////////////////////////
```

The procedural always block is used and is level sensitive to changes on a_in, b_in, cin_in and an op_code_in.

The functionality is coded by using the case construct.

case construct is used to infer the parallel logic and it infers the logic without latches as all case conditions are included.

Please refer the operational table of the arithmetic unit.

Example 4 Verilog RTL of the 8-bit arithmetic unit

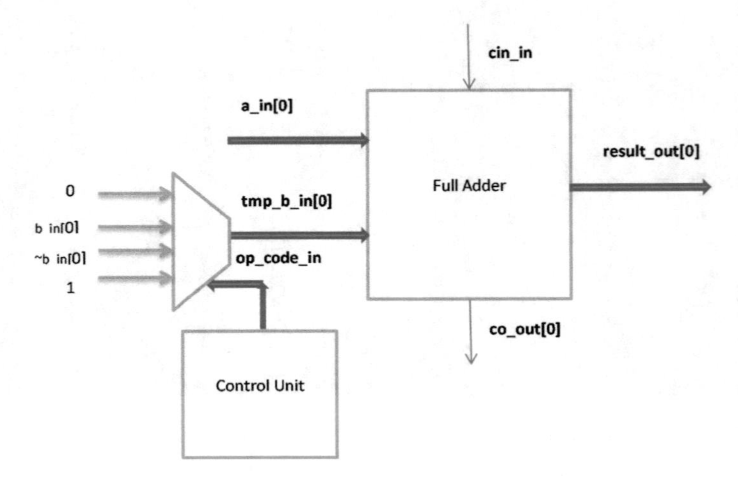

Fig. 12.8 Synthesis result of one-bit arithmetic unit

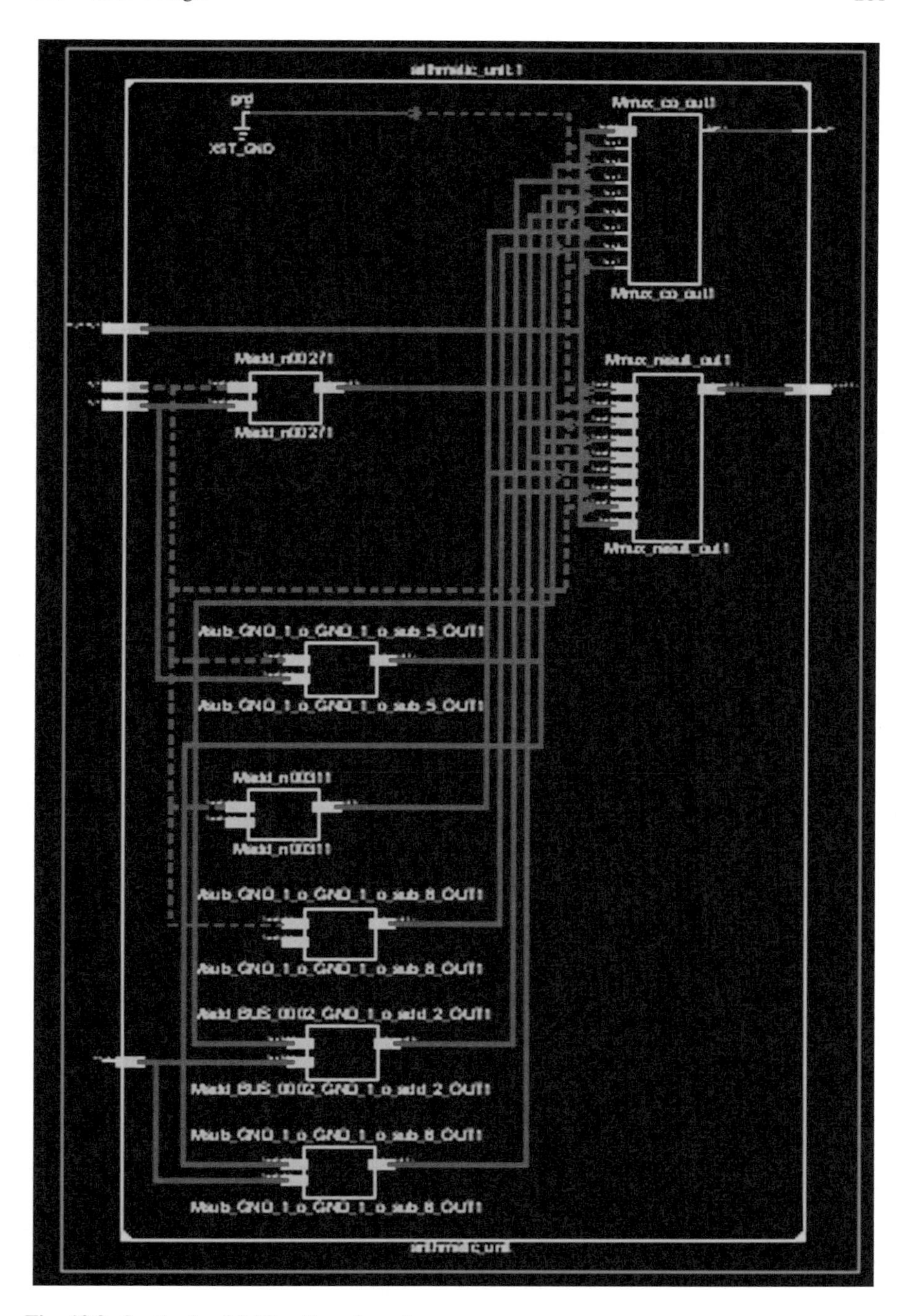

Fig. 12.9 Synthesis of 8-bit arithmetic unit

12.1.3 Arithmetic and Logic Unit

Figure 12.10 shows the ALU with the associated logic circuit to perform the operation on two 8-bit numbers a_in, b_in. For logic operations, the carry input (cin_in) is ignored, and the output result_out is generated depending on the operational code of the instruction. Depending on the operational code, ALU is used to perform either arithmetic or logic operation. During arithmetic operations, if result is greater than 8-bit, then carry output co_out is set to logic 1 that indicates carry propagation outside to MSB.

Table 12.6 indicates the number of bits required at inputs and outputs for the ALU to perform execution of the 11 instructions. It performs only one operation at a time depending on the status of the operational code. The table has the seven arithmetic instructions and four logic instructions. The pin or signal description is shown in Table 12.5.

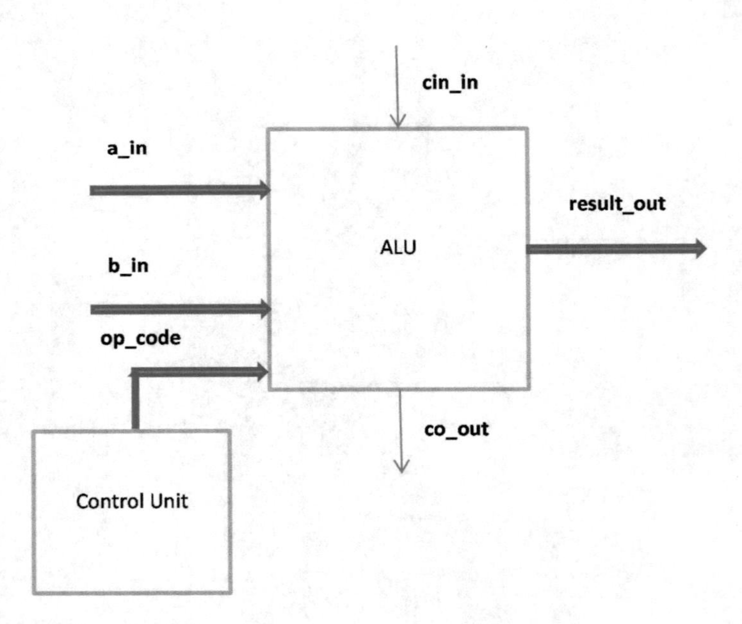

Fig. 12.10 ALU top-level diagram

Table 12.5 Signal or pin description of 8-bit ALU

Signal or pin name	Size (bits)	Description
a_in	8	An 8-bit operand
b_in	8	An 8-bit operand
cin_in	1	Carry input to a ALU
op_code_in	4	4-bit opcode of instruction
result_out	8	An 8-bit output from ALU
co_out	1	One-bit output carry from ALU

Table 12.6 Operational table for 8-bit ALU

Operational code	Instruction	Description
0000	Transfer a_in	Generate an output a_in + 0 + 0
0001	Addition without carry	a_in + b_in + 0
0010	Addition with carry	a_in + b_in + 1
0011	Subtract without borrow	a_in − b_in
0100	Subtract with borrow	a_in − b_in − 1
0101	Increment a_in by 1	a_in + 1
0110	Decrement a_in by 1	a_in − 1
1000	a_in OR with b_in	a_in OR b_in
1001	a_in XOR with b_in	a_in XOR b_in
1010	a_in AND with b_in	a_in AND b_in
1011	Complement a_in	Not a_in

An efficient Verilog RTL using the synthesizable case constructs to infer the parallel logic is shown in Example 5. For the op_code_in[3] = 0, it performs the arithmetic operation, and when op_code_in[3] = 1, it performs the logic operation (Fig. 12.11).

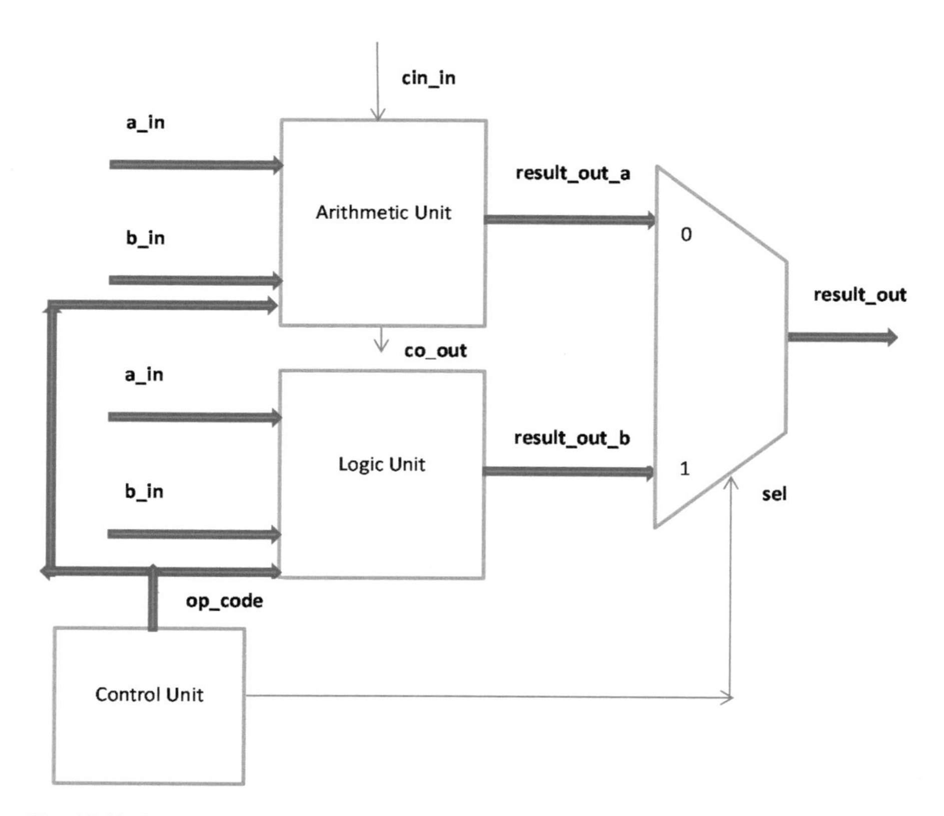

Fig. 12.11 Expected synthesis result of -bit ALU

```
////////////////////////////////////////////////////////////////////////
module arithmetic_logic_unit # (parameter data_size=8,
parameter opcode_size=4)
(

input [data_size-1:0] a_in,
input [data_size-1:0] b_in,
input cin_in,
input [opcode_size-1:0] op_code_in,
output reg [data_size-1:0] result_out,
output reg co_out );

always @ ( *)
begin
  if (~op_code_in[3])
      begin
   case (op_code_in[2:0])

      3'b000 : {co_out,result_out} = {1'b0,a_in} ;
      3'b001 : {co_out,result_out} = a_in + b_in;
      3'b010 : {co_out,result_out} = a_in + b_in + cin_in ;
      3'b011 : {co_out,result_out} = a_in - b_in ;
      3'b100 : {co_out,result_out} = a_in - b_in - cin_in;
      3'b101 : {co_out,result_out} = a_in + 1'b1;
      3'b110 : {co_out,result_out} = a_in - 1'b1;
      default : {co_out,result_out} = 9'b0_0000_0000 ;
      endcase
   end
     else
      begin
      case ( op_code_in [2:0])
     3'b000 : {co_out,result_out} = {1'b0, (a_in | b_in) };
     3'b001 : {co_out,result_out} = {1'b0, (a_in ^ b_in) };
     3'b010 : {co_out,result_out} = { 1'b0, (a_in & b_in) };
     3'b011 : {co_out,result_out} = { 1'b0, ~a_in };
      default :  {co_out,result_out} = 9'b0_0000_0000 ;
   endcase
   end

      end
endmodule
////////////////////////////////////////////////////////////////////////
```

The procedural always block is used and is level sensitive to changes on a_in, b_in, cin_in and an op_code_in.

The functionality is coded by using the case construct.

case construct is used to infer the parallel logic and infers the logic without latches due to use default condition.

Please refer the operational table for the arithmetic logic unit.

The procedural block is level sensitive to changes on a_in, b_in and op_code_in.

case construct is used to infer the parallel logic for the desired functionality.

Please refer the functional table of the logical operations.

All case conditions are covered and hence the case construct is full case. It infers the parallel logic without latches.

Example 5 Verilog RTL for 8-bit ALU

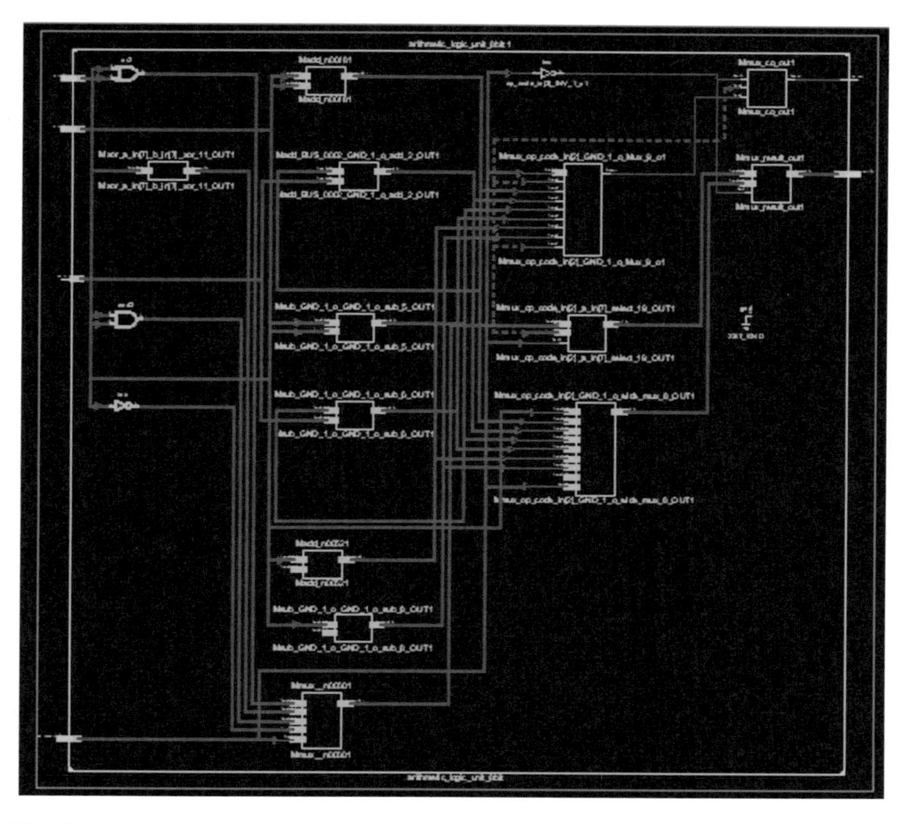

Fig. 12.12 Actual synthesis result of ALU

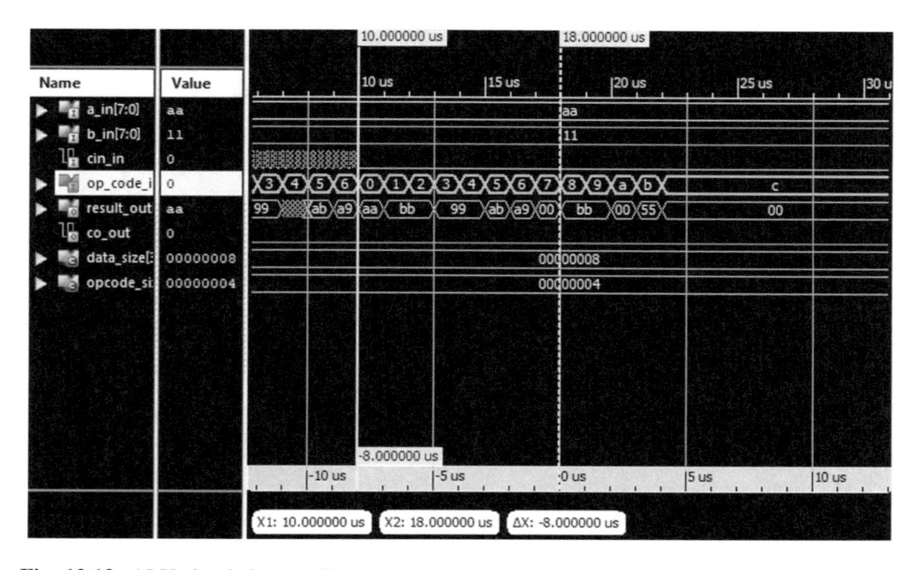

Fig. 12.13 ALU simulation result

The synthesis result of the 8-bit ALU is shown in Fig. 12.12. As shown in the figure, it consists of the parallel logic and used to perform the arithmetic operations and logic operations. Using the multiplexer tree at the output, either arithmetic or logic operation result is generated. The logic does not use the concept of resource sharing and hence has maximum area and power. Readers are requested to refer Chap. 13 for efficient RTL designs and to understand about the recommended tweaks. Refer the simulation result (Fig. 12.13).

12.2 Functions and Tasks

Task and functions are used in the Verilog to describe the commonly used functionality. Instead of coding the same RTL at the different places, it is good and common practice to use the functions or tasks depending on the requirement. For easy maintenance of the code, it is better to use the functions or tasks like the subroutine.

12.2.1 Counting Number of 1's from the Given String

Example 6 shows the ***task*** used to count 1's from the given string. The following are important points need to remember while using the task.

1. Task can consist of the time control statements and even delay operators.
2. Task can have input, output declarations.
3. Task can consist of function calls, but function cannot consist of the task.
4. Task can have output argument and not used to return the value when called.
5. Task can be used to call other tasks.
6. It is not recommended to use the task while coding the synthesizable Verilog RTL.
7. Tasks are used while coding the behavioral or simulation model.

Example 6 is the description to count number of 1's from the given string. In this example, ***task*** is used with arguments data_in, out. The name of ***task*** is *count_1-s_in_byte*. In most of the protocol descriptions, it is required to perform some operations on the input string. In this example, the string is 8-bit input data_in, and output result is 4-bit out. It is not recommended to use the task to generate synthesized logic.

The simulation result is shown in Fig. 12.14, and as shown for the 8-bit binary input data_in = 0000_0000, it generates result that is out = 0 as number of 1's is equal to 0. For data_in = 11101110, it generates out = 6 as six 1's are in the input string.

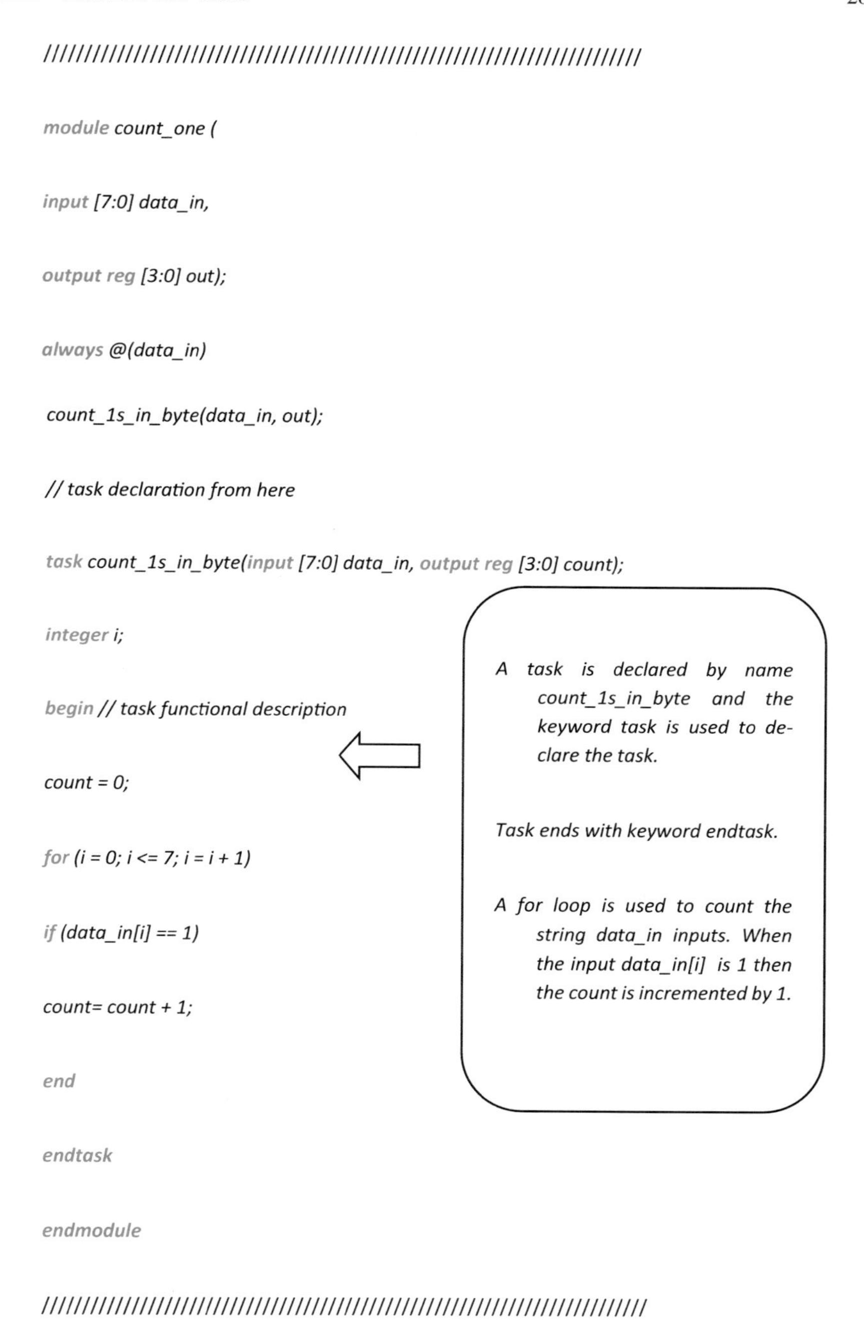

```
/////////////////////////////////////////////////////////////////////////

module count_one (

input [7:0] data_in,

output reg [3:0] out);

always @(data_in)

count_1s_in_byte(data_in, out);

// task declaration from here

task count_1s_in_byte(input [7:0] data_in, output reg [3:0] count);

integer i;

begin // task functional description

count = 0;

for (i = 0; i <= 7; i = i + 1)

if (data_in[i] == 1)

count= count + 1;

end

endtask

endmodule

/////////////////////////////////////////////////////////////////////////
```

A task is declared by name count_1s_in_byte and the keyword task is used to declare the task.
Task ends with keyword endtask.
A for loop is used to count the string data_in inputs. When the input data_in[i] is 1 then the count is incremented by 1.

Example 6 Verilog RTL using the task

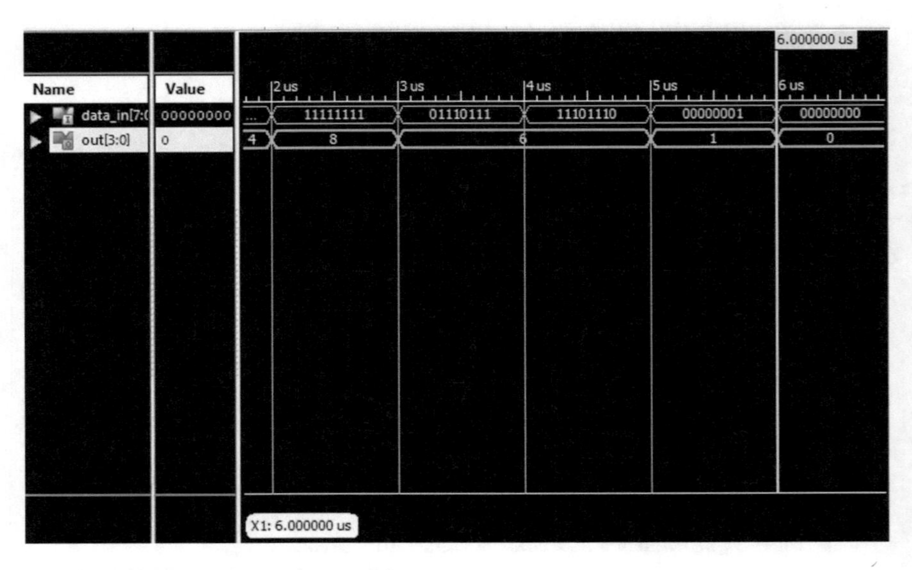

Fig. 12.14 Simulation result for Example 6

12.2.2 RTL Design Using function to Count Number of 1'S

Example 7 uses the function to count 1's from the given string. The following are important points need to remember while using the ***function***.

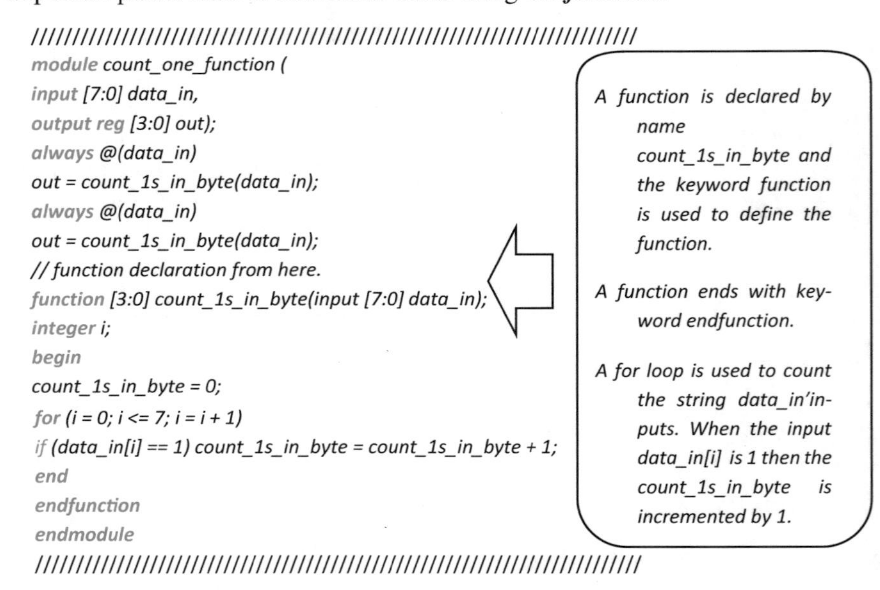

```
//////////////////////////////////////////////////////////////////////
module count_one_function (
input [7:0] data_in,
output reg [3:0] out);
always @(data_in)
out = count_1s_in_byte(data_in);
always @(data_in)
out = count_1s_in_byte(data_in);
// function declaration from here.
function [3:0] count_1s_in_byte(input [7:0] data_in);
integer i;
begin
count_1s_in_byte = 0;
for (i = 0; i <= 7; i = i + 1)
if (data_in[i] == 1) count_1s_in_byte = count_1s_in_byte + 1;
end
endfunction
endmodule
//////////////////////////////////////////////////////////////////////
```

A function is declared by name count_1s_in_byte and the keyword function is used to define the function.

A function ends with keyword endfunction.

A for loop is used to count the string data_in'inputs. When the input data_in[i] is 1 then the count_1s_in_byte is incremented by 1.

Example 7 Verilog RTL with function calls

1. Function cannot consist of the time control statements and even delay operators.
2. Function can have at least one input argument declarations.
3. Function can consist of function calls, but function cannot consist of the task.
4. Function executes in zero simulation time and returns single value when called.
5. It is not recommended to use the function while coding the synthesizable Verilog RTL.
6. Functions are used for coding the behavioral or simulatable model.
7. Functions should not have non-blocking assignments.

Example 7 is the description to count number of 1's from the given string. In this example, function is used with arguments 'data_in'. The name of function is 'count_1s_in_byte'. In most of the protocol descriptions, it is required to perform some operations on the input string. In this example, the string is 8-bit input 'data_in', and output result is 4-bit 'out'. It is not recommended to use the function to generate synthesized logic.

The simulation result is shown in Fig. 12.15, and as shown for the 8-bit binary input data_in = 0000_0000, it generates result that is out = 0 as number of 1's is equal to 0. For data_in = 10101011, it generates out = 5 as 5, 1's are in the input string.

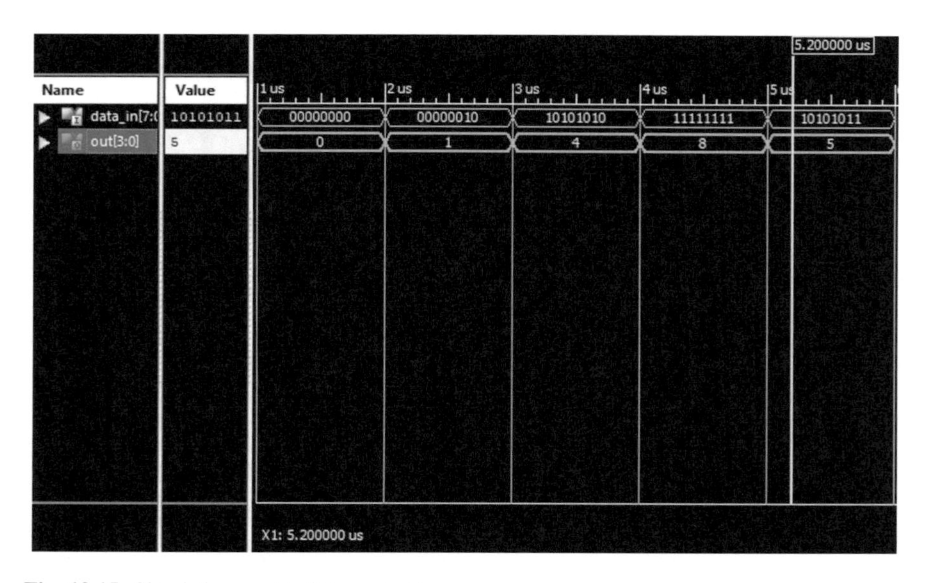

Fig. 12.15 Simulation result of Example 7

12.3 Synthesis Result of RTL Using function

Most of the time depending on the design requirements, we can think about using function to infer the combinational logic. The RTL is coded (Example 8) to infer the XOR gate having registered output. In this example, the function xor_logic is used to implement the XOR gate.

As shown in the synthesis result (Fig. 12.16), the inferred logic has XOR gate and D flip-flop.

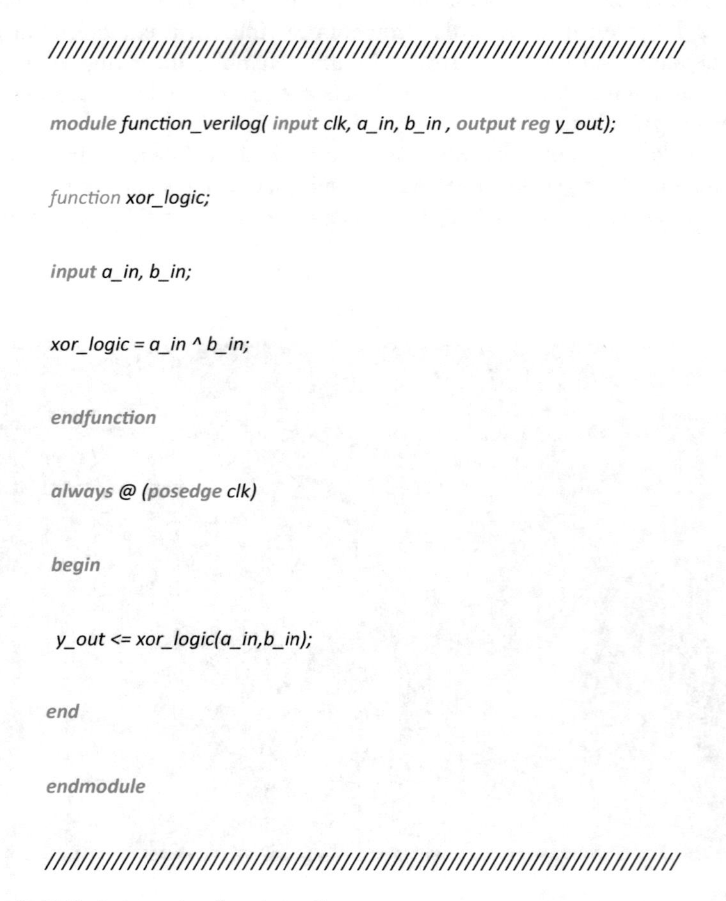

```
//////////////////////////////////////////////////////////////////////////////
module function_verilog( input clk, a_in, b_in , output reg y_out);
function xor_logic;
input a_in, b_in;
xor_logic = a_in ^ b_in;
endfunction
always @ (posedge clk)
begin
 y_out <= xor_logic(a_in,b_in);
end
endmodule
//////////////////////////////////////////////////////////////////////////////
```

Example 8 RTL design using function call

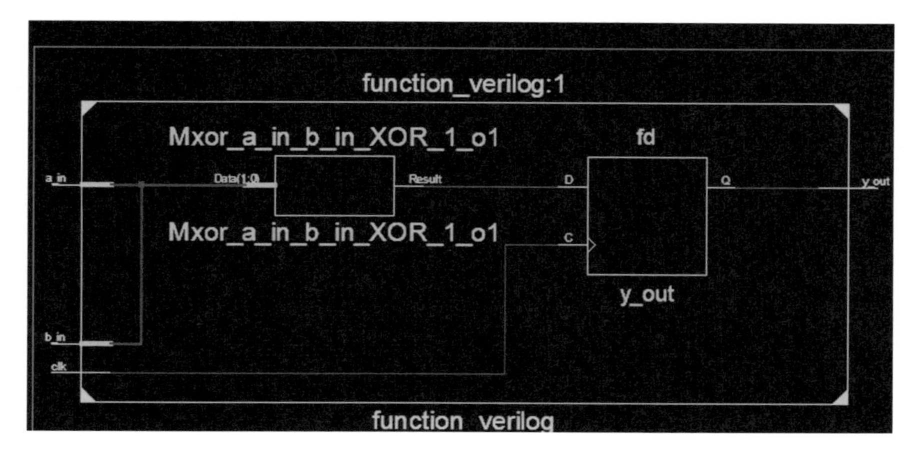

Fig. 12.16 Synthesis result of Example 8

12.4 Synthesis Result of RTL Using task

The RTL is coded (Example 9) to infer the XOR gate having registered output. In this example, the task xor_logic is used to implement the XOR gate.

As shown in the synthesis result (Fig. 12.17), the inferred logic has XOR gate and D flip-flop.

```
///////////////////////////////////////////////////////////////////////////////

module task_verilog( input clk,a_in, b_in , output reg y_out);

task xor_logic;

input a, b;

output y;

y = a ^ b;

endtask

always @ (posedge clk)

begin

xor_logic ( a_in, b_in, y_out);

end

endmodule

///////////////////////////////////////////////////////////////////////////////
```

Example 9 RTL design using task

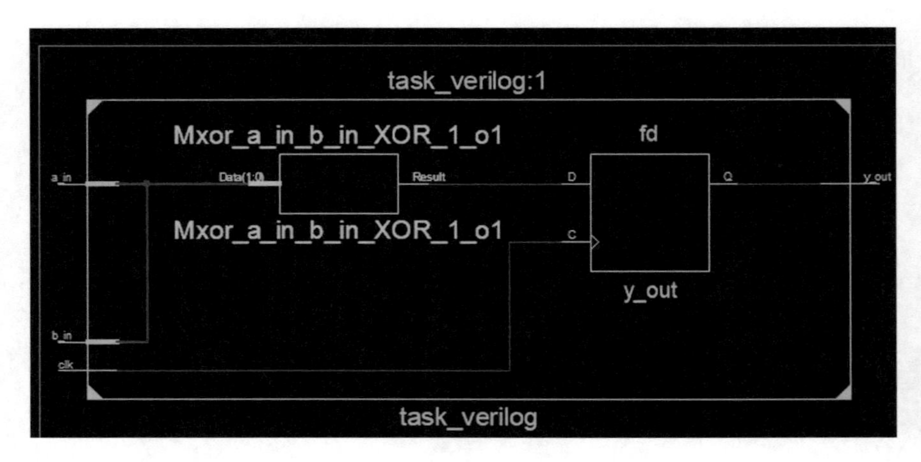

Fig. 12.17 Synthesis result of Example 9

12.5 Exercises

The exercises are based on the understanding of synthesizable Verilog constructs. Complete the exercises for better understanding and application of Verilog constructs.

1. *Code the RTL using synthesizable constructs to implement 8-bit barrel shifter.*
2. *The logic inferred by the following code is*

```
module design_logic ( input  a_in, b_in, c_in, output reg result-
     out, carry_out);

always @ (*)

begin

if (~c_in)

 {carry_out, result_out} = a_in+b_in+c_in;

else

 {carry_out, result_out} = a_in+(~b_in)+c_in;

end

endmodule
```

   a. *Adder and subtractor*
   b. *Adder, subtractor and 2:1 mux*
   c. *Only adder*
   d. *Syntax error and RTL should be tweaked*

3. *Tweak the RTL shown in the exercise question 2 to infer only minimum number of adders.*
4. *Tweak the RTL shown in the Example-5 to have registered inputs and registered output.*

12.6 Summary

The following are important points to conclude this chapter

1. The design partitioning can give the good and clear visibility of the data and control paths during the RTL design.
2. The Verilog RTL for the complex design should have the separate modules for the data paths and control paths.
3. Use the resource sharing concepts while coding for the logic unit. All the logical operations can be performed by using full adder with additional combinational logic.
4. Do not use the function and task while coding the RTL design.
5. Function does not consist of delays or timing control constructs.
6. Task can consist of timing control and delay constructs.
7. Use the pipelining for better performance of the design.
8. Code the RTL design using minimum resources.
9. Use the synthesizable constructs to implement the complex designs and have the better design partitioning at functional level.

Chapter 13
RTL Tweaks and Performance Improvement Techniques

The concept of the RTL tweaks to improve the performance of the design is discussed in this chapter. The chapter discusses about the area, speed and power improvement basics and useful during the RTL design and synthesis stage to improve the design performance.

As discussed in Chap. 12 for any design, the goal is to achieve the area, speed, and power. In such scenario, the optimization techniques play important role, and the following few section discusses the performance improvement of the design using RTL tweaks. The powerful techniques such as resource sharing, pipelining, and clock gating are discussed and are useful during the RTL design phase.

13.1 Arithmetic Resource Sharing

Most of the time we experience the need of the resource sharing. During the RTL design of the arithmetic and logic unit, we can improve the performance of the design by sharing the common resources such as adders to have the better data and control path optimization!

Example 1 shows the implementation of addition without resource sharing. The intended design functionality is to design the combinational logic and shown in Table 13.1.

As shown in the synthesized logic in Fig. 13.1, it uses three full adders and two multiplexers. The synthesized logic is inefficient as all the addition operations are performed simultaneously and multiplexer output is control signal dependent. So, there is wastage of more power, and the RTL coded is inefficient as per as area utilization is concern.

Even the data and control path optimization is not used while coding the RTL.

V. Taraate, *Digital Logic Design Using Verilog*,
https://doi.org/10.1007/978-981-16-3199-3_13

```
////////////////////////////////////////////////////////////////////////////

module logic_without_resource_sharing(

input  a_in,
input  b_in,
input  c_in,
input  d_in,
input  e_in,
input  f_in,
input s1_in;,
input s2_in,
output reg  y_out,
output reg  z_out );
always @ ( a_in, b_in, c_in, d_in, s1_in)
begin
  if ( s1_in )
      y_out = a_in + b_in;
      else
      y_out = c_in + d_in;
end

always @ ( a_in, b_in, e_in, f_in, s2_in)
begin
  if ( s2_in )
      z_out = e_in + f_in;
      else
      z_out = a_in + b_in;
end


endmodule
////////////////////////////////////////////////////////////////////////////
```

The combinational design is coded by using multiple always blocks.

The addition operation is performed by using operator +. Due to if-else construct it infers the 2:1 MUX.

Example 1 Verilog RTL without using the concept of resource sharing

Table 13.1 Functional table description

s1_in	y_out
1	a_in + b_in
0	c_in + d_in
s2_in	z_out
0	a_in + b_in
1	e_in + f_in

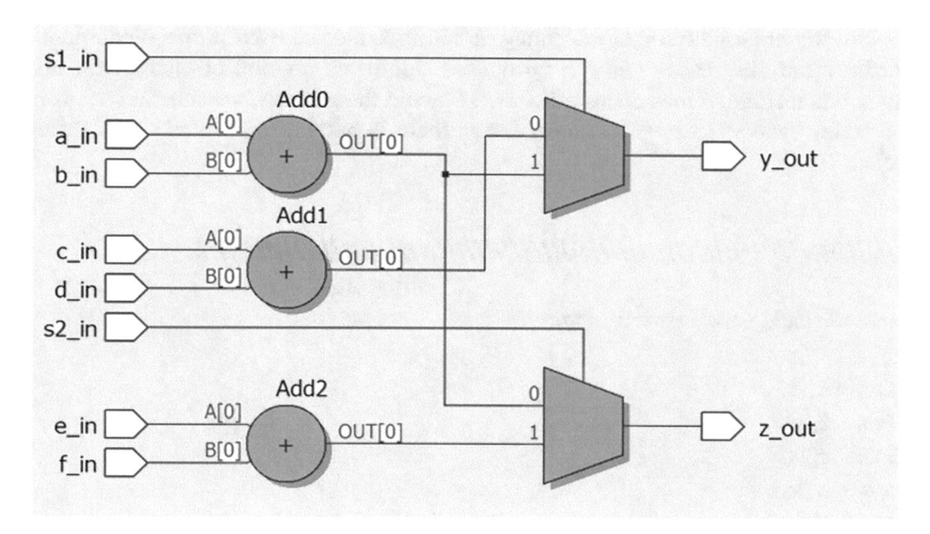

Fig. 13.1 Synthesized logic without resource sharing

13.1.1 RTL Design Using Resource Sharing to Have Area Optimization

Resource sharing is one of the efficient techniques used during the RTL design to share the common resources. As coded in Example 2, the multiple adders at the output are generating the results simultaneously and wait for data from the multiplexer tree which is sensitive to the control signals either s1_in or s2_in.

As shown in Fig. 13.2, the inferred logic is optimized and has the adders at the output and chain of multiplexer to pass the desired input depending on the status of select inputs. The resource sharing has minimized the single adder for 1-bit addition. For the 8-bit addition, this technique is useful to improve the area by eliminating eight adders.

13.2 Gated Clocks and Dynamic Power Reduction

Gated clock signals are used to turn on or turn off the switching at the clock net for the multiple clock domain designs and uses the clock enable inputs. When enable input is high, the clock switching is on, and when enable input is low, the clock switching is off. The clock gating logic is used to control the clock turn on or turn off. Clock gating is an efficient technique used in the ASIC design to reduce the switching power at the clock input of register or flip-flop. By using the clock gating cell, the clock switching can be controlled as and when required according to the design functional requirements. The clock gating is useful to reduce the dynamic power for the design.

But the issue with the clock gating is that it cannot be used in the synchronous designs, and the reason being it introduces significant amount of clock skew and even this technique introduces glitches. To avoid the glitches, special care needs to be taken by ASIC design engineer hence there is need of dedicated clock gating cells.

```
//////////////////////////////////////////////////////////////////////

module logic_with_resource_sharing (

input  a_in,
input  b_in,
input  c_in,
input  d_in,
input  e_in,
input  f_in,
input s1_in,
input s2_in,
output reg  y_out,
output reg  z_out );
reg temp1,temp2,temp3,temp4;

always @ ( a_in, b_in, c_in, d_in, s1_in)
begin
  if ( s1_in )
      begin
        temp1 = a_in;
        temp2 = b_in;
      end
      else
      begin
        temp1 = c_in;
        temp2 = d_in;
      end
end

always @ ( a_in, b_in, e_in, f_in, s2_in)
begin
if ( s2_in )
```

The combinational design is coded by using multiple always procedural blocks.

In this the functionality is described using multiple number of multiplexers at the input side to sample the desired inputs.

Example 2 Verilog RTL with resource sharing technique

```
    begin
     temp3 = e_in;
     temp4 = f_in;
    end
    else
    begin
     temp3 = a_in;
     temp4 = b_in;
    end
end

always @ ( temp1, temp2, temp3, temp4 )
begin
 y_out = temp1 + temp2;
 z_out = temp3 + temp4;
end

endmodule
//////////////////////////////////////////////////////////////////////////////
```

Adders are used as common resources to perform addition depending on the inputs sampled by using multiple multiplexers.

Example 2 (continued)

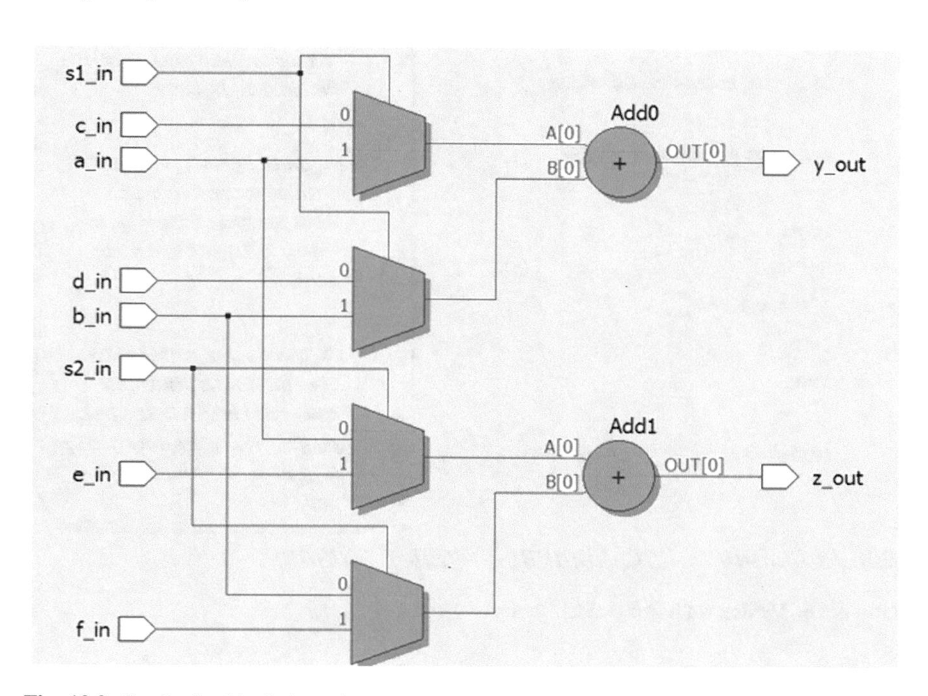

Fig. 13.2 Synthesized logic by using common resources

Verilog RTL is coded and shown in Example 3 and uses enable input to control the clock switching activity. For 'enable = 1', the clock input 'clk' toggles, and for 'enable = 0', clock input is permanently active low so no switching at clock input.

The synthesized logic is shown in Fig. 13.3 where clock is gated by using AND logic. The logic is prone to glitches, and it is recommended to use the dedicated clock gating cell. Refer Chap. 24 to have more understanding about the low power design.

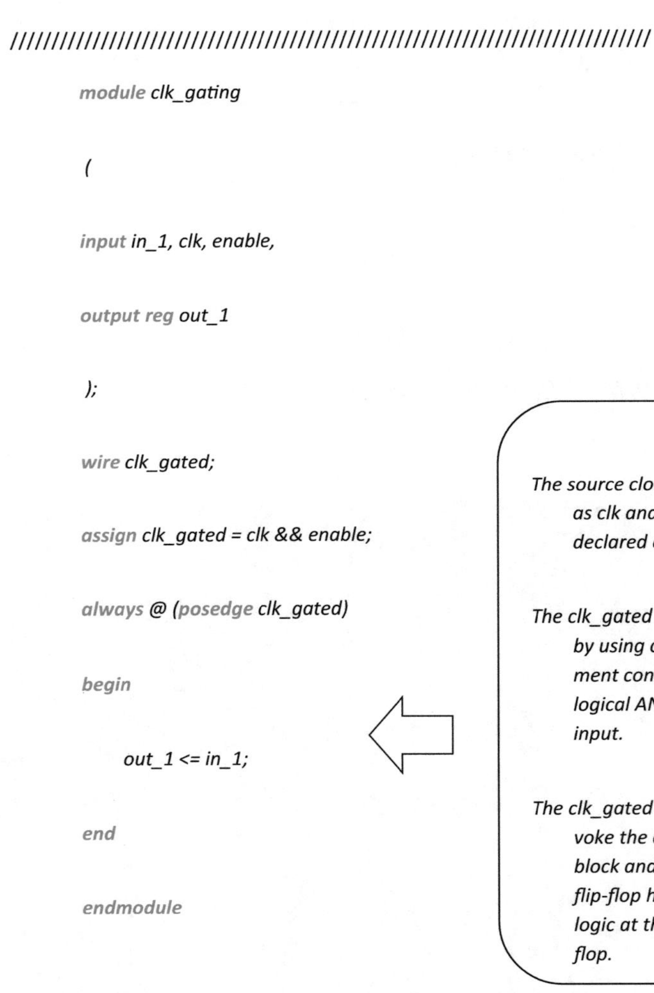

//
module clk_gating
(
input in_1, clk, enable,
output reg out_1
);
wire clk_gated;
assign clk_gated = clk && enable;
always @ (posedge clk_gated)
begin
out_1 <= in_1;
end
endmodule
The source clock signal is declared as clk and gated clock signal is declared as clk_gated.
The clk_gated signal is generated by using continuous assignment construct assign and it is logical AND of clk and enabl' input.
The clk_gated signal is used to invoke the always procedural block and hence it infers the D flip-flop having the clock gating logic at the clock input of flip-flop.
//

Example 3 Verilog RTL using clock gating

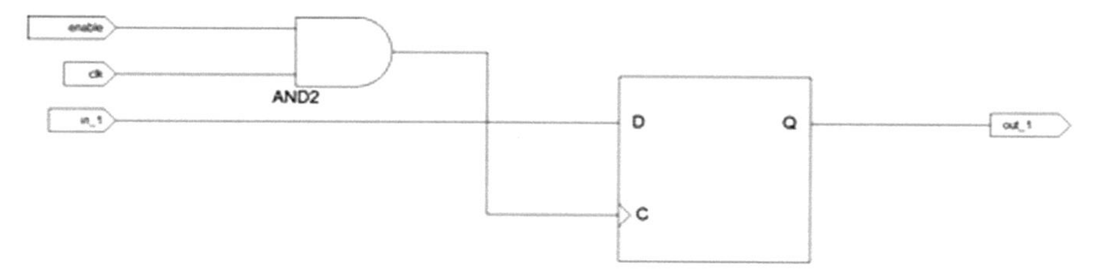

Fig. 13.3 Synthesized clock gating logic

13.3 Use of Pipelining in Design

Pipelining is one of the powerful techniques used to improve the performance of the design at the cost of latency. This technique is used in many processor designs and many ASIC design applications to perform multiple tasks at a time. This section discusses the design without pipelining and design with pipelining.

13.3.1 Design Without Pipelining

During the initial stage of the design, most of the designs are coded by using synthesizable Verilog constructs without the use of the pipelined logic. If the desired speed, that is, design performance is not met, then ASIC designer can tweak the RTL design. One of the best approaches is pipelining by inserting the register according to the clock latency and data rate requirements.

Example 4 describes the design functionality using Verilog synthesizable constructs without the use of any pipelined logic.

The synthesized logic is shown in Fig. 13.4 and consists of two flip-flops sensitive to the common clock source clk.

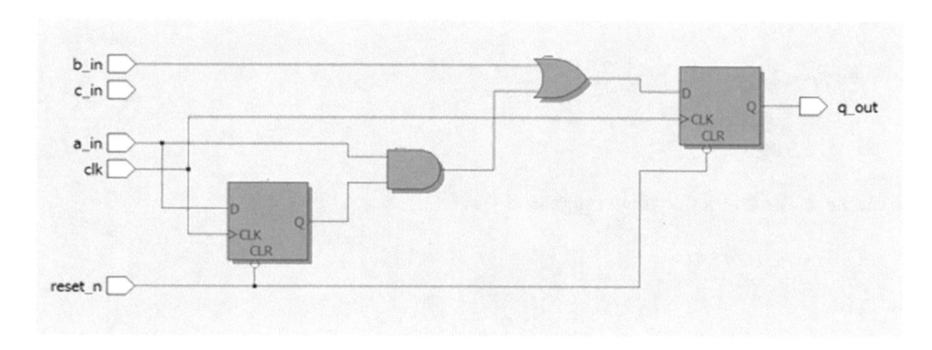

Fig. 13.4 Synthesized logic without pipelined stage

```
//////////////////////////////////////////////////////////////////////////////

module design_without_pipeline (

input a_in,

input b_in,

input clk,

input reset_n,

output reg q_out ) ;

reg q1_out;

always @ (posedge clk , negedge reset_n)

begin

    if (~reset_n)

        begin
```

The always procedural block is sensitive to changes on the rising edge of clock clk and active low reset reset_n.

This block is used to infer the sequential logic sensitive to positive edge of clock clk. The reset is an asynchronous active low reset reset_n.

Example 4 Verilog RTL without pipelined stage

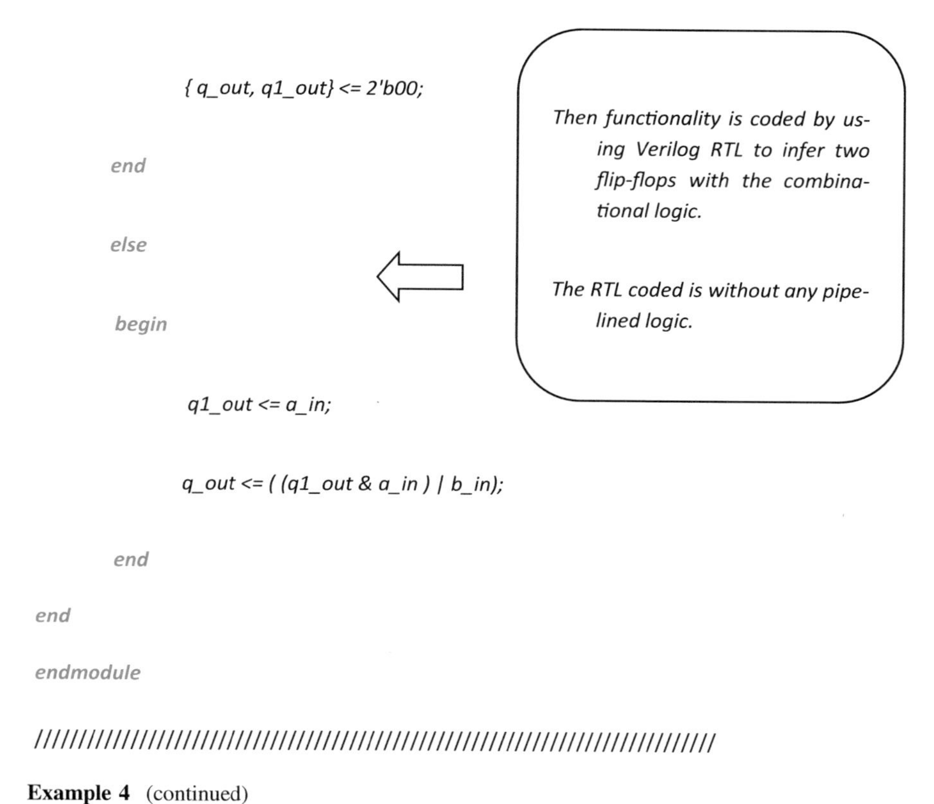

{ q_out, q1_out} <= 2'b00;
end
else
begin
q1_out <= a_in;
q_out <= ((q1_out & a_in) | b_in);
end
end
endmodule
//
Then functionality is coded by using Verilog RTL to infer two flip-flops with the combinational logic.
The RTL coded is without any pipelined logic.

Example 4 (continued)

13.3.2 *Speed Improvement Using Register Balancing or Pipelining*

To improve the design performance, the combinational logic AND output can be given to the additional pipelined register, and the output of the pipelined register can drive one of the input of OR logic.

This technique will improve the overall performance of the design at the cost of one clock latency. The improvement in the design performance is due to the reduction in the combinational delay in the register-to-register path. This improves the data arrival time for the reg-to-reg path!

Verilog RTL is shown in Example 5, and by using additional register as the pipelined logic, the register balancing is achieved.

The synthesized logic is shown in Fig. 13.5 and consists of three flip-flops sensitive to the common clock source clk.

```
/////////////////////////////////////////////////////////////////////////////////

module design_with_pipeline (

input a_in,

input b_in,

input clk,

input reset_n,

output reg q_out );

reg q1_out, q2_out;

always @ (posedge clk , negedge reset_n)

begin

    if (~reset_n)

begin

        { q_out, q2_out, q1_out} <= 3'b000;
```

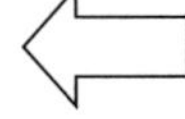

The always procedural block is sensitive to changes on the rising edge of clock clk and active low reset reset_n.

This block is used to infer the sequential logic sensitive to positive edge of clock clk. The reset is an asynchronous active low reset reset_n.

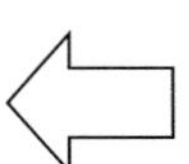

The functionality is coded by using Verilog RTL to infer three flip-flops with the combinational logic. To improve the speed of the design the combinational logic is spitted using additional register.

The structure described is RTL with single stage pipelined logic.

Example 5 Verilog RTL with pipelined stage

```
    end
    else
     begin
          q1_out <= a_in;
          q2_out <=  (q1_out & a_in )
          q_out <= (q2_out| b_in);
     end
end
endmodule
////////////////////////////////////////////////////////////////////////////////
```

Example 5 (continued)

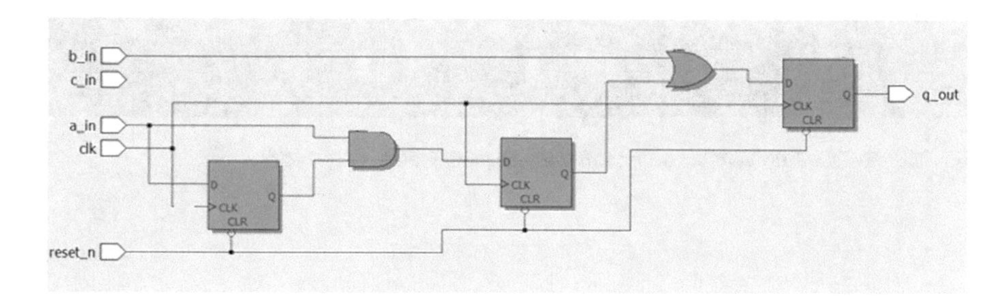

Fig. 13.5 Synthesized logic with pipelined stage

13.4 Counter Design and Duty Cycle Control

As discussed in Chaps. 9 and 10, we need to have the efficient counters, shift registers, and memories in the ASIC design. The MOD-3 counter which has three states 00,01,10 is coded using the Verilog synthesizable construct and is shown in Example 6. Due to three states the on duty cycle is 33.33%, where duty cycle is $(T_{on}/(T_{on} + T_{off}))$, where T_{on} is on time and T_{off} is off time.

The synthesis result of the MOD-3 counter is shown in Fig. 13.6 and has the two flip-flops and next state logic as the adder and multiplexer. It generates the sequence as 00,01,10,00,01,10....

The simulation result of MOD-3 synchronous up counter is shown in Fig. 13.7 and generates an output sequence as 00,01,10,00,01,10.... During reset_n = 0, an output is equal to 00. The counter is sensitive to rising edge of the clock.

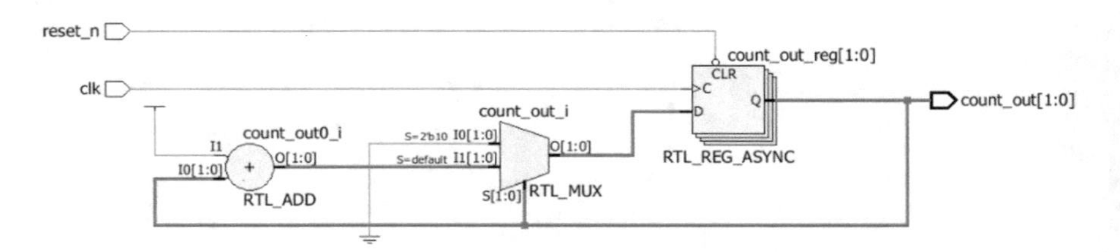

Fig. 13.6 Synthesis result of MOD-3 up counter

Fig. 13.7 MOD-3 up counter waveform without having 50% duty cycle

```
////////////////////////////////////////////////////////////////////////////////

module modulo_3_up_counter(

                    input clk, reset_n,

                     output reg [1:0] count_out

                     );

 always @ (posedge clk, negedge reset_n)

 begin

   if ( ~reset_n)

   count_out <= 2'b00;

   else

     if ( count_out == 2'b10)

       count_out <= 2'b00;

       else

       count_out <= count_out + 1;

 end

 endmodule

////////////////////////////////////////////////////////////////////////////////
```

Example 6 MOD-3 up counter RTL design

13.5 MOD-3 Counter RTL Design to Have 50% Duty Cycle

The MOD-3 counter which has three states 00,01,10 is coded using the Verilog synthesizable construct and is shown in Example 7. The on duty cycle is 50.00%, where duty cycle is ($T_{on}/(T_{on} + T_{off})$), where T_{on} is on time and T_{off} is off time. Both time durations are same. That is for three half cycles, counter MSB output is high, and for three half cycles, the MSB of counter output is low.

The strategy used is using the positive edge-sensitive flip-flops and negative edge-sensitive flip-flops in the parallel path to have the output with 50% duty cycle.

The synthesis result of the MOD-3 counter having 50% duty cycle is shown in Fig. 13.8 and has the two positive edge-sensitive flip-flops, single negative level-sensitive flip-flop, and next state logic as the adder, OR gate, and multiplexer. It generates the sequence with 50% duty cycle and counter output as 00,01,10,00,01,10….

The simulation result of MOD-3 synchronous up counter is shown in Fig. 13.9 and generates an output sequence as 00,01,10,00,01,10…. During reset_n = 0, an output is equal to 00. The counter is sensitive to rising edge of the clock. Check during the period highlighted using two vertical yellow markers the output of MOD-3 counter is having 50% duty cycle. That is on time is equal to off time.

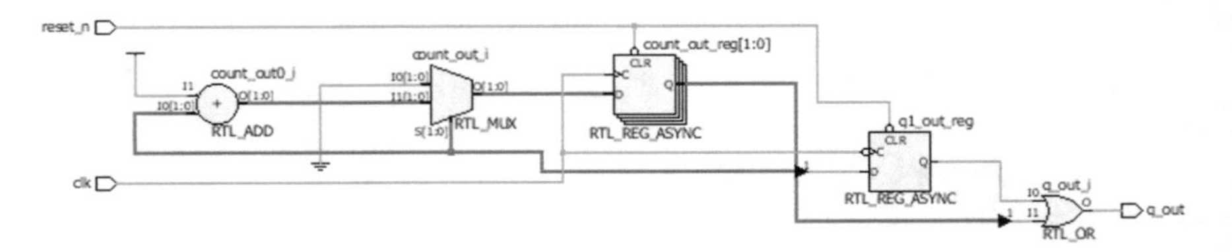

Fig. 13.8 Synthesis result of MOD-3 counter having 50% duty cycle

Fig. 13.9 Waveform of MOD-3 counter having 50% duty cycle output

```
////////////////////////////////////////////////////////////////////////////

module MOD_3_duty_cycle_control(

                input clk, reset_n,

                output q_out

                );

 reg [1:0] count_out;

 reg q1_out;

always @ (posedge clk, negedge reset_n)

begin

  if ( ~reset_n)

  count_out <= 2'b00;

    else

    if ( count_out == 2'b10)

        count_out <= 2'b00;

    else

     count_out <= count_out + 1;

end
```

Example 7 RTL tweak to get the 50% duty cycle at the MOD-3 counter output

```
always @ (negedge clk, posedge reset_n)

if (~reset_n)

      q1_out <= 1'b0;

else

      q1_out <= count_out[1];

assign q_out = q1_out | count_out[1];

endmodule

/////////////////////////////////////////////////////////////////////////////////////
```

Example 7 (continued)

13.6 Exercise

The exercises are based on the understanding of synthesizable Verilog constructs. Complete the exercises for better understanding and application of Verilog constructs.

1. *Code the RTL for MOD-5 up-counter. Consider the counter is sensitive to the falling edge of the clock and has the asynchronous active low reset.*
2. *Tweak the RTL in Exercise-1 to get the MOD-5 counter having 50% duty cycle output.*
3. *Code the RTL using resource sharing for the following*

s1_in	y_out
1	a_in * b_in
0	c_in * d_in
s2_in	z_out
1	a_in * b_in
0	e_in * f_in

13.7 Summary

The following are important points to conclude this chapter:

1. The resource sharing technique is useful to improve the design performance.
2. Using the resource sharing, common arithmetic resources can be shared.
3. Use the pipelining concept to improve the speed of the design.
4. The pipelined stages increase the latency but reduce the reg-to-reg path data arrival time.
5. The dynamic power can be optimized using the clock gating technique.
6. Use the dedicated clock gating cells to optimize for the power.
7. Use the duty cycle control strategies to have the T_{on} equal to T_{off} at the output of sequential design.

Chapter 14
Finite State Machines Using Verilog

The RTL design should have better performance and should have the better data and control path optimization. Finite-state machines are used to design the control and timing logic and even to detect the sequence. Finite-state machines can be coded by using different encoding styles. These encoding styles are binary, gray, and one-hot encoding and are discussed in this chapter.

The finite-state machine (FSM) is especially important design block for any of the ASIC design. Most of the ASIC designs and controllers need the efficient and synthesizable state machines. The FSMs can be designed and coded very efficiently by using the Verilog synthesizable constructs. During the design phase, the RTL design team should use the synthesizable Verilog constructs to implement the state machines to have the better performance!

Basically, FSMs are useful to detect the desired sequences or the preordered or defined events and are source synchronous designs. FSMs can be coded efficiently for the better synthesis outcome using the multiple or single procedural block. In the practical scenario, it is recommended to use the multiple procedural blocks to code the state machines. One of the procedural blocks can describe the combinational logic and level sensitive to the inputs or the states. Whereas the other procedural block can be edge sensitive to positive edge of clock or to the negative edge of clock.

The multiple procedural block FSM is better for the readability and can generate efficient synthesis results. The main objective of this chapter is to code the efficient FSM for better performance using synthesizable Verilog constructs.

V. Taraate, *Digital Logic Design Using Verilog*,
https://doi.org/10.1007/978-981-16-3199-3_14

14.1 Moore Versus Mealy Machines

FSMs are classified as Moore and Mealy machines. In the Moore machines, the output is function of the present state or current state, and in the Mealy machine, the output of FSM is function of the present or current state as well as present input.

In the Moore machine, an output is stable for one clock cycle, and hence, output is glitch or hazard free. In the Mealy machines, an output may or may not be stable for one clock cycle as it is function of current state or change in the input.

The timing analysis for the Moore machine is quite simple due to clean register to register path, but for the Mealy machine, there might be chances of glitches or hazards as the output is function of the input changes and current state.

But the disadvantage of Moore machine is that it needs a greater number of states as compared to Mealy machine. Practical scenario is that Mealy machine has one state less as compared to Moore machine.

Fig. 14.1 describes the internal structure for the machine, and it consists of combinational block as next state logic whose output is dependent upon the changes in the Current_state and an input, state register block which is dependent on the Next_state, and an output logic block which is purely combinational in nature and depends upon the changes in the Current_state. As discussed, earlier in the Moore machine, output is function of Current_state and hence stable for one clock cycle.

Fig. 14.2 describes the internal structure for the machine, and it consists of combinational block as next state logic whose output is dependent upon the changes in the Current_state and an input, state register logic which is dependent on the Next_state, and an output logic block which is purely combinational in nature and depends upon the Current_state as well changes in the input. As discussed earlier, Mealy machine output is function of Current_state as well as changes in the inputs and hence may or may not be stable for one clock cycle. Due to this, the Mealy machines are prone to glitches!

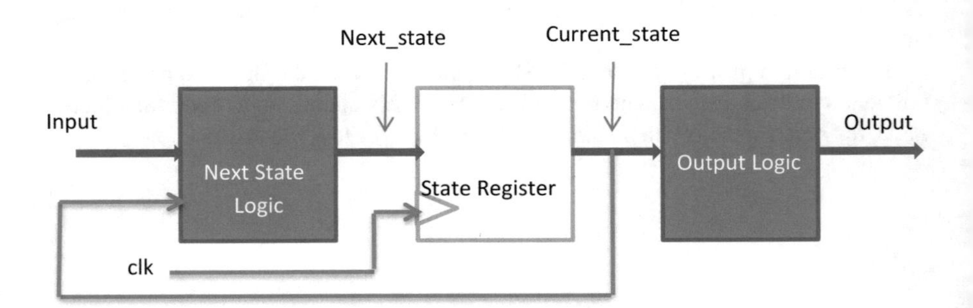

Fig. 14.1 Block diagram of Moore machine

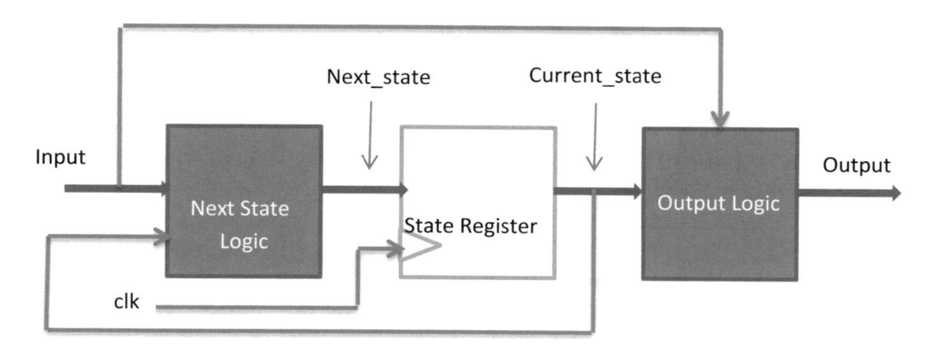

Fig. 14.2 Block diagram of Mealy machine

How to code an efficient FSM is one of the important points to discuss about! As a RTL design engineer, the overall performance of the design is dependent upon the efficient RTL coding. Most of the inexperienced RTL engineers use single procedural ***always*** block for coding the behavior of the FSM. But single ***always*** block FSM always results into inefficient coding and has issue while synthesizing the design and even during timing analysis.

In the practical scenarios, two or three ***always*** procedural block FSMs should be used. In this chapter, I have recommended to use the three procedural block FSMs. Multiple procedural block FSM increases the number of lines of code but most efficient during synthesis and timing analysis. Even this improves the overall readability and reusability during the reviews and design cycle. This also improves the overall design performance!

Table 14.1 Differences between Moore and Mealy machines

Moore machine	Mealy machine
Outputs are function of current state only	Outputs are function of the current state and inputs also.
As output is function of current state, it is stable for one clock cycle	Output is function of current state and inputs so may change if input changes and hence may or may not be stable for one clock cycle.
Output is stable for one clock cycle and not prone to glitches or spikes	Output may change multiple times depending on changes in the input and hence prone to glitches or hazards.
It requires a greater number of states as compared to Mealy machine	Mealy machine needs at least one state less as compared to Moore machine.
STA is easy as combinational paths between the registers are shorter	STA is complex as combinational paths are relatively larger area as compared to Moore machine.
Higher operating frequency as compared to Mealy machine	Lower operating frequency as compared to Moore machine

1. One of the procedural ***always*** blocks is used to describe the functionality for the next state logic, and it is level sensitive to changes on the inputs and Current_state.
2. Another procedural ***always*** block is used to describe the state register logic and sensitive to positive or negative edge of clock and hence used to infer the state register sequential logic.
3. Third procedural ***always*** block is sensitive to the changes on Current_state and used to infer the combinational logic. This is true for the Moore machine.
4. For the Mealy machine, third procedural ***always*** block is sensitive to the changes on Current_state as well as input and used to infer the combinational logic.

Table 14.1 illustrates the important differences between Moore and Mealy machines.

The template shown in Fig. 14.3 is useful and gives information about the steps and declaration for the FSM coding.

The practical FSM for the toggle flip-flop is designed using the three procedural block ***always*** blocks. The Example 1 describes the efficient Verilog RTL for state transition Table 14.2. The state table is used to describe the state transition on the active clock edge.

Table 14.2 State transition table for the toggle flip-flop

current_state	next_state
0	1
1	0

```
/////////////////////////////////////////////////////////////////////////////

1. Declare module having the FSM name and list the inputs and
   outputs.
   module FSM_NAME( // input output list);

2. Declare the state variables using parameter, for
   FSM has two states then declare
   parameter s0;
   parameter s1;

3. Declare the intermediate variable net data type reg for the next
   state and current state. For Example, for one bit data type declare:
   reg current_state;
   reg next_state;

//Use non-blocking assignment to code the sequential state register
    logic

4. Code the state register logic sensitive to the edge, for example:
   always@ (posedge clk , negedge reset_n)
   begin

    //Functionality of state register logic

    end

   //Use the blocking assignments to code the next state combina-
   tional logic

5. Code the next state logic, which is sensitive to level, for example
   always @ (*)
   begin
    case (current_state)
```

Fig. 14.3 Steps for efficient FSM Verilog RTL coding

```
//Functionality of the next state logic
endcase
end
//Use the blocking assignment to code the combinational output
logic

6. Code  the output logic which is sensitive to level, for example
always @ (current_state)// For Moore machine
always@(current_state, input) // For Mealy machine
begin
 case (current_state)
 //Functionality of the output  logic
 endcase
 end
/////////////////////////////////////////////////////////////////////////////
```

Fig. 14.3 (continued)

```
/////////////////////////////////////////////////////////////////////////////

//FSM of the toggle flip-flop
module toggle_flip_flop_fsm (
input clk,
input reset_n,
output  reg y_out );
```

The input and outputs are declared for the toggle flip-flop.

Inputs are named as clk, reset_n and an output is named as y_out.

State parameters are declared as s0, s1. By using the net data type reg the 'current_state and next_state are declared.

```
parameter s0=0;
parameter s1=1;
reg current_state;
reg next_state;

//State register logic
always@ (posedge clk , negedge reset_n)
begin
 if (~reset_n)
   current_state <= s0;
 else
   current_state <= next_state;
 end
```

The state register sequential block is sensitive to positive edge of clock.

During reset condition current_state is assigned to 's0' and during normal operation the next_state is assigned to current_state

```
//Next state combinational logic
always @ (current_state)
begin
 case (current_state)
 s0 : next_state = s1;
 s1 : next_state =s0;
 default : next_state =s0;
 endcase
 end
```

The next state logic combinational block is sensitive to the current_state for the toggle flip-flop.

The functionality is described by using the case construct. Please refer the state transition table

Example 1 Verilog RTL for toggle flip-flop

```
//Output combinational logic
always@ (current_state)
case ( current_state)
s0 : y_out = 1'b0;
s1 : y_out =1'b1;
default : y_out=1'b0;
endcase
endmodule
```

The output combinational logic is function of the current_state and the functionality is described using the case construct. During state s0 the output y_out' is assigned to logic 0 and during state s1 the output y_out is assigned to logic 1. So, it infers the toggle flip-flop.

//

Example 1 (continued)

The synthesized logic for the toggle flip-flop is shown in Fig. 14.4, and it infers the state register triggered on the positive edge of clock and having active low asynchronous reset 'reset_n'. Due to use of the ***case*** construct, the decoding logic is inferred, and the decoding logic is a NOT gate. The output is available at y_out, and the output toggles on every positive edge of clock clk.

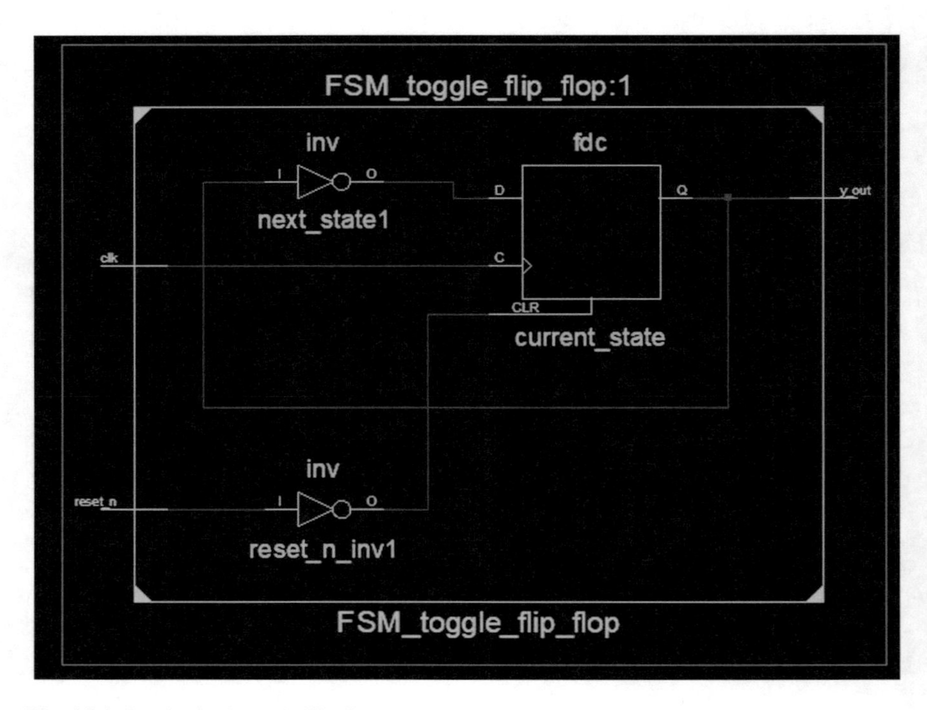

Fig. 14.4 Synthesized toggle flip-flop

14.1.1 Level to Pulse Converter

The level-to-pulse converter partial state transition diagram is shown in Fig. 14.5; Partial State transition diagram for mealy level to pulse converter gives information about the state transition. As shown in the figure in the state s0, the data input data_in is logic 0, and in the state s1, the data input data_in is logic 1. The state transition table for the Mealy level-to-pulse converter is shown in Table 14.3.

The synthesizable Verilog RTL using three procedural ***always*** blocks is coded and shown in the Example 2.

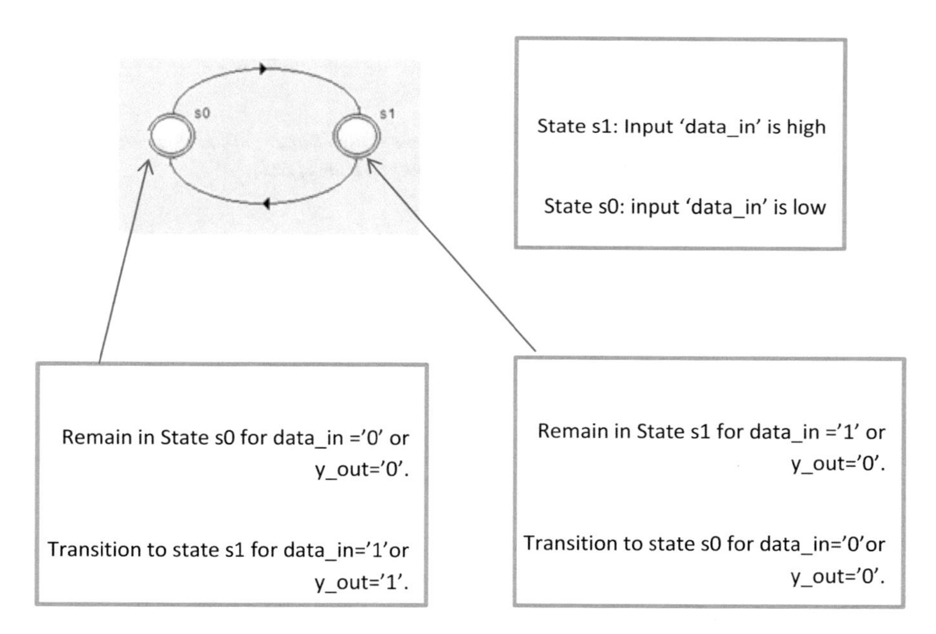

Fig. 14.5 Partial state transition diagram for Mealy level-to-pulse converter

Table 14.3 State transition table for the Mealy level-to-pulse converter

data_in	current_state	next_state	y_out
0	s0	s0	0
1	s0	s1	1
0	s1	s0	0
1	s1	s1	0

```
/////////////////////////////////////////////////////////////////////////
//Synthesizable RTL for Mealy level to pulse converter

module level_pulse_converter (
input clk,
input reset_n,
input data_in,
output  reg y_out);

parameter s0=0;
parameter s1=1;

reg current_state;
reg next_state;

//State register logic
always@ (posedge clk , negedge reset_n)
begin
 if (~reset_n)
   current_state <= s0;
 else
   current_state <= next_state;
 end

//Next state logic combinational block
always @ (*)
begin
 case (current_state)

 s0 : if (data_in) next_state = s1;
        else next_state=s0  ;
```

The input and outputs are declared for the level to pulse converter.

Inputs are declared as clk, reset_n and data_in. The output is declared as y_out.

State parameters are declared as s0, s1. By using the net data type reg the current_state and next_state are declared.

The state register sequential block is sensitive to positive edge of clock.

During reset condition current_state'is assigned to s0 and during normal operation the next_state is assigned to current_state

The next state logic combinational block is sensitive to the current_state for the level to pulse converter.

The functionality is coded by using the cas' construct. Please refer the state transition table

Example 2 Verilog RTL for level-to-pulse converter

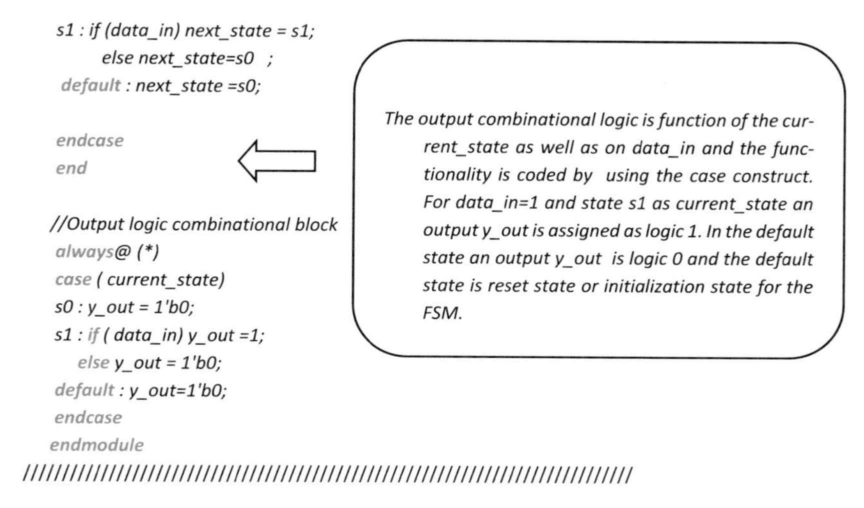

```
s1 : if (data_in) next_state = s1;
       else next_state=s0  ;
 default : next_state =s0;

endcase
end

//Output logic combinational block
always@ (*)
case ( current_state)
s0 : y_out = 1'b0;
s1 : if ( data_in) y_out =1;
    else y_out = 1'b0;
default : y_out=1'b0;
endcase
endmodule
////////////////////////////////////////////////////////////////////////////////
```

The output combinational logic is function of the current_state as well as on data_in and the functionality is coded by using the case construct. For data_in=1 and state s1 as current_state an output y_out is assigned as logic 1. In the default state an output y_out is logic 0 and the default state is reset state or initialization state for the FSM.

Example 2 (continued)

As shown in the Verilog RTL, the output of level-to-pulse converter is function of changes in an input data_in and current_state. The Verilog RTL infers the logic structure as shown in Fig. 14.6.

The synthesized logic for the Mealy level-to-pulse converter is shown in Fig. 14.7, and it infers the register logic with the combinational structure at output. Thus, the output of Mealy level-to-pulse converter is function of the current_state and an input data_in.

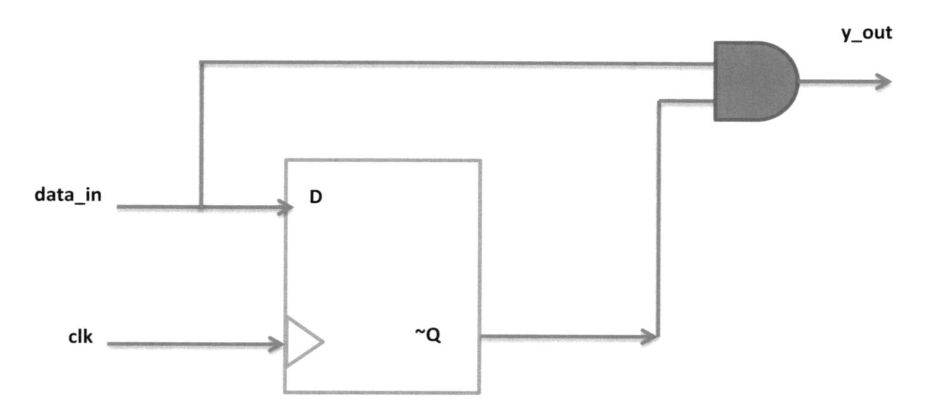

Fig. 14.6 Level-to-pulse converter logic diagram

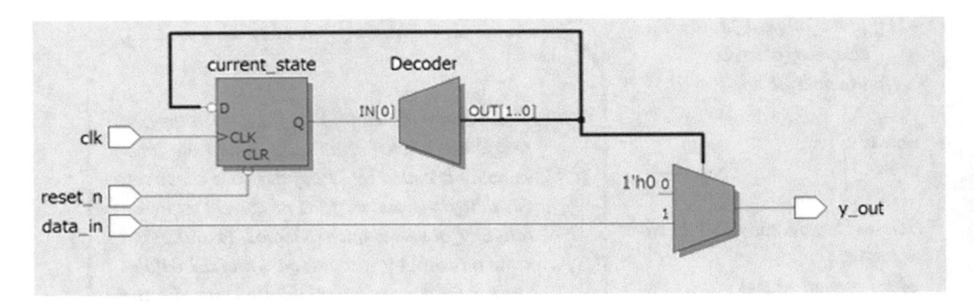

Fig. 14.7 Synthesized logic for level-to-pulse converter

14.2 FSM Encoding Styles

FSM can be coded by using many styles, and practically, there are three popular encoding styles used to code the FSMs. These styles are named as follows:

a. **Binary Encoding**: FSM can be described by using binary encoding styles, and by using this style, the number of flip-flops or 1-bit registers used is equal to $\log_2 N$. Where N are number of states. Consider an FSM has four states then the number of flip-flops equal to $\log_2 4$ which is equal to 2.
b. **Gray Encoding**: FSM can be efficiently described by using gray encoding technique, and in this style, the gray codes are used to represent the states. The number of flip-flops or 1-bit registers used is equal to $\log_2 N$. Where N are number of states. Consider an FSM has four states then the number of 1-bit registers equal to $\log_2 4$ which is equal to 2.
c. **One-hot encoding**: FSMs can be efficiently coded by using one-hot encoding style. One hot indicates that only one bit is active at a time or hot at a time. The number of flip-flops or 1-bit registers used is equal to N. Where N are number of states. Consider an FSM has four states then the number of flip-flops required are equal to 4. This style requires more area, but advantage is that it has clean register to register path, and it makes STA quite simple. If FSM has 16 states, then one-hot encoding needs 16 flip-flops.

The comparison of different FSM encoding styles for state machine having 16 states is shown in the following Table 14.4

The encoding styles for four state FSMs is shown in Table 14.5.

Table 14.4 FSM encoding style comparison

FSM Encoding	Binary	Gray	One hot
Representation	Binary number	Gray number	16 bit one-hot number
No. of 1-bit registers (flip-flops)	4	4	16
Area	Same as gray encoding	Same as binary encoding	More

Table 14.5 FSM state representation

FSM states	Binary	Gray	One hot
S0	00	00	0001
S1	01	01	0010
S2	10	11	0100
S3	11	10	1000

14.2.1 Binary Encoding

As discussed earlier, the binary encoding style can be used if the area optimization is the goal. In this encoding style, state parameters for the binary encoding are represented in the binary format.

14.2.1.1 Two-Bit Binary Counter FSM

Two-bit binary counter FSM is coded and shown in the Example 3. As described in the example, the number of states is equal to 4, and it needs four state variables s0, s1, s2, and s3. The number of flip-flops used to represent the functionality of counter is equal to 2.

The state transition table is shown below in Table 14.6. The state transition diagram is shown in Fig. 14.8. The transition from one state to another state occurs on the positive edge of clock. Default state is s0, and it is reset state.

Table 14.6 State transition table for binary encoding

current_state	next_state
s0 = 00	s1 = 01
s1 = 01	s2 = 10
s2 = 10	s3 = 11
s3 = 11	s0 = 00

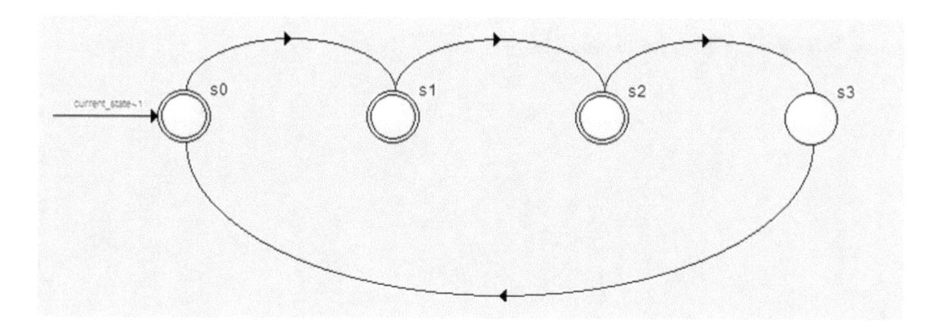

Fig. 14.8 State transition diagram for two-bit binary counter

//

```
//FSM of 2-bit binary up-counter
module binary_2_bit_counter (
input clk,
input reset_n,
output reg [1:0] y_out );
parameter s0=2'b00;
parameter s1=2'b01;
parameter s2 =2'b10;
parameter s3 = 2'b11;

reg [1:0] current_state;
reg [1:0] next_state;

//State register logic

always@ (posedge clk , negedge reset_n)
begin
 if (~reset_n)
   current_state <= s0;
 else
   current_state <= next_state;
 end

//Next state combinational logic
always @ (current_state)
begin
 case (current_state)
 s0 : next_state = s1;
 s1 : next_state =s2;
 s2 : next_state =s3;
 s3 : next_state =s0;
 default : next_state =s0;
 endcase
 end
```

The state parameters are declared as so, s1, s2, s3.

The binary encoding style is used and the reg data type is used to declare the current_state and next_state.

The state register sequential block is sensitive to positive edge of clock.

During reset condition current_state is assigned to s0 and during normal operation the next_state is assigned as current_state

The next state logic combinational block is sensitive to the current_state for the two-bit binary counter

The functionality is coded by using the case construct. Please refer the state transition table. Default state is s0.

Example 3 Verilog RTL for two-bit binary counter

```
//Output combinational logic
always@ (current_state)
case ( current_state)
s0 : y_out = 2'b00;
s1 : y_out = 2'b01;
s2 : y_out = 2'b10;
s3 : y_out = 2'b11;
default : y_out=2'b00;

 endcase
endmodule
```

The output combinational logic is function of the current_state and the functionality is described using the case construct. For example, state s1 as current_state an output y_out is assigned as binary 01. In the default state an output y_out is binary 00 and the default state is reset state or initialization state for the FSM.

//

Example 3 (continued)

The synthesized logic for the two-bit binary counter is shown in Fig. 14.9. As shown in the figure, the state register is triggered on the positive edge of the clock and has active low asynchronous reset reset_n. The output combinational logic is decoding structure due to the use of ***case*** construct'.

The state machine coded works on the positive edge of the clock and generates an output sequence as 00, 01, 10, 11, 00.... Fig. 14.10 shows the output sequence advanced on the rising edge of the clock during reset_n = 1 duration.

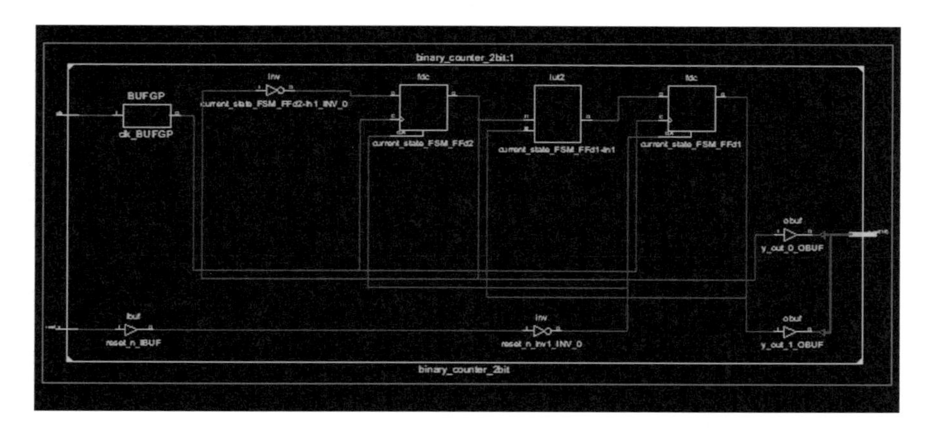

Fig. 14.9 Synthesized logic for two-bit counter

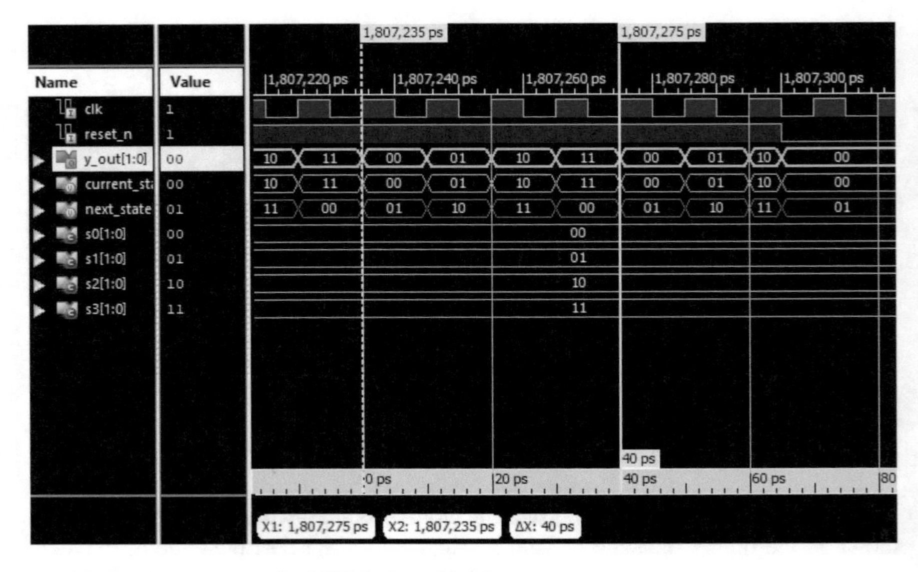

Fig. 14.10 Simulation result of FSM of two-bit binary counter

14.2.2 Gray Encoding

As discussed earlier, the gray encoding style can be used if the area requirement is a constraint on the design. In this encoding style, state parameters are represented in the gray format.

14.2.2.1 Two-Bit Gray Counter FSM

Two-bit gray counter FSM is coded using synthesizable Verilog constructs and shown in the Example 4. As shown in the example, the number of states is equal to 4 and has four state variables s0, s1, s2, and s3. The number of flip-flops used to design the functionality of counter is equal to 2.

The state transition table is shown below in Table 14.7. The state transition diagram is shown in Fig. 14.11. The transition from one state to another state occurs on the positive edge of clock. Default state is s0, and it is reset state.

Table 14.7 State transition table for gray encoding

current_state	next_state
s0 = 00	s1 = 01
s1 = 01	s3 = 11
s3 = 11	s2 = 10
s2 = 10	s0 = 00

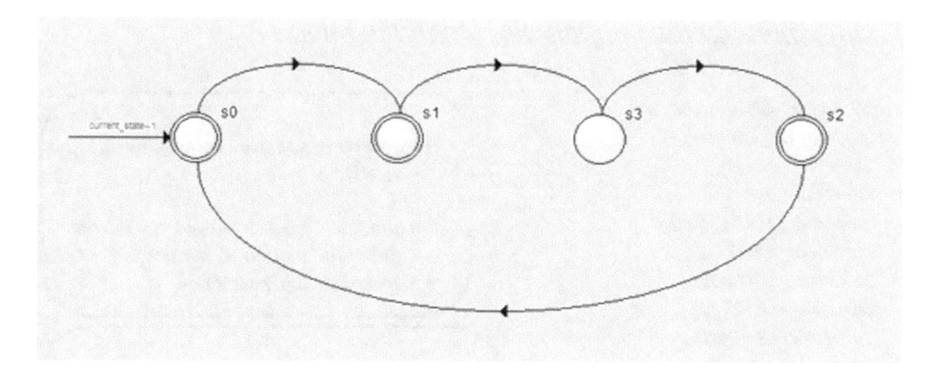

Fig. 14.11 State transition diagram for gray encoding

The synthesized logic for the two-bit gray counter is shown in Fig. 14.12. As shown in the figure, the state register is sensitive to the positive edge of the clock and has active low asynchronous reset reset_n. The output combinational logic is decoding structure and inferred due to the use of the ***case*** construct'.

The state machine coded works on the positive edge of the clock and generates an output sequence as 00, 01, 11, 10, 00.... Fig. 14.13 shows the output sequence advanced on the rising edge of the clock to get the next gray number during reset_n = 1 duration.

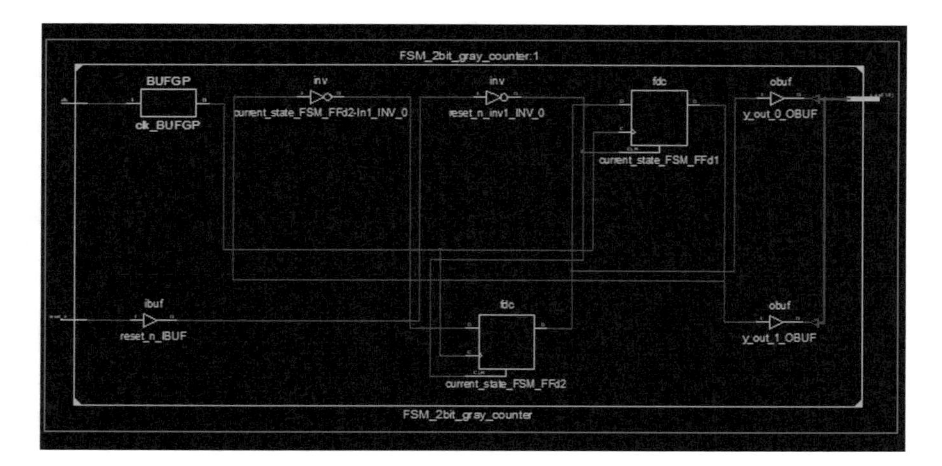

Fig. 14.12 Synthesized logic for the gray encoding style

//

```
//FSM for 2-bit gray counter
module gray_counter (
input clk,
input reset_n,
output  reg [1:0] y_out );
parameter s0=2'b00;
parameter s1=2'b01;
parameter s2 =2'b10;
parameter s3 = 2'b11;
reg [1:0] current_state;
reg [1:0] next_state;

//State register logic
always@ (posedge clk , negedge reset_n)
begin
 if (~reset_n)
   current_state <= s0;
 else
   current_state <= next_state;
 end


//Next state combinational logic
 always @ (current_state)
 begin
  case (current_state)
  s0 : next_state = s1;
  s1 : next_state =s3;
  s3 : next_state =s2;
  s2 : next_state =s0;
  default : next_state =s0;
  endcase
  end
  //Output combinational logic
  always@ (current_state)
  case ( current_state)
  s0 : y_out = 2'b00;
  s1 : y_out = 2'b01;
  s3 : y_out = 2'b11;
  s2 : y_out = 2'b10;
  default : y_out=2'b00;
  endcase
 endmodule
```

The state parameters are declared as so, s1, s2, s3.

The gray encoding style is used and the reg data type is used to declare the current_state and next_state.

The state register sequential block is sensitive to positive edge of clock.

During reset condition current_state is assigned to s0 and during normal operation the next_state is assigned to current_state

The next state logic combinational block is sensitive to the current_state for the two-bit gray counter

The functionality is coded by using the case construct. Please refer the state transition table. Default state is s0.

The output combinational logic is function of the current_state and the functionality is coded by using the case construct. For example, state s3 as current_state an output y_out is assigned to binary 11. In the default state an output y_out is binary 00 and the default state is reset state or initialization state for the FSM.

//

Example 4 Verilog RTL with gray encoding style

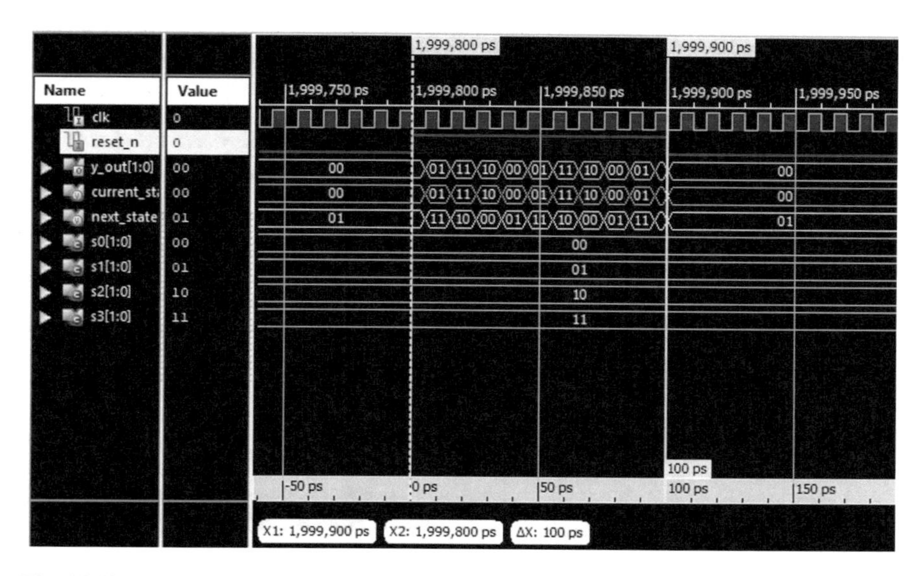

Fig. 14.13 Simulation result of FSM of two-bit gray counter

14.3 One-Hot Encoding

Two-bit counter FSM using one-hot encoding is coded and shown in the Example 5. As shown in the example, the number of states is equal to 4, and it needs four state variables, namely s0, s1, s2, and s3. The number of flip-flops used to design the functionality of counter is equal to 4 as one-hot encoding style is used.

The state transition is shown in Table 14.8. The transition from one state to another state occurs on the positive edge of clock. Default state is s0, and it is reset state.

The synthesized logic for the two-bit binary counter using one-hot encoding is shown in Fig. 14.14. As shown in the figure, the state register is sensitive to the positive edge of the clock and has active low asynchronous reset reset_n. This encoding method uses the four flip-flops to infer the design. The output combinational logic is decoding structure to generate 2-bit output using the ***case*** construct'.

The state machine coded works on the positive edge of the clock and generates an output sequence as 00, 01, 10, 11, 00.... Fig. 14.15 shows the output sequence advanced on the rising edge of the clock during reset_n = 1 duration. The encoding style used is one-hot encoding.

```
/////////////////////////////////////////////////////////////////////////////

//FSM for one-hot encoding
module one_hot_encoding (
input clk,
input reset_n,
output reg [1:0] y_out);
parameter [3:0] s0=4'b0001;
parameter [3:0] s1=4'b0010;
parameter [3:0] s2 =4'b0100;
parameter [3:0] s3 = 4'b1000;
reg [3:0] current_state;
reg [3:0] next_state;

//State register logic
always@ (posedge clk , negedge reset_n)
begin
 if (~reset_n)

   current_state <= s0;
  else
   current_state <= next_state;
  end

//next state logic
always @ (current_state)
begin
 case (current_state)
 s0 : next_state =s1;
 s1 : next_state =s2;
 s2 : next_state =s3;
 s3 : next_state =s0;
 default : next_state =s0;
 endcase
 end

//Output combinational logic
 always@ (current_state)
 case ( current_state)
 s0 : y_out = 2'b00;
 s1 : y_out = 2'b01;
 s2 : y_out = 2'b10;
 s3 : y_out = 2'b11;
 default : y_out=2'b00;
 endcase
endmodule
/////////////////////////////////////////////////////////////////////////////
```

The state parameters are declared as so, s1, s2, s3.

The one-hot encoding style is used and the reg data type is used to declare the current_state and next_state. The reg is 4 bit wide.

The state register sequential block is sensitive to positive edge of clock.

During reset condition current_state is assigned to s0 and during normal operation the next_state is assigned to current_state

The next state logic combinational block is sensitive to the current_state for the two-bit binary counter

The functionality is coded by using the case construct. Please refer the state transition table. Default state is s0.

The output combinational logic is function of the current_state and the functionality is coded by using the case construct. For example, state s3 as current_state an output y_out is assigned to binary 11. In the default state an output y_out is binary 00 and the default state is reset state or initialization state for the FSM.

Example 5 Verilog RTL for one-hot encoding

Table 14.8 State transition table for one-hot encoding

current_state	next_state
s0 = 0001	s1 = 0010
s1 = 0010	s2 = 0100
s2 = 0100	S3 = 1000
s3 = 1000	s0 = 0001

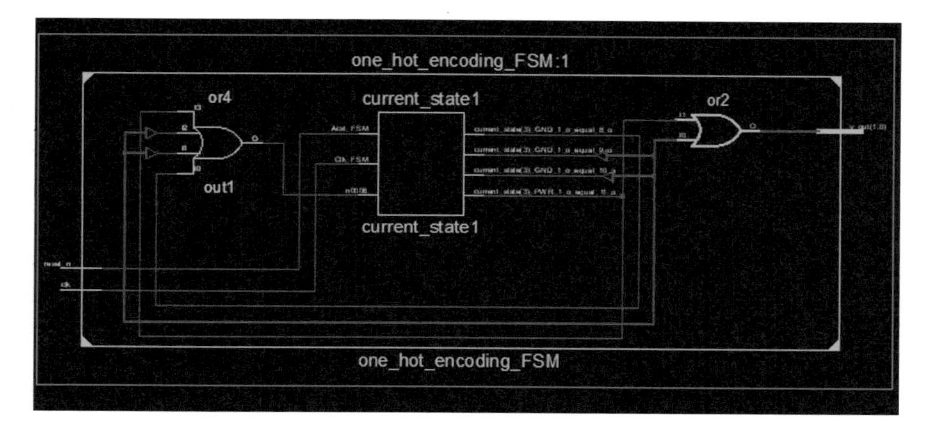

Fig. 14.14 Synthesized logic for one-hot encoding

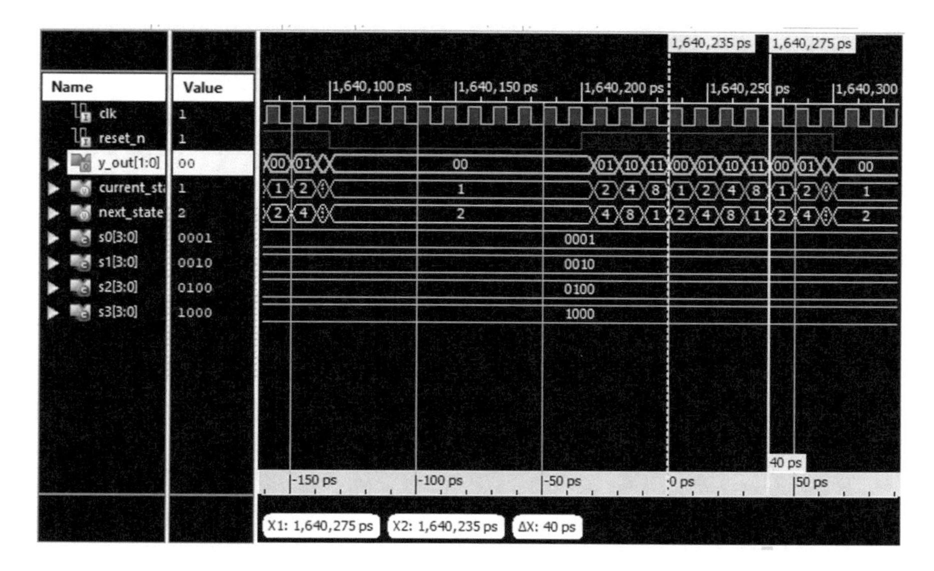

Fig. 14.15 Simulation result of FSM of two-bit binary counter using one-hot encoding

14.4 Sequence Detectors Using FSMs

FSMs are used to code the functionality of sequence detectors. The efficient RTL coding using Verilog is used in the practical scenario to get the correct output for the desired sequence. Depending on the requirements, either Moore or Mealy machines can be used to detect the correct sequence.

14.4.1 Mealy Sequence Detector Using Two always Procedural Blocks

The state transition table for the sequence detector is shown below with the desired output (Table 14.9).

The synthesizable Verilog RTL is shown in the Example 6 to detect the sequence 1010.

The state machine coded works on the positive edge of the clock and has active low asynchronous reset and used to detect the sequence 1010. Fig. 14.16 shows the output sequence to detect the sequence 1010 during reset_n = 1 duration. The encoding style used is binary encoding.

The synthesizable Mealy machine sequence detector infers the state register sequential logic having two flip-flops and combinational decoding logic. In the design of Mealy sequence detector an output is function of the state, and input changes. The synthesized logic is shown in Fig. 14.17.

Table 14.9 State transition table for sequence detector

Current_state	Input	Next_state	output
S0 = 00	1	S1 = 01	2'b00
S1 = 01	0	S2 = 10	2'b01
S2 = 10	1	S3 = 11	2'b10
S3 = 11	0	S0 = 00	2'b11

```
////////////////////////////////////////////////////////////////////////////

//sequence detector to detect 1010
module sequence_detector_1010 (
input clk, data_ in, reset_n,
output reg [1:0] y_ out );
reg [1:0] state; // Declare state register
parameter S0 = 0, S1 = 1, S2 = 2, S3 = 3; // Declare states

always @ (posedge clk , negedge reset_n)
begin
if (~reset_n)
state <= S0;
else
case (state)
S0: if (data_in)
begin
state <= S1;
end
else begin
state <= S0;
end
S1: if (~data_in)
begin
state <= S2;
end
else
begin
state <= S1;
end
S2: if (data_in)
begin
state <= S3;
end
else
begin
state <= S2;
end
S3: if (~data_in)
begin
state <= S0;
end
else
begin
state <= S3;
```

The state register sequential block is sensitive to positive edge of clock.

During reset condition state is assigned to S0 and during normal operation the state is assigned to desired state. Please refer the state transition table

Example 6 Verilog RTL for the sequence detector

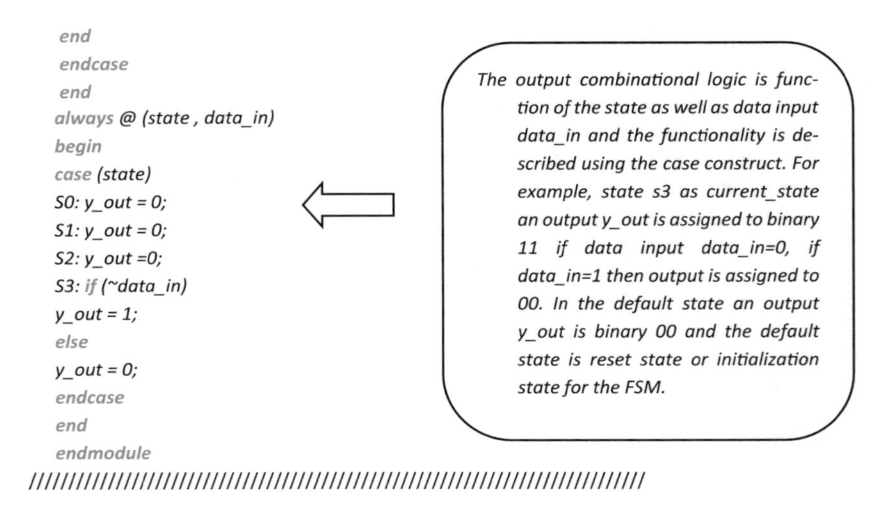

```
    end
    endcase
    end
  always @ (state , data_in)
  begin
  case (state)
  S0: y_out = 0;
  S1: y_out = 0;
  S2: y_out =0;
  S3: if (~data_in)
  y_out = 1;
  else
  y_out = 0;
  endcase
  end
  endmodule
//////////////////////////////////////////////////////////////////////////////
```

The output combinational logic is function of the state as well as data input data_in and the functionality is described using the case construct. For example, state s3 as current_state an output y_out is assigned to binary 11 if data input data_in=0, if data_in=1 then output is assigned to 00. In the default state an output y_out is binary 00 and the default state is reset state or initialization state for the FSM.

Example 6 (continued)

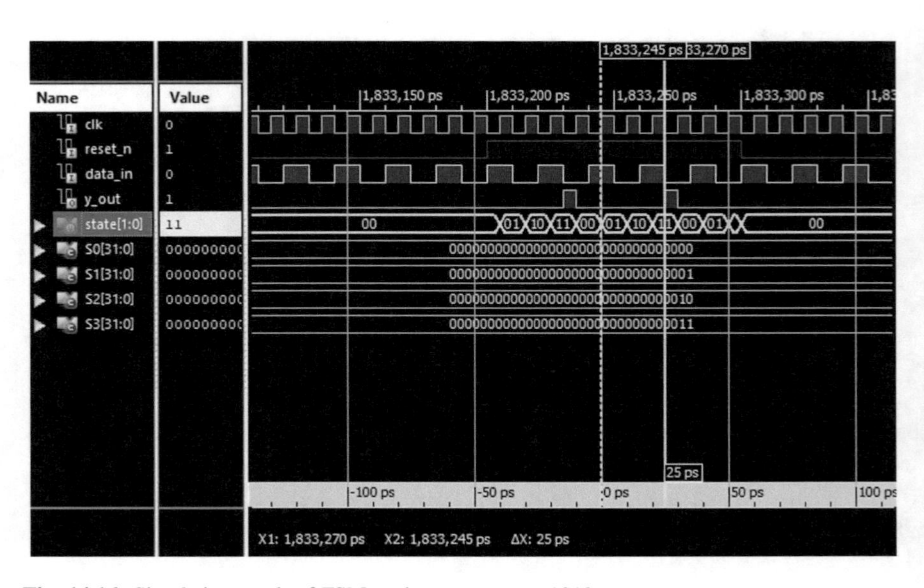

Fig. 14.16 Simulation result of FSM to detect sequence 1010

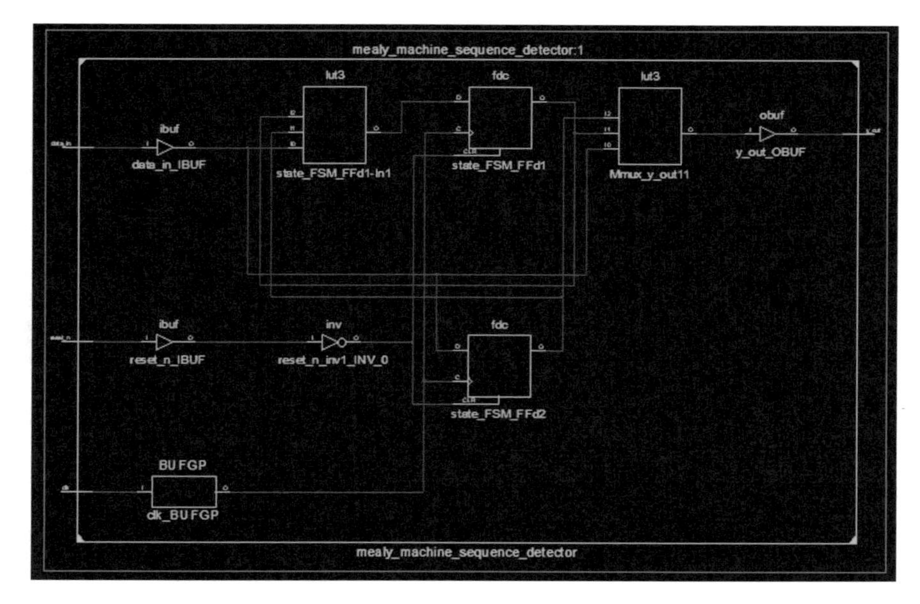

Fig. 14.17 Synthesized logic for sequence detector

14.4.2 Mealy Machine: Sequence Detector to Detect 101 Overlapping Sequence

Another Mealy machine sequence detector to detect the overlapping sequence 101 is shown in the state diagram Fig. 14.18. For the overlapping sequence of 101 also, this works and generates an output y_out = 1 after detecting the sequence 101. The state transition table is shown in Table 14.10.

As shown in Table 14.10, the Mealy machine output y_out is active high when input sequence 101 is detected. Output is function of the current state and changes in the input.

The Verilog RTL for the Mealy sequence detector is shown in the Example 7.

The inferred logic uses the two flip-flops and LUTs for the combinational logic and shown in Fig. 14.19.

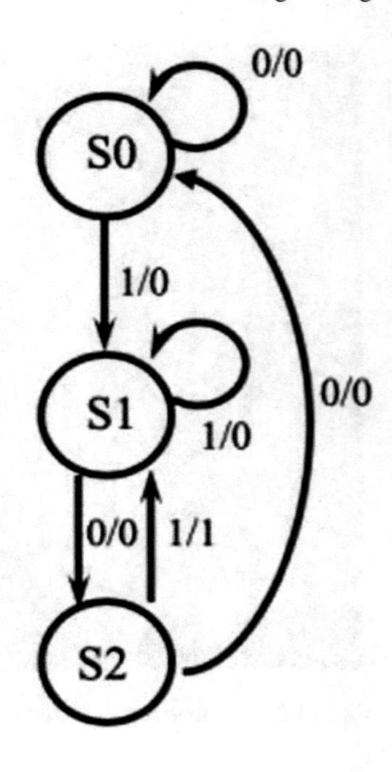

Fig. 14.18 Mealy machine state diagram for 101 sequence

Table 14.10 State transition table for sequence detector 101

Input	Current _state	Next_state	output
1	S0	S1	0
0	S1	S2	0
1	S2	S1	1

The state machine coded works on the positive edge of the clock and has active low asynchronous reset and used to detect the overlapping sequence 101. Fig. 14.20 shows the output to detect the sequence 101 during reset_n = 1 duration. The encoding style used is binary encoding.

```
///////////////////////////////////////////////////////////////////////////
//Mealy FSM to detect 101 overlapping sequence
module sequence_detector_101 (

    input clk,
    input reset_n,
    input data_in,
    output reg  y_out);
    parameter s0=2'b00;
    parameter s1=2'b01;
    parameter s2 =2'b10;
    reg [1:0] current_state;
    reg [1:0] next_state;

    //State register logic

    always@ (posedge clk , negedge reset_n)
    begin
     if (~reset_n)
       current_state <= s0;
     else
       current_state <= next_state;
     end

    //Next state combinational logic
    always @ (current_state, data_in)
    begin
     case (current_state)
     s0 : if (data_in) next_state = s1;
          else next_state=s0;
     s1 : if (~data_in) next_state =s2;
          else next_state=s1;
     s2 : if (data_in) next_state =s1;
          else next_state=s0;
     default : next_state =s0;
     endcase
     end

     //Output combinational logic
     always@ (current_state, data_in)
     case ( current_state)
     s0 : y_out = 0;
     s1 : y_out = 0;
     s2 : if (data_in) y_out=1;
```

The state register sequential block is sensitive to positive edge of clock.

During reset condition current_state is assigned to 's0' and during normal operation the next_state is assigned to current_state

The next state logic combinational block is sensitive to the current_state as well as data_in for the sequence detector.

The functionality is coded by using the case construct. Please refer the state transition table. Default state is s0.

The output combinational logic is function of the current_state and the functionality is coded by using the case construct. For example, state s2 as current_state then if data_in is logic 1 then output y_out is assigned as logic 1. The default state is reset state or initialization state for the FSM.

Example 7 Sequence detector example to detect overlapping 101 sequence

```
    else y_out=0;
  default : y_out=2'b0;
  endcase
 endmodule
```

//

Example 7 (continued)

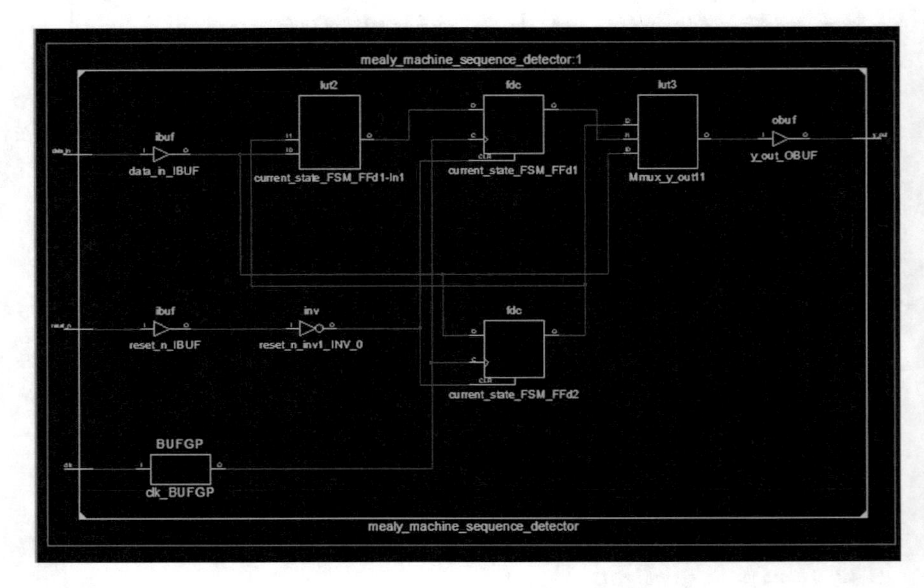

Fig. 14.19 Synthesis result of sequence detector 101 for mealy FSM

14.5 Improving the Design Performance for FSMs

The objective or goal during the FSM coding is efficient synthesis and better design timing. The reusability and use of the state encoding is another important point RTL designer needs to focus. Even the coding style should be compact as well as readable.

The following are important guidelines used to improve the FSM performance.

a. Do not use the single ***always*** block FSM. As the issue is in readability and it doesn't yield in the efficient synthesis results.
b. Use multiple procedural block FSMs. In practical ASIC designs, two or three ***always*** block FSMs are used as it improves the readability and reusability, and even it yields into the efficient synthesis results.

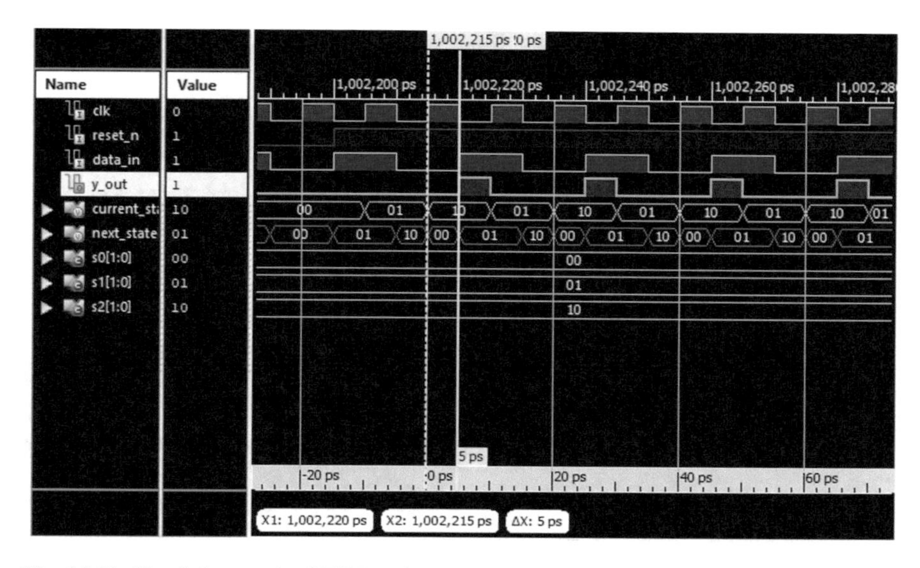

Fig. 14.20 Simulation result of FSM to detect sequence 101

c. Declare the state parameters according to the required state encoding, and then, declare the next_state and current_state.
d. Use non-blocking assignments for coding the state register logic.
e. Use blocking assignments for coding the next state combinational logic.
f. Use blocking assignments for coding the output combinational logic.
g. Include the ***default*** condition in the ***case*** construct to avoid the unintentional latches.
h. While using the ***if-else*** construct, the number of transitions in the state diagram should be same as number of if-else conditions.
i. Register the FSM outputs as it ensures that an output is glitch free.
j. For better and efficient synthesis outcome, use the one-hot encoding method.

14.6 Exercises

The exercises are based on the understanding of synthesizable Verilog constructs. Complete the exercises for better understanding and application of Verilog constructs.

1. *Design the Moore FSM to detect the sequence 101*. Use three always block FSMs.
2. *Perform the simulation to check for the functional correctness of the design.*
3. Perform the synthesis to find out the resources used.

14.7 Summary

As discussed in this chapter, the following are important points to conclude the chapter.

1. FSMs are coded very efficiently by using Verilog RTL for better synthesis outcome.
2. FSMs are of two types: Moore and Mealy.
3. In the Moore type FSMs, the output is function of current state only.
4. In the Mealy FSMs, the output is function of current state as well as inputs.
5. FSM encoding styles are binary, gray, and one hot.
6. One-hot encoding style is used for glitch-free outputs and yields the better and clean synthesis.
7. In ASIC designs, two or three always block FSMs are used to generate efficient synthesis.
8. Register the FSM outputs as it ensures that an output is glitch free.
9. For better and efficient synthesis outcome, use the one-hot encoding method.
10. Use non-blocking assignments for coding the state register logic.
11. Use blocking assignments for coding the next state combinational logic.
12. Use blocking assignments for coding the output combinational logic.
13. Include the ***default*** condition in the ***case*** construct to avoid the unintentional latches.

Chapter 15
Non-synthesizable Verilog Constructs and Testbenches

The chapter discusses about the inter-delay, intra-delay assignments and other non-synthesizable constructs useful during the testbenches. The chapter is useful to understand about the non-synthesizable constructs and how to check for the functional correctness of the design.

As discussed in the previous chapters, we have used the synthesizable Verilog constructs during the RTL design. During the RTL verification phase, the objective is to check for the functional correctness of the design using non-synthesizable constructs. The following few sections are useful to understand the use of non-synthesizable constructs during design verification.

15.1 Intra-delay and Inter-delay Assignments

The RTL design using Verilog should be verified to check for the functional correctness of the design. To verify the RTL design functionality, the testbench needs to be coded using the non-synthesizable Verilog constructs. Please refer Appendix III for the non-synthesizable Verilog constructs.

15.1.1 Simulation for Blocking Assignments

Consider the Verilog code having the blocking assignments without the delay and shown in Example 1.

V. Taraate, *Digital Logic Design Using Verilog*,
https://doi.org/10.1007/978-981-16-3199-3_15

```
///////////////////////////////////////////////////////////////////////////////

module test_design;
reg clk;
reg [7:0] a,b,c,d;
always #10 clk = ~clk;
always@(posedge clk)
begin
    a=b+c;
    b=a+d;
    c=a+b;
end
initial
begin
    clk=0;
    a=8'h2;
    b=8'h3;
    c=8'h4;
    d=8'h5;
end
endmodule

///////////////////////////////////////////////////////////////////////////////
```

Example 1 Simulation of Verilog blocking assignment

In Example 1, the procedural always block executes every time on the event on the clock clk. The initial block executes only once and used to update the values of a, b, c, and d. The simulation result is shown in the following Waveform 1.

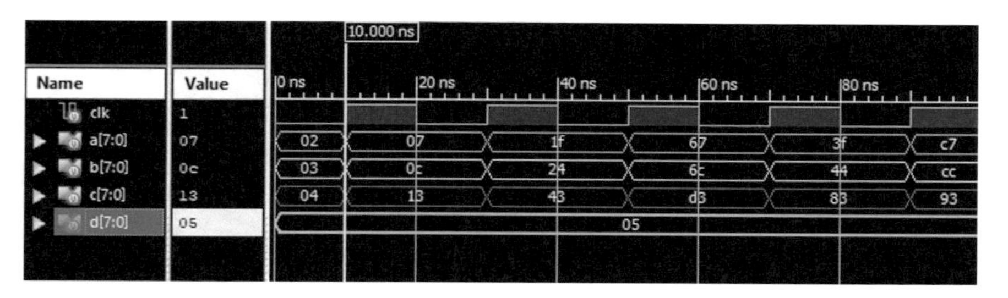

Waveform 1 Simulation result for Verilog blocking assignment

15.1.2 Simulation of Non-blocking Assignments

Consider the Verilog code having the non-blocking assignments shown in Example 2.

```
/////////////////////////////////////////////////////////////////////////////////

module test_design;
reg clk;
reg [7:0] a,b,c,d;
always #10 clk = ~clk;
always@(posedge clk)
begin
    a <=b+c;
    b <=a+d;
    c <=a+b;
end
initial
begin
   clk=0;
   a=8'h2;
   b=8'h3;
   c=8'h4;
   d=8'h5;
end
endmodule
/////////////////////////////////////////////////////////////////////////////////
```

Example 2 : Simulation for Verilog non-blocking assignment

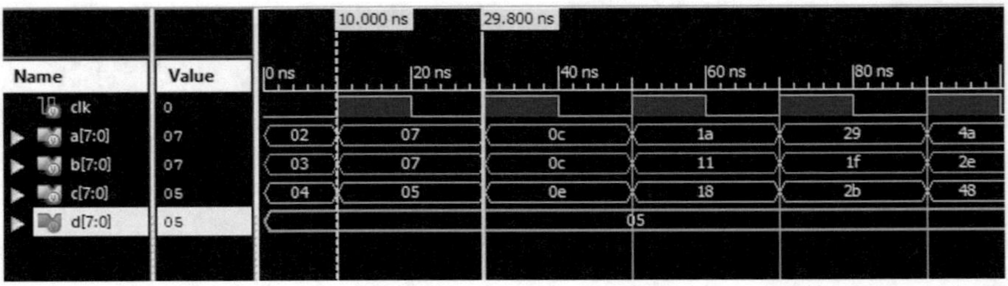

Waveform 2 Simulation result for Verilog non-blocking assignment

Table 1 Difference between initial and always block

Initial	Always
In this block assignment executes in the 0-simulation time and continues for the next specified sequence	In this block assignments continues to execute in simulation time 0 and repeats forever depending on the sensitivity list event
This block is executed only once, and the simulation stops at the end of this block	The simulation in this block continues forever. If wait construct is there, then it will be held during simulation session
It is non-synthesizable construct	It is synthesizable construct

The simulation result for Example 2 is shown in the following Waveform 2.

15.2 The always and initial Procedural Block

We have discussed about the use of the always procedural block to code the RTL design. The initial procedural block is used in the testbenches to generate stimulus at various time stamp.

The difference in the initial and always block is documented in Table 1.

15.2.1 Blocking Assignments with Inter-assignment Delays

In the inter-assignment delays with the blocking assignment, it delays both the evaluation of the assignment and update of the assignment.

Consider the Verilog code shown in Example 3.

```
//////////////////////////////////////////////////////////////////////////
module test_design;
reg clk;
reg [7:0] a,b,c,d;
always #10 clk = ~clk;
always@(posedge clk)
begin
  b=a+a;
#3 c=b+a;
#1 d=c+a;
end
initial
begin
    clk =0;
    a=4;
    b=3;
    c=2;
   d=5;
end
endmodule
//////////////////////////////////////////////////////////////////////////
```

Example 3 Verilog blocking assignment with inter-assignment delay

The following Waveform 3 is useful to understand the simulation results for the blocking assignment having the inter-assignment delays.

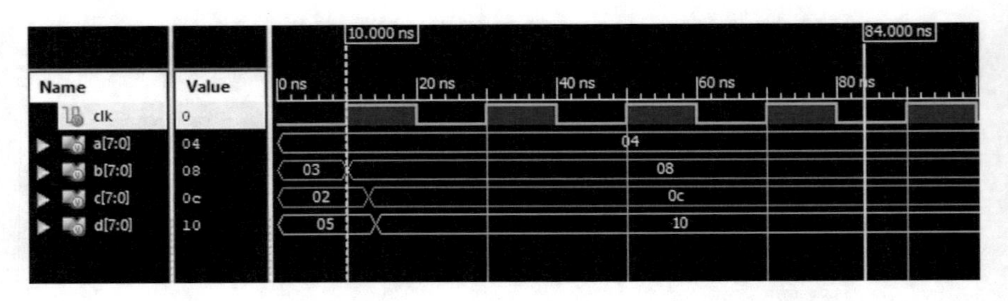

Waveform 3 Simulation result for the Verilog blocking assignment with inter-assignment delay

15.2.2 Blocking Assignments with Intra-assignment Delays

In the intra-assignment delays with the blocking assignment, it delays the evaluation of the assignment but not the update of the assignment.

Consider the Verilog code shown in Example 4.

The Waveform 4 gives the simulation results for the blocking assignment with the intra-assignment delays.

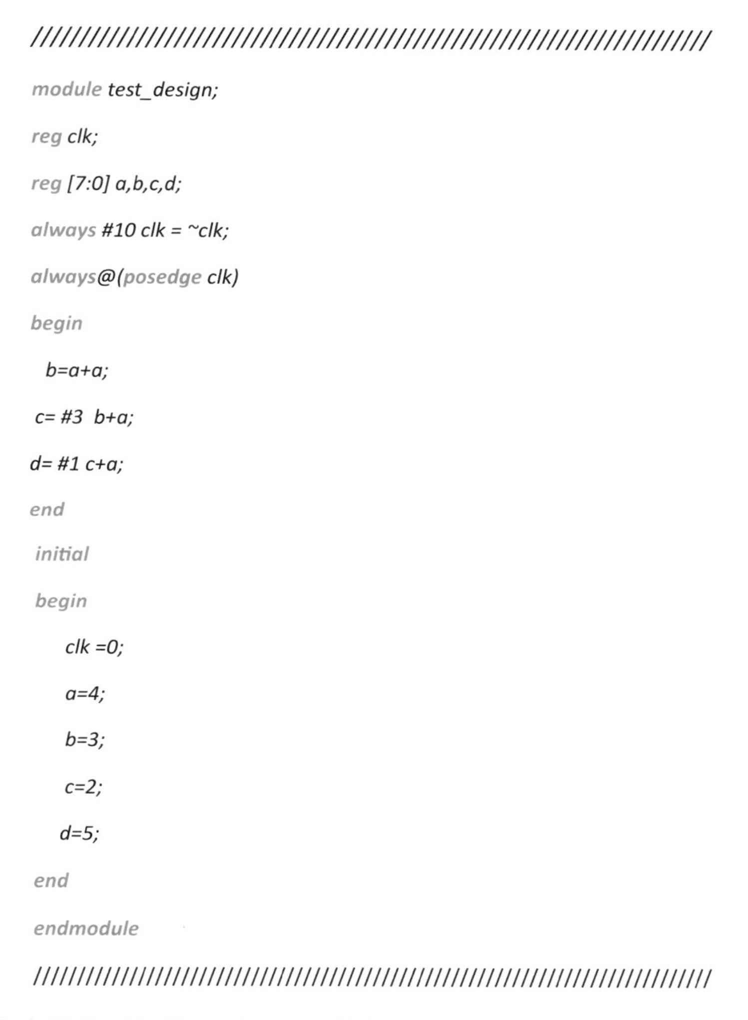

```
//////////////////////////////////////////////////////////////////////////////
module test_design;
reg clk;
reg [7:0] a,b,c,d;
always #10 clk = ~clk;
always@(posedge clk)
begin
  b=a+a;
 c= #3  b+a;
d= #1 c+a;
end
 initial
 begin
     clk =0;
     a=4;
     b=3;
     c=2;
    d=5;
 end
 endmodule
//////////////////////////////////////////////////////////////////////////////
```

Example 4 Verilog blocking assignment with intra-assignment delay

Waveform 4 Simulation result for the Verilog blocking assignment with intra-assignment delay

15.2.3 Non-blocking Assignments with Inter-assignment Delays

Using the intra-assignment delays with the non-blocking assignment, it delays both the evaluation of the assignment and the update of the assignment.

Consider the Verilog code shown in Example 5.

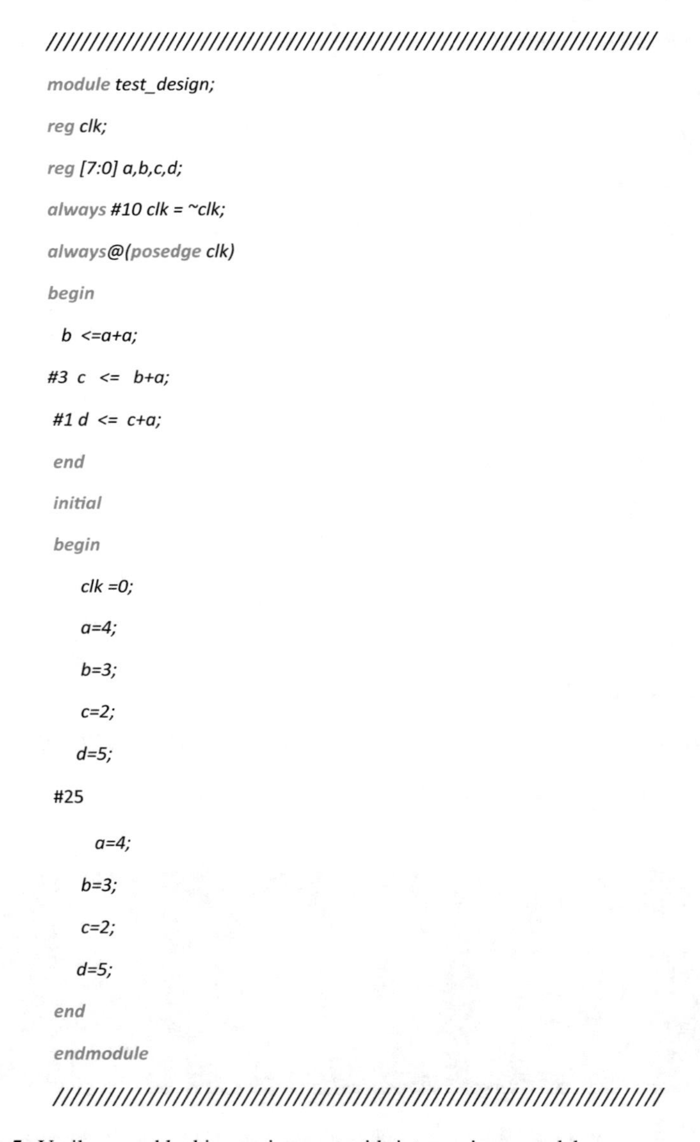

```
//////////////////////////////////////////////////////////////////////
module test_design;
reg clk;
reg [7:0] a,b,c,d;
always #10 clk = ~clk;
always@(posedge clk)
begin
  b <=a+a;
#3 c  <=  b+a;
 #1 d <=  c+a;
 end
 initial
 begin
    clk =0;
    a=4;
    b=3;
    c=2;
   d=5;
 #25
      a=4;
    b=3;
    c=2;
   d=5;
 end
 endmodule
//////////////////////////////////////////////////////////////////////
```

Example 5 Verilog non-blocking assignment with inter-assignment delay

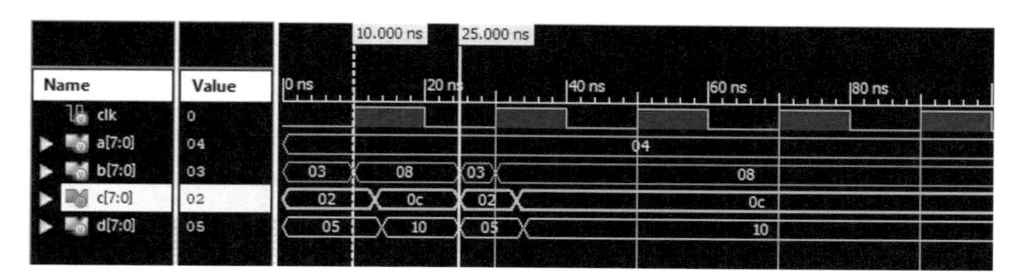

Waveform 5 Simulation result for the Verilog non-blocking assignment with inter-assignment delay

The Waveform 5 gives the simulation results for the non-blocking assignment with the inter-assignment delays.

15.2.4 Non-blocking Assignments with Intra-assignment Delays

In the intra-assignment delays with the blocking assignment, it delays the update of the assignment but not the evaluation of the assignment.

Consider the Verilog code shown in Example 6.

The Waveform 6 gives the simulation results for the non-blocking assignment with the intra-assignment delays.

```
//////////////////////////////////////////////////////////////////////////
module test_design;
reg clk;
reg [7:0] a,b,c,d;
always #10 clk = ~clk;
always@(posedge clk)
begin
  b <=a+a;
 c  <= #3  b+a;
d  <= #1 c+a;
end
initial
begin
    clk =0;
    a=4;
    b=3;
    c=2;
   d=5;
#25
      a=4;
    b=3;
    c=2;
   d=5;
end
endmodule
//////////////////////////////////////////////////////////////////////////
```

Example 6 Verilog non-blocking assignment with intra-assignment delay

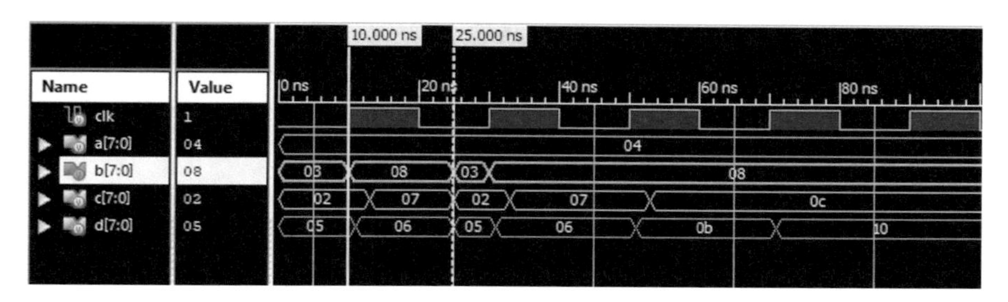

Waveform 6 Simulation result for the Verilog non-blocking assignment with intra-assignment delay

15.3 Role of Testbenches

In Chaps. 1, 2, 3, 4, 5, 6, 7, 8, 9, 10, 11, 12, 13 and 14, we have discussed about the RTL design synthesis and logic inferred. Verilog HDL is powerful for the simulation of the design. By using non-synthesizable constructs, the Verilog Design Under Verification (DUV) can be verified to find out functional correctness of the design. Consider Verilog Design of BCD up–down counter having inputs as clk and reset_n. The counter has four-bit output q_out [3:0]. The RTL description of BCD counter is shown in Example 7.

The testbench to test the functional correctness of the BCD up–down counter should generate the stimulus at the clk, reset_n, and at up_down. The objective is to monitor the count at the output. The testbench using the non-synthesizable constructs is shown in Example 8 and uses to pass the stimulus to the UUT, where UUT is Unit Under Test.

The testbench generates the results shown in the following Waveform 7. As shown the counter counts from 0 to 9 for up_down = 1 and from 9 to 0 for up_down = 0.

As discussed above, the basic simulation can be carried out by coding the testbench which can force the stimulus to the design under test. For the moderate complex FPGA designs this approach can work. But for large density SOC designs it is essential to use the sophisticated self-checking testbenches. It is essential for the verification engineer to understand about the creation of the test cases, test plans, and test vectors. Even the best industry practice is use of the verification architectures which has components like drivers, monitors, and checkers. This discussion is out of scope and reader can refer the RTL verification books and other literature.

```
////////////////////////////////////////////////////////////////////////////////

module bcd_up_down_counter (
                input clk, reset_n,
                input up_down ,
                output reg [3:0] count
                 );
always @ (posedge clk, negedge reset_n)
begin
  if ( ~reset_n)
  count <= 4'b0000;
  else
    if ( up_down)
      if ( count == 4'b1001)
      count <= 4'b0000;
      else
      count <= count + 1;
    else
      if ( count == 4'b0000)
          count <= 4'b1001;
          else
          count <= count - 1;
 end
endmodule
////////////////////////////////////////////////////////////////////////////////
```

Example 7 Four-bit BCD up–down counter Verilog RTL

```
/////////////////////////////////////////////////////////////////////////////

//Verilog testbench to pass the data to up_down,  reset_n and clk to BCD
counter

`timescale 1ns/1ps

module test_BCD_counter;

reg clk;

 reg reset_n;

 reg up_down;

 wire [3:0] count;

 bcd_up_down_counter UUT (

                    .clk (clk),

                    .reset_n(reset_n),

                    .up_down(up_down) ,

                    .count(count)

                    );

 always  #10 clk= ~clk;

 always  #200 up_down = ~up_down;

 initial

 begin

  clk=0;

  reset_n=0;

  up_down = 0;

  #25;

  reset_n = 1;

  #400;

  reset_n = 0;

 end
endmodule

/////////////////////////////////////////////////////////////////////////////
```

Example 8 Testbench to check for the functional correctness of the BCD up–down counter

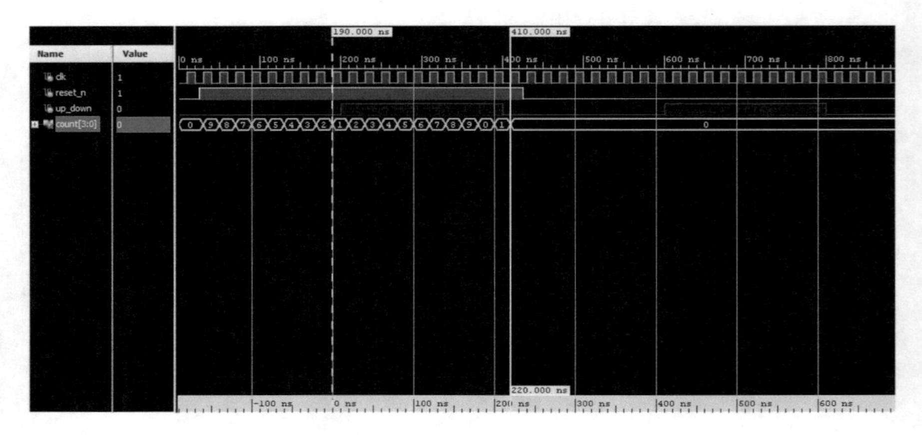

Waveform 7 Simulation result of BCD up–down counter

15.4 Multiple Assignments Within the begin–end

If multiple assignments need to be included within the procedural block, then we need to have the ***begin..end***. All the blocking assignments within ***begin..end*** are executed sequentially.

Consider the testbench (Example 9) which has the multiple assignments within the initial and always procedural block.

As shown in Example 9 a = 1 at 30 ns and b = 1 at 20 ns, these blocking assignments are executed sequentially as they are included within ***begin..end***. The simulation waveform is shown in Fig. 15.1, as first rising edge of clock is at 10 ns, the a = 1 at 40 ns and b = 1 at 40 + 20 = 60 ns.

```
////////////////////////////////////////////////////////////////////////////////
module multiple_assignments_begin_end;
reg clk;
reg a;
 reg b ;
always #10 clk=~clk;
  initial
  begin
    clk=0;
     a = 0;
     b = 0;
  end
  always @(posedge clk)
 begin
     #30 a = 1;
     #20 b = 1;
 end
endmodule
////////////////////////////////////////////////////////////////////////////////
```

Example 9 Verilog testbench which uses begin–end within always procedural blocks

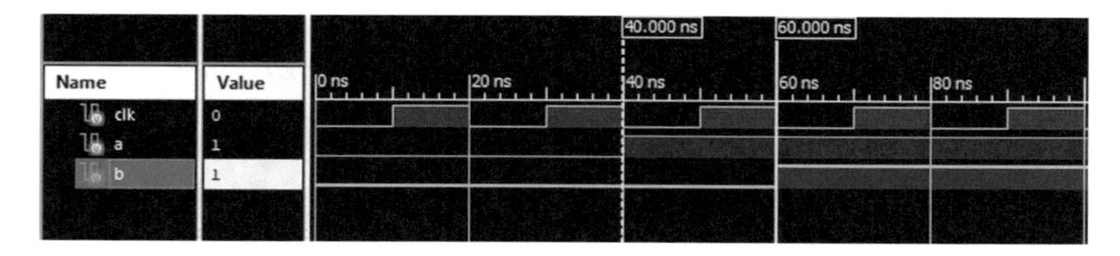

Fig. 15.1 Simulation result for Example 9

15.5 Multiple Assignments Within the fork–join

If multiple assignments need to be included within the procedural block then if we have the ***fork..join***. All the blocking assignments within ***fork..join*** are executed concurrently.

Consider the testbench (Example 10) which has the multiple assignments within the initial and always procedural block.

```
/////////////////////////////////////////////////////////////////////////////
module fork_join;
reg clk;
reg a;
reg b ;
always #10 clk=~clk;
  initial
  begin
     clk=0;
     a = 0;
     b = 0;
  end
  always @(posedge clk)
 fork
     #30 a = 1;
     #20 b = 1;
 join
endmodule
/////////////////////////////////////////////////////////////////////////////
```

Example 10 Verilog testbench which uses fork–join within always procedural blocks

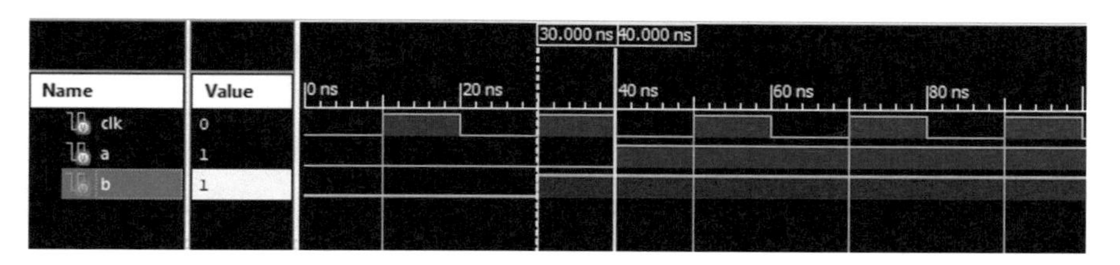

Fig. 15.2 Simulation result of Example 10

As shown in Fig. 15.2, a = 1 at 30 ns and b = 1 at 20 ns so these blocking assignments are executed concurrently as they are included within ***fork..join.*** The simulation waveform is shown in Fig. 15.2, as first rising edge of clock is at 10 ns, the b = 1 at 30 ns and a = 1 at 40 ns.

15.6 Display Tasks

If the RTL design has hundreds of the inputs and outputs, then using the waveforms it becomes difficult and time consuming to check for the functional correctness of the design. In such scenarios, we need to use the display tasks such as ***$monitor***, ***$display,*** and ***$strobe*** to print the result. The testbench which uses these display tasks is shown in Example 11.

```
////////////////////////////////////////////////////////////////////////////////
module tb_design;
 reg  a, b, c, y;
 initial
begin
   // let us initialize all signals to zero at time unit 0ns
  a <= 0;
  b <= 0;
  c <= 0;
  y <= 0;
  // at the end of the 10ns let us update the a=1 and c=1
  #10  a <= 1;
         c <= 1;
// $display the display task prints the value of the a,b,c,y in the order they
appear
$display (" time=%0t a=%0b b=%0b c=%0b y=%0b", $time, a, b, c, y);

// to print the values at the end of the current time stamp
$strobe (" time=%0t a=%0b b=%0b c=%0b y=%0b", $time, a, b, c, y);

  // wait for the 10ns and after that assign
 #10  y <=  a & b ^ c;
// To display the values automatically when variable or the expression
value changes
$monitor (" time=%0t a=%0b b=%0b c=%0b y=%0b", $time, a, b, c, y);
 end
 endmodule
////////////////////////////////////////////////////////////////////////////////
```

Example 11 Use of display tasks to print the values at various time stamp

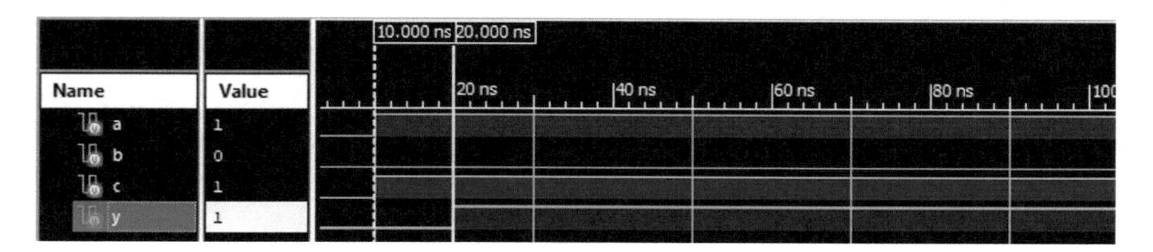

Fig. 15.3 Simulation result of Example 11

The ***$display*** is used to print the value immediately.
The ***$strobe*** is used to print the value at the current time stamp.
The ***$monitor*** to display the changes in the values at the end of every time stamp.

The result of example is shown below, and the simulation waveform is shown in Fig. 15.3.

```
time=10000 a=0 b=0 c=0 y=0
time=10000 a=1 b=0 c=1 y=0
time=20000 a=1 b=0 c=1 y=1
```

15.7 Exercises

The exercises are based on the understanding of non-synthesizable Verilog constructs. Complete the exercises for better understanding and application and use of Verilog constructs.

1. *Find the output values a,b,c,d at various time stamp*

```
//////////////////////////////////////////////////////////////////////////

module test_design;

reg clk;

reg [7:0] a,b,c,d;

always #10 clk = ~clk;

always@(posedge clk)

begin

  b <=a+a;

 c  <= #5  b+a;

d  <= #2 c+a;

end

initial

begin

    clk =0;

    a=3;

    b=4;

    c=2;
```

```
    d=5;
#25
       a=4;
     b=3;
     c=2;
    d=5;
end
endmodule
//////////////////////////////////////////////////////////////////////////
```

2. Find a,b at various time stamp

```
//////////////////////////////////////////////////////////////////////////
module multiple_assignments_begin_end;
reg clk;
reg a;
 reg b ;
always #10 clk=~clk;
   initial
   begin
     clk=0;
      a = 0;
      b = 0;
   end
   always @(posedge clk)
```

```
 begin
     #20 a = 1;
     #40 b = 1;
  end
endmodule
//////////////////////////////////////////////////////////////////////////////

     3.   Find a,b at various time stamp
//////////////////////////////////////////////////////////////////////////////
module multiple_assignments_begin_end;
reg clk;
reg a;
 reg b ;
always #10 clk=~clk;
   initial
   begin
     clk=0;
      a = 0;
      b = 0;
   end
   always @(posedge clk)
fork
      #20 a = 1;
      #40 b = 1;
join
endmodule
//////////////////////////////////////////////////////////////////////////////
```

15.8 Summary

Following are few of the important points to conclude the chapter.

1. To verify the RTL design functionality, the testbench need to be coded using the non-synthesizable Verilog constructs.
2. Testbench is driver to drive the stimulus to design under test.
3. During the simulation, we can use inter- or intra-delay assignments.
4. The initial block is executed only once, and the simulation stops at the end of this block.
5. The simulation in the always block continues forever. If wait construct is there, then it will be held during simulation session.
6. The display tasks such as $monitor, $display, and $strobe and used to print the result
7. Blocking assignments included within begin..end executes sequentially.
8. Blocking assignments within the fork..join executes concurrently.

Chapter 16
FPGA Architecture and Design Flow

Programmable logic devices are used to realize the complex logic. Due to programmable features, the modern high-density FPGAs are used to prototype the complex ASICs and SOCs. This chapter discusses about the FPGA architecture, design flow, and the simulation using the FPGA.

Most of the time we use the FPGA as a programmable logic to realize the complex ASICs and SOCs. The chapter is useful to understand about the FPGA design flow. The chapter is useful to understand about the PLD classification and programmable features and their use during the prototype phase.

16.1 Introduction to PLD

During the past decade, the programmable logic devices (PLD) market has grown, and the demand of the PLDs has increased as they are useful to realize high-density logic and to prototype the new ideas. The chip which has programmable features and can be programmed is treated as PLD. The PLD is also named as filed programmable device (FPD). FPDs are used to implement the complex digital logic, where the functionality of the integrated circuit can be configured by the user to realize the designs. The programming functionality of such integrated circuit is accomplished by using the vendor-specific or EDA tools.

The first programmable chip introduced in the market was programmable read-only memory (PROM). PROM has number of address lines and data lines. Address lines are used as logic circuit inputs, and data lines are used as logic circuit output. The PROM has inefficient architecture and cannot be used to realize the complex digital logic.

The first programmable device introduced later to PROM during 1970s is PLA which has two levels of logic and used to realize the small density logic. After evolution of PLA, the real evolution of programmable logic device took place.

V. Taraate, *Digital Logic Design Using Verilog*,
https://doi.org/10.1007/978-981-16-3199-3_16

After PLA, the SPLD, CPLD, and FPGA architectures are evolved during early 1980s. Early programmable logic device is shown in Fig. 16.1.

The PLD classification is shown in Table 16.1.

Following are the key terminology used to understand the field programmable devices.

PAL: A relatively small density field programmable device (FPD) which has programmable AND plane followed by fixed OR plane is called as programmable array logic (PAL).

PLA: A relatively small density field programmable device (FPD) which has programmable AND plane followed by programmable OR plane is called as programmable logic array (PLA). The PLA structure has two levels of logic and can be available as the full-custom chip.

SPLD: Any structure which is like PAL or PLA is called as simple programmable logic device (SPLD). SPLDs are used to realize small gate count state machines due to the better and clean timing performance.

CPLD: The structure which consists of multiple SPLD like blocks on the same chip with interconnection logic is called as **complex programmable logic device (CPLD)**. CPLD is also called as mega PAL, super PLA, or enhanced PLD (EPLD). In the practical scenario, CPLDs are used to realize gate count state machines due to improved timing performance as compared to the SPLDs. The CPLD structure is shown in Fig. 16.2.

Fig. 16.1 Early programmable logic device

Table 16.1 PLD classification

PLD	SPLD	CPLD	FPGA
Logic Cell	PAL or PLA	SPLD	CLB
Density	Few hundred logic gates	Few thousand logic gates	Few lakh logic gates
Type	Gate rich logic	Gate rich logic	Flip-flop-rich logic
Application	Small density FSM	Moderate gate count FSMs	Complex FSMs

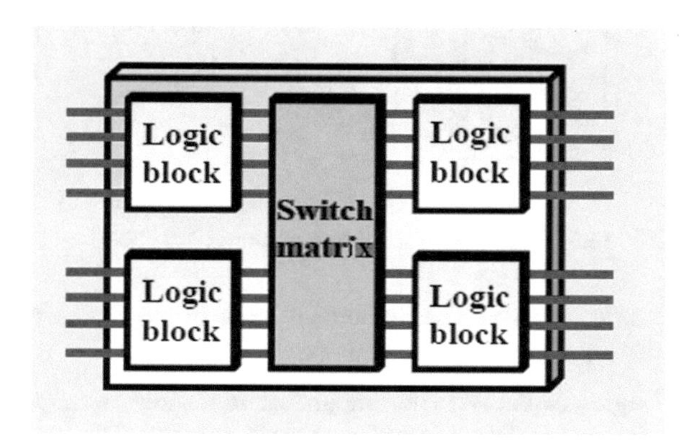

Fig. 16.2 Block diagram of CPLD

FPGA: It is the programmable logic which consists of the greater number of resources like flip-flops and configurable or programmable logic blocks to realize the high-density logic! As they have programmable features and can be programmed using the vendor-specific EDA tools at field, they are called as field programmable gate array (FPGA). FPGA is also called as programmable ASIC and consists of the configurable logic blocks (CLB), IO blocks (IOB), clocking resources, and programmable interconnects. Modern FPGAs even consist of the multipliers, block RAMs, DSP blocks, high-speed interfaces, and processor cores. The FPGA having the important functional blocks, logic blocks, IO blocks, and programmable interconnect is shown in the following Fig. 16.3.

Interconnect: The routing resource in the field programmable device is called as a interconnect.

Programmable Switch: It is the switch used to route the connections between various functional blocks.

Configurable Logic Block (CLB): The logic block which can be configured for the desired combinational and sequential logic functionality is called as CLB. While implementing the logic on the FPGA, the logic is decomposed into small density logic blocks and mapped on the multiple CLBs.

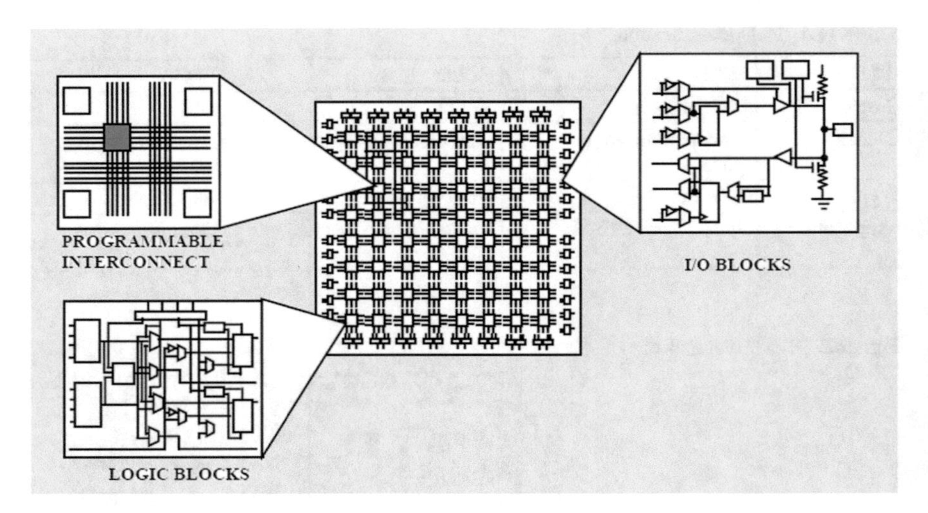

Fig. 16.3 Basic FPGA architecture (Source: XILINX)

Logic Density: The amount of logic in the FPGA per unit area is called as logic density.

Logic Capacity: The amount of logic that is mapped into the single filed programmable device is called as logic capacity. The logic capacity is given in the form of the number of logic gates in the gate array. The logic capacity can be thought in the form of number of two input NAND gates or universal gates.

Performance: The maximum operating frequency of the field programmable device is measure of the performance for the sequential logic. For the combinational logic, the longest path in the design decides the performance.

The comparison of the structured ASIC and FPGA design is listed in the following Table 16.2.

Table 16.2 Comparison of structured ASIC with FPGA (Source: XILINX)

Selection Criteria	Structured ASICs*	EasyPath FPGAs
Time to Prototype Samples	4–8 weeks	0 weeks
Total Time to Volume Production	12–15 weeks	8 weeks
Vendor NRE/Mask Costs	$100–$200 K	$75 K
Design Costs for Conversion	$250–$300 K	$0
Additional Cost of Tools for Conversion	$100–$200 K	$0
Unit Costs	Low	Low
Risk	High	Low
Flexibility to Make Changes In System	Inflexible	Flexible
Design Conversion from Prototype to Production	Additional Engineering	Conversion Free

16.2 FPGA as Programmable ASIC

Modern FPGAs are named as programmable ASICs and used in various applications which include the prototyping of the ASIC and SOC designs. The FPGA classification using various techniques is discussed in this section.

16.2.1 SRAM Based FPGA

Most of the FPGAs in the market are based on the SRAM technology. They store the configuration bit file in the SRAM cells designed using latches. As the SRAM is volatile, they need to be configured at the start. There are two modes for programming, and they are master and slave. The SRAM memory cell is shown in Fig. 16.4.

In the master mode, FPGA reads configuration data from the external source, and that source can be flash.

In the slave mode, FPGA is configured by using the external master device such as processor. The external configuration interface can be JTAG and useful during the boundary scan.

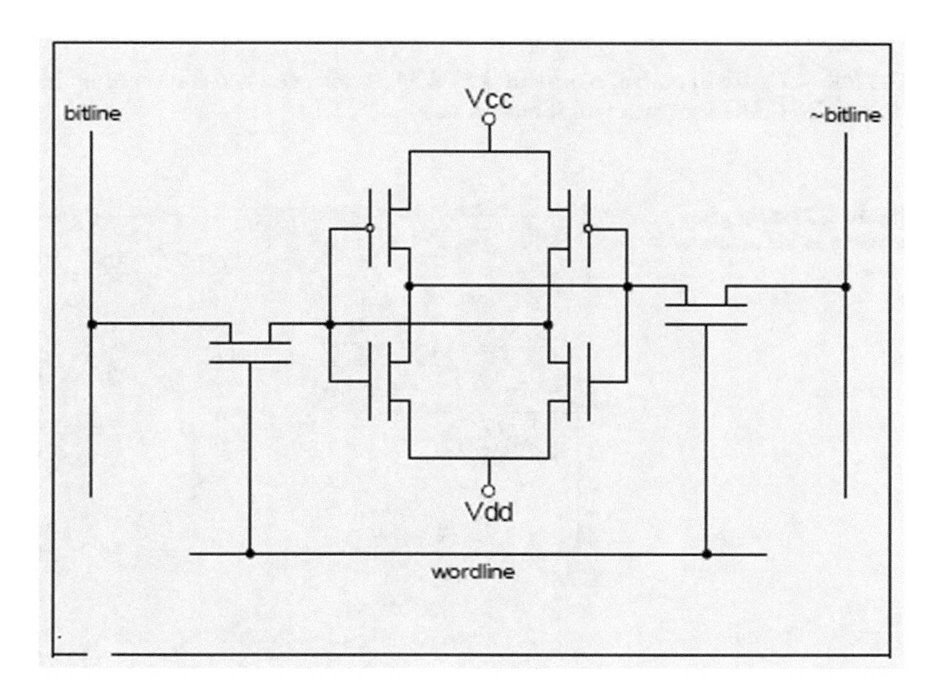

Fig. 16.4 SRAM cell

16.2.2 Flash Based FPGA

In these types of FPGAs, the flash memory is used to store the configuration data. So, the primary resource for this FPGA is the flash memory. So, these kinds of FPGAs have the lower power consumption, and they are less tolerant for the radiation effects. In the SRAM-based FPGAs, the internal flash is only used during power-up to load the configuration file. The floating gate transistor used in the flash memory is shown in Fig. 16.5.

16.2.3 Antifuse FPGAS

These types of FPGAs are used to programme only once, and they are different as compared to previous two types of FPGAs. Antifuse is opposite to fuse, and initially at the start, they does not conduct current but can be burned to conduct current. Once they are programmed, there is no any way to reprogramme as burned antifuse cannot be forced to the initial state (Fig. 16.6).

16.2.4 Important FPGA Blocks

The following are important blocks in the FPGA architecture and discussed in this section. The FPGA architecture is shown in Fig. 16.7.

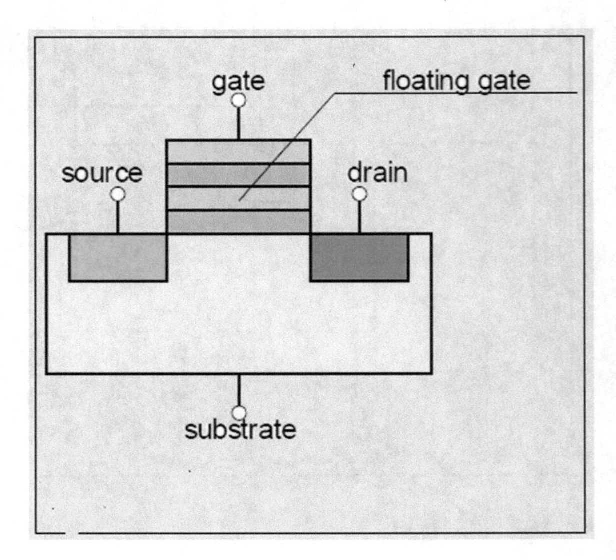

Fig. 16.5 Floating gate transistor in flash memory

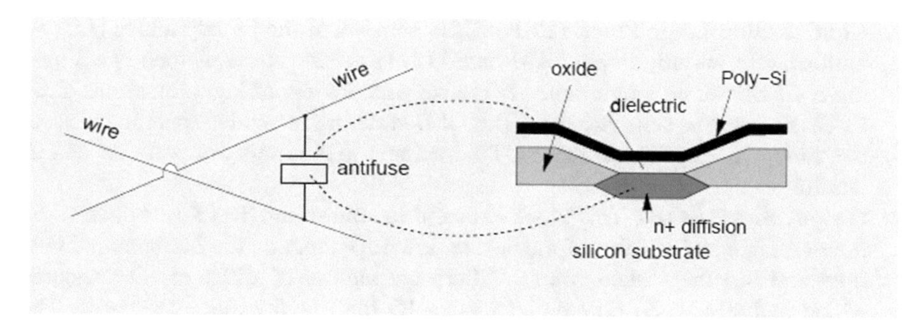

Fig. 16.6 Antifuse structure

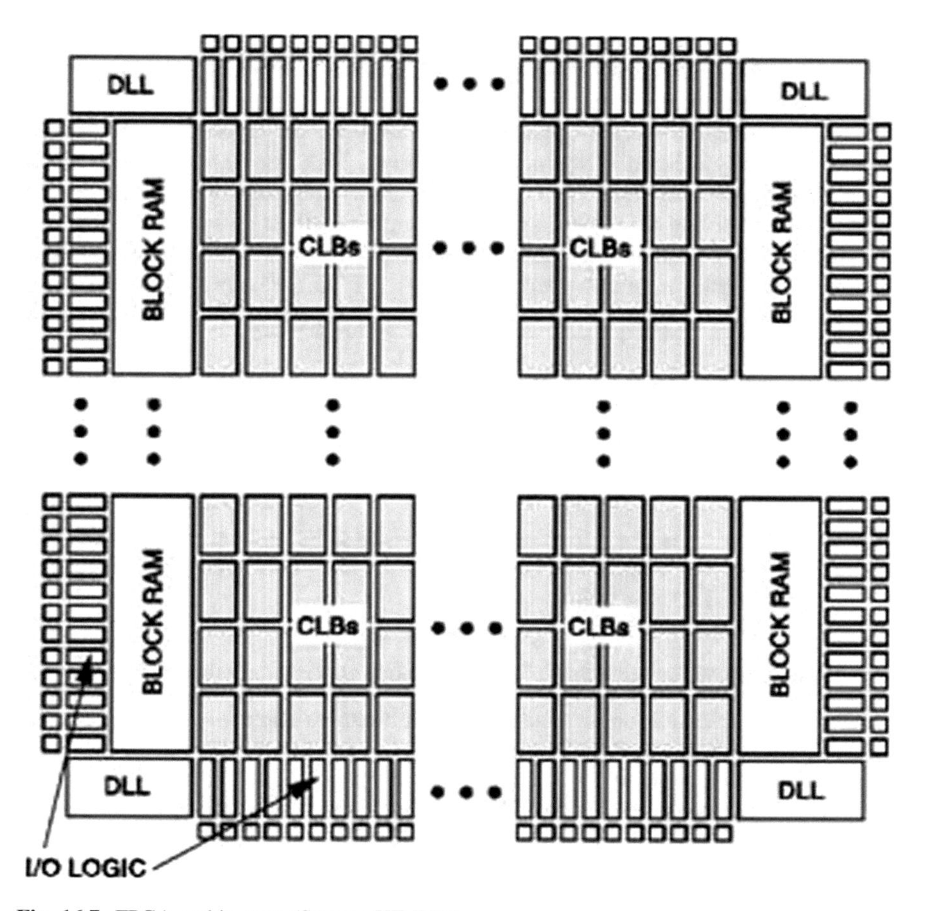

Fig. 16.7 FPGA architecture (Source: XILINX)

1. **Configurable Logic Block (CLB)**: CLB consists of the lookup tables (LUTs), multiplexers, and flip-flops. RAM-based LUTs are used to implement the digital logic. CLBs can be programmed to realize wide variety of logic functions. Even CLBs are used to store the data. The CLB structure is vendor-specific. Most of the time we experience the LUTS having 4,6,9 inputs for various FPGA families.
2. **Output Block (IOB)**: This block is useful to control the data flow between the internal logic and IO pins of the device. Each IO supports the bidirectional data flow and has the tristate control. There are almost 24 different IO standards which includes seven differential special IO high perforamnce standards. The double data rate registers are also provided with the digitally controlled impedence feature.
3. **Block RAM (BRAM)**: They are used to store the larger amount of the data and available in the form of dual port RAM, for example, 18 Kbit dual port block RAM. FPGA can consist of such multiple BRAM blocks depending on the device architecture.
4. **Digital Clock Managers (DCMs)**: They are used for clock management and provides fully calibrated digital clock solution. They are used to distribute the clock with uniform clock skew. Even they are useful to delay the clock signals, multiply or divide the clock signals with uniform clock skew.
5. **Multipliers**: Dedicated multiplier block is used to perform the multiplication of two n bit digital numbers. Depending on the FPGA device family, the n can vary. If n = 18, then the dedicated block is used to perform the multiplication of two 18-bit numbers.
6. **DSP Blocks**: They are embedded DSP blocks used to realize the DSP functions such as filtering anddata processing. These blocks are used to improve the overall performance of the FPGA while processing the larger amount of data during the DSP applications.

16.3 FPGA Design Flow

FPGA design flow includes the following important steps and shown in Fig. 16.8.

1. Design specification understanding and requirement capture
2. Design Simulation and Synthesis
3. Design Implementation
4. Device Programming.

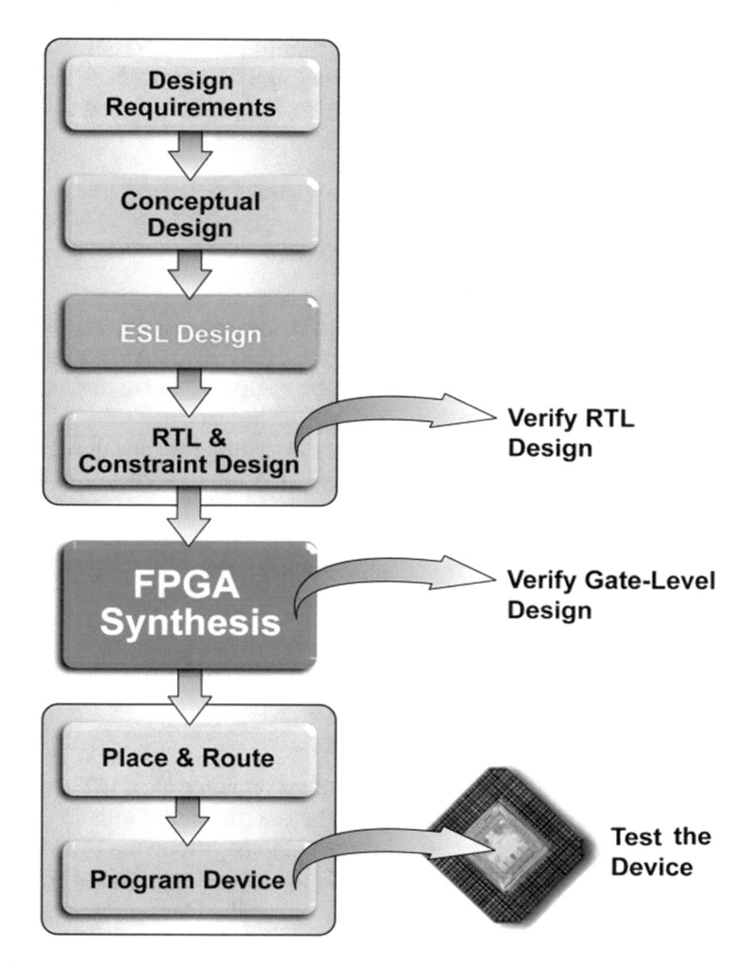

Fig. 16.8 FPGA design flow

The design steps are elaborated in the following section.

16.3.1 Design Entry

Before the design entry, the design planning needs to be completed by understanding the design requirements and the design specifications. The design specifications need to be documented in the form of block and sub-block level design and called as architecture and micro-architecture document. The design architecture and micro-architecture should include the overall design partitioning into smaller modules to get clarity about the intended design functionality and data flow.

During the architecture design phase, the requirement of memory, area, speed, and power needs to be estimated. Depending on the requirement, the FPGA device needs to be selected for the implementation.

RTL design can be coded by using either Verilog (.v) or VHDL (.vhd) or by using SystemVerilog (.sv). After the design entry, the design needs to be simulated to check for the functional correctness of the design. This is called as functional simulation.

16.3.2 Design Simulation and Synthesis

During functional simulation, the set of inputs are applied to the design with intention to check for the functional correctness of the design. Although the timing or area, power issues can crop up during the later design cycle but design team is at least confident about the functional correctness of the design.

The major goal of the FPGA design engineer is to infer the intended design logic using FPGA! The synthesis is the process of getting the lower level of the design abstraction from the higher level. During the logical synthesis, the RTL is used as one of the inputs to get the lower level of abstraction as the gate-level netlist. The netlist is device independent and can be in the standard format like electronic design interchangeable format (EDIF).

16.3.3 Design Implementation

The design passes through the various steps during implementation. These steps are translate, map, and place and route. During the design implementation, the EDA tool translates the design into the desired format and maps it on the FPGA fabric by considering the overall area requirements. The mapping is performed by the EDA tool and functionality uses the logic cells or macrocells. During the mapping process, the EDA tool uses the macrocells, programmable interconnects, and the IO blocks. The special dedicated blocks like multipliers, DSP, and BRAMs are also mapped using vendor tools. The blocks are placed on the predefined geometry inside the FPGA and routed by using the programmable interconnects to get the intended functionality. The step is called as place and route.

To check for the design timing performance and whether the constraints are met or not, the timing analysis is performed, and it is called as post-layout STA. During the STA, the timing paths are checked with the delays associated with the programmable interconnects. The intention is to find out how many timing paths have set up or hold violations? Extracting the RC delays and using them by timing analyzer is called as back-annotation.

16.3.4 Device Programming

The FPGA is programmed by using the vendor-specific or proprietary bit-stream file. Bit-stream is binary data file needs to be loaded into the FPGA to program the device.

Depending on the use of the various PFGA resources for the design, the EDA tool generates device utilization summary. Please refer Appendix III for the XILINX Spartan and Virtex series devices.

16.4 Logic Realization Using FPGA

The architecture of FPGA consists of the array of CLBs, block RAMs, multipliers, DSPs, IOBs, and digital clock managers. Delay Locked Loop (DLL) are used to generate the clock. The floor plan for the XILINX Spartan series FPGA is shown in Fig. 16.9.

16.4.1 Configurable Logic Block

As shown in the following, basic CLB consists of the LUTs, flip-flop, and multiplexer logic. The configuration data is held in the latch. The CLB architecture is vendor-specific and can consist of multiple LUTs, flip-flops, multiplexers, and

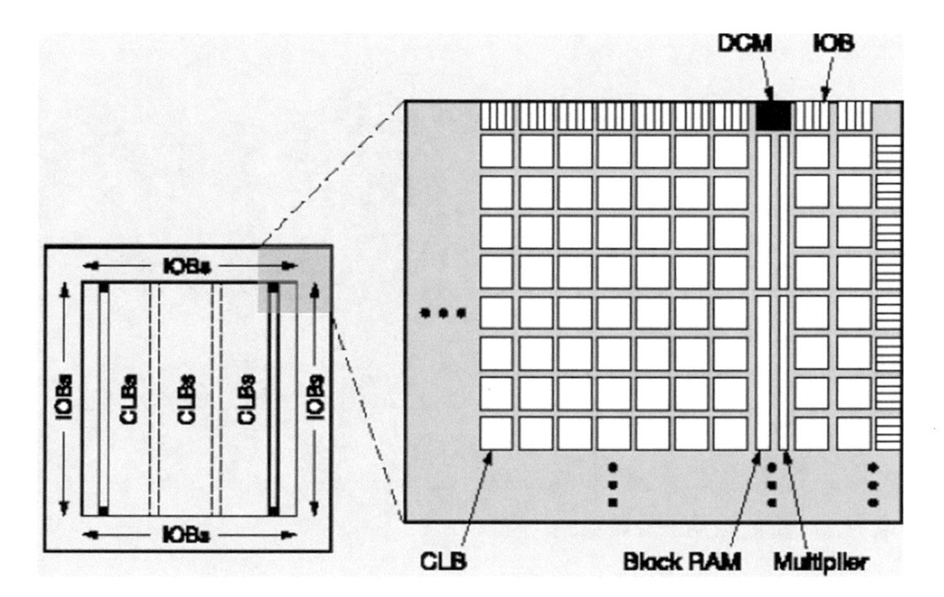

Fig. 16.9 Xilinx Spartan series device

latches. The following Verilog code is realized by using the single four input LUT without flip-flop, and the output is called as combinatorial output.

```
always@(*)
   begin
          y_out= a_in && b_in;
   end
```

The following Verilog RTL uses single LUT with single flip-flop to realize the logic. The logic is sequential logic as output is function of the present input and past output.

```
always@(posedge clk)
   begin
          y_out= a_in && b_in;
   end
```

The CLB shown in Fig. 16.10 can be also useful to implement 16-bit shift register. The LUTs can be cascaded to infer the longer size shift register, or it can be used to introduce the pipelining to improve the design performance.

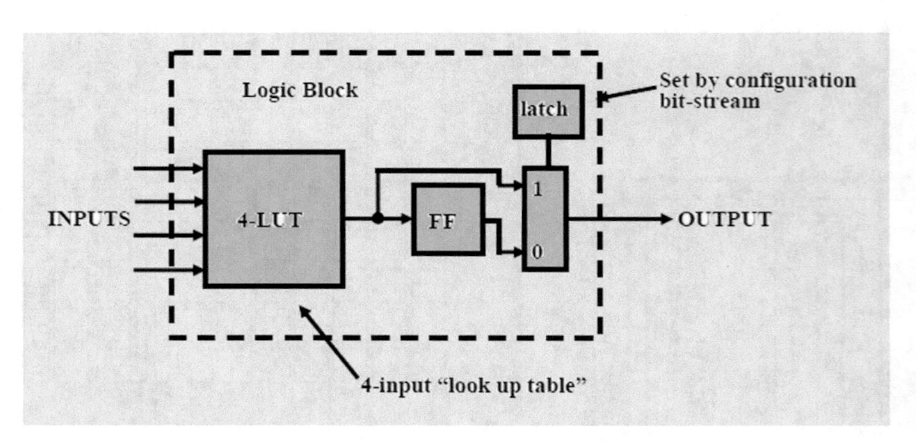

Fig. 16.10 Xilinx basic CLB structure

16.4.2 Input Output Block (IOB)

An input–output block is used to establish the communication of the logic residing on the FPGA fabric with outside world and consists of the number of flip-flops and buffers with the tristate control mechanism. The IO block can be used to have a registered input and registered output. The IOB structure of modern FPGA is complex and can consist of many IO control features which may include DDR, special-purpose high-speed interfaces. The basic IO block structure is shown in Fig. 16.11.

16.4.3 Block RAM

XILINX Spartan-3 family supports 200 MHz block RAM organized in the four columns to get the synchronous configurable 18 Kb blocks. Each block RAM contains 18,432 bits, among them, 16 Kb are allocated for the data storage, and remaining 2 Kb are allocated for the parity. Block RAM can be used as single port memory or dual port memory and has independent port access. Each port is synchronous and has independent clock, clock enable, and write enable. Read

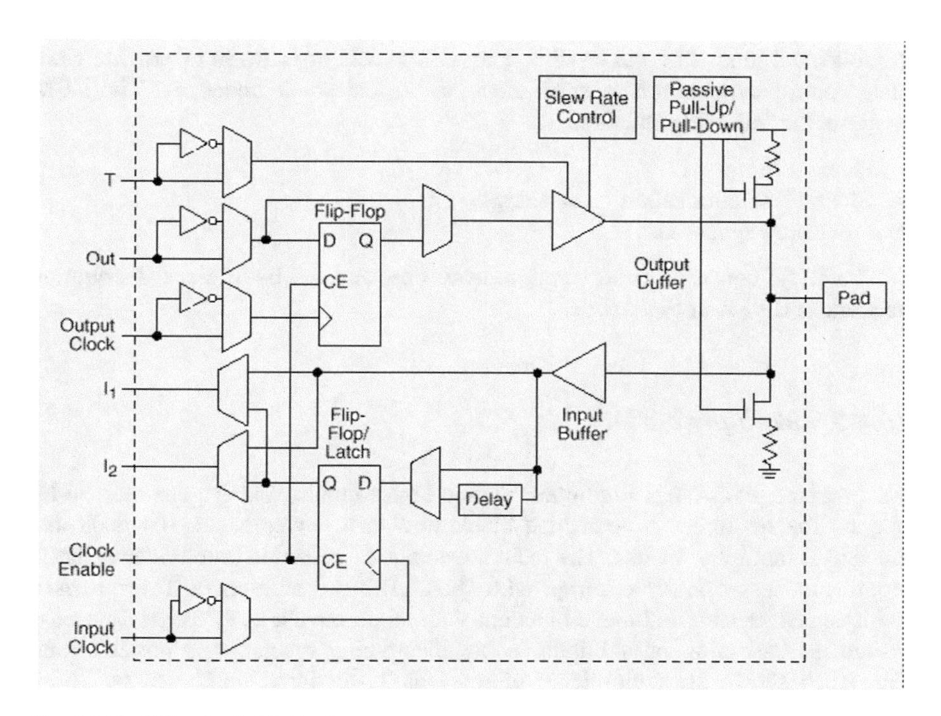

Fig. 16.11 Xilinx basic IO block

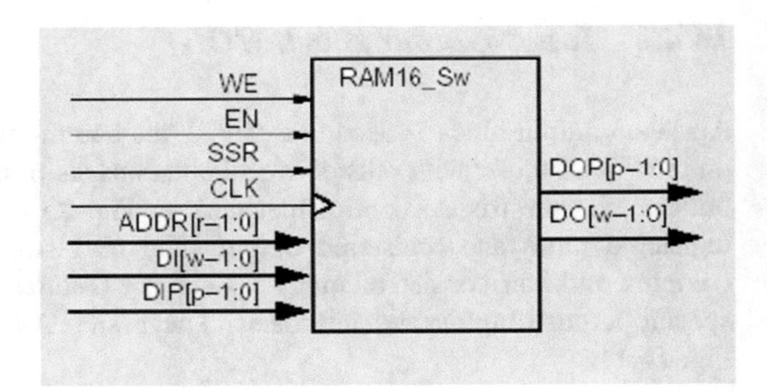

Fig. 16.12 Xilinx single port BRAM

operations are also synchronous in nature and may require the clock enable. The applications of Block RAM is to store the data, BRAMs are used during the FIFO designs, or as a memory buffers, or to implement stacks and they are even useful to hold larger amount of data in the design of complex state machines. Single port RAM is shown in Fig. 16.12.

16.4.4 Digital Clock Manager (DCM) Block

The Xilinx device family uses the delay-locked loop (DLL), and Altera uses the phase-locked loop (PLL) as clock manager. The role of DCM is to provide complete control over the phase shift, clock skew, and clock frequency. The DCM supports the following functions.

- Phase shifting
- Clock skew elimination or balancing
- Frequency synthesis.

The DCM consists of the variable delay line, and the basic clock distribution network is shown in Fig. 16.13.

16.4.5 Multiplier Block

All Spartan3 FPGA has multiplier having 18-bit inputs, and it generates 36-bit output. The multiplier is embedded block, and each device has 4–104 dedicated embedded multiplier blocks. The main advantage of embedded multiplier is that it requires the lesser power as compared to the CLB-based multipliers. They are used to implement the fast arithmetic functions with minimum use of the general-purpose resources. Cascading of multiplier using the routing resources is possible, and Fig. 16.14 shows the multiplier configured as 22-bit input multiplied by 16-bit input to infer the multiplier having the 38-bit output. The multiplier can be used

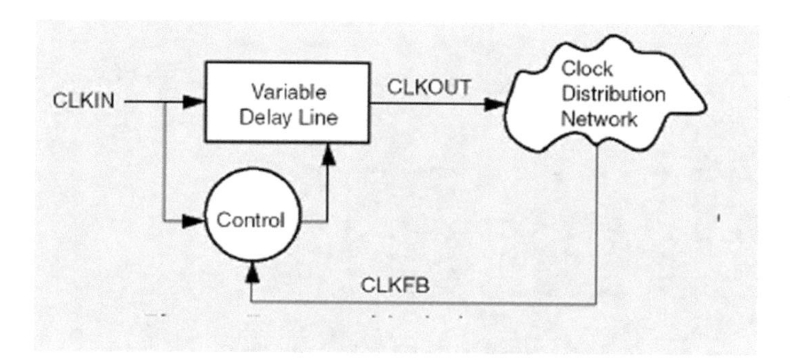

Fig. 16.13 Xilinx basic DLL block

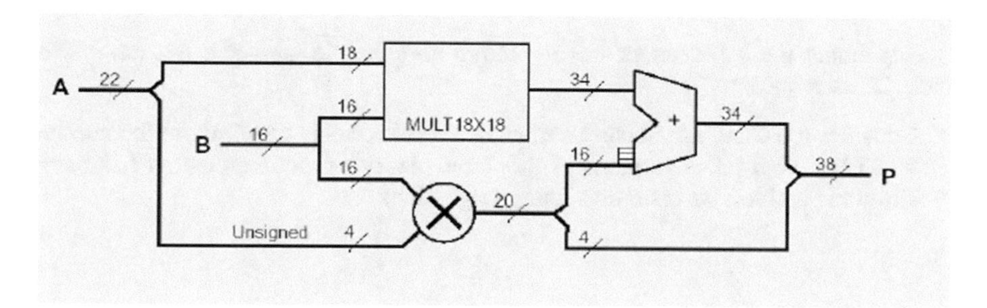

Fig. 16.14 Xilinx basic multiplier block

during the signed or unsigned number multiplication. The multipliers are extensively used in the DSP applications. The basic block is shown in Fig. 16.14.

In this section, only few basic blocks are documented and discussed. As stated earlier in the above few sections, the modern FPGA architecture has other dedicated blocks such as DSP blocks, processor cores, and high-speed interfaces. Readers can go through the data sheets of the desired FPGA family and can understand the architectures to implement the high-density designs.

16.5 Exercises

Complete the following exercises to understand the interpretation of the RTL design at FPGA fabric level.

1. Consider the FPGA has the CLB architecture shown in the following figure.

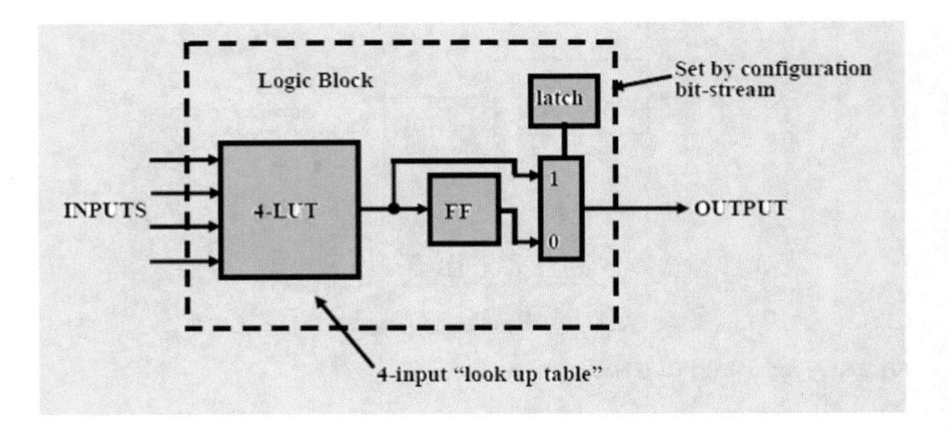

Implement the 3:8 decoder having active high enable, and find out number of such CLBs required.

2. Consider the CLB architecture shown in the exercise-1, and find out the number of CLBs required to implement the 2-bit binary up counter which has asynchronous active low reset and rising edge clock.

16.6 Summary

Following are few important points to summarize this chapter.

1. PLDs are classified into three main categories and are SPLD, CPLD, and FPGAs.
2. PAL, PLA are also called as SPLDs and used to realize small gate count designs.
3. CPLDs are moderate density FPDs and are used to design small gate count FSMs due to good timing performance.
4. FPGAs are used to design the complex gate count FSMs and are called as flip-flop-rich logic.
5. The design specifications need to be documented in the form of block and sub-block level design and called as architecture and micro-architecture document.
6. During functional simulation, the set of inputs are applied to the design with intention to check for the functional correctness of the design.
7. The synthesis is the process of getting the lower level of the design abstraction from the higher level.
8. The design passes through the various steps during implementation. These steps are translate, map, and place and route.

9. The FPGA is programmed by using the vendor-specific or proprietary bit-stream file.
10. The architecture of FPGA consists of the array of CLBs, block RAMs, multipliers, DSPs, IOBs, and digital clock managers. Delay Locked Loop (DLL).
11. Modern FPGA consists of the dedicated multipliers, DSP blocks, and high-speed interfaces with the processor cores.
12. FPGA designer needs to use the design guidelines while using the FPGAs.

Chapter 17
FPGA Design and Guidelines

As discussed in the previous few chapters, the design guidelines play important role during the FPGA design implementation. In such scenario, the chapter discusses about the design guidelines for FPGA-based designs. How to use the design guidelines is explained with the RTL designs coded using the synthesizable Verilog constructs.

The design and RTL coding guidelines are useful to improve the design performance. The area, speed, and power requirements of the design should be met, and the efficient RTL should yield into the better performance of the design. The following few section discusses about the design and coding guidelines useful during the FPGA design.

17.1 Design Guidelines for FPGA Based Designs

The RTL design concepts discussed in Chaps. 1, 2, 3, 4, 5, 6, 7, 8, 9, 10, 11, 12, 13, 14, 15 and 16 can be used to design the logic for desired FPGA family. The coding guidelines during the RTL design phase and the design guidelines for the FPGA are always useful to achieve the better performance of the design. Following are the few coding and design guidelines need to be followed during logic design and implementation using FPGA.

17.1.1 Verilog Coding Guidelines

Guidelines while using Verilog to have efficient RTL design are listed in this section, and it is always recommended to use these guidelines during RTL design phase! Among these, few guidelines are based on the concepts explained using stratified event queue in the Chap. 7.

V. Taraate, *Digital Logic Design Using Verilog*,
https://doi.org/10.1007/978-981-16-3199-3_17

17.1.1.1 *Blocking Versus Non-blocking Assignments: (Please Refer* Chaps. 7 and 11)

I. It is recommended to use ***blocking assignments*** while modeling the ***combinational design***.
II. It is recommended to use ***non-blocking assignments*** while modeling ***sequential design.***
III. It is recommended to use the ***non-blocking*** assignments while modeling the ***latches.*** While implementing RTL design, it is essential to overcome the potential unintentional latches. Unintentional latches are inferred due to missing ***else*** or due to incomplete ***case*** conditions.
IV. It is recommended, ***not to mix the blocking and non-blocking assignments*** in the same always block.

The sequential design logic uses non-blocking assignments and is shown in the Example 1, and it results into the desired intended synthesis result (Fig. 17.1. So, it is recommended to use the non-blocking assignments in the sequential designs!

```
////////////////////////////////////////////////////////////////////////////////

module sequential_design_using_NBA(input  d_in, clk, reset_n, output
reg y);

reg tmp1;

always @ (posedge clk, negedge reset_n)

begin

if (~ reset_n)

        {y,tmp1} <= 2'b00;

else

begin

        tmp1 <= d_in;

        y <= tmp1;

end

end

endmodule

////////////////////////////////////////////////////////////////////////////////
```

Example 1 Sequential design using non-blocking assignments

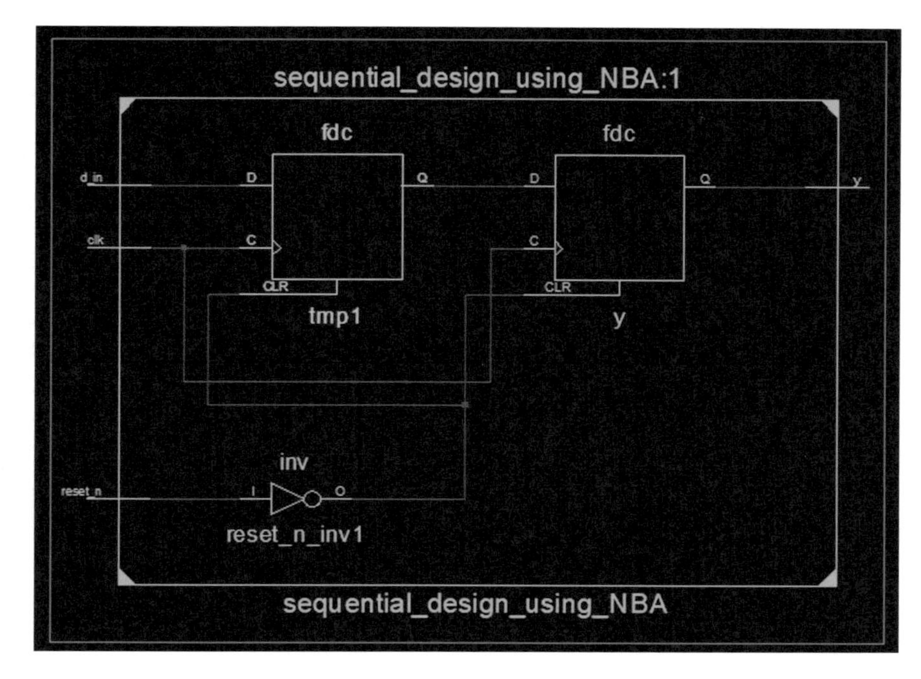

Fig. 17.1 Synthesis result of sequential design using non-blocking assignments

The sequential design logic uses blocking assignments and is shown in the Example 2, and it results into the wrong synthesis result (Fig. 17.2). So, it is not recommended to use the blocking assignments in the sequential designs!

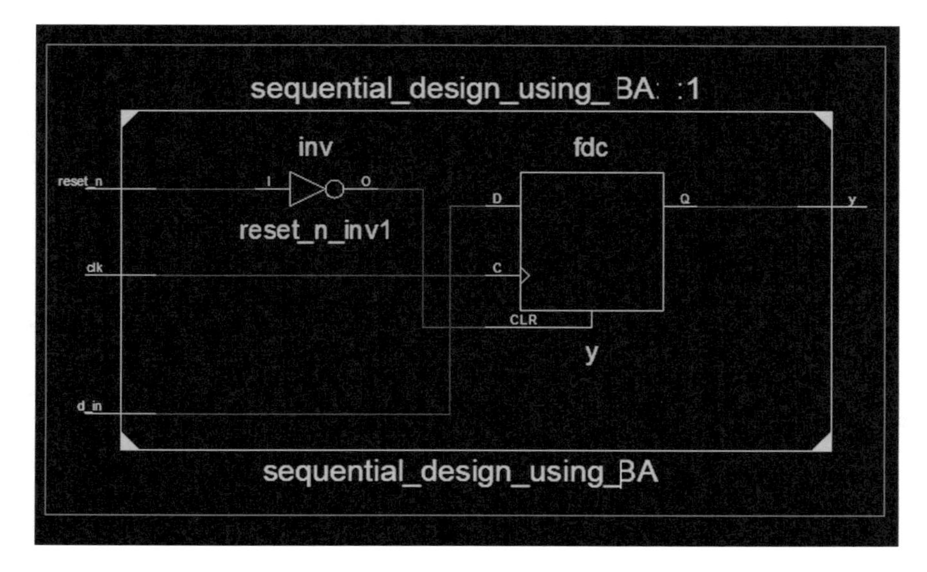

Fig. 17.2 Synthesis result of sequential design using blocking assignments

```
////////////////////////////////////////////////////////////////////////////////
module sequential_design_using_BA(input  d_in, clk, reset_n, output reg
y);

reg tmp1;

always @ (posedge clk, negedge reset_n)

begin

if (~ reset_n)

        {y,tmp1} = 2'b00;

else

begin

        tmp1 = d_in;

        y = tmp1;

end

end

endmodule

////////////////////////////////////////////////////////////////////////////////
```

Example 2 Sequential design using blocking assignments

17.1.1.2 Priority Versus Parallel Logic

I. It is recommended to use ***if-else*** construct to design the priority logic. Priority encoder or priority interrupt control logic can be modeled by using the ***nested if-else*** statements. It is recommended to use ***case*** construct to infer the parallel logic. Priority logic infers the design having the longer combinational path due to ***nested if-else constructs.*** It is always recommended to use ***case*** construct to infer the parallel logic.

The parallel logic using if-else construct to infer the 4:1 mux is shown in the Example 3 and infers the gate-level structure as shown in Fig. 17.3.

```
////////////////////////////////////////////////////////////////////////////////
module priority_logic( input x,y,z,w, input [1:0] sel, output reg q );

always@*

begin

 if (sel==2'b00)

        q= x;

 else if (sel==2'b01)

        q= y;

else if (sel==2'b10)

        q= z;

else

        q= w;
end
 endmodule
////////////////////////////////////////////////////////////////////////////////
```

Example 3 RTL design of 4:1 mux using nested if-else

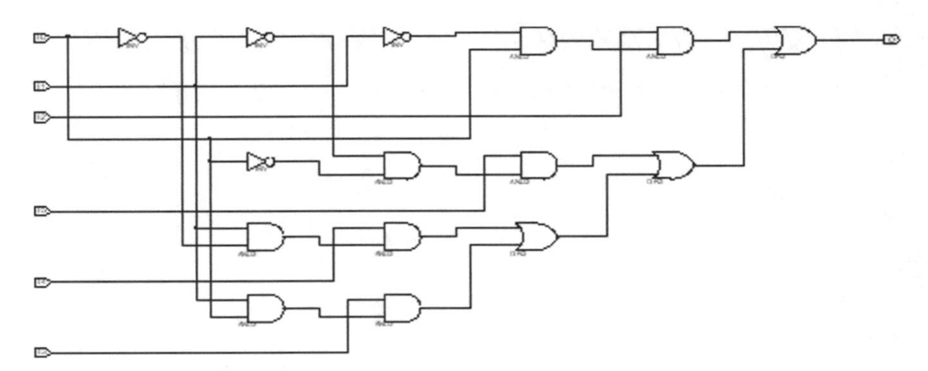

Fig. 17.3 Gate-level schematic of 4:1 mux

The parallel logic using case construct to infer the 4:1 mux is shown in the Example 4 and infers the gate-level structure as shown in Fig. 17.4.

```
////////////////////////////////////////////////////////////////////////////
module parallel_logic( input x,y,z,w, input [1:0] sel, output reg q );
always@*
begin
 case(sel)

      2'b00 : q =x;
      2'b01 : q= y;
      2'b10 : q= z;
      default : q=w;
      endcase
end
endmodule
////////////////////////////////////////////////////////////////////////////
```

Example 4 RTL design of 4:1 mux using case

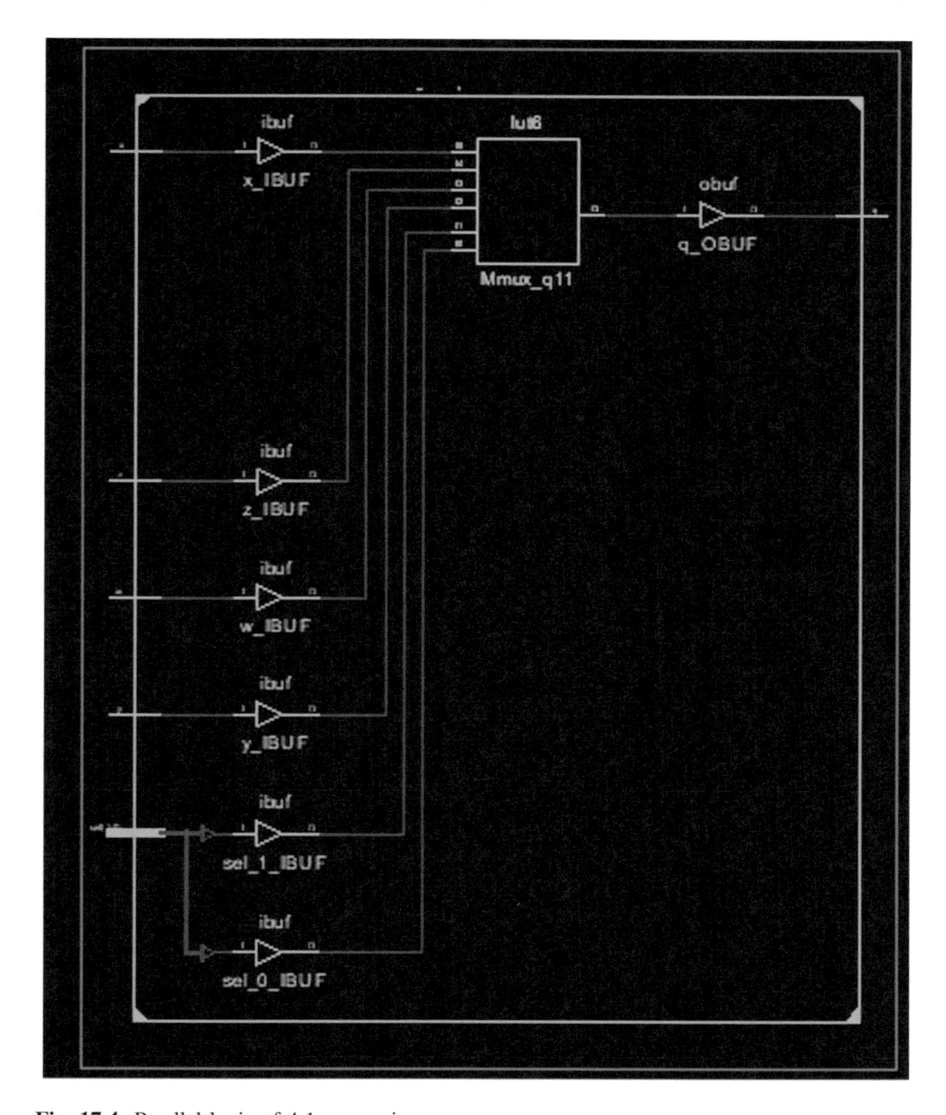

Fig. 17.4 Parallel logic of 4:1 mux using case

17.1.2 FSM Guidelines

I. Binary encoding techniques are efficient for a design having 16 or fewer states. As number of states increases, the next state combinational logic performs slower operation.

II. One-hot encoding technique is efficient and reliable as compared to the binary encoding due to glitch-free behavior. One-hot encoding requires low density next state logic and is useful in design of larger FSM blocks. But the main drawback of one-hot encoding is that it uses a greater number of flip-flops!

III. While designing FSM, designer needs to take care of the following important points.

   a. Do not leave any unused states. Initialize the unused states to reset value or use the ***default*** assignment.
   b. Do not implement the FSM with combination of flip-flops and latches. Avoid the unintentional latches in the FSM design to improve the reliability.
   c. Model the FSM blocks by using ***case*** constructs as it infers the parallel logic.
   d. Have the separate always procedural block to code the next state combinational logic, output combinational logic, and state register logic. This improves the speed of FSM.
   e. Register FSM output as it preserves the hierarchy.
   f. Use the look ahead Mealy machines for better design performance.

17.2 Combinational Design and Combinational Loops

I. It is recommended to use ***continuous assignment*** construct to design the ***combinational logic.***

II. While designing the combinational logic, it is essential to avoid the combinational loops. Combinational loop causes instability in digital designs as it violates the synchronous design concepts due to infinite looping. The combinational loop generates the oscillatory output, and the period of the oscillatory output signal is mainly dependent on the delay introduced by the combinational elements in the feedback path.

17.3 Grouping the Terms

I. Use the signal grouping to improve the performance of FPGA-based design. For example, if the expression $q = (x + y + z + w)$ is to be implemented using FPGA, then cascade structure is inferred as shown. But using grouping of the few terms and by modifying the expression as $q = (x + y) + (z + w)$, the logic infers the parallel structure. Even due to grouping, the timing performance of design is improved.

The Example 5 shown does not use the grouping of terms and infers the cascade or priority logic as shown in Fig. 17.5. The cascade logic has the 3*tpd delay.

```
////////////////////////////////////////////////////////////////////////////////
module without_grouping_terms( input x,y,z,w, output q );

assign q = x & y & z & w;

endmodule

////////////////////////////////////////////////////////////////////////////////
```

Example 5 Expressions without grouping

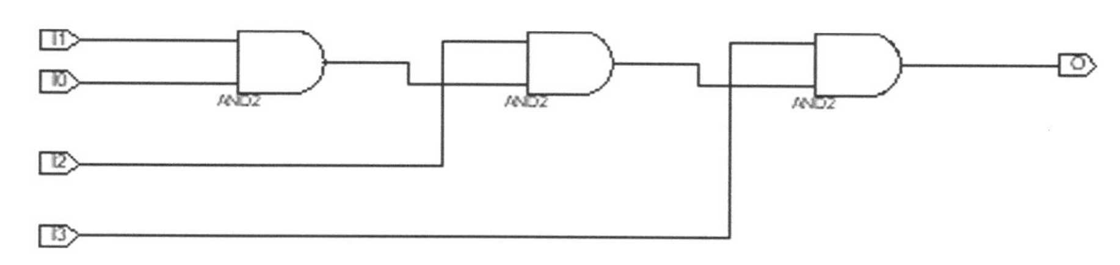

Fig. 17.5 Cascade logic

The Example 6 shown uses the grouping of terms and infers the parallel logic as shown in Fig. 17.6. The parallel logic has the 2*tpd delay.

```
////////////////////////////////////////////////////////////////////////////////
module grouping_terms( input x,y,z,w, output q );

assign q = ((x & y) & (z & w));

endmodule

////////////////////////////////////////////////////////////////////////////////
```

Example 6 Expressions with grouping the terms

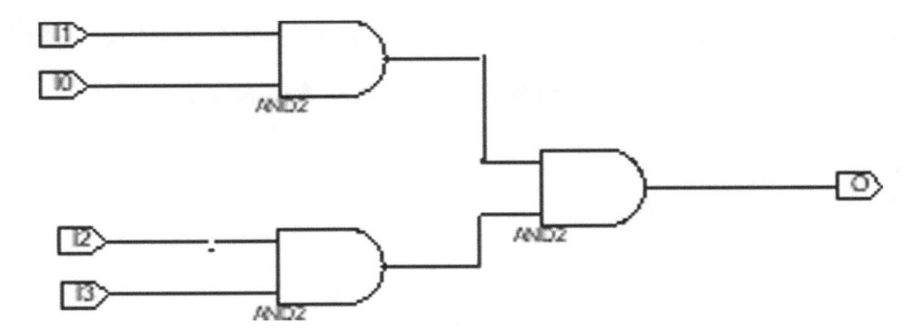

Fig. 17.6 Two-stage parallel input logic

17.3.1 Assignments

I. It is recommended ***not to have the assignments to same variable or output port within multiple always block.*** It gives error as multiple drivers drive to the same net or reg.
II. It is recommended ***not to have assignments using #0 delay***.

17.4 Simulation and Synthesis Mismatch

Most of the synthesis tools ignore the sensitivity list of always procedural block used for combinational, but during simulation, the procedural block invokes, only when there is an event on one of the signals included in the sensitivity list. Due to incomplete sensitivity list, it creates the logic which has simulation synthesis mismatch.

17.4.1 Post-synthesis Verification

It is highly recommended to perform the post-synthesis verification for the FPGA-based design. Post-synthesis verification with the SDF assures the correct intended behavior of the gate-level netlist. There should not be mismatch between the functional verification of the design and the post-synthesis verification!

17.5 Guidelines for Area Optimization

FPGAs have finite number of resources, so it is recommended to follow the design guidelines to optimize the area. The area optimization techniques are resource sharing, logic duplication. [Note: Many times, it has been observed that logic duplication can even increase area and the use of logic duplication technique is dependent on the design scenarios!].

17.5.1 Resource Sharing

Always it is observed that adders consume more area as compared to multiplexers. The resource sharing is powerful technique to share the common resources to minimize the area. It is essential for the FPGA design team to consider resource sharing while using arithmetic operators. The resource sharing is one of the powerful area optimization techniques. But it is recommended that not to share resources from different modules or from different hierarchy. Resources can be shared from the same module or from the same hierarchies. The resource sharing improves the overall data and control path for the design!

17.5.2 Logic Duplication

It is the powerful technique to reduce the net delay by allowing the placement tool to place the replicated logic in various areas of die [2]. The major drawback of this technique is that it increases the area of the design while replicating the flip-flop or sequential logic.

As per the FPGA area optimization is concerned, logic duplication can act as very efficient technique but depends on the design specific scenarios!

FPGA-specific design scenario
Consider example of implementing 8:256 decoder using single ***case*** construct. If FPGA architecture has CLB which has two, 4 input LUTs and a single two-input LUT at the output (Fig. 17.7) then to realize the single bit output the tool uses 3 LUTs. So, for 256-bit output, 768 LUTs are utilized.

By using multiple case construct, let us implement two 4:16 decoders, and duplicate the AND array at the output. This technique we can call as logic duplication ! By using logic duplication, if two 4:16 decoders are used with 256 AND gate array, then the overall device utilization is just 288 LUTs to implement the 8:256 decoder, and it reduces the device utilization by around 480 LUTs. That is huge reduction in the overall area.

For the 8:256 decoder, the logic duplication uses the four input LUTs and two input LUTs to get the single output; the structure of CLB used to get single output is shown in the (Fig. 17.7). We need to have such type of 256 CLBs.

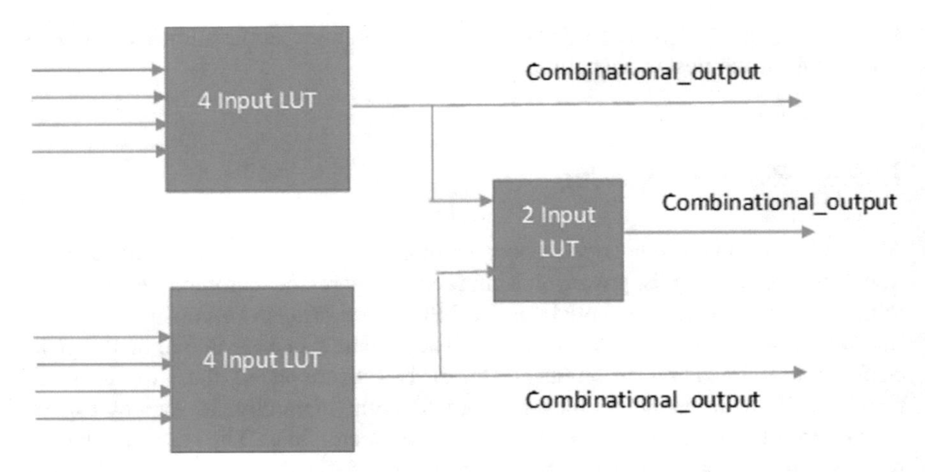

Fig. 17.7 Xilinx basic LUT structure

17.6 Guidelines for Clock

The performance and reliability of an FPGA-based design are based upon the clocking schemes. For the FPGA-based design and implementation, it is recommended that

a. ***Use single global clock.***
b. ***Avoid use of gated clocks.***
c. ***Avoid mixed use of positive and negative edge-triggered flip-flops.***
d. ***Avoid use of internally generated clock signals.***
e. ***Avoid ripple counters and asynchronous clock division.***

It is recommended by most of the FPGA vendors that do not use the internal generated clocks as it causes the functional and timing issues in the design. If internal generated clocks are required in the design, then use DLL or PLL to generate the clocks. The internal generated clocks by using combinational logic are prone to glitches, and it creates the functionality issues in the design. Due to the combinational delays, it creates the timing issues in the FPGA designs. The major problem for using the internal generated clocks is the issue during synthesis and timing analysis. Xilinx provides the library component global clock buffers BUFGCTL and BUFGMUX to generate internal clocks.

To avoid glitches, it is recommended to register the output of the internal generated clocks. It is recommended to use the clock generation logic. For low power designs, it is essential to use the clock gating, but it is prone to glitches. So, it is recommended to use the clock gating cells for low-power FPGA-based design.

It is recommended not to use the asynchronous pulse generator circuit. As shown in Figure 17.8 a, it is asynchronous way of pulse generation. This technique should be avoided as it is prone to glitches and difficult to synthesize and place and route.

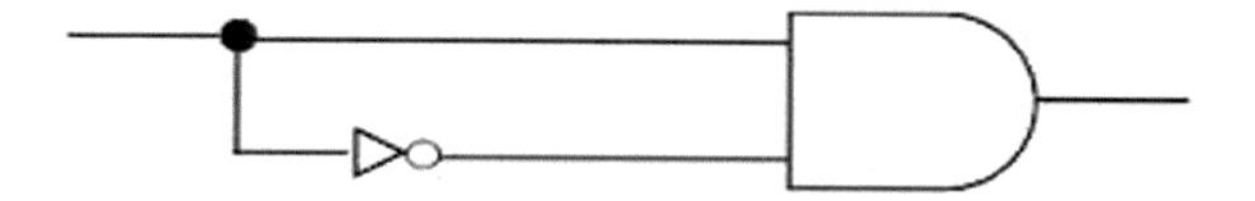

Fig. 17.8 Asynchronous pulse generator

Depending on the pulse width requirement, replace the inverter shown in figure by chain of odd number of inverters.

Figure 17.9 is the recommended pulse generator where the pulse width is dependent on the clock period. It is recommended to use two-level synchronizer at the input of pulse generator to avoid the metastability issues.

17.7 Synchronous Versus Asynchronous Designs

In synchronous design, the data input is sampled on every active edge of clock, and clock signal controls the data transfer from input to output. Figure 17.10 is register to register path in which the combinational logic (CL) drives the data to the input of flip-flop. For the desired operation of the design, it is essential that the data input should be stable during the setup time and hold time window for the flip-flop. The propagation delay of combinational logic limits the maximum operating frequency of the design.

To meet the timing requirement, use the pipelining feature to improve the performance of synchronous design. As FPGA is flip-flop-rich logic, pipelining can be used for improvement of the speed of the design at the cost of clock latency.

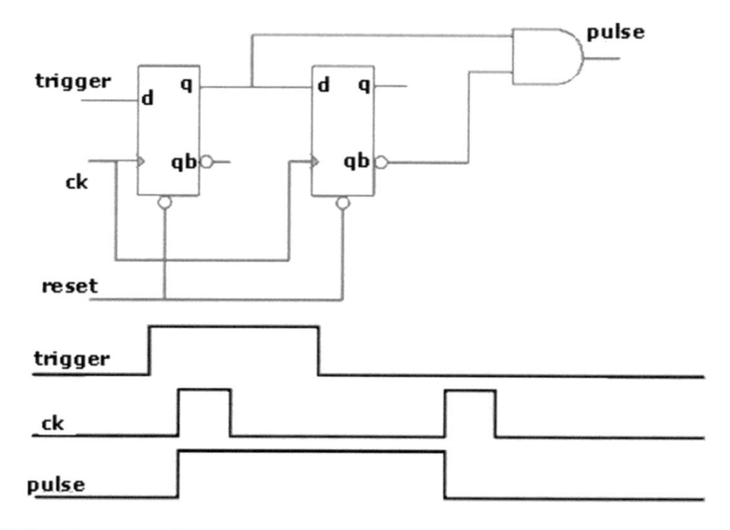

Fig. 17.9 Synchronous pulse generator

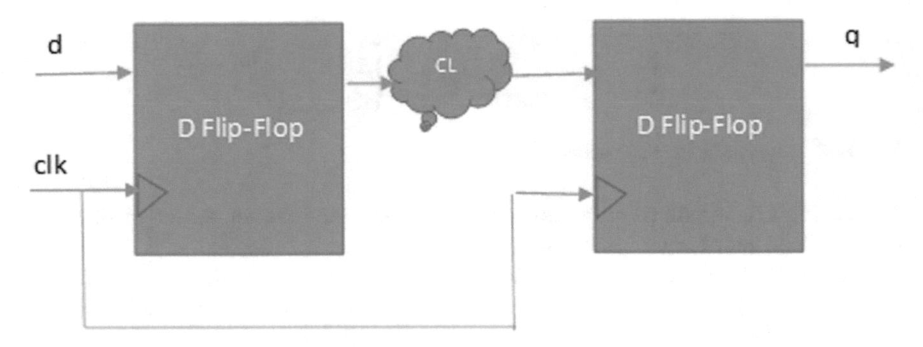

Fig. 17.10 Synchronous Logic

On the other hand, an asynchronous design does not have common clock (e.g., ripple counters) and is prone to glitches or spikes. It is difficult to specify the timing of asynchronous design by using timing constraints. Many times, an asynchronous design infers the logic which has the glitches or short time duration pulses shorter than the clock period. If the glitches are passed through the combinational logic, then the output leads to an incorrect value. Figure 17.11 describes an asynchronous logic prone to glitches.

Many times, it has been observed that an asynchronous logic reduces the device resources but prone to hazards. So, it is recommended to use the synchronous logic while implementing the sequential design. Synchronous logic always makes STA easy [4]!

17.8 Guidelines for Use of Reset

Resets are classified as synchronous and asynchronous resets. Asynchronous resets are easy to implement as they do not depend on the clock. But STA becomes difficult and complex while using asynchronous resets. At the same time, automatic insertion of the test structure is difficult.

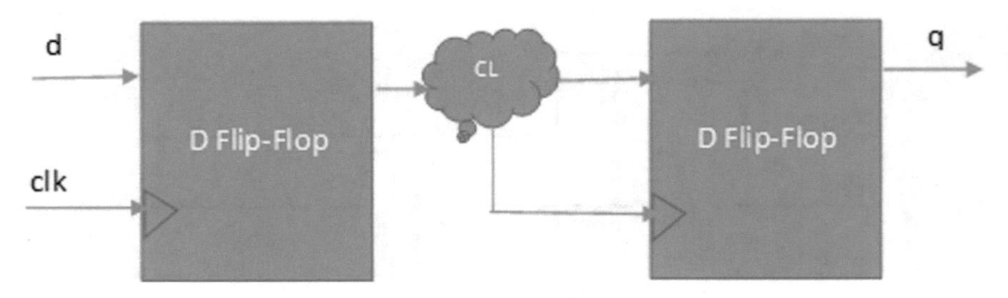

Fig. 17.11 Asynchronous logic

On the other hand, synchronous resets are difficult to implement as it requires more resources, and they are dependent on the clock. Synchronous resets slowdown the design performance. It is recommended that FPGA designer should avoid internally generated conditional resets.

It has been observed during FPGA-based designs that, reset deassertion circuit may be required while using asynchronous reset. If reset signal is deasserted and if does not pass the setup and hold timing check then flip-flop goes into metastable state and it can lead to potential functional issues in the design.

It is recommended to use the synchronized asynchronous resets. That is asynchronously asserted and synchronously deasserted. Figure 17.12 is the recommended scheme to pass asynchronous active low reset (reset_n) through the two-level synchronizer.

For large density or complex FPGA-based designs with multiple hierarchies, it is essential to use the Linting tool which can provide desired information about the reset and clock trees.

17.9 Guidelines for CDC

It is impossible to verify the clock domain crossing (CDC) during the functional verification, and even it is impossible to verify CDC by using timing analysis tool due to asynchronous nature of clock path. The major problem encountered in CDC is functionality failure due to metastability. To avoid metastability, it is recommended to use the multi-flop level synchronizer while passing signals from one clock domain to another.

Linting tools are used to ensure the use of synchronizer chain in the clock domain crossing paths. Use two- or three-flop-level synchronizer as shown in Fig. 17.13 to transfer the signals from one clock domain to another. This will avoid metastability in the design.

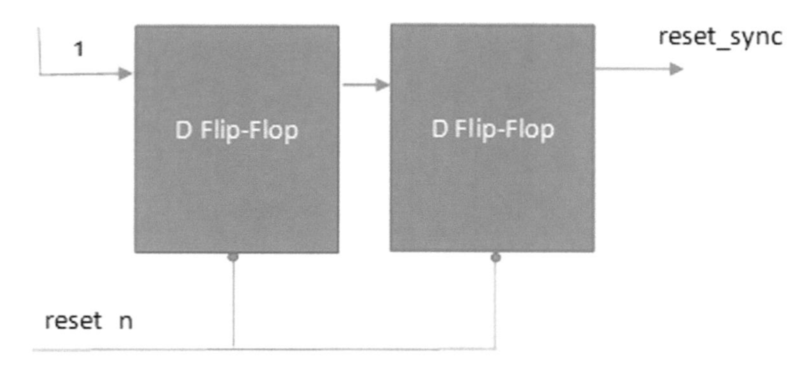

Fig. 17.12 Reset generation logic

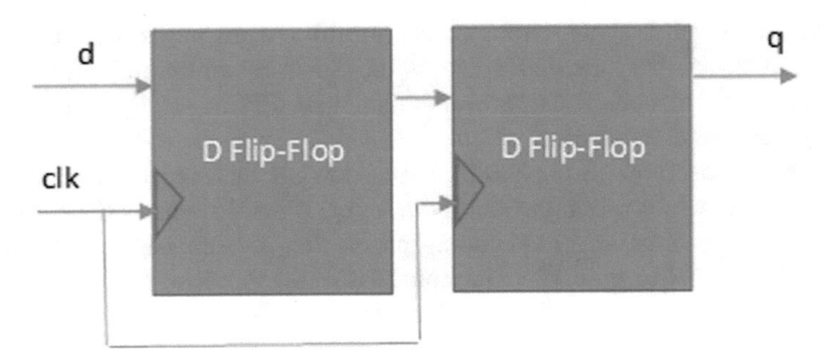

Fig. 17.13 Two-flop-level synchronizer

17.10 Guidelines for Low Power Design

Reducing the power for many applications is critical, and due to complexity of designs, only use of power-efficient FPGA devices or architecture is not sufficient! It is essential for designer to understand the features supported by the EDA tools to optimize the dynamic power. The recommendation by many FPGA vendors is to reduce the switching activity in the sequential logic and clock routing.

For the low power design, it is recommended to use the gated clocks or the low-power clock gating cells. Dynamic power of a cell is dependent on voltage, load capacitance, and on clock frequency. Due to switching at the clock input, it has been observed that the dynamic power increases. So, to reduce dynamic power, it is recommended to use clock gating cells. Figure 17.14 shows the clock gating cell.

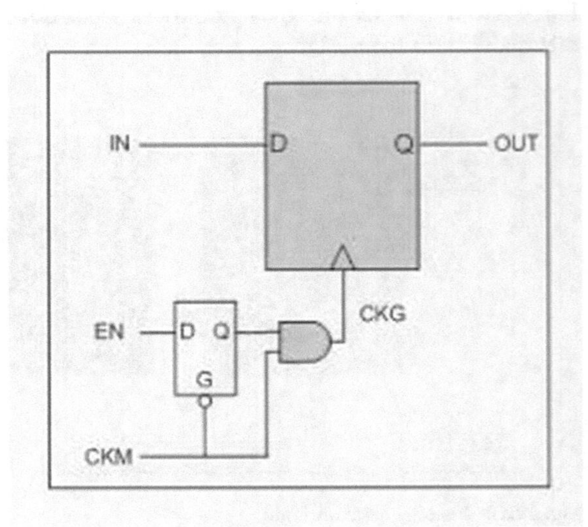

Fig. 17.14 Low-power clock gating cell

17.11 Guidelines for Use of Vendor-Specific IP Blocks

It is always recommended by the FPGA vendor to have the brief and detailed understanding of the FPGA device family and the architecture of FPGA device. It is recommended to use the vendor-specific design and coding guidelines to improve the performance of design. It is highly recommended to encrypt the IP by using standard security standards.

During synthesis phase it is recommended to infer the micro-functions such as multipliers, shift registers, memories, and DSP blocks to ensure the optimized results.

For the better performance, it is recommended to use the desired timing constraints and analyze the timing constraints by using the timing analyzer. It is even recommended to use the place and route effort level while implementing the design. The place and route effort level allows the EDA tool to use the algorithm to improve the design performance, and even it improves the design placement. It is also recommended to use the IOB resources and speed grade during design implementation stage. While using the synchronous interface, it is recommended to use the single clock synchronous RAM (read and write in the same clock domain), and while using asynchronous interfaces, use the dual port RAM.

17.12 Summary

Following are important points to summarize this chapter.

1. It is recommended to use ***blocking assignments*** while modeling the ***combinational design***.
2. It is recommended to use ***non-blocking assignments*** while modeling ***sequential design.***
3. It is recommended ***not to mix the blocking and non-blocking assignments*** in the same always block.
4. It is recommended to use ***if-else*** construct to design the priority logic.
5. It is recommended to use ***case*** construct to infer the parallel logic.
6. Priority logic infers the design having the longer combinational path due to ***nested if-else constructs.***
7. Have the separate always procedural block to code the next state combinational logic, output combinational logic, and state register logic. This improves the speed of FSM.
8. Use the signal grouping to improve the performance of FPGA-based design.
9. To avoid metastability, it is recommended to use the dual or three-flop-level synchronizer while passing signals from one clock domain to another.
10. For the low power design, it is recommended to use the gated clocks or the low-power clock gating cells.

Chapter 18
ASIC Design

ASIC is an Application Specific Integrated Circuit and designed for the specific applications. For ASIC design engineer it is required to have good understanding of the ASIC synthesis and optimization. The chapter discusses about the ASIC types, basics of ASIC design flow.

In the previous few chapters, we have discussed the RTL design using Verilog and FPGA-based designs. The chapter focuses on various types of ASICs and ASIC flow and important terms which can be useful during the ASIC design.

18.1 What Is ASIC?

ASIC is an application-specific integrated circuit. Integrated circuits are made up of silicon wafer, and each silicon wafer consists of thousands of cells. If any integrated circuit is designed for specific application, then it is called as an ASIC. The examples are chip designed for the car controller, chip designed for satellite communication, and interfacing chips to establish communication between the CPU and memory. The microprocessors and memories are general-purpose integrated circuits which are not treated as an ASIC. The following are main types of ASIC, and classification is shown in Fig. 18.1.

18.1.1 Full Custom ASIC

In such type of ASIC, the design starts from the scratch. The ASIC design team creates the ASIC logic cells and layout required for the logic. The analog and digital design can be implemented by using full-custom ASICs. In such type of ASICs, predefined standard cells or gates are not used to describe the functionality of the design.

V. Taraate, *Digital Logic Design Using Verilog*,
https://doi.org/10.1007/978-981-16-3199-3_18

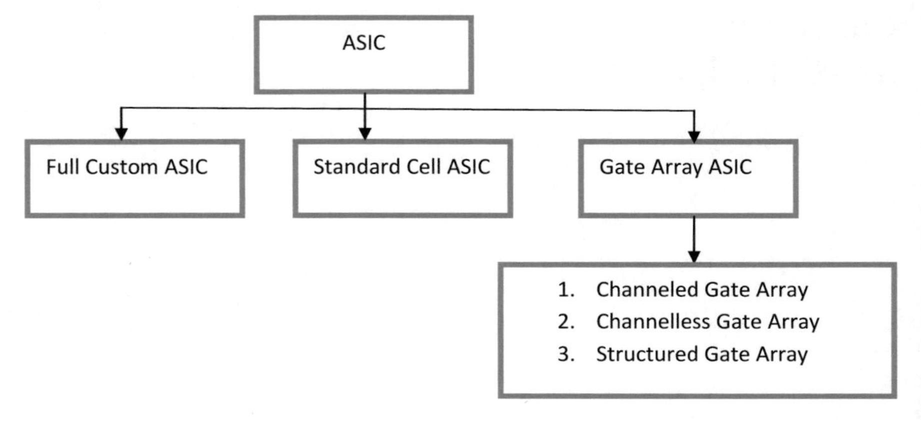

Fig. 18.1 Type of ASICs

18.1.2 Standard Cell ASIC

In such type of ASIC, the design team uses the predefined logic cells which are also called as standard cells. Few of the standard cells are logic gates, mux, flip-flops, or latches. These standard cells are predefined and pretested, so it saves a lot of time of design team and money and there is less risk while using these standard cells. These types of ASIC designs are flexible like full-custom ASIC designs but reduce overall risk. The standard cell libraries are designed by using full-custom design flow.

18.1.3 Gate Array ASIC

In such type of ASICs, the array consists of number of transistors which are predefined or prefabricated on the silicon wafer. The array is also called as base or basic array, and the transistor cell is called as basic cell or base cell. The interconnects between the cell and within the inside structure of the cell are customized and hence improve the programmability. The types of these types of ASICs are

a. Channeled gate array
b. Channel-less gate array
c. Structured gate array.

While designing the ASIC, the following are important objectives.

1. **Speed of an ASIC**: Whether the ASIC is working at the desired speed or not?
2. **Area of an ASIC**: What is the maximum area of an ASIC?
3. **Power of an ASIC**: What is the leakage and dynamic power dissipation in the best-case and worst-case scenarios?
4. **Time to Market for an ASIC**: What is the time to market for an ASIC?

18.2 ASIC Design Flow

To design an ASIC, the design team needs to have in-depth understanding of the important steps from specification to the layout. These important steps are used during the design phases. Figure 18.2 shows the ASIC design flow with the information about the important steps used during the design cycle.

As shown in Fig. 18.2, an ASIC design flow consists of the important design steps and can be considered as design milestones. Every ASIC design starts with the basic idea, and the idea to develop chip functionality is outcome of the in-depth market research! After the idea is finalized for the desired design functionality, the actual ASIC design implementation cycle starts with the specification extraction. The following section discusses the important steps during the ASIC design cycle.

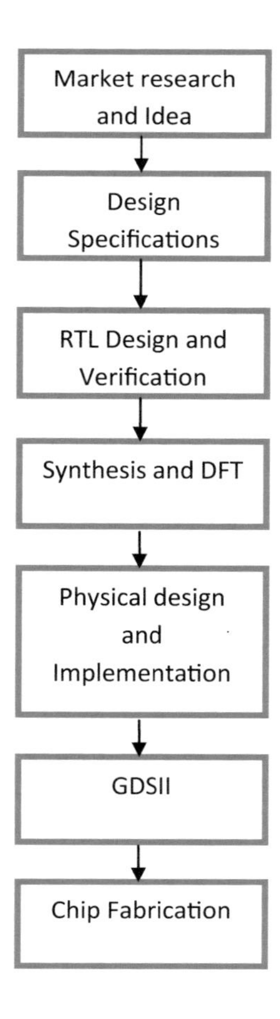

Fig. 18.2 Basic ASIC design flow

18.2.1 Design Specification

The input used to extract and finalize the design specification is the data collected during the market research for the feasible ideas or products. The following are the important points needed to be documented in the specification document.

a. Functionality of the design. That is what the chip exactly does?
b. Design goals and constraints for the design
c. Performance constraints like the speed, power, and area for the design
d. Technology constraints like the physical dimensions, area, and size of the cell
e. Fabrication techniques for the ASIC design
f. Vendor-dependent constraints and third-party IPs
g. Memories and macros used for the design
h. The data rate and the interface definitions for the design
i. Packaging information and the testing or verification planning for the design
j. Risk and dependability and time to market for the design.

The above specifications are described in the form of block diagrams, and this phase is called as architecture-level design. Most of we know that, the architecture is the block level representation of an ASIC design. For example, if 16-bit processor needs to be designed, then the architecture can consist of ALU, control logic, instruction decoder and encoder, interrupt logic, serial IO controller, bus arbitration logic, counters, and pointer logic. All these functional blocks are interconnected together to get the desired architecture required to perform the specific application. The chip architect designs the multiple architectures, and the best suitable architecture for an ASIC is finalized depending on the requirement of speed, power, and design resources. This architecture document is used in the ASIC design cycle to document the functionality of each and every functional block, and it is block-level representation of the design.

After the architecture for an ASIC is finalized, the architecture blocks are divided into the form of the sub-blocks which has the interfaces and logic details, and this is called as micro-architecture design. The micro-architecture for every functional block is useful to understand about the intended design functionality. The chip architect with good amount of experience can design the viable and feasible micro-architecture by understanding the functional, timing, and power requirements for the chip. Most of the ASIC micro-architecture uses the low power design techniques, DFT-friendly design details, information about the area requirements, partition for the multiple clock domain designs, and the timing details of the interfaces. In the micro-architecture, the software and hardware design partitioning should be included with the technology-dependent component details.

18.2.2 RTL Design and Verification

The important milestone is to design the functionality of an ASIC using synthesizable constructs by using Verilog, SystemVerilog, or VHDL and is treated as the RTL design milestone. RTL stands for the Register Transfer Level and can be efficiently described by using HDL. RTL design team uses the micro-architecture document as an input to design the functionality. The objective of the RTL design team is to describe the design functionality to realize the intended logic. The functionalities can be processor implementations, pipelined features, state machine coding, data transfer modules, memories, etc.

The RTL design is used as an input by the verification team, and the verification team uses this for early detection of the functional bugs. The verification is useful to check for the functional correctness of the design. The RTL verification for any ASIC is important milestones, and objective is to check for the functional correctness and to improve the overall coverage for the design. Most of the ASIC design flow uses the verification methodologies and languages like Verilog or SystemVerilog during the verification of the ASIC design.

18.2.3 ASIC Synthesis

Once the RTL verification is completed and the coverage goals are met, the next important milestone is logic synthesis and objective is to get the gate-level netlist. The process of getting the gate-level netlist is called as logic synthesis. The ASIC synthesis tool uses the RTL design coded using Verilog, SystemVerilog, or VHDL, design constraints, and the ASIC library as an input and generates the gate-level netlist as an output. Figure 18.3 shows the inputs used by the synthesis tool and output as a gate-level netlist.

The popular synthesis tools used in the industry are Synopsys Design Compiler (DC), Cadence RTL Compiler, etc. The synthesis tool considers the time, power,

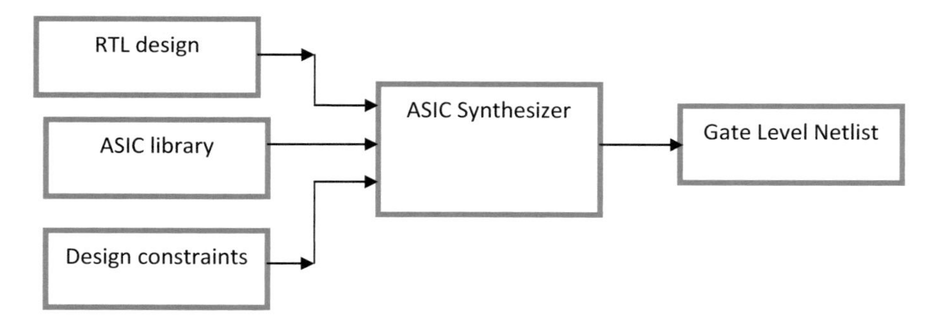

Fig. 18.3 ASIC synthesis **input and output**

and testability as the major important factors to generate the gate-level netlist. Synthesis tool tries to meet the constraints by calculating the cost of various implementations. The gate-level netlist is the structural description having only standard cells. The gate-level netlist is verified for the functional correctness of the design, and this phase is called as gate-level verification.

After successful gate-level verification of the RTL design, the objective is to check for the timing violations. This phase is called as pre layout STA. During this milestone, the objective of STA team is to find out the timing violations for the design. During this phase, STA is performed without considering the parasitic (RC) effect. The objective is to fix the setup time violations in the design and to improve the overall performance of the design. In most of the ASICs, the hold time violations for various timing paths are fixed after CTS and routing that is during the physical design flow.

Before physical implementation of an ASIC design, the design for testability (DFT) should be checked to understand about the DRC violations. Even the objective is to find out the various design faults. As discussed above, the RTL should be DFT-friendly for the efficient scan chain insertions and to find out the overall fault coverage for the design. The DFT techniques and processes are out of scope as per as this book is concerned.

18.2.4 Physical Design and Implementation:

The next milestone in the ASIC design flow is physical design and implementation. In this phase, the gate-level netlist is processed and passes through various phases to have geometric representations. The geometric representation can be treated as the layout of the design. The discussion on the physical design flow is out of scope,

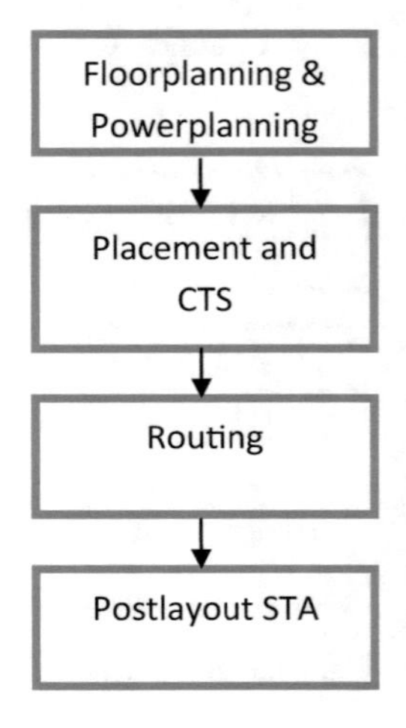

Fig. 4 Physical design flow important steps

and readers are requested to refer the physical design and synthesis books. Few of the important steps during physical design flow are shown in Fig. 4.

The gate-level netlist is used by the physical design tool to get the layout. The important steps in the physical design flow are floor planning, power planning, placement of standard cells and macros, clock tree synthesis, routing, and post-layout STA to get the final layout. The design is basically converted from the gate-level abstraction to the switch-level abstraction by using standard cells and macros. The netlist after place and route phase is given as an input to the STA tool to fix the timing violations, and this process is called as post-layout STA. The post-layout STA is by considering the routing delays.

The physical design and implementation tool uses the design rule library to get the GDSII file. The design rule library consists of the guidelines based on the fabrication processes. GDSII file is used by the foundry to fabricate the integrated circuit. The industry leading tool for the physical design and implementation is IC Compiler from Synopsys or Encounter from Cadence.

The physical verification needs to be performed to verify the intended design functionality and to make sure that the layout is designed according to the rules! After the physical verification and timing analysis, the layout is ready for the fabrication. In this phase, the layout data is converted into the photolithography masks. After the fabrication process, the wafer is diced into various individual chips and packaged as well as tested.

18.3 ASIC Design and Synthesis Strategies

Consider the 8-bit processor logic which has the functional blocks as shown in Fig. 18.5. The RTL design team should use the strategy to code the block-level design and to check for the functional correctness of the design.

During synthesis, we need to have the block-level synthesis using the block-level constraints and top-level synthesis using the constraints. The main optimization constraints are area and speed, and during logic synthesis, the objective is to achieve the design performance for the block-level and top-level design.

The block-level constraints met do not mean that top-level constraints will also meet. It depends upon the complexity of the design and on the overall timing associated for the design. More about the ASIC synthesis strategies and STA, refer Chaps. 19–21.

If the constraints are not met, then the design team should use the RTL tweaks or architecture and micro-architecture tweaks and are discussed in Chap. 21.

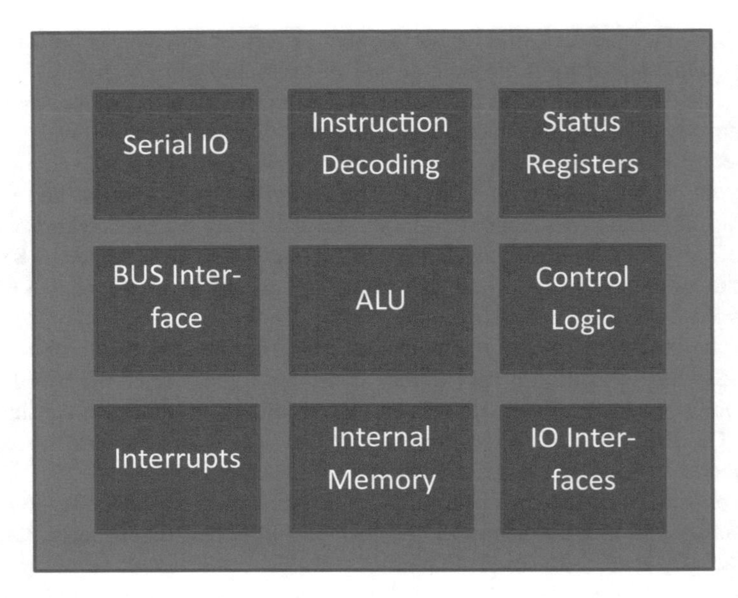

Fig. 5 Top-level processor design

18.4 Summary

The following are the important points to conclude this chapter.

1. ASIC is an application-specific integrated circuit and used to design the chips for the specific applications.
2. In the full-custom ASIC designs, the design does not use the predefined library cells. The design is done from the scratch.
3. In the semi-custom ASIC design, the design uses the predefined and pretested standard cell library components.
4. RTL synthesis is process of getting the lower level of design abstraction from higher-level design.
5. The logic synthesis tool uses the Verilog RTL, libraries, and constraints as an input.
6. The physical design flow has the steps like floor planning, power planning, CTS, placement and routing, post-layout STA, back-annotation, and layout.
7. The optimization can be achieved by tweaking the RTL design or by using the tool-based synthesis optimization algorithms.

Chapter 19
ASIC Synthesis and SDC Commands

During the ASIC synthesis the objective is to get the gate level netlist. The synthesis tool uses the optimization constraints, and these constraints are specified by using the SDC commands. The chapter discusses about the ASIC synthesis and important SDC commands used during synthesis.

As discussed in Chap. 18, the ASIC synthesis tool uses the Verilog RTL, constraints, and ASIC library and performs the synthesis and optimization to generate the gate-level netlist. The chapter discusses the ASIC synthesis flow and the SDC commands.

19.1 ASIC Synthesis Using Design Compiler

This section only focuses on the logic synthesis using the Design Compiler to get the gate-level netlist. As discussed previously, the logic synthesis tool uses the RTL design either Verilog (.v) or VHDL (.vhd) files, the design constraints (.sdc), and library(.lib) as an input and generates the optimized gate-level netlist using standard cells available in the library. The gate-level netlist is technology dependent and can change if process node varies. Depending on the functionality, the gate-level netlist for the 40 nm can be different as compared to gate-level netlist generated for the lower process nodes like 20 nm or 14 nm process node. ASIC synthesis tool performs few steps to generate the gate-level netlist. The important steps during the ASIC synthesis are translate, map, and optimize. The important steps during the FPGA synthesis are translate, optimize, and map. Figure 19.1 gives the brief information about the ASIC synthesis steps to generate the gate-level netlist.

1. ***Read Library***: During the logic synthesis, the synthesis tool reads the DesignWare libraries, technology libraries, and symbol libraries. The DesignWare library consists of the complex cells like adders, comparators, multipliers, etc. The technology library consists of the logic gates and flip-flops, latches, etc. While synthesizing the synthesis, optimization algorithms

V. Taraate, *Digital Logic Design Using Verilog*,
https://doi.org/10.1007/978-981-16-3199-3_19

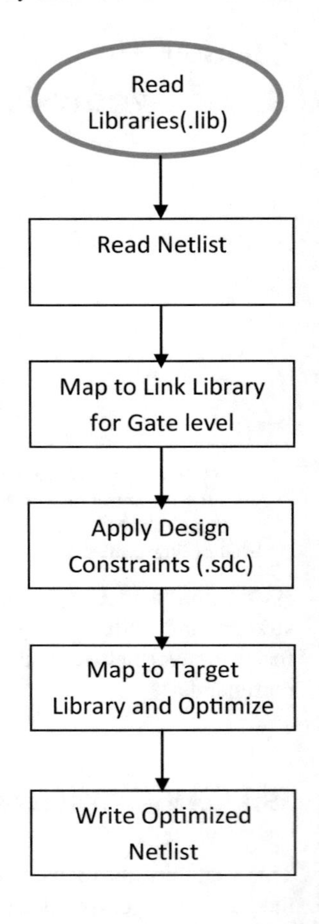

Fig. 19.1 ASIC synthesis important **steps**

automatically determine when to use the technology library cells and when to use the DesignWare library components. These library cells are used efficiently to generate the gate-level netlist.

2. The next step is to read the RTL description described by using either Verilog or VHDL.
3. The synthesis tool after reading the libraries and the RTL performs many required steps like optimization, conversion to unoptimized Boolean logic, and technology-independent optimization and finally maps the logic using the technology library. The above process is called as linking the logic to the desired target library.
4. The synthesis tool uses the design constraints like area, speed, and power while optimizing the design using the standard cells according to the target library. So basically, link library can be IO library, cell library, or macrolibrary and used to link the design and target library is used while optimizing the design.

5. For efficient RTL coding, it is required that RTL design engineer should have good understanding of the target standard cell library. After the design is optimized, then the design is ready for the design for testability (DFT) that is to detect early faults in the design. During RTL design stage only, the DFT-friendly RTL needs to be described to enable quick scan insertions and testing for various faults in the design.
6. The optimized netlist can be in the Verilog (.v) format or in the database (.ddc) format and used by the placement and routing tool. Based on the routing, the back-annotation can be performed with actual routing delays for accurate timing analysis. If timing goals are not met, then the design can be resynthesized to meet the timing goals.

19.2 ASIC Synthesis Guidelines

The Synopsys Design Compiler reads the startup file from the current working directory. The startup file is ***synopsys_dc.setup.*** There should be two startup files: One should be in the current working directory, and another should be in the root directory where the Design Compiler is installed. To use the tool, the following important parameters need to be set up.

1. ***search_path***: This parameter is used to search for the synthesis technology library for reference during synthesis.
2. ***target_library***: This parameter is used by the DC while mapping the logic cells during synthesis. The target library consists of the logic cells.
3. ***symbol_library***: All the logic cells have symbolical representation. This parameter is used to point the library that contains the visual information for the logic cells present in the technology synthesis library.
4. ***link library***: The tool uses the cells from the target_library for mapping the desired functionality, and this parameter is used to point to the reference of the logic gates in the synthesis technology library.

The above four parameters for.*synopsys_dc.setup* are described by using the following.

```
set search_path " ./synopsys/libraries/syn/cell_library/syn"
set target_library "tcbn65lpwc.db, tcbn65lpbc.db"
set link_library " $target_library $symbol_library"
set symbol_library " standard.sldb dw_foundation.sldb"
```

Once the above variable or parameters are set up for the desired process node library, then the synthesis tool can be invoked at the command prompt.

The design objects are documented in Table 19.1 and used during synthesis. Every design is the description of the logic circuit to perform some of the logical operations. The design can be single system description or can consist of the multiple sub-systems. The design objects are described in Table 19.1.

19.3 Constraining Design Using Synopsys DC

The design is described using VHDL, Verilog languages using the synthesizable constructs. This design needs to be used as an input by Design Compiler. Table 19.2 describes the important commands used by DC.

19.3.1 Reading the Design

It is essential for the ASIC design team to understand about the difference between the read command and analyze, elaborate command? The following are important highlights:

1. The analyze and elaborate commands are used to pass required parameters while elaborating the design.
2. The read command is used while entering for the pre-compiled designs or netlists in DC.
3. Using analyze and elaborate commands, the different architectures can be specified during elaboration for the same analyzed design.
4. The read command does not allow the use of the different architectures.

Table 19.1 Design objects used by Synopsys DC

Design object	Description
Cell	Cell is also called as instance. The instantiated name of the sub-design is called as cell
Reference	It is original design to which cell or instance refers. For example, instantiated sub-design must refer to the design which consists of the functional description of the sub-design
Ports	The primary inputs and outputs or IOs of the design are called as ports
Pins	The primary inputs, outputs, and IOs of cells in the design are called as pins
Net	Wires used for the connection between ports of the pins of the different designs are called as net
Clock	The input port or pin used as clock source is called as clock
Library	The technology-specific cells used for targeting for synthesis, linking, or reference are called as library

Table 19.2 Commands used to read the design

Command	Description
read –format <format_type> <filename>	Used to read the design. For example, to read the Verilog module processor.v the command can be **read-format verilog processor.v**
analyze –format <format_type> <list of file names>	Used to analyze the design to find the syntax errors and to perform the translation before building the generic logic. The generic logic is part of the Synopsys generic technology-independent library. The components are named as GTECH. This logic is unmapped representation of the Boolean functions. The command can be used as **analyze –format verilog processor.v**
elaborate –format <list of module names>	Used to elaborate the design and can be used to specify the different architectures during design elaboration for the same analyzed design. The command can be **elaborate –library work processor**

Table 19.3 Command used to check the design

Command	Description
check_design	Used to check the design problems like shorts, opens, multiple connections, and instantiations with the no connections **check_design**

19.3.2 Checking of the Design

After the design has been read using the Synopsys DC, the check_design is used. Table 19.3 describes the command used to check the errors in the design.

19.3.3 Clock Definitions

The clock needs to be created using the command create_clock, and this is used as reference by timing analysis tool. Table 19.4 describes the clock definition commands.

Table 4 Commands for clock definition

Command	Description
create_clock –name <clock_name> -period <clock_period> <clock_pin_name>	Used to create the clock for the design and used as reference during timing analysis. The clock is always associated with the clock pin of the design. If design does not have clock, then it will be treated as virtual clock. The command can be used to generate 200 MHz clock with 50% duty cycle and is **create_clock –name clock -period 5 master_clock**

If the design requirement is to have the clock for variable duty cycle with rising edge at 1 ns and clock period of 5 ns, then the same command can be modified as

```
create_clock -name clock - period 5 -waveform {1,5} -name master_clock
```

If the design does not have the clock pin, then the virtual clock can be created using the following commands.

This command generates virtual clock of frequency 200 MHz with 50% duty cycle.

```
create_clock -name clock -period 5
```

This command generates virtual clock of frequency 200 MHz with variable duty cycle with rising edge at 1 ns and falling edge at 5 ns.

```
create_clock -name clock -period 5 -waveform {1,5}
```

19.3.4 Skew Definition

As we know that the skew is difference between arrivals of the clock signal at various pins of the flip-flop, it is essential to design the clock network to have balanced skew. If clock at the source flip-flop is delayed with reference to the destination flip-flop, then the skew is called as negative clock skew and useful for the hold. If clock at the destination flip-flop is delayed with reference to the source flip-flop, then the skew is called as positive clock skew and useful for the setup. The reason for positive clock skew is the clock at the destination flip-flop is delayed and hence the data can arrive late.

Table 19.5 Commands used for the skew definitions

Command	Description
set_clock_skew –rise_delay <rising_clock_skew> -fall_delay <falling_clock_skew> <clock_name>	This command is used to define the clock skew for the design. This can be described as **set_clock_skew –rise_delay 2 –fall_delay 1 master_clock**

Table 19.6 Commands used for the Input, output delay definitions

Command	Description
set_input_delay –clock <clock_name> <input_delay> <input_port>	Used to specify the input port delay with reference to the clock. To specify 1 ns delay with reference to clock, the command can be used as **set_input_delay –clock master_clock 1 data_in**
set_output_delay –clock <clock_name> <output_delay> <output_port>	Used to specify the output port delay with reference to the clock. To specify 1 ns delay with reference to clock, the command can be used as **set_output_delay –clock master_clock 1 data_out**

The Synopsys DC will not be able to synthesize the clock tree; so to overcome the problem, the clock skew can be used to model the propagation delay that exists in the clock tree.

Table 19.5 describes the commands used by Synopsys DC while defining clock skew for the design.

19.3.5 Specifying the Input and Output Delay

The input and output delays can be specified by using set_input_delay and set_output_delay, respectively. Table 19.6 describes the command used with the required parameter definition.

19.3.6 Specify the Minimum (min) and Maximum (Max) Delay

The input and output delays can be specified as min or max depending on the requirements. Table 19.7 describes the min and max delay definitions.

Table 19.7 Commands used for min and max IO delay definitions

Command	Description
set_input_delay –clock <clock_name> -max <delay> <input_port>	Used to specify the max input port delay with reference to the clock. To specify 2 ns delay with reference to clock, the command can be used as **set_input_delay –clock master_clock –max 1 data_in**
set_input_delay –clock <clock_name> -min <delay> <input_port>	Used to specify the min input port delay with reference to the clock. To specify the 1 ns delay with reference to clock, the command can be used as **set_input_delay –clock master_clock –min 1 data_in**
set_output_delay –clock <clock_name> -max <delay> <output_port>	Used to specify the max output port delay with reference to the clock. To specify 2 ns delay with reference to clock, the command can be used as **set_output_delay –clock master_clock –max 2 data_out**
set_output_delay –clock <clock_name> -min <delay> <output_port>	Used to specify the min output port delay with reference to the clock. To specify 1 ns delay with reference to clock, the command can be used as **set_output_delay –clock master_clock –min 1 data_out**

19.3.7 Design Synthesis

Using the command compile, the design can be synthesized; prior to synthesis, the design constraints need to be given to the design. The design can be synthesized using the different effort levels like low, medium, and high.

Table 19.8 describes the compile command.

Table 19.8 Command used for compiling the design

Command	Description
compile –map_effort <map_effort_level>	This command is used to synthesize the design with different effort levels like low, medium, and high. The command for the medium effort level can be **compile –map_effort medium**

19.3.8 Command to Save the Design

The design can be saved by using write command in various formats using Design Compiler. The format can be Verilog or database format (ddc).

Table 19.9 describes the command used to save the design.

Table 19.9 Command used to save the gate-level netlist

Command	Description
write –format < format_type > -output < file_name >	This command is used to save the output generated by synthesis tool in various formats. For the Verilog format, the command can be **write –format verilog -output processor_netlist.v**

19.4 Synthesis and Optimization Techniques

Before discussion on the synthesis reports and timing reports, let us understand the different synthesis techniques used during the optimization. The optimization can be performed at the RTL, architecture level, or during synthesis. The fully optimized design is that which has met the area and timing requirements. The optimization at the RTL level can be achieved by few tweaks to meet the intended functionality. In such type of optimizations, care needs to be taken that the optimized code should have the same simulation results before synthesis and after synthesis. There are few standard techniques used in the real practical scenarios to have better synthesis optimizations and results. Few of such techniques are discussed in this section.

19.4.1 Resource Allocation

This is used for the better synthesis results, and this optimization technique uses the sharing of common resources.

Consider the Verilog procedural block shown below

```
always@(*)
begin
if(a_in==1)
        y_out= b_in+c_in;
else
        y_out = b-in+d_in;
end
```

The above functionality infers the two adders: one to perform addition of a_in and b_in and another to perform addition of b_in and d_in. It also infers the 2:1 mux to select the output of one of the adders. The synthesis result is shown in Fig. 19.2.

In the synthesis result shown without use of the resource allocation, the common input b_in is not shared. If the RTL shown above is tweaked to have only one adder, then the synthesis optimization is better due to minimum area. Figure 19.3 shows the logic having minimum resources.

The modified optimized RTL description is shown in the following example

```
always@(*)
begin
if(a_in==1)
        y_tmp= c_in;
else
        y_tmp= d_in;
end
assign y_out = b_in + y_tmp;
```

So, prior to the sharing of the resources the area was more but by using the resource sharing technique the area has improved.

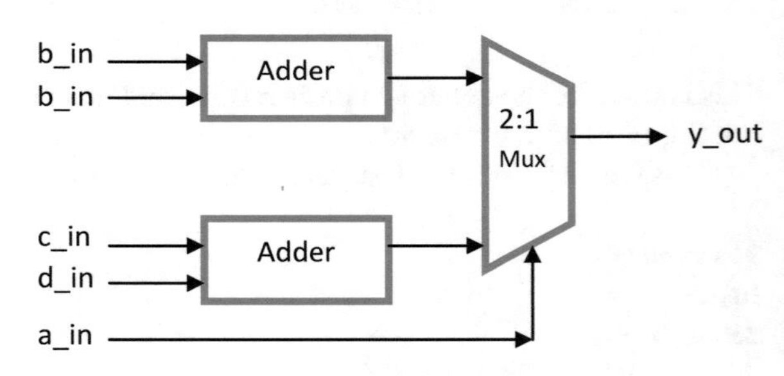

Fig. 19.2 Synthesis result without resource allocation

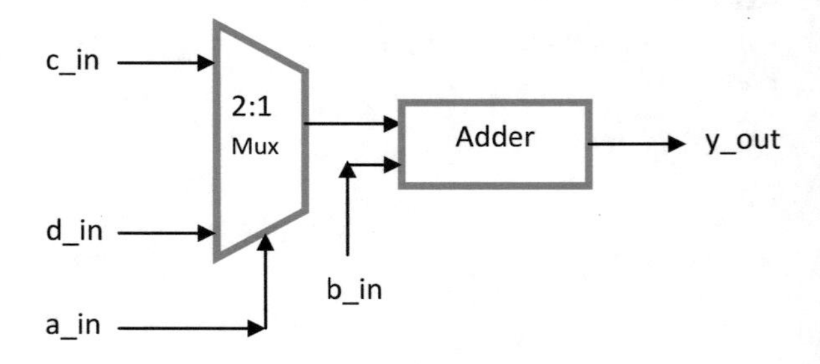

Fig. 19.3 Synthesis result with resource allocation

19.4.2 Common Factors and Sub-expression Use During Optimization

In most of the RTL designs, it is essential to use the expressions or sub-expression. Most of the time, the sub-expressions are not reused. If the sub-expression computed values are reused, then the synthesis tool will be able to perform the better optimization.

Consider the example shown below. In the example, b_in + c_in is used during the multiple assignments.

```
assign y_tmp = b_in + c_in;
assign z_out = d_in – ( b_in + c_in);
```

The following tweak while using the continuous assignment can give the better logic using minimum resources.

***assign** z_out = d_in – y_tmp;*

Consider another RTL code, the common factor can be reused while coding an efficient Verilog RTL.

```
always@(*)
begin
if (a_in)
        y_out = b_in & ( c_in + d_in);
else
        z_out = e_in ^ (c_in + d_in);
end
```

In the above example, the common factor is (c_in + d_in) and can be reused. The above code can be modified as

```
always@(*)
tmp_add = c_in + d_in;
begin
if (a_in)
        y_out = b_in & (tmp_add);
else
        z_out = e_in ^ (tmp_add);
end
```

These minor modifications in the Verilog RTL can be useful to have more optimized logic.

19.4.3 Moving the Piece of Code

In most of the RTL designs, the expressions are used while using the ***for*** or ***while*** loops. These expression values may or may not change during every iteration. Those statements used within ***for*** or ***while*** loops whose value will not change can be handled by using the tweaks in the code. The synthesis tool during optimization phase handles such scenario, but there are chances of redundant logic generation. This can be avoided by moving the expression outside of the loop. Consider the following Verilog RTL.

```
//The value of y_tmp in the range of 0 to 9.
assign y_tmp = a_in + b_in;
for ( y_tmp = 0; y_ymp < 9; y_tmp = y_tmp + 1)
z_out = y_tmp − 9;
```

In the above example, it is assumed that y_out is not assigned with the new value within the loop and the above expression remains constant for every iteration within the loop. The synthesis tool infers the logic having the subtractors during synthesis, and this occupies more area. The above Verilog RTL functionality can be modified to avoid the unnecessary logic.

//The value of y_tmp in the range of 0 to 9

```
assign y_tmp = a_in + b_in;
assign tmp = y_tmp-6
for ( y_tmp = 0; y_ymp < 9; y_tmp = y_tmp + 1)
z_out = tmp;
```

19.4.4 Constant Folding

Consider the use of constants in the RTL design. Instead of declaring the constants, use the direct computed or required value for the y_out. The piece of RTL code is shown in the following example.

```
integer c_in = 3;
assign y_out = c_in *3;
```

Instead of using the above Verilog RTL, the better way is use the value 9 and assign the value to y_out and this technique is called as constant folding.

19.4.5 Dead Zone Elimination

The section of the code which is never executed is called as dead zone code. The dead zone code elimination technique can be used for the better synthesis results.

The piece of Verilog RTL is shown below

```
integer c_in = 3;
integer b_in = 2;
always@(*)
if (b_in > c_in)
        y_out = 1;
else
        y_out = 0;
end
```

In the above RTL code the condition is always false and hence if construct always executes the false condition assignment. The synthesis tool during synthesis will perform such kind of optimizations. But if the code is modified, then it will reduce the compilation time during synthesis.

19.4.6 Use of Parentheses

In most of the RTL designs if parentheses are used efficiently, then the synthesis results can be optimized.

For example, if the assign construct is used in the RTL without any parentheses then it infers the logic which has more propagation delay (Fig. 19.4).

```
assign y_out = a_in + b_in – c_in –d_in;
```

If the above RTL is modified as shown below, then it gives the clear timing and data path (Fig. 19.5).

```
assign y_out= (a_in+b_in) – (c_in+d_in);
```

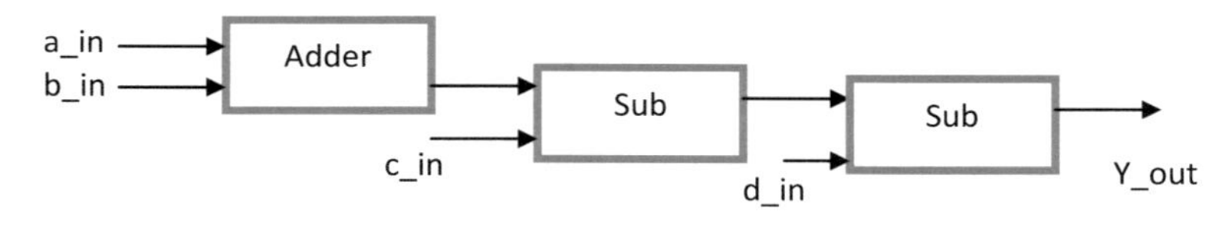

Fig. 19.4 Synthesis result without use of parentheses

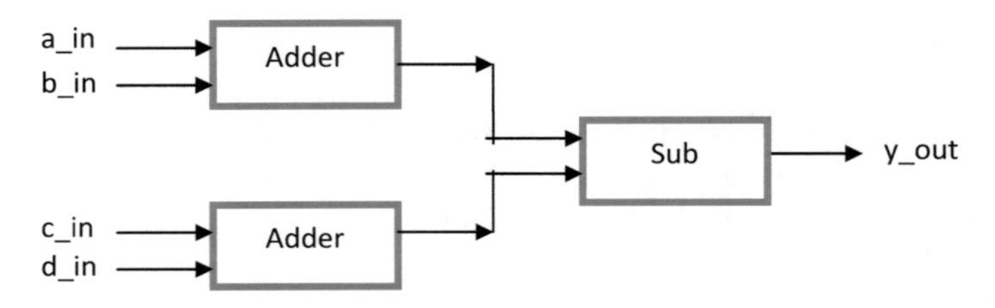

Fig. 19.5 Synthesis result with use of parentheses

19.4.7 Partitioning and Structuring the Design

The design needs to be structured and partitioned for the better synthesis outcome. It is the practical reality that the design which is better partitioned generates better synthesis results and even it reduces the synthesis runtime. The following are important guidelines recommended for the better and efficient design partitioning

1. Partition the design for the design reuse.
2. To describe the functionality, use the different modules.
3. Use the combinational logic in the same block.
4. Use the separate block or structure logic for the random logic.
5. Partition the design at the top level.
6. Do not use the glue logic at the top level.
7. Use the separate module for state machines that is isolating the state machines form the other logic.
8. Limit the logic size to maximum 10 K gates for every block.
9. Avoid use of the multiple clocks in the same block.
10. Isolate the synchronizers for the multiple clock domain designs.

19.5 Summary

The following are the few important points to conclude this chapter.

1. RTL design engineer should have good understanding of the target standard cell library.
2. The optimization can be performed at the code level or during synthesis.
3. The fully optimized design is that which has met the area and timing requirements.
4. If the sub-expression computed values are reused, then the synthesis tool will be able to perform the better optimization.
5. The resource allocation or sharing of common resources is the better technique to improve the area of the design.

6. Use the constant folding and dead zone elimination to improve the area optimization.
7. The design which is better partitioned generates better synthesis results, and even it reduces the synthesis runtime.

Chapter 20
Static Timing Analysis

STA is non-vectored approach used in the timing closure. STA is used to find whether all the timing paths are met or not. For RTL design engineer it is essential to have good understanding of different timing paths. This chapter discusses about the STA concepts and their use in the timing closure.

In the previous chapters, we have discussed the important RTL concepts and synthesis in detail. But we have not discussed the timing parameters for the ASIC design. The timing analysis is especially an important phase for any ASIC design, and it can be performed during various design phases. Timing analysis can be performed before design layout stage and after design layout stage. So, it is essential and important to understand important timing parameters and considerations for an ASIC design.

Before layout, the timing analysis is performed on gate-level netlist of the design with goal to fix the setup time. Timing analysis tool uses the design constraints and the timing libraries to perform the timing analysis for the design. Timing analysis is of types of static and dynamic. Static timing analysis (STA) is performed without using any set of vectors, and dynamic timing analysis is performed by using set of vectors for the design. The goal is to fix the setup and hold time violations for the design.

For any sequential element, two important timing parameters are setup and hold time.

If setup time or hold time is violated, then the design goes into metastable state. So, it is essential to find out timing issues and fix the timing violations for the design and this process is performed by the timing analysis tool. Popular timing analysis tool is Synopsys Prime Time (Synopsys PT). This chapter focuses on the important timing considerations and their importance and use during timing closure.

V. Taraate, *Digital Logic Design Using Verilog*,
https://doi.org/10.1007/978-981-16-3199-3_20

20.1 Setup Time

The amount of time for which the input signal D of the flip-flop should maintain stable value either logic 0 or logic 1 before arrival of an active edge of the clock is called as setup time.

The setup time considerations are especially important for the design when the design is overconstrained. There can be many setup violations. The designer can perceive that the violations in the design are due to the tight constraints in the design.

To meet the setup time, it is required that the data should arrive at the input of D flip-flop before certain amount of time before arrival of the active clock edge. For example, if we consider design operated with 200 MHz clock frequency (clock cycle time = 5 ns) and have set up time requirement of 1 ns, then it is required that data should arrive at least at 4 ns so that the required setup time of 1 ns can be met.

Consider Fig. 20.1 consisting of combinational logic at the input of the register. If setup time is t_{su}, then the data should arrive at the D input to meet the setup time.

So, the required time to travel data at D input is $T_{clk} - t_{su}$. The data arrival time is t_{comb} that is delay of combinational logic.

Figure 20.2 shows the valid setup time region with the necessary condition to meet the desired setup time. Let us consider the positive edge of the clock, the data arrives at the D input of flip-flop prior to setup time window. So, there is no any setup violation in the design.

Data arrival time is the amount of time required to arrive the data at the data input of the D flip-flop. It is given by

Data arrival time = Propagation delay of flip-flop + combinational delay
Data required time = time duration of clock cycle − setup time

The difference in between data arrival time and data required time is called as setup slack, and to meet setup time, the slack should be positive.

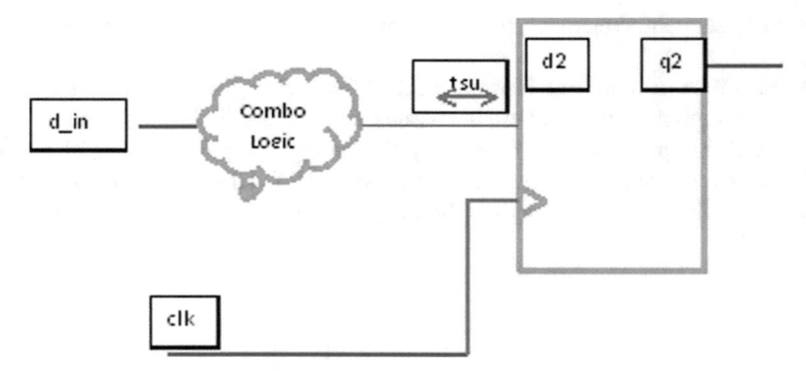

Fig. 20.1 Input–to-register path

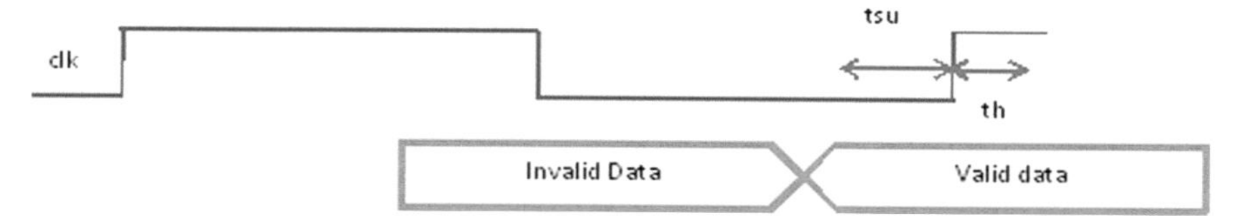

Fig. 20.2 Valid setup time region timing sequence

20.2 Hold Time

The amount of time for which the input signal D of flip-flop should maintain the stable value either logic 0 or logic 1 after arrival of an active edge of the clock is called as hold time.

Hold time is an important timing parameter consideration in the design. For most of the design constrained at high frequency, it is critical to meet the hold time. During the STA at ASIC layout stages, most of the hold violations are reported and fixed. The hold violations in the design are due to the fact that data is arriving slowly as compared to the required time.

For example, consider the scenario in Fig. 20.3. The design is constrained at 200 MHz operating frequency; that is, clock cycle time is 5 ns. If hold time requirement is 1 ns and data arrived at D input of flip-flop changes during the 1 ns window after arrival of active clock edge, then there is hold violation in the design.

As shown in Fig. 20.3, the valid data is present at the D input of the flip-flop. Both setup and hold times are met for the design; hence, there is no any timing violation in the design.

Data arrival time = Propagation delay of flip-flop + combinational delay should be greater than hold time of flip-flop.

If propagation delay of flip-flop is 3 ns and combinational delay is 1 ns for the design, then data will never change during the 1 ns window so there is no any chance of hold violation in the design.

But consider the design scenario, for the design the flip-flop propagation delay is 0.8 ns, hold time is 1 ns and there is no any combinational logic in the data path, in such scenario the hold violation occurs in the design.

So, it is important to note that the data should be stable at the D input of flip-flop during setup and hold time window.

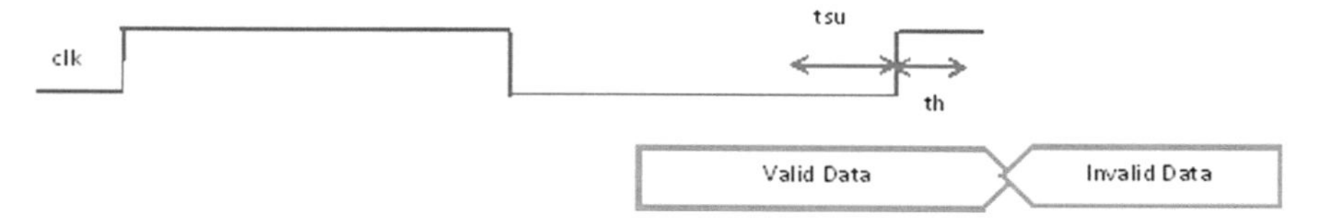

Fig. 20.3 Valid hold time region timing sequence

20.3 Clock to Q Delay

The amount of time required for the flip-flop to generate valid output either logic 0 or logic 1 after arrival of an active clock edge is called as propagation delay of flip-flop. Propagation delay of flip-flop is also called as clock to q delay.

The amount of time required for the flip-flop to generate valid output either logic 1 or logic 0 after arrival of the active clock edge is called as propagation delay of flip-flop. The propagation delay of flip-flop is also called as clock to output delay or clock to q delay of flip-flop.

Consider t_{su} is setup time of flip-flop, t_h is hold time of flip-flop, and t_{pff} is propagation delay of flip-flop. Figure 20.4 shows the various timing parameters for the register.

20.3.1 Frequency Calculations

As shown in Fig. 20.4, the timing parameters of flip-flop(reg1) are given as t_{pff1}, t_{su1} and the timing parameters for the register 2 are given as t_{pff2}, t_{su2}. The combinational logic design delay in the data path is given as t_{comb}.

These timing parameters are used to find the maximum operating frequency for the design.

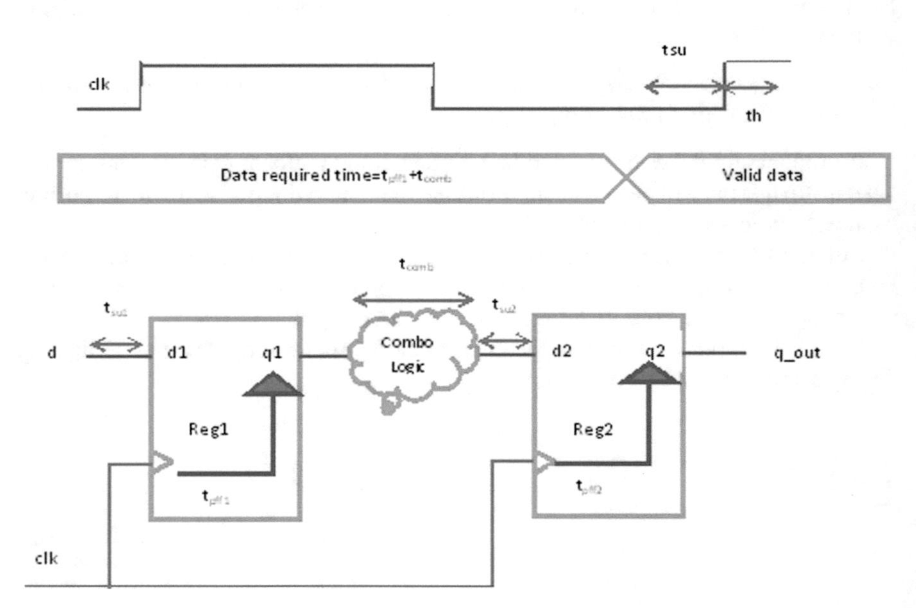

Fig. 20.4 Register-to-register path

To find the maximum operating frequency of the design, find out the data required time and data arrival time. The data required time is the addition of all the delays in the register-to-register path.

Therefore, the data required time is given by: $t_{pff1} + t_{comb}$.

The data arrival time is given by: $T_{clk} - t_{su2}$ where T_{clk} is the clock cycle time and t_{su2} is the setup time of second flip-flop.

So, the maximum frequency is calculated by equating the data required time and data arrival time.

$$t_{pff1} + t_{comb} = T_{clk} - t_{su2}$$
$$T_{clk} = t_{pff1} + t_{comb} + t_{su2}$$
$$F_{max} = 1/(t_{pff1} + t_{comb} + t_{su2})$$

Consider both flip-flops have same timing parameter values, that is, $t_{pff1} = t_{pff2} = 2$ ns, $t_{su1} = t_{su2} = 1$ ns, and $t_{comb} = 2$ ns. Then, the maximum operating frequency is

$$F_{max} = 1/(2 + 2 + 1)ns = 200\ MHz$$

20.4 Skew in Design

Consider the example shown in Fig. 20.5. In this example, the flip-flop1(Reg1) is triggered early and flip-flop2(Reg2) is triggered late. Flip-flop1 is called as launch flip-flop, and flip-flop2 is called as capture flip-flop. As the launch flip-flop is triggered first and capture flip-flop is triggered last, there is skew in the clock pulse and it is called as positive clock skew.

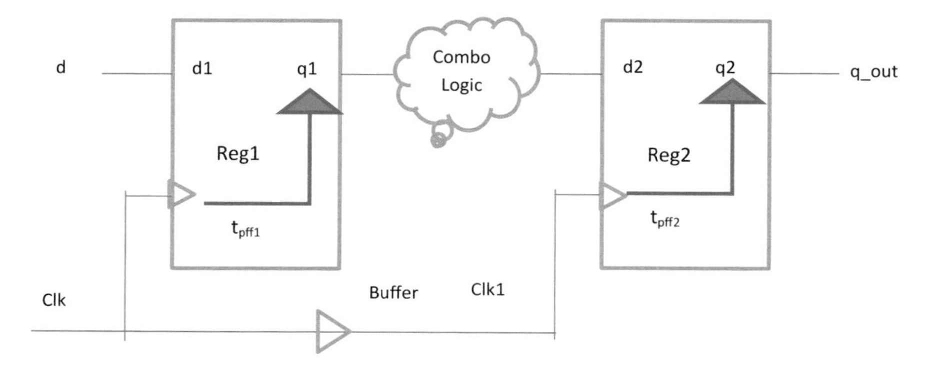

Fig. 20.5 Positive clock skew in the design

In the above example, clock and data travel in the same direction, and due to buffer delay, the clk1 is delayed by buffer delay as compared to clk input.

To find the maximum operating frequency for the above design, find out the data required time and data arrival time. The data required time is the addition of the delays in the register-to-register path.

Therefore, the data arrival time is given by: $t_{pff1} + t_{comb}$.

The data required time is given by: $T_{clk} - t_{su2} + t_{buf}$ where T_{clk} is the clock cycle time and t_{su2} is the setup time of second flip-flop where t_{buf} is the buffer delay of the buffer in the clock path.

So, the maximum frequency is calculated by equating the data required time and data arrival time.

$$t_{pff1} + t_{comb} = T_{clk} - t_{su2} + t_{buf}$$
$$T_{clk} = t_{pff1} + t_{comb} + t_{su2} - t_{buf}$$
$$F_{max} = 1/(t_{pff1} + t_{comb} + t_{su2} - t_{buf})$$

Consider both flip-flops have same timing parameters, that is, $t_{pff1} = t_{pff2} = 2$ ns, $t_{su1} = t_{su2} = 1$ ns, $t_{buf} = 1$ ns, and $t_{comb} = 2$ ns. Then, the maximum operating frequency is

$$F_{max} = 1/(2 + 2 + 1 - 1)ns = 250\ MHz$$

So, from the above discussion, positive clock skew is good to improve the performance of design. In the above example due to the buffer delay of 1 ns, the clock at flip-flop2 is delayed by 1 ns time as compared to the clk at flip-flop1. So, the time of 1 ns delayed clock can be compensated by setup time and hence increases frequency by 50 MHz.

Let us consider another example shown in Fig. 20.6. In this example, source flip-flop is triggered last and destination flip-flop is triggered first. In the other way, one can perceive that the clock and data are traveling in the opposite direction.

To find the maximum operating frequency for the above design, find out the data required time and data arrival time. The data required time is the addition of the delays in the register-to-register path.

Therefore, the data arrival time is given by: $t_{pff1} + t_{comb} + t_{buf}$.

The data required time is given by $T_{clk} - t_{su2}$ where T_{clk} is the clock cycle time and t_{su2} is the setup time of second flip-flop where t_{buf} is the buffer delay of the buffer in the clock path.

So, the maximum frequency is calculated by equating the data required time and data arrival time.

$$t_{pff1} + t_{comb} + t_{buf} = T_{clk} - t_{su2}$$
$$T_{clk} = t_{pff1} + t_{comb} + t_{su2} + t_{buf}$$

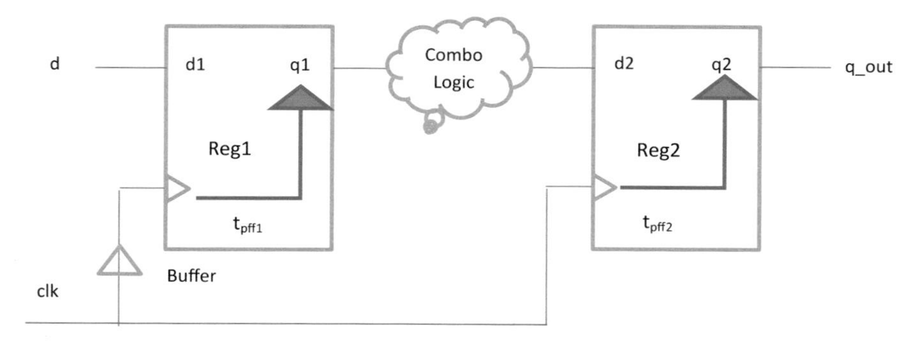

Fig. 20.6 Negative clock skew in the design

$$F_{max} = 1/(t_{pff1} + t_{comb} + t_{su2} + t_{buf})$$

Consider both flip-flops have same timing parameters, that is, $t_{pff1} = t_{pff2} = 2$ ns, $t_{su1} = t_{su2} = 1$ ns, $t_{buf} = 1$ ns, and $t_{comb} = 2$ ns. Then, the maximum operating frequency is

$$F_{max} = 1/(2 + 2 + 1 + 1)ns = 166.66\ MHz$$

So, from the above discussion negative clock skew degrades the performance of design. In the above example due to the buffer delay of 1 ns, the clock at flip-flop1 is delayed by 1 ns time as compared to the clk at flip-flop2. So the time of 1 ns buffer delay is added in the data path with the flip-flop delay and hence reduces the clock frequency for the design.

20.5 Timing Paths in Design

As discussed in the above section, the STA is a non-vectored approach to check the timing violation and the performance of the ASIC design! The STA tool uses the algorithm to check for the violations in all possible timing paths.

Timing paths in design start at start point. The clock port of the flip-flop or input port of the design is treated as start point. Timing path terminates or ends at the end point. The data input of flip-flop or an output port is treated as end point.

For any RTL design, there can be four timing paths and they are named as

- Input-to-register path (input to reg path)
- Output-to-register path (output to reg path)
- Register-to-register path (reg to reg path)
- Input-to-output path (combinational path).

So, timing analyzer checks for the worst possible delays through the timing paths but ignores the logical operations. As timing analyzer ignores the logic operations, it is a non-vectored approach and faster as compared to the simulator. But reader needs to understand that the timing analysis is used to check for the timing correctness of the design but not used to check for the logical functional correctness of the design.

This following section discusses the different timing paths in the design.

20.5.1 Input-to-Register Path

Input-to-register path has start point input port q1 and end point data input d2 of the flip-flop. This path is also called as input–register path group. Figure 20.7 shows the input port q1 and combinational logic (combo logic), and the path from q1 to d2 through combo logic is treated as input-to-register path.

20.5.2 Register-to-Output Path

Register-to-output path has start point clock input port clk and end point data output q_out of the register element. This path is also called as output–register path group. Figure 20.8 shows the start point port clk, and data d travels through the register through combinational logic, hence named as register-to-output path.

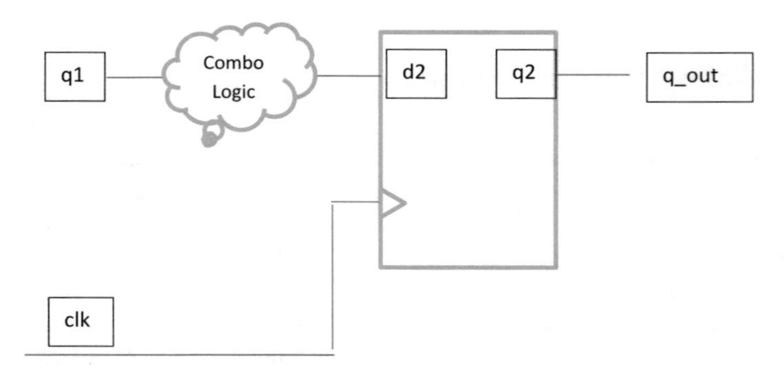

Fig. 20.7 Input-to-register path

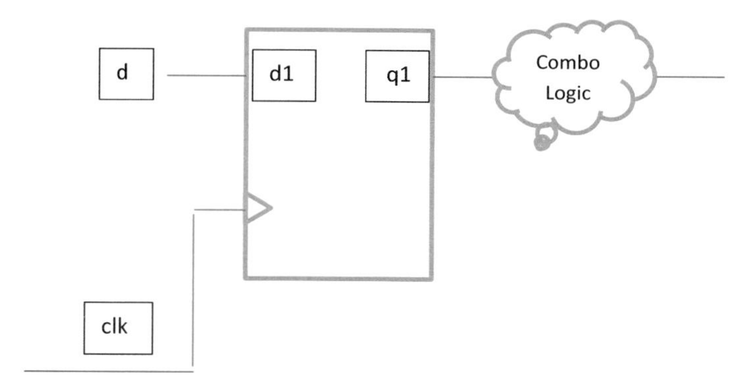

Fig. 20.8 Register-to-output path

20.5.3 Register-to-Register Path

Register-to-register path has start point clock input port clk, and first flip-flop acts as a launch register, end point data input d2 of the second flip-flop. This path is also called as clock path group. Figure 20.9 shows the clock port clk, and launched data by flip-flop1 passes through the combinational logic (combo logic) and arrives at the data input d2 of the second flip-flop. This path decides the maximum operating frequency of the design.

20.5.4 Input-to-Output Path

Input-to-output path has start point input port d and end point data output q1_out. This path is also called as combinational path group. Figure 20.10 shows the input port d, and the data passes through the combinational logic (combo logic) to generate an output q1_out.

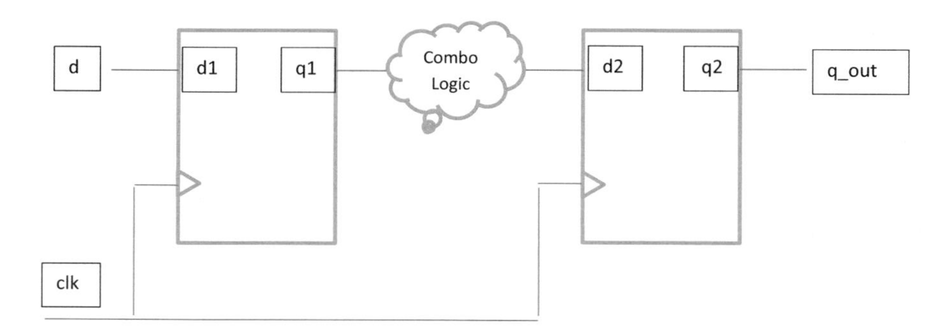

Fig. 20.9 Register-to-register path

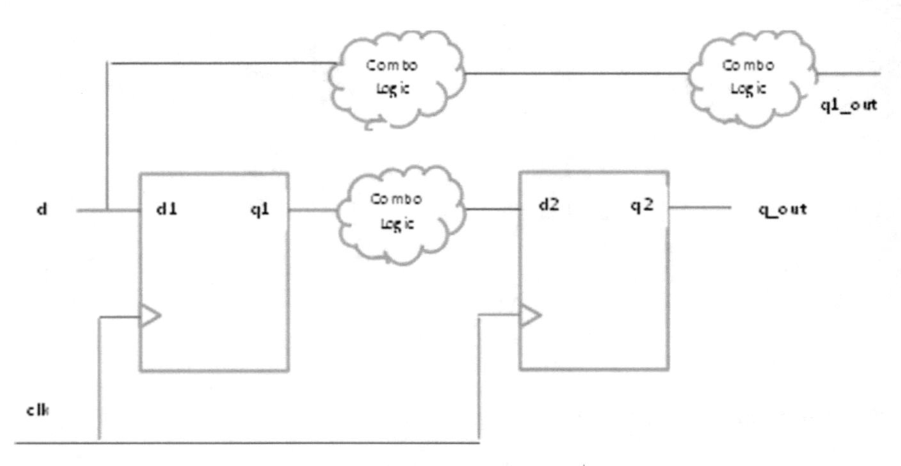

Fig. 20.10 Input-to-output combinational path

20.6 Timing Goals for the Design

In the practical scenarios, the design timing goals are described using the clock definitions for the design and by specifying the IO timing with respect to the clock. The reason for all this definition for the synchronous designs is because data arrives from the clocked device and the data goes to the clocked device.

The template shown in Fig. 20.11 describes the definitions to specify the timing goals for the design.

Use the SDC commands to define the clock, input delays, output delays, and clock skew.

The SDC commands to specify the timing goals are listed in Fig. 20.12.

20.7 Min–Max Analysis for ASIC Design

So, from the above discussion the setup time is good due to faster clock arrival and slow data arrival. To overcome the setup violations, the data should arrive fast, launch clock should arrive fast, and capture clock should arrive slowly.

Hold time violation is because data arrival is fast, capture is slow, and data arrival is fast. The hold time can be fixed using the strategy to have the data arrival slow, launch is slow, and capture is fast.

In the practical scenarios, the min–max corner analysis can be performed by using minimum value of timing parameters and by using maximum value of timing parameters. During setup time analysis, consider the maximum data path delays and minimum delays in the clock path. During hold analysis, consider minimum delays in the data path and maximum delays in the clock path.

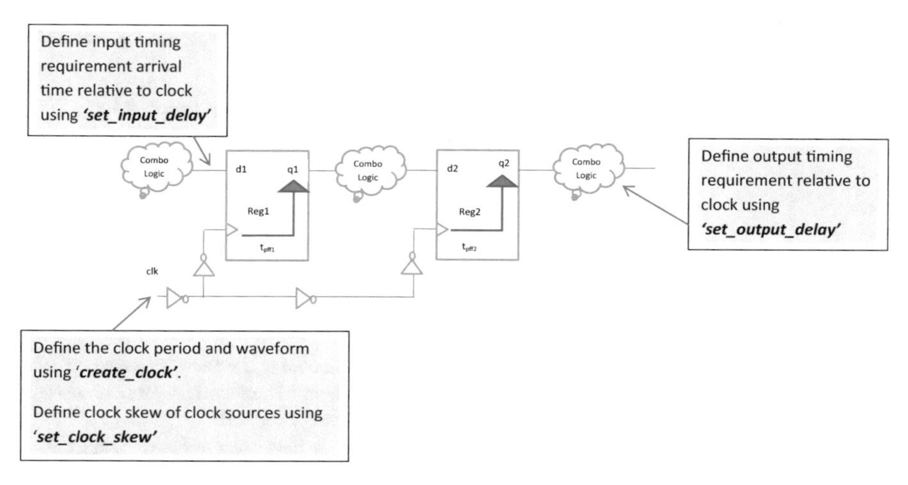

Fig. 20.11 Timing goals for synchronous design

The example shown in Fig. 20.13 is used to describe the minimum and maximum analysis for the design.

In this example, the minimum delays are considered in the clock path and maximum delays are considered in the data path. Consider the timing parameters of flip-flop1 and flip-flop2.

Consider the first flip-flop delay is (1.35, 1.5) ns, the second flip-flop delay is (1.65, 1.75) ns, and the combinational path delay is 2 ns. NOT gate propagation delay is (0.75, 0.8), and setup time of both the flip-flops is (0.6, 0.65).

Skew in the design is due to the NOT gates in the clock path. This skew is calculated as follows. By using minimum delay analysis, the skew in the design is 1.2 − 0.6 = 0.6 ns. This skew is due to additional delay of NOT gate for the capture flop.

$$\text{Data Arrival time} = T_{pff1} + T_{combo} = 1.5 + 2 = 3.5\ ns.$$

Data required time is equal $t_{clk} + t_{skew} - t_{su} = t_{clk} + 0.6 - 0.6$. Then, the maximum operating frequency is as follows.

Therefore, the minimum time period of design is

$$T_{pff1} + T_{combo} = t_{clk} + t_{skew} - t_{su}$$
$$t_{clk} = T_{pff1} + T_{combo} - t_{skew} + t_{su} = 1.5 + 2 - 0.6 + 0.6 = 3.5\ ns$$
$$F_{max} = 1/(3.5\ ns) = 285.71\ MHz$$

Fig. 20.12 SDC commands to specify timing goals

- *Define the clock for 200MHz operating frequency and having 50% duty cycle by using*

  create_clock –period 5.00 –name clk [get_ports {clk}]

  The above SDC command generates the clock of 200MHz with the 50% duty cycle that is on time is equal to off time.

- *Specify the clock latency. For example, if the clock latency is of 1ns then use the command 'set_clock_latency'*

  set_clock_latency –source 1.00 [get_clocks clk]

- *Timing analyzer uses the longest or shortest path during timing analysis. The longest delay path is specified by –late and shortest delay path is specified by the –early path.*
- *During the setup analysis the timing analyzer uses the late clock latency for the data arrival path and early clock latency for the clock arrival path. The clock latency for setup is defined with reference to rising (-rise) or falling (-fall) clock transitions.*
- *During the hold analysis the timing analyzer uses the early clock latency for the data arrival time and late clock latency for the clock arrival time.*
- *The definitions for the clock latency can be specified by the following SDC*

  set_clock_latency –source –early –rise -0.5 [get_clocks clk]

  set_clock_latency –source –early –fall -0.45 [get_clocks clk]

- *Specify the separate clock uncertainty for the setup (-setup) and for the hold (-hold)*

  set_clock_uncertainty –setup 1.0 [get_clocks clk]

  set_clock_uncertainty –hold 0.5 [get_clocks clk]

- *Specify the minimum and maximum input delays for the design using set_input_delay*

  set_input_delay –clock clk –max 2.0 find (port d1)

  set_input_delay –clock clk –min 1.4 find (port d1)

- *Specify the minimum and maximum output delays for the design using set_output_delay*

  set_output_delay –clock clk –max 1.8 find (port q_out)

  set_output_delay –clock clk –min 1.2 find (port q_out)

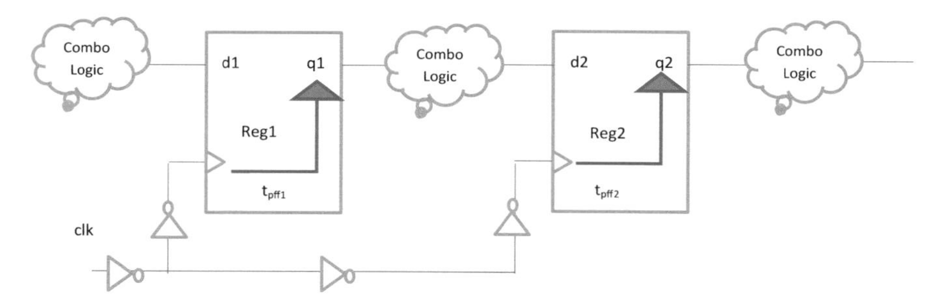

Fig. 20.13 Min–max delay analysis

20.8 Fixing Design Violations

The following are few important techniques used to fix the design violations.

20.8.1 Tweaks at the Architecture Level

To fix the design violations, the last option is to make the required and necessary changes at the architecture level of design. But the architecture-level changes are not recommended for the design as it can have significant impact on the design and implementation cycle. But after incorporating changes at the micro-architecture of the design or during optimization if the timing constraints are not met, then sometimes it is essential to incorporate the changes at the architecture level. The designer needs to suggest the chief architect about the required changes in the architecture. The chief architect needs to take care of the design functionality as the changes in the architecture can affect the design functionality. It is essential to make the desirable changes by keeping the same design functionality.

20.8.2 Tweaks at Micro-architecture Level

If the design optimization fails to meet the required timing, then it is essential to make the necessary and required changes at the micro-architecture level. The micro-architecture document is the golden reference document for the RTL design, and due to that, the designer has insight about it. The greater detailed understanding of the micro-architecture always plays a crucial and significant role during the RTL design stage.

20.8.3 Optimization During Synthesis

Synthesis tool used during logic synthesis is more efficient due to the inbuilt synthesis and optimization algorithms. They are driven by the coding and design styles adopted at the synthesis level. If design does not meet the timing, then optimization and performance strategies need to be used. To meet the desired timing goals, the designer can use the optimization concepts like pipelining, register duplications, register balancing, etc. Consider the scenario, if the design needs to be optimized to eliminate the 100 timing violations, and among them 20 to 30 timing violations are not possible to fix by using synthesis optimizations, then the better approach can make the necessary and required changes in the RTL code and fix these violations. Here, performance improvement techniques are useful.

The reader needs to ask themselves that why it is challenging to fix the timing violations in the design? As the design complexity increases from block level to chip level due to multiple hierarchies in the design, and hence the propagation delay between registers increases due to inefficient design partitioning. This has significant impact on register-to-register path timing. It may be possible that the multiple timing paths can be violated due to non-meeting of the setup and hold time parameters.

It is general observation that, at the block level if design meets the timing goals then the design does not have any timing violations at the block level. But at the top-level design due to integration of multiple blocks, there exist possibilities of several timing violations. At the top level, these violations can be fixed by minimizing the logic density between the registers. If data required time is greater than the data arrival time, then it is treated as clean register-to-register path due to positive slack. This indicates that there is no any setup violation in the design at top level.

20.9 Fixing Setup Violations in the Design

The following are few techniques used to fix the setup time violations:

1. Logic duplication
2. Encoding methods
3. Late arrival signal fixes
4. Register balancing.

20.9.1 Logic Duplication

This technique increases the effective area but generates two independent paths during synthesis. This technique is effective to fix setup time violation. For

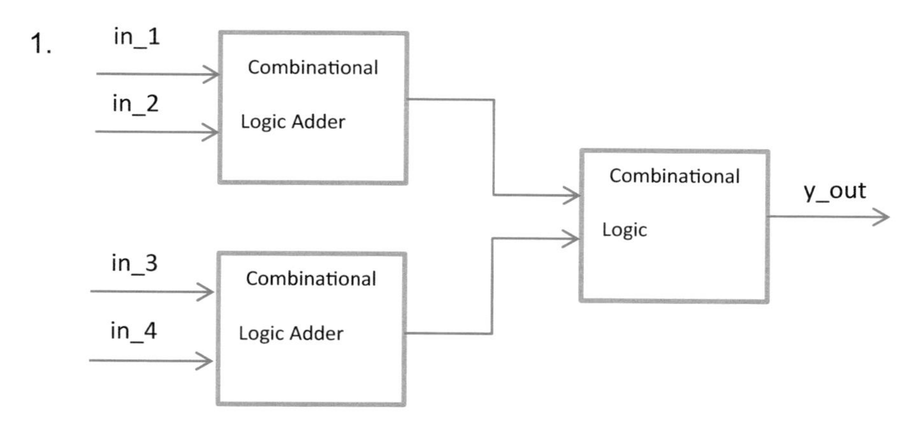

Fig. 20.14 ASIC logic without duplication

example, consider Fig. 20.14. Consider inputs in_1, in_2, in_3, and in_4 are registered inputs, and the combinational logic is in the register-to-register path. If every adder has propagation delay of 3 ns, then overall combinational path delay is 6 ns. But due to logic duplication, the two independent paths can be optimized to improve the timing.

As shown in Fig. 20.15, the two independent paths have been created using logic duplication technique and hence the optimization for these two independent paths is possible by retaining same functionality. Logic duplication technique increases the area.

20.9.2 Encoding Methods

The popular used encoding techniques are ***priority encoding*** and ***multiplexed encoding***. Consider the continuous assignments used to code the combinational logic.

```
assign y_out = a_in && b_in && c_in && d_in && e_in && f_in && g_in && h_in;
```

The above assignment infers the priority logic, where a_in has highest priority over any other input signal. The inferred logic is shown in Fig. 20.16.

In the priority encoding method, the overall delay is of 7 * t_{pd}; if t_{pd} is equal to 1 ns, then the overall propagation delay is of 7 ns.

To improve the design performance, it is essential to reduce the propagation delay of combinational logic and hence multiplexed encoding technique can be efficient as compared to the priority encoding technique. Figure 20.17 shows the multiplex encoding by using the continuous assignment.

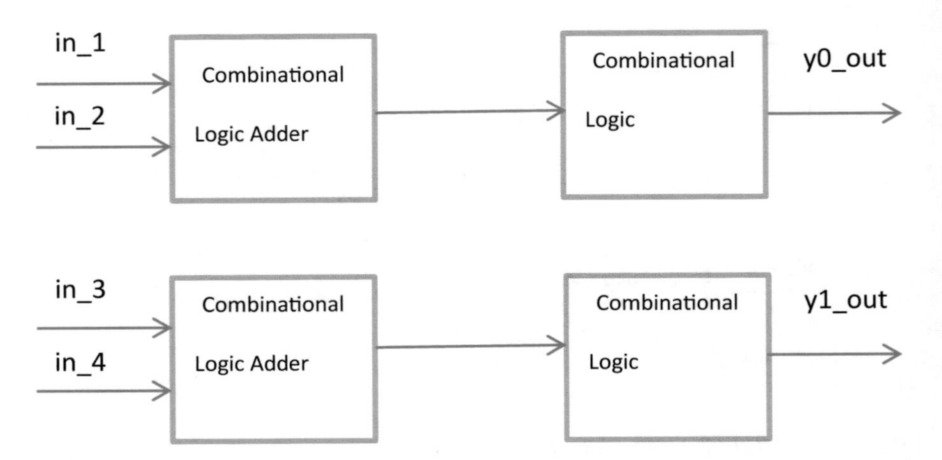

Fig. 20.15 ASIC logic with logic duplication

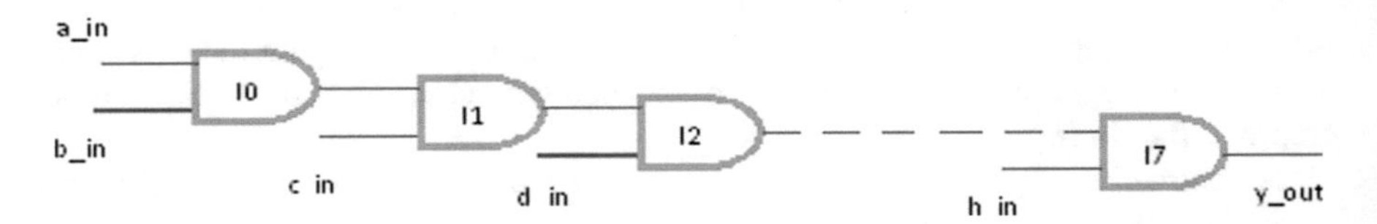

Fig. 20.16 Priority encoding logic

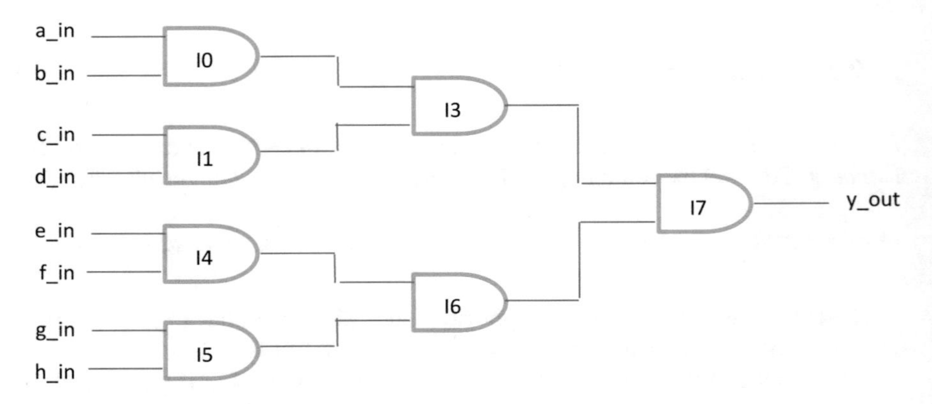

Fig. 20.17 Sequential or multiplexed encoding logic

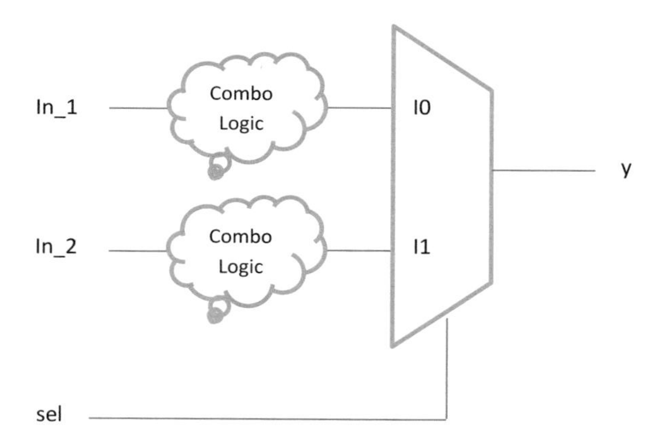

Fig. 20.18 Logic with late arrival of signals

```
assign y_out= ((a_in && b_in) && (c_in && d_in)) && ((e_in && f_in) &&
(g_in && h_in).
```

As shown in Fig. 20.17, the number of levels has been reduced from seven to three and hence the overall propagation delay for the multiplexed encoding is only $3 * t_{pd}$. If the t_{pd} is 1 ns, then overall propagation delay for the multiplexed encoding is only three-stage delay, that is, 3 ns. So, this technique has improved performance as compared to the priority encoding technique.

20.9.3 Late Arrival Signals

For any design if control signals are late arriving, then it has significant impact on the design timing. Due to late arrival of the control signal, setup time is violated.

In the example shown in Fig. 20.18, in_1 and in_2 are multiplexer inputs and arrive quickly but sel_in is select line of multiplexer and arrives late. The select input sel_in is late arrival signal. This signal has significant impact on the setup time of design.

To improve the timing and to avoid the setup time violation, the combinational logic can be pushed ahead toward output side, and the multiplexer logic can be pushed toward input side. The combinational logic can be duplicated at the input of multiplexers. This technique increases area but improves the overall design performance by compensating the time required for the combinational logic and late arrival signal. Another important point to understand is this technique allows the logic partitioning efficiently into two groups and is useful for further improvement in the timing.

20.9.4 Register Balancing

To fix the setup time and to improve the design performance, register balancing is one of the powerful techniques. Consider the operating frequency of the design as 200 MHz; that is, clock period is on 5 ns. The register-to-register path has high combinational delay due to which the data arrival time is greater than the data required time. In such scenario, the slack is negative, and it violates the setup time for the design.

Consider the example shown in Fig. 20.19; register1-to-register2 path has combinational logic and has delay of 3 ns. If we consider the setup time of register as 1 ns, propagation delay of flip-flop as 2 ns, and hold time as 0.5 ns, then the data arrival time for register1-to-register2 path is 5 ns and data required time is $T_{clk} - 1$ ns. So, the clock time period is $T_{clk} = 6$ ns. This violates the setup time of design for the given design constraints of 5 ns.

For register2-to-register3 path, the combinational delay specified is 2 ns, and if we consider same timing parameters of the register, then the data required time is T_{clk}-1 ns and data arrival time is 3 ns. This meets the timing constraints for the design. For register2-to-register3 path, the data is arrived at the D input of register3 at 3 ns and waiting for the clock which is arriving after 2 ns. So, there is additional time margin of 1 ns, this can be used to improve the design performance, and this technique is called as balancing the timing between two registers.

This can be achieved by splitting the combinational logic between the register1 and register2 into two paths and pushing the combinational logic having delay of 1 ns to the register2-to-register3 path.

This will give the clean timing for all register-to-register paths as the data arrival time for both the paths will be 4 ns. This meets the design constraints, and the operating frequency for the design meets the target of 200 MHz.

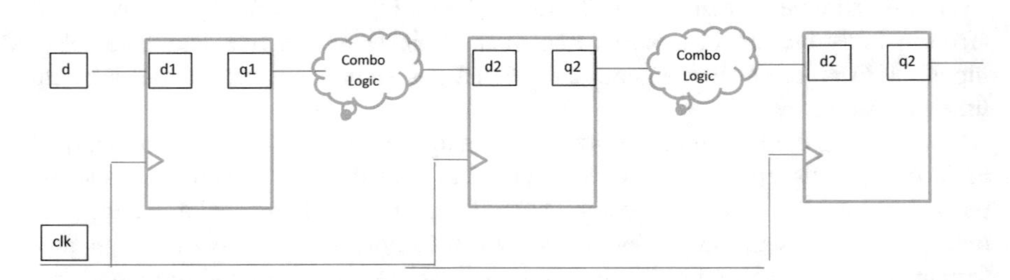

Fig. 20.19 Register balancing example

20.10 Hold Violation Fix

Hold time violation occurs in the design if the data at the D input of register changes fast. For example, consider the design shown in Fig. 20.7; if combinational logic delay is less and due to that if data at D input of register changes fast, then there exists hold violation for the design. During the hold time window, if the data changes, then there is hold violation.

To fix the hold violation for the design, it is recommended to insert the buffers in the data path, but care needs to be taken that this should not violate the setup time requirements for the design. Inserting buffers in the data path increases the time required to change data at the D input of register and is useful to fix the hold violation. The logic after inserting the buffers in the data path is shown in Fig. 20.20.

20.11 Timing Exceptions in the Design

There are two main timing exceptions, and they are named as false paths and multi-cycle paths. These timing exceptions need to be reported to timing analyzer using SDC commands.

20.11.1 Asynchronous and False Paths

If the changes on any one of the signals or ports do not affect the output of design, then the path needs to be reported as false path. False path is basically timing exception and needs to be notified to the synthesis tool. For example, consider the following expression.

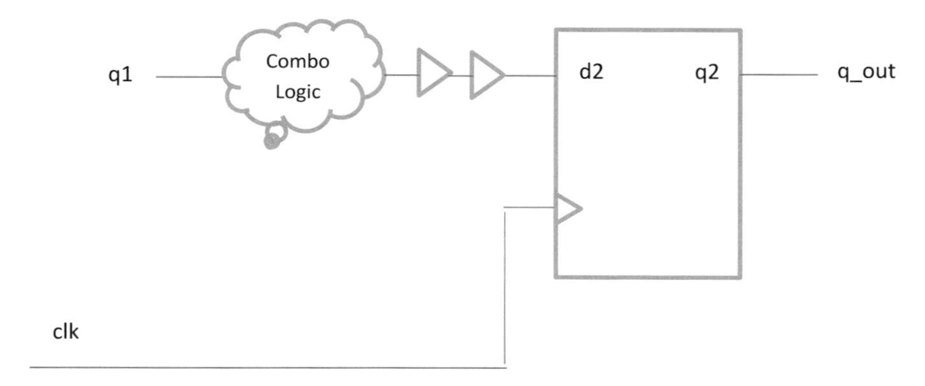

Fig. 20.20 Hold violation fix

```
assign y_out = (a_in + b_in) + (c_in + d_in)
```

In this example, if the d_in is set to zero due to some reason then the logical output depends on only a_in, b_in, c_in inputs and the path from d_in to y_out will be considered as false path.

Asynchronous path: Asynchronous path needs to be notified to the synthesis tool, and these path violations need to be treated as false violations and need to be ignored.

Figure 20.21 describes the false path, and this needs to be reported to the timing analyzer. The SDC command discussed in Chap. 10 can be used to specify the false path.

```
set_false_path –from [ get_ports {a b} ] –to [ get_ports c_d ]
The above SDC command indicates the changes on input ports a, b will not
affect the output c,_d and need to be treated as false path
set_false_path –from [ get_ports {c d} ] –to [ get_ports a_b ]
The above SDC command indicates the changes on input ports c, c will not
affect the output a_b and need to be treated as false path
```

20.11.2 Multi-cycle Paths

If any path in the design has delay of more than one clock cycle, then the path is treated as multi-cycle path. Consider the following design scenario where register (FF4) to register (FF5) delay is of 40 ns and clock period is of 5 ns. To update the d input of register with new value, the number of clock pulses required is equal to 8.

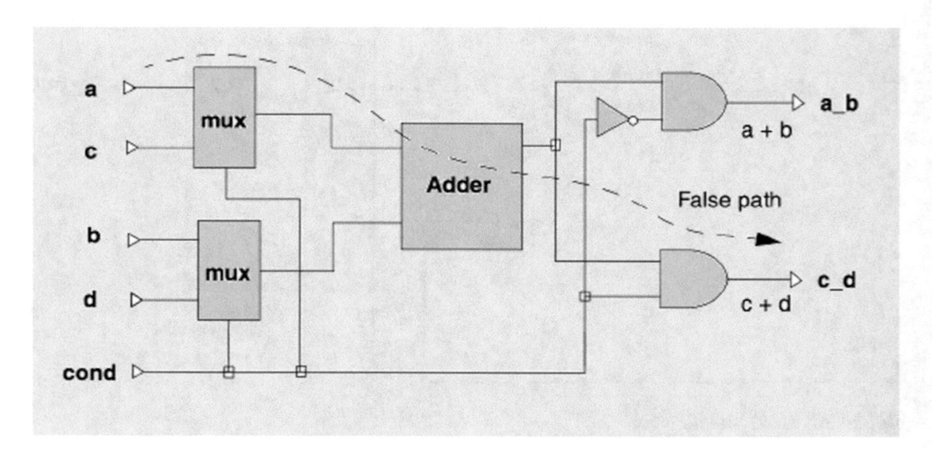

Fig. 20.21 False path example [Synopsys timing constraints and optimization user guide, version D-2010.03]

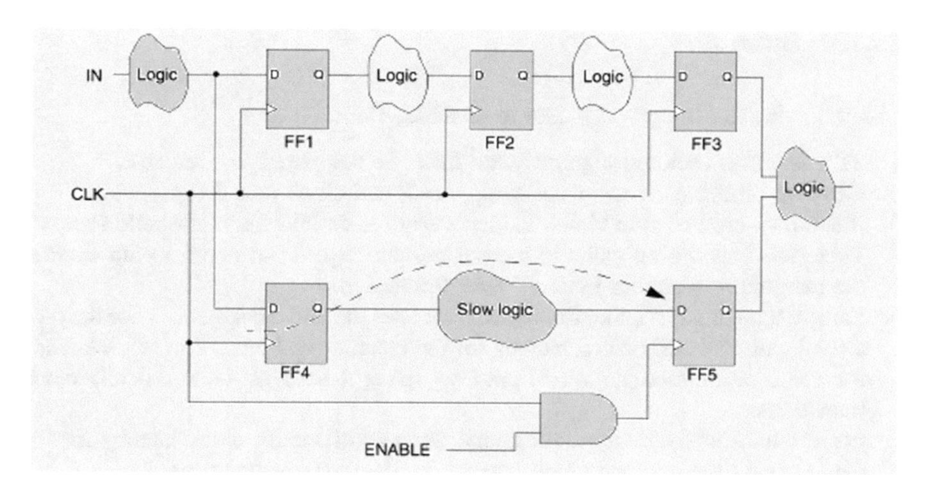

Fig. 20.22 Multi-cycle path example [Synopsys timing constraints and optimization user guide, version D-2010.03]

This needs to be informed to the tool so that setup and hold check can be pushed according to the requirement. The multi-cycle path is a timing exception.

The SDC command discussed in Chap. 19 can be used to specify the multi-cycle path (Fig. 20.22).

```
set_multicycle_path -setup 2 -from [ get_cells FF4 ] \
-to [ get_cells FF5 ]
The above SDC command indicates that the path specified is multi-cycle
path and the setup is pushed by 2 clock cycles
set_multicycle_path -hold 1 -from [ get_cells FF4 ] \
-to [ get_cells FF5 ]
The above SDC command indicates that the path specified is multi-cycle
path and the setup is pushed by 1 clock cycles
```

20.12 Pipelining and Performance Improvement

The design performance for the design can be improved by adding the multiple stage pipelining in the ASIC design. The overall latency to get an output data is dependent upon the number of pipelined stages. Pipelining will increase the area as register utilization for multiple bits increases.

Due to use of pipelining, the overall performance of the design also improves. Readers are requested to refer Chap. 11 for better understanding of the pipelining.

20.13 Summary

The following are important points to conclude the chapter.

1. STA is a non-vectored approach and faster as compared to simulator.
2. Flip-flop timing parameters are setup, hold, and clock to q delay.
3. If setup or hold time is violated, then design goes into the metastable state.
4. There are four timing paths in the design, and register-to-register path decides the maximum operating frequency for the design.
5. For the setup analysis, the timing analyzer uses the late clock latency for the data arrival path and early clock latency for the clock arrival path. The clock latency for setup is defined with reference to rising (-rise) or falling (-fall) clock transitions.
6. For the hold analysis, the timing analyzer uses the early clock latency for the data arrival time and late clock latency for the clock arrival time.
7. The multi-cycle paths and false paths are the timing exceptions.

Chapter 21
Design Constraints And Optimization

Synopsys Design Compiler is industry leading logic synthesis tool and popular as Synopsys DC. Most of the leading ASIC design companies uses the Synopsys DC during the logic synthesis and Synopsys PT for the timing analysis and timing closure. The chapter focuses on the design constraints and optimization using Synopsys DC.

The optimization using the Synopsys DC for mainly the area, speed using various SDC commands, is discussed in this chapter. The chapter discusses the use of the design constraints, ASIC synthesis, and optimization strategies useful during the ASIC designs. The following sections are useful to understand the design constraints and optimization using the Synopsys DC.

21.1 Introduction to Design Constraints

Modern ASIC SOCs are extraordinarily complex in the nature and consist of more than millions of logic gates. Design complexity has grown exponentially in the past decade due to the demand of the sophisticated and intelligent devices. In such scenario, there is additional overhead and cost during the design synthesis and timing closure. As discussed in Chaps. 18 and 19, the ASIC design passes through various phases which include architecture design micro-architecture design, design entry using HDL, simulation, and synthesis. The Synopsys DC is the leading EDA tool used to perform the logic synthesis and optimization, and Synopsys PT is used for the timing closure.

As a ASIC design engineer, it is essential to have exposure about the design synthesis and timing analysis. These concepts are covered in Chaps. 18–20. The understanding of the design constraints and the commands used to constrain the design for the area, speed, and power is very much useful during chip design various phases. This chapter discusses how to specify the design constraints using Synopsys DC and how to optimize the design.

V. Taraate, *Digital Logic Design Using Verilog*,
https://doi.org/10.1007/978-981-16-3199-3_21

The design constraints are classified as design rule constraints and optimization constraints. The classification is shown in Fig. 21.1.

Synthesis flow is discussed in Chap. 19 with the important SDC commands. For better understanding, the synthesis flow is shown in Fig. 21.2. The flow includes various steps useful during the synthesis of any kind of logic. The compilation strategy can be chosen as top-down or bottom-up. The commands used during each phase are discussed in the subsequent sessions.

```
/* read the design object */
read -format verilog full_adder.v
/* specify the technology requirements */
target_library = my_library.db
symbol_library = my_library.sdb
link_library = "*" + target_library
/* define the design environment */
set_load 2.0 sum_out
set_load 1.2 carry_out
set_driving_cell -cell FD1 all_inputs()
set_drive 0 clk_name
/* set the design constraints */
set_input_delay 1.25 -clock clk {a_in, b_in}
set_input_delay 3.0 -clock clk c_in
set_max_area 0
/* synthesize the design */
compile
/* generates reports */
report_constraint
report_area
/* save the design database */
write -format db -hierarchy -output full_adder.db
```

Example 1 Important steps during synthesis and compilation

1. **Read Design Object**: Design object is Verilog RTL code which is simulated to check for the functional correctness. The commands used are

```
analyze, elaborate, read
```

2. **Specify Technology Requirements**: In these steps, the design rules and libraries required need to be specified. The commands used are

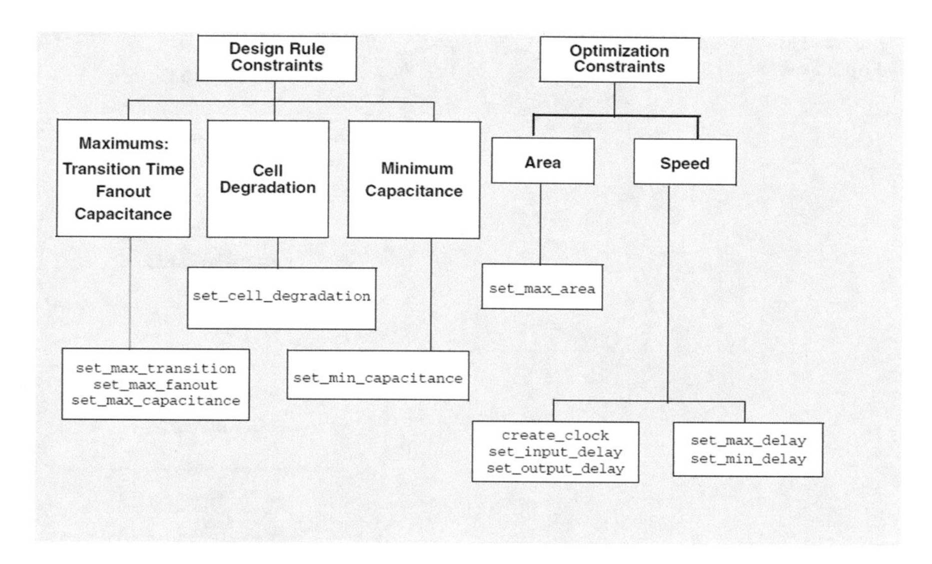

Fig. 21.1 Constraint classification

Library Objects

```
link_library
target_library
symbol_library
```

Design Rules

```
set_max_transition
set_min_transition
set_max_fanout
set_min_fanout
set_max_capacitance
set_min_capacitance
```

3. **Design Environment Definitions**: The design environment includes the process, temperature, voltage conditions, drive strength, and effect of load driving the design. The commands used are

```
set_operating_conditions
set_wire_load
set_drive
```

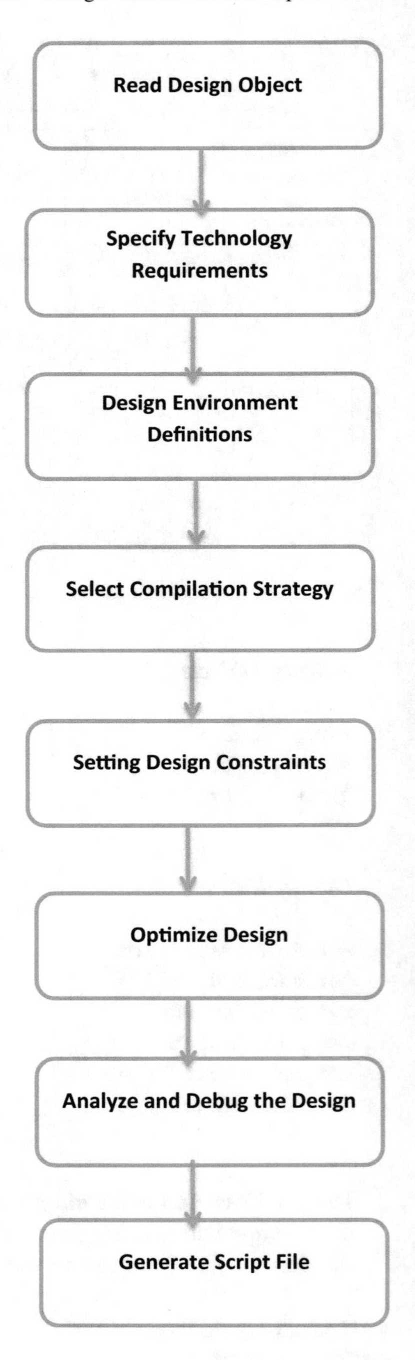

Fig. 21.2 Flow for synthesis and optimization

```
set_driving_cell
set_load
set_fanout_load
```

4. **Select Compilation Strategy**: The strategies used for optimizing hierarchical design includes top-down, bottom-up, and compile-characterize. The advantages and disadvantages of each strategy are discussed in the subsequent section
5. **Setting Design Constraints**: The constraints need to be set for the design optimization and for the timing analysis. The commands used are

```
create_clock
set_clock_skew
set_input_delay
set_output_delay
set_max_area
```

6. **Optimize Design**: Synthesize the design to generate technology-specific gate-level netlist. The command used is
compile.
7. **Analyze and debug the**: This step is important to understand the potential problems in the design by generating various reports. The commands used are

```
check_design
report_area
report_constraint
report_timing
```

8. **Generate Script file**: The design database is stored in the form of script file.

Consider the top-level design as full adder having inputs a_in, b_in, c_in, and outputs sum_out, carry_out. The top-down compilation strategy is used, and the script is shown below and can be used in the practical scenario. Refer Chap. 19 for the SDC commands. To synthesize the design and to compile, use the script shown in Example 1.

21.2 Compilation Strategy

The strategy used during the compilation of any design can be top-down or bottom-up compilation. Each compilation strategy has its own advantages and disadvantages.

21.2.1 Top-Down Compilation

The top-down compilation uses the top-level design constraints and is easier as compared to the bottom-up compilation approach. Following are the advantages and disadvantages of the top-down compilation

Advantages

1. Optimization engines work on full design, complete paths
2. Usually get best optimization result
3. No iteration required
4. Simpler constraints
5. Simpler data management.

Disadvantages

1. Longer runtime
2. More memory requirements.

The commands used for the top-down compilation are

```
dc_shell> current_design TOP
dc_shell> compile –timing_high_effort_script
```

21.2.2 Bottom-Up Compilation

The bottom-up compilation uses the sub-module-level compilation first, and then it moves towards top level. The care must be taken by the synthesis team to set ***'set_dont_touch'*** attribute on the sub-modules to avoid recompilation of the sub-modules. The synthesis team needs to know the timing information of the inputs and outputs for each of the sub-module. The advantages and disadvantages are documented below

Advantages

1. Faster as compared to top-down compilation
2. Less compilation time required per run
3. Less memory requirement.

Disadvantages

1. Optimization works on the sub-module or sub-design
2. More iterations are required
3. More hierarchies to be maintained.

Consider the design has two sub-modules. The commands used for the bottom-up compilation are

```
dc_shell> current_design submodule1
dc_shell> compile -timing_high_effort_script
dc_shell> set_dont_touch submodule1
dc_shell> current_design submodule2
dc_shell> compile -timing_high_effort_script
dc_shell> set_dont_touch submodule2
dc_shell> current_design TOP
dc_shell> compile -timing_high_effort_script
```

21.3 Area Optimization Techniques

There are several techniques used to optimize the overall area of the design. The highest priority of the synthesis team is to optimize for the timing followed by area. There are several efficient area optimization techniques at the RTL level. In the previous few chapters, we have discussed the resource sharing. Following are the important guidelines used to optimize for the area

1. Avoid the use of the combinational logic as individual block or module
2. Do not use the glue logic between two modules
3. Use ***set_max_area*** attribute while synthesizing the design.

21.3.1 Avoid Use of Combinational Logic as Individual Block

It is recommended that, do not use the combinational logic as individual block. If the individual combinational module is used, then DC will not be able to optimize the individual block. This is not a good design partitioning technique. The hierarchy of the module is fixed, and Design Compiler will not be able to modify the hierarchy of the design. Consider the design scenario shown in Fig. 21.3. It has module I and module II, and module II is individual combinational block so the Design Compiler will not be able to optimize module II, as Design Compiler does not optimize the port interfaces.

If the design is partitioned efficiently, then the overall optimization will boost the design performance. A better partitioned ASIC design should have combined

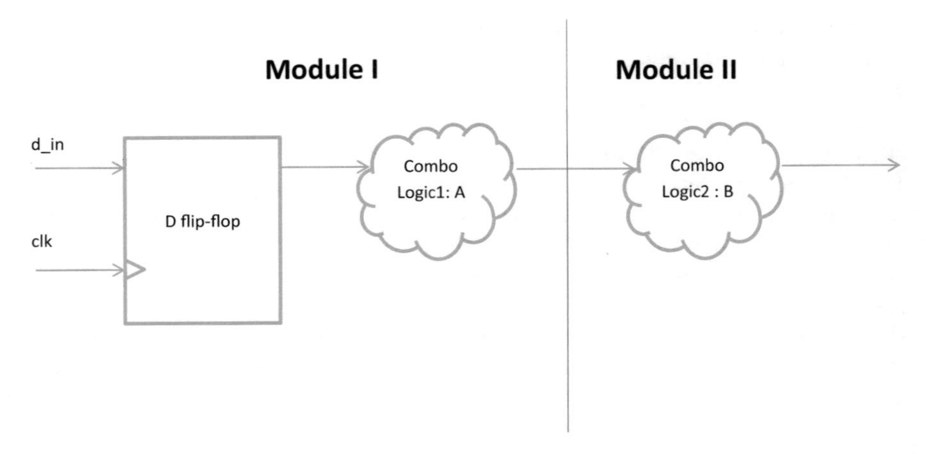

Fig. 21.3 Combinational logic as individual module

functionality of module I and module II. The functionality of A + B in the using module is shown in Fig. 21.4 and results in faster optimization during the synthesis.

21.3.2 Avoid Use of Glue Logic Between Two Modules

If the module II in Fig. 21.3 is replaced by glue logic that is instance of logic gate, then it glues between the different modules as shown in Fig. 21.5. Such type of design partitioning is not good, the reason being the logic gate cannot be optimized by the Design Compiler as design is not partitioned properly. To avoid this type of scenario, it is recommended to use the ***group*** command. Either group the glue logic with the module I or module II. Following command used to group the glue logic with module I.

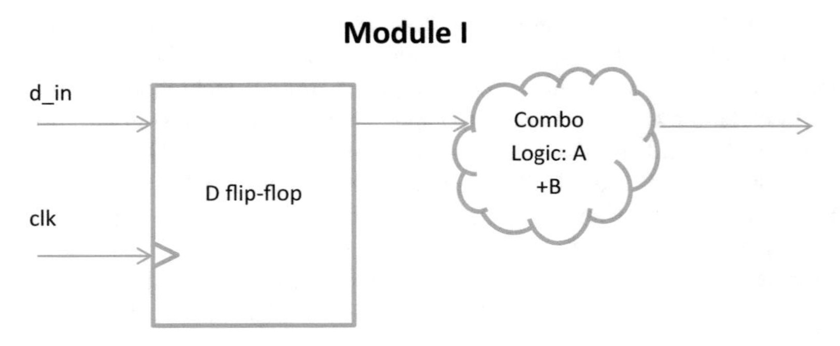

Fig. 21.4 Eliminating individual combinational module

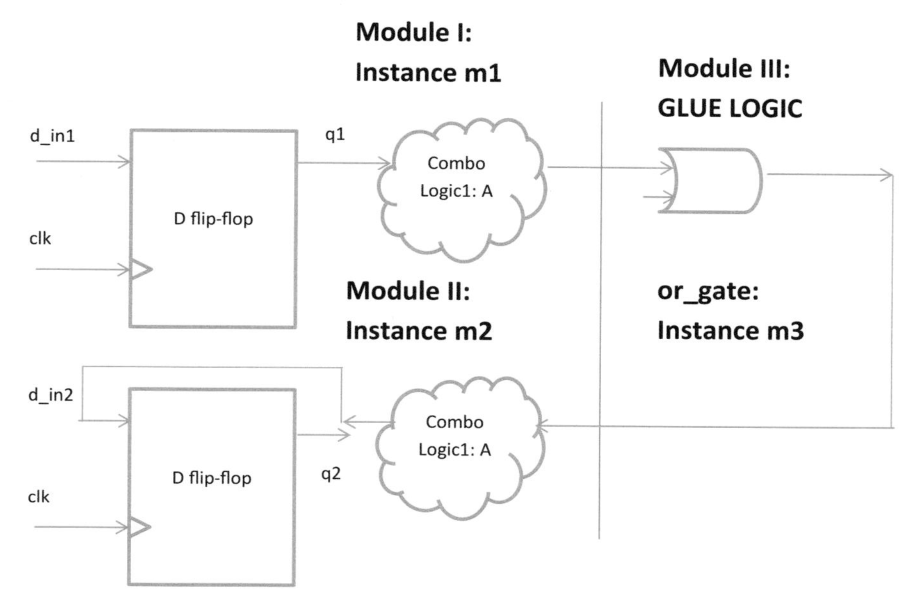

Fig. 21.5 Glue logic between two blocks

```
dc_shell> group {m1, m3} –design_name moduleIII cell_name or_gate
```

Following command used to group the glue logic with module II.

```
dc_shell> group {m2, m2} –design_name moduleIII
                                    cell_name or_gate
```

21.3.3 Use of Set_max_area Attribute

To obtain the least possible area, it is recommended to use the attribute ***set_-max_area***. This attribute is effective during the optimization of the design. Design Compiler gives the highest priority to the timing optimization. If timing is met, then only the area optimization phase can start. The priorities for the design optimization are listed below

1. Design rule constraints (DRC)
2. Timing
3. Power
4. Area.

21.3.4 Area Report

The area report is generated by the Design Compiler using ***report_area*** command. The sample area report is shown Example 2. The area report for any design consists of the number of ports, nets, references. It also gives information about the combinational, sequential, and total cell area.

21.4 Timing Optimization and Performance Improvement

During optimization, the timing has the highest priority as compared to the power and area. During the first phase of optimization, the DC checks for the design rule constraints (DRC) violations, then the timing violations and the power constraints, and finally the area constraints. This section discusses the few timing optimization commands supported by the Design Compiler.

```
Number of ports:                 3
Number of nets:                  8
Number of cells:                 7
Number of references:            2
Combinational area:              100.349998
Non combinational area:          125.440002
Net Interconnect area: undefined (Wire load has zero net area)
Total cell area:                 225.790009
Total area:                      undefined
```

Example 2 Area report

21.4.1 *Design Compilation Using* Map_effort High

Most of the time synthesis teams use the option as ***map_effort*** medium while performing the synthesis. It is advisable that during synthesis of the first phase synthesis team can use the option as map_effort medium as it reduces the compilation time. If the design constraints are not met, then the designer can go for the incremental compilation with the option as map_effort high. This can improve the design performance by at least 5–10%.

The sdc command is shown below.

```
dc_shell> compile –map_effort_high –incremental_mapping
```

21.4.2 *Logical Flattening*

The design hierarchy can be broken by using logical flattening. The option allows all the logic gates of the design at the same level of hierarchy. This allows the compiler to have better performance and better area utilization for the design. If the hierarchical design is complex, then this option may not work. If number of hierarchies in the design increases, then compiler will take the larger amount of time during the design optimization.

Use the following command to achieve the logical flattening for the design

```
dc_shell> ungroup –all -flatten
dc_shell> compile –map_effort high –incremental mapping
dc_shell> report_timing –path full –delay max –max_path 1 –nworst 1
```

21.4.3 *Use of Group_path Command*

The design performance can boost upto 10% by using the map_effort high option. But if timing is not met with the incremental compilation, then it is essential to group the critical timing paths and use the weight factor to improve the design performance. This command is useful to improve the timing performance. The command is shown below

```
dc_shell> group_path –name critical1 –from <input_name> –to <out-
put_name> –weight <weight factor>
```

Consider the design scenario where the setup violation of 0.38ns. The setup violation is the difference between the data required time and data arrival time. So, the slack is negative and setup time is violated.

```
dc_shell> read -format Verilog combinational_design.v
dc_shell> create_clock -name clk -period 15
dc_shell> set_input_delay 3 -clock clk in_a
dc_shell> set_input_delay 3 -clock clk in_b
dc_shell> set_input_delay 3 -clock clk c_in
dc_shell> set_output_delay 3 -clock c_out
dc_shell> current_design = combinational_design
dc_shell> compile -map_effort medium
dc_shell> report_timing -path full -delay max -max_path 1 -nworst 1
```

After the design synthesis, it is successful use the ***report_timing*** command while performing the timing analysis. The timing report for the synthesized design can be obtained using the multiple options as listed in the above script and is shown in Example 3

To fix the setup time violation and to improve the design performance use the group_path with the weight factor. More the weight factor, more is the compilation time!

```
dc_shell> group_path -name critical1 -from c_in -to c_out -weight 8
dc_shell> compile -map_effort high -incremental mapping
dc_shell> report_timing -path full -delay max -max_path 1 -nworst 1
```

The above-listed commands are useful to generate the timing report with positive slack and remove setup violation and is shown in Example 4.

As shown in the timing report (Example 4) during the max analysis with the compile_map high option and weight factor of 5 the setup slack is met.

```
Startpoint: c_in (input port)
Endpoint: c_out (output port)
Path Group: clk
Path Type: max
Point                              Incr        Path
----------------------------------------------------------------
input external delay               0.00        0.00 f
c_in (in)                          0.00        0.00 f
U19/Z (AN2)                        0.87        0.87 f
U18/Z (EO)                         1.13        2.00 f
add_8/U1_1/CO (FA1A)               2.27        4.27 f
add_8/U1_2/CO (FA1A)               1.17        5.45 f
add_8/U1_3/CO (FA1A)               1.17        6.62 f
add_8/U1_4/CO (FA1A)               1.17        7.80 f
add_8/U1_5/CO (FA1A)               1.17        8.97 f
add_8/U1_6/CO (FA1A)               1.17        10.14 f
add_8/U1_7/CO (FA1A)               1.17        11.32 f
U2/Z (EO)                          1.06        12.38 f
C_out (out)                        0.00        12.38 f
data arrival time                              12.38 f

clock clk (rising edge)            15.00       15.00
clock network delay (ideal)        0.00        15.00
output external delay              -3.00       12.00
data required time                             12.00
----------------------------------------------------------------
Data required time                             12.00
Data arrival time                              -12.38

Slack (violated)                               -0.38
```

Example 3 Timing report with negative slack

```
Startpoint: c_in (input port)
Endpoint: c_out (output port)
Path Group: max
Path Type: max
Point                                  Incr         Path
-------------------------------------------------------------------
input external delay                   0.00         0.00 f
c_in (in)                              0.00         0.00 f
U19/Z (AN2)                            0.87         0.87 f
U18/Z (EO)                             1.13         2.00 r
add_8/U1_1/CO (FA1A)                   2.27         4.27 f
add_8/U1_2/CO (FA1A)                   1.17         5.45 f
add_8/U1_3/CO (FA1A)                   1.17         6.62 r
add_8/U1_4/CO (FA1A)                   1.17         7.80 f
add_8/U1_5/CO (FA1A)                   1.19         8.99 r
add_8/U1_6/CO (FA1A)                   1.15         10.14 f
add_8/U1_7/CO (FA1A)                   0.79         10.93 f
U2/Z (EO)                              1.06         11.99 f
C_out (out)                            0.00         11.99 f
data arrival time                                   11.99 f

clock clk (rising edge)                15.00        15.00
clock network delay (ideal)            0.00         15.00
output external delay                  -3.00        12.00
data required time                                  12.00
-------------------------------------------------------------------
Data required time                                  12.00

Data arrival time                                   -11.99
___________________________________________________________________
Slack (met)                                   0.01
```

Example 4 Timing report with the positive slack

21.5 Sub-module Characterizing

In the practical ASIC designs, the design can have multiple hierarchies. Consider that the top-level design consists of sub-modules X, Y, Z. If individually these sub-modules are synthesized and optimized, they may meet the timing requirements individually! When these sub-modules are instantiated in the higher level of hierarchy top, then there may be possibility that they may or may not meet the timing. The reason for this may be the glue logic used in between the sub-modules X, Y, Z or the tight constraints at the top-level hierarchy.

Under such circumstances to meet the design constraints it is advisable to use the ***characterize*** command. This command allows the capturing of the boundary conditions for the sub-module which is based on the top-level hierarchy environment. Each sub-module can be compiled and characterized independently.

Following is the script which can enable the characterize of the individual sub-modules. Consider the sub-module X, Y, Z instance names as I1, I2, and I3.

```
dc_shell> current_design=TOP
dc_shell> characterize I1
dc_shell> compile -map_effort high -incremental mapping
dc_shell> current_design=TOP
dc_shell> characterize I2
dc_shell> compile -map_effort high -incremental mapping
dc_shell> current_design=TOP
dc_shell> characterize I3
```

```
dc_shell> compile -map_effort high -incremental mapping
dc_shell> current_design=TOP
```

21.6 Register Balancing

Register balancing is efficient and powerful command to split the combinational logic from one pipelined stage to another pipelined stage. This technique improves the design performance by moving the logic and hence reduces the register-to-register delay. Consider the pipelined design shown in Fig. 21.6 and consists of the three flip-flops and combinational logic. In the first pipelined stage, the combinational logic is 4-variable function, and the second pipelined stage has combinational logic as 8-variable function and has more propagation delay as compared to the 4-variable combinational logic. Due to the different propagation delays in two different pipelined stages, the design performance is based on the register-to-register timing path which has more delay.

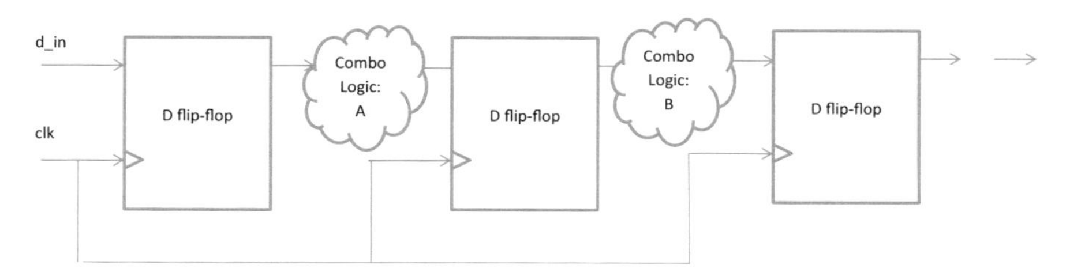

Fig. 21.6 Pipelined stages

Under such circumstances, the register balancing can be used to split the combinational logic from one of the pipelined stage to another pipelined stage without affecting the functionality of the design. This is achieved by compiler by using the following set of commands.

```
dc_shell> balance_registers
dc_shell> report_timing -path full -delay max -max_path 1 -nworst 1
```

21.7 FSM Optimization

For the optimization of the finite-state machines, the FSM Compiler is used. The use of FSM compiler is to achieve the area optimization and to improve the design performance. In the practical ASIC designs, the state machines are coded as an independent block. The design which has state machines, and the other logic cannot be considered as good design partitioning. The reason being, if the other logic is isolated from the state machine logic, then the designer can choose for the best-suited encoding style while coding for the state machines. So always use the separate sub-module for the logic and for the state machines to achieve better design performance.

The script shown in Example 5 can be used for the FSM extraction and optimization.

```
/* read the design object */
dc_shell> read -format verilog state_machines.v
/* Map the design */
dc_shell> compile –map_effort medium
/* if the design is not partitioned then group the logic */
dc_shell> set_fsm_state_vector { <flip_flop_name>, <flip_flop_name>,...}
dc_shell> group –fsm –design_name <fsm_design_name>
/* extract the state machine from netlist in the state machine table for-
mat */
dc_shell> set_fsm_state_vector { <flip_flop_name>, <flip_flop_name>,...}
dc_shell> set_fsm_encoding { "state0=0", "state1=1", .......}
dc_shell>extract
/* write the design in the FSM format */
dc_shell>write –format  st –output state_machine.st
/* if the design is already in the state machine format then read the de-
sign */
dc_shell>read –format  st  state_machine.st
/* define the order of the state */
dc_shell>set_fsm_order {state0,state1,....}
/* define the encoding style */
dc_shell> set_fsm_encoding_style <encoding_style>
/* compile the design */
dc_shell> compile –map_effort high
```

Example 5 FSM extraction script

21.8 Fixing Hold Violations

Fixing of the hold violations is quite easy as compared to the setup violations provided that the design is not complex. For complex designs, it may be time-consuming and challenging phase to fix these violations. To fix the setup violations, it is essential to modify the architecture of the design, and in turn, it has a greater impact on the RTL coding of the design. The setup violations are fixed during the pre-layout STA and hold violations can be fixed during pre-layout or post layout STA phase. Design Compiler is efficient to fix the hold violations automatically. Use the following command to fix the hold violation.

```
dc_shell> set_fix_hold clk1
dc_shell> compile –map_effort_high- incremental_mapping
```

21.9 Report Command

Following are few commands used to generate reports.

```
Timing Path Group 'clk1'
-------------------------------------------------------------------------
Levels of Logic:                 6.00
Critical Path Length:            3.64
Critical Path Slack:             -2.64
Critical Path Clk Period:        11.32
Total Negative Slack:            -55.45
No. of Violating Paths:           59.00
No. of Hold Violations:          1.00
-------------------------------------------------------------------------
Timing Path Group 'clk2'
-------------------------------------------------------------------------
Levels of Logic:                 10.00
Critical Path Length:            3.59
Critical Path Slack:             -0.29
Critical Path Clk Period:        22.65
Total Negative Slack:            -2.90
No. of Violating Paths:          11.00
No. of Hold Violations:           0.00
-------------------------------------------------------------------------
Cell Count
-------------------------------------------------------------------------
Hierarchical Cell Count:         1736
Hierarchical Port Count:                   114870
Leaf Cell Count:                 323324
-------------------------------------------------------------------------
```

Example 6 : qor report

21.9.1 Report_qor

This is used to generate report which consists of timing summary of all the path groups. This gives overall information about the timing for the design. Example 6 shows the sample report with multiple timing path groups using ***report_qor*** command.

21.9.2 Report _constraints

This command is useful to generate the reports which have the information about the user constraints and the actual design values. Example 7 is generated using the ***report_constraints*** command.

```
Weighted
Group (max_delay/setup)      Cost        Weight      Cost
-------------------------------------------------------------------
CLK                          0.00        1.00        0.00
default                      0.00        1.00        0.00
-------------------------------------------------------------------
max_delay/setup              0.00
Constraint                   Cost
-------------------------------------------------------------------
max_transition               0.00 (MET)
max_fanout                   0.00 (MET)
max_delay/setup              0.00 (MET)
critical_range               0.00 (MET)
min_delay/hold               0.40 (VIOLATED)
max_leakage_power            6.00 (VIOLATED)
max_dynamic_power            14.03 (VIOLATED)
max_area                     48.00 (VIOLATED)
-------------------------------------------------------------------
```

Example 7 Report constraints

21.9.2.1 Report_contraints_all

This command is used to show all the timing and DRC violations. Example 8 is report generated using the ***report_constraints_all*** command.

max_delay/setup ('clk1' group)

Endpoint	*Required Path Delay*	*Actual Path Delay*	*Slack*
data[15]	*1.00*	*3.64 f*	*-2.64 (VIOLATED)*
data[13]	*1.00*	*3.64 f*	*-2.64 (VIOLATED)*
data[11]	*1.00*	*3.63 f*	*-2.63 (VIOLATED)*
data[12]	*1.00*	*3.63 f*	*-2.63 (VIOLATED)*

Example 8 All constraint report

Table 21.1 Constraint validation

Command	Description
check_design	Used to check for the design consistency and reports the unconnected nets, ports, etc.
check_timing	Used to verify the timing setup is complete

21.10 Constraint Validation

Following are the important commands used to validate the design (Table 21.1).

21.11 Commands Used for the DRC, Power, and Optimization

Following are the important commands used to specify the design rules, power, and optimization constraints (Table 21.2).

Table 21.2 DRC, power, and optimization definition

Command	Type	Description
set_max_transition	DRC	Used to define the largest transition time
set_max_fanout	DRC	Used to set the largest fanout for the design
set_max_capacitance	DRC	Used to set the maximum capacitance allowed for the design
set_min_capacitance	DRC	Used to set the minimum capacitance allowed for the design
set_operating_conditions	Optimization constraints	Used to set the PVT conditions as it affects on timing
set_load	Optimization Constraints	Used to model load on output port
set_clock_uncertainty	Optimization Constraints	Used to define the estimated network skew
set_clock_latency	Optimization Constraints	Used to define the estimated source and network delays
set_clock_transition	Optimization Constraints	Used to define the estimated input skew
set_max_dynamic_ power	Power constraints	Used to set the maximum dynamic power
set_max_leakage_ power	Power constraints	Used to set the maximum leakage power
set_max_total_ power	Power constraints	Used to set the maximum total power
set_dont_touch	Optimization Constraints	It is used to prevent the optimization of mapped gates

```
/* set the clock */
set clock clk
/* set clock period */
set clock_period 2
/* set the latency */
set latency 0.05
/* set clock skew */
set early_clock_skew [expr $ clock_period/10.0]
set late_clock_skew [expr $ clock_period/20.0]
/* set clock transition */
set clock_transition [expr $ clock_period/100.0]
/* set the external delay */
Set external_delay [expr $ clock_period*0.4]
/* define the clock uncertainty*/
set_clock_uncertainty –setup $ early_clock_skew
set_clock_uncertainty –hold$ late_clock_skew

Name the above script as clock.src, and Source the above script

/* report clock and timing*/
dc_shell> report_timing
dc_shell> report_clock
dc_shell> report_timing
dc_shell> report_constraints –all_violations
```

Example 9 Sample script for constraining design at 500 MHz

Example 9 is the sample script and can be used to constrain the design at operating frequency of 500 MHz.

21.12 Summary

Following are the important points to conclude this chapter.

1. Constraints are classified as optimization, design rule, and environmental
2. Avoid the use of the combinational logic as individual block or module
3. Do not use the glue logic between two modules
4. Use set_max_area attribute while synthesizing the design
5. The top-down compilation uses the top-level design constraints and is easier to execute as compared to the bottom-up compilation approach.

6. The design hierarchy of the design can be broken by using logical flattening of the design.
7. Register balancing is very efficient and powerful command to move the combinational logic from one pipelined stage to another pipelined stage.
8. For the optimization of the finite-state machines the FSM compiler is used. The use of FSM compiler is to achieve the small area optimization and to improve the design performance.
9. The setup violations are fixed during the pre-layout STA and hold violations can be fixed during pre-layout or post-layout STA phase.
10. The use of FSM compiler is to achieve the area optimization and to improve the design performance.

Chapter 22
Multiple Clock Domain Design

Multiple clock domain design understanding is essential for ASIC design engineer. In most of the practical design scenarios the multiple clock domain designs are used and it is essential to understand need of the synchronizers for passing control signals from one of the clock domain to another clock domain. This chapter discusses about the multiple clock domain design techniques and the control and data path synchronizers and their use!

Most of the ASIC designs has the multiple clock domains and the RTL design engineers should be familiar with the concepts of the multiple clock domain and the issues or challenges in the multiple clock domain designs! The chapter is useful to understand the issues in the clock domain crossing and the data and control path synchronizers. Following few sections discusses the synchronizers used in the multiple clock domain designs and the strategies during the RTL design!

22.1 What Is Multiple Clock Domain?

It is quite simple to design single clock domain design logic. If all the flip-flops in the design are clocked by single clock source, then the design is said to be synchronous. If the flip-flops are triggered by the different clock sources, then the design is said to be asynchronous design. In the modern ASIC or SOCs, the design can have multiple clock sources of different frequencies. For example, consider Fig. 22.1, as shown the flip-flop regA is triggered by CLK1 and flip-flop regB is triggered by the CLK2.

V. Taraate, *Digital Logic Design Using Verilog*,
https://doi.org/10.1007/978-981-16-3199-3_22

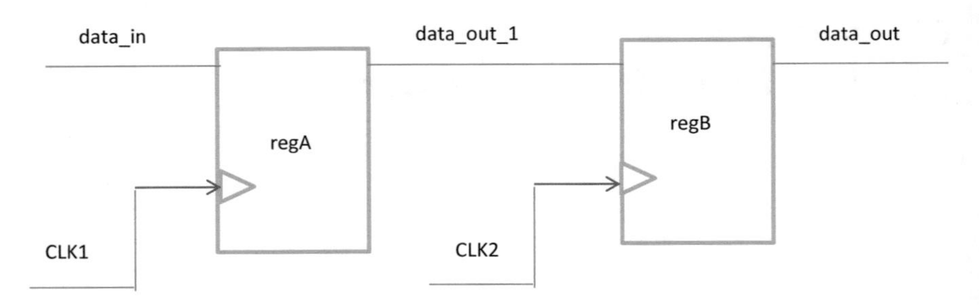

Fig. 22.1 Multiple clock domain logic

In Fig. 22.1, the data is sampled by in the clock domain using the clock source CLK1, and output from the clock domain 1 is data_out_1. The flip-flop is named as regA in clock domain 1 and named as regB in the clock domain 2. The regB has clock input as CLK2 and samples the data output from clock domain 1 on the rising edge of CLK2. The output from clock domain 2 is data_out. The difference between the single clock domain and multiple clock domain design is phase difference between arrivals of the clock signals. The clock sources CLK1 and CLK2 are derived from different clock source or may have the same or different frequency. The design is considered as multiple clock domain design. The data is launched from one clock domain and captured in another clock domain.

22.2 What Is Clock Domain Crossing (CDC)

The data transfer can be from slower clock domain to faster clock domain or from faster clock domain to slower clock domain. The data or control signal crosses from one of the clock domain to another clock domain, and it is treated as clock domain crossing.

If we consider the single clock domain design coded using the synthesizable Verilog constructs (Example 1), then there is no any issue in the data integrity due to valid output from both the flip-flops.

The synthesis result of Example 1 is shown in Fig. 22.2 and as shown it infers two cascade flip-flops!

```
//////////////////////////////////////////////////////////////////////////////

// The  Verilog RTL of level synchronizer

module level_synchronizer (

input data_in,

input clk,

output reg data_out);

reg data_out_1;

always@( posedge clk)

begin

        data_out_1<= data_in;

        data_out <= data_out_1;

end

endmodule

//////////////////////////////////////////////////////////////////////////////
```

Example 1 Verilog RTL for single clock domain design

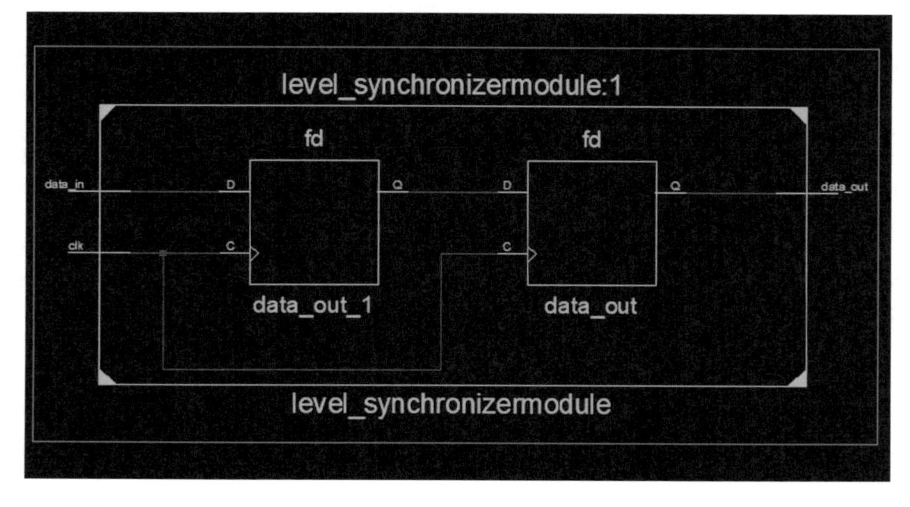

Fig. 22.2 Two-stage level synchronizer

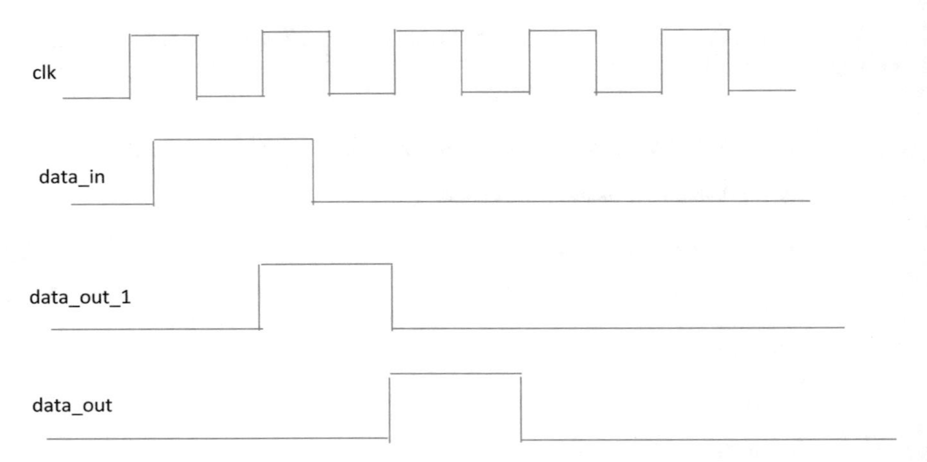

Fig. 22.3 Timing sequence for two-stage level synchronizer

The timing sequence of Example 1 is shown in Fig. 22.3.

If we consider the multiple clock domain designs shown in Example 2, then there is a data integrity issue due to invalid output or metastable output.

The synthesis outcome of the Verilog RTL code (Example 2) is shown in Fig. 22.4.

The timing sequence of Example 2 is shown in Fig. 22.5.

As shown in Fig. 22.5, the output from regB that is data_out is in the metastable state for one clock cycle. Metastability is the scenario in the design due to occurrences of multiple events close to each other, and the design has setup and hold time violations. The scenario results into the synchronization failure between multiple clock domain designs. It is due to different clock frequencies and different phases of the clock in the design. It is essential for the designer to think about why design goes into metastable state? The reasons are every flip-flop has setup and hold time, and if the data changes during the setup time and hold time window, then the design has timing violations and results into the invalid output. Metastable state of the design is not a stable state of the design so if the data output data_out is fed to the other design module, then the output from that module is unpredictable state or invalid logic state. So, to avoid the metastability issues in the design, it is essential to have synchronizers in the data path and control path of the design.

The issue of metastability can be resolved by deploying the level synchronizers while passing the control signals from one of the clock domain to another clock domain. Figure 22.6 shows the multiple clock domain design with the two-stage level synchronizer logic.

```
//////////////////////////////////////////////////////////////////////////////
//The RTL design having multiple clock sources
module multiple_clock_domain_design (
input clk1, clk2,
input data_in,
output reg data_out);
reg data_out_1;

// the procedural block sensitive to rising edge of clk1
always@( posedge clk1)
begin
        data_out_1<= data_in;
end
// the procedural block sensitive to rising edge of clk2
always@(posedge clk2)
begin
        data_out<=  data_out_1;
end
endmodule
//////////////////////////////////////////////////////////////////////////////
```

Example 2 Verilog RTL for multiple clock domain design

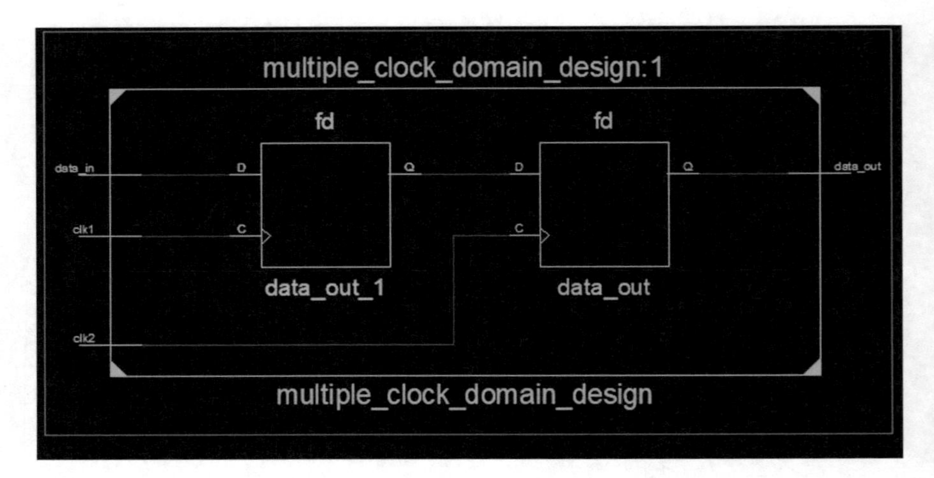

Fig. 22.4 Synthesis result for the multiple clock domain design logic

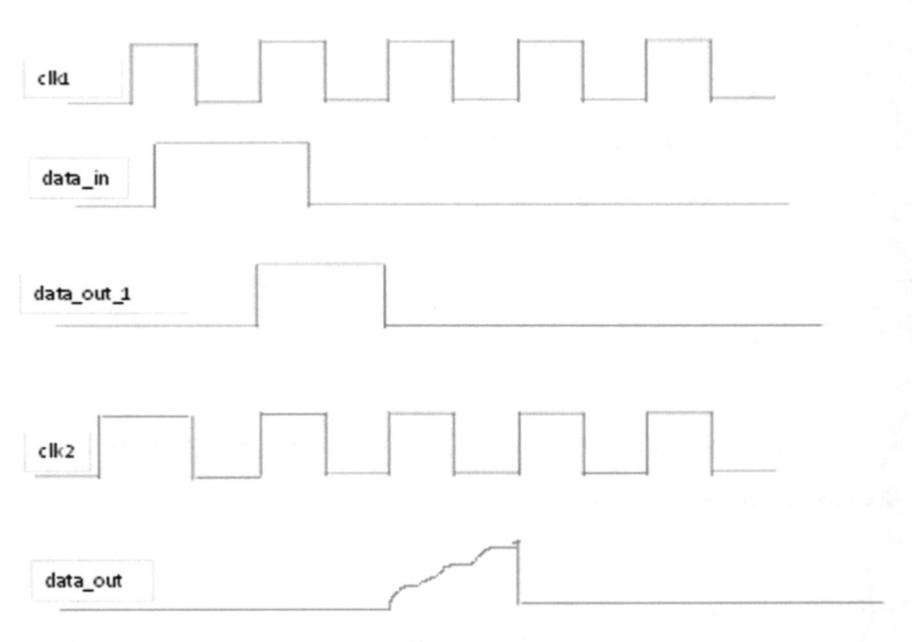

Fig. 22.5 Timing sequence for the metastable output

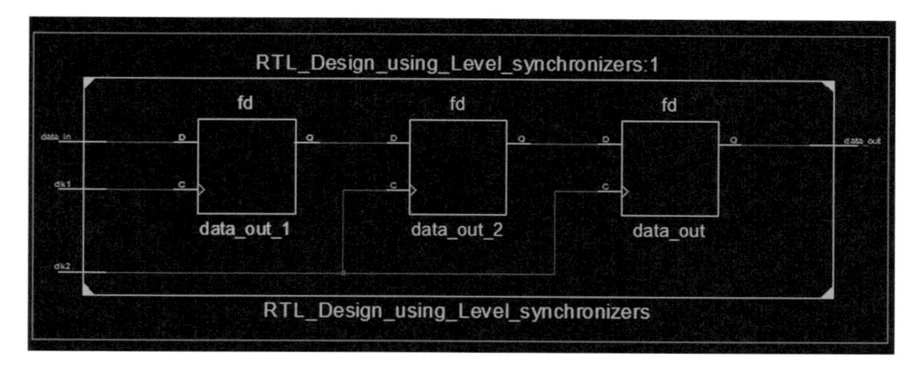

Fig. 22.6 Two-stage level synchronizer in the control path

```
////////////////////////////////////////////////////////////////////////////////
module RTL_Design_using_Level_synchronizers
(
input clk1, clk2,
input data_in,
output reg data_out);
reg data_out_1, data_out_2;
// the clock domain 1 output logic
always@( posedge clk1)
begin
        data_out_1<= data_in;
end

//Use of the two-flip-flop level synchronizer in second clock domain
always@(posedge clk2)
begin
        data_out_2<=data_out_1;
        data_out<= data_out_2;
end
endmodule
////////////////////////////////////////////////////////////////////////////////
```

Example 3 Verilog RTL for use of two-stage level synchronizer in the control path

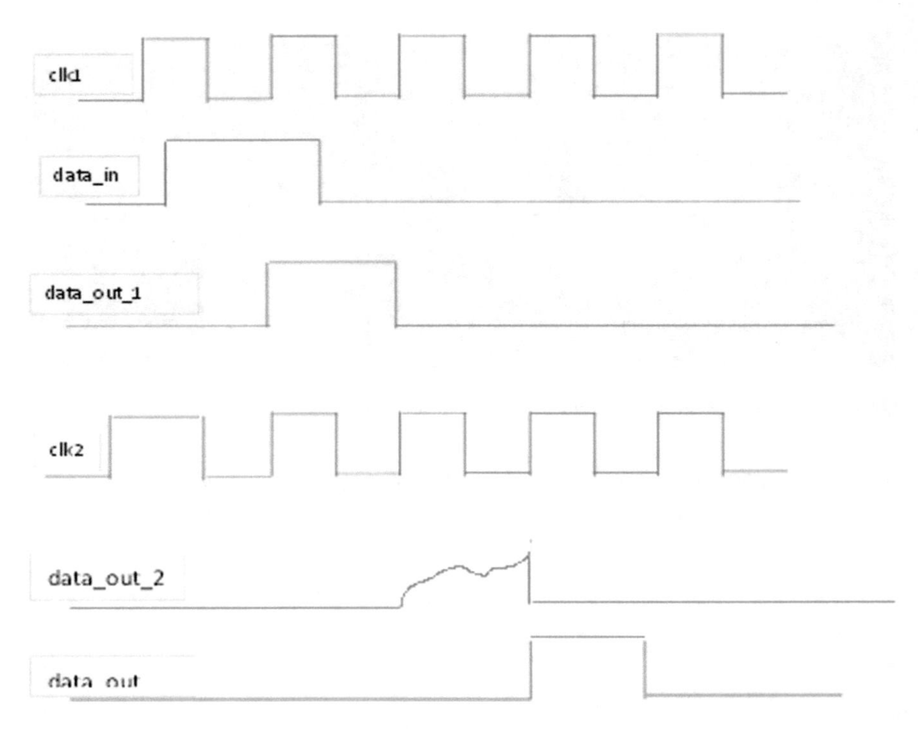

Fig. 22.7 Timing sequence with use of the two-stage synchronizer

As shown in Fig. 22.6, the level synchronizer is used in the second clock domain. The level synchronizer is designed by using regC, regB and used to sample the data data_out_1 from the clock domain 1. The first flip-flop in the second clock domain has metastable output, but the output register in the second clock domain generates the stable legal output data_out on the next clock edge. The Verilog RTL is shown in Example 3.

The timing sequence for Example 3 is shown in Fig. 22.7.

As shown in Fig. 22.7, during second rising edge of clk1, the output data_out_1 is sampled and the output of flip-flop in the first clock domain is logic 1. On the third clock edge of clk2, the output of flip-flop data_out_2 goes to the metastable state due to violation of either setup or hold time. But the flip-flop having an output data_out meets the timing and samples the valid data value of logic 1 on the fourth clock edge of clk2. Hence, the output data_out is valid or legal value state value, and it is logic 1. In most of the scenarios, it is true that the output data_out_2 is in the metastable state due to violation of the setup and hold time. So, during the timing analysis, it is essential to set the false path from output of regA to the output of regC. The SDC command for setting the false path is

set_false_path –from regA/q –to regC/q

22.3 Level synchronizers

The level synchronizers are used to pass the control signal information from one of the clock domain to another clock domain. In the practical scenario, either two-stage or three-stage synchronizers are used. In the two-stage level synchronizer, the number of registers (flip-flops) used is two, and three-stage level synchronizers are designed by using three registers (flip-flops). The latency of control information transfer is dependent on the number of flip-flops. The two-stage level synchronizer is shown in Fig. 22.8.

As discussed in the previous section, the functionality of the two-stage level synchronizer is coded using the synthesizable Verilog constructs and shown in Example 4.

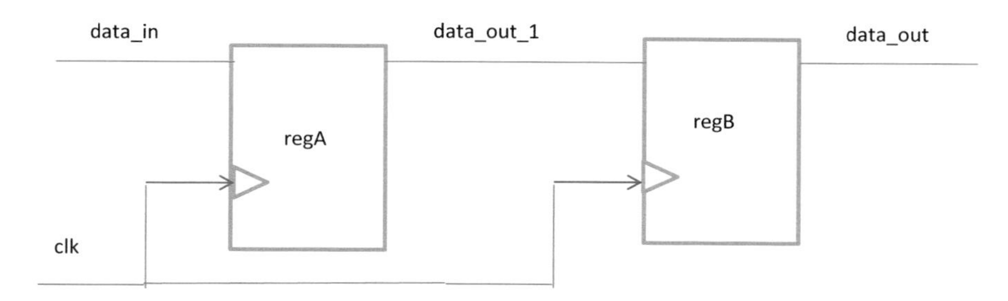

Fig. 22.8 Level synchronizer logic diagram

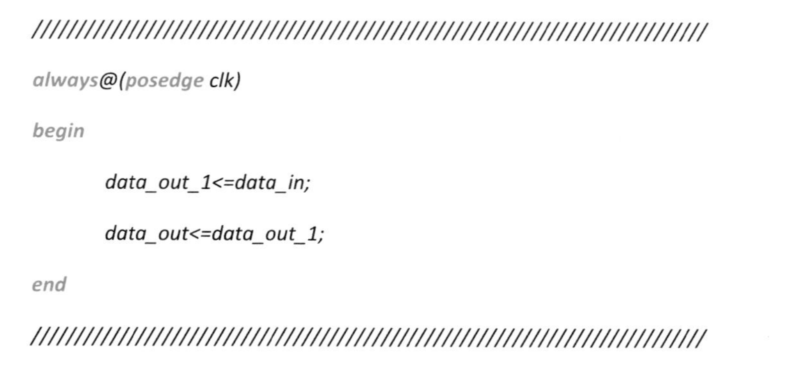

```
/////////////////////////////////////////////////////////////////////////////
always@(posedge clk)
begin
        data_out_1<=data_in;
        data_out<=data_out_1;
end
/////////////////////////////////////////////////////////////////////////////
```

Example 4 Verilog functional description for two-stage synchronizer

```
////////////////////////////////////////////////////////////////////
// the Verilog RTL of three stage level synchronizers
module level_synchronizer (
input data_in,
input clk,
output reg data_out);
reg data_out_1;
reg data_out_2;
// three flip-flop cascade stages
 always@( posedge clk)
 begin
          data_out_1<= data_in;
          data_out_2 <= data_out_1;
          data_out <= data_out_2;
 end
 endmodule
 ///////////////////////////////////////////////////////////////////////////
```

Example 5 Verilog functional description for three-stage level synchronizer

The three-stage level synchronizer is coded by using Verilog and shown in Example 5.

The synthesis outcome of the three-stage level synchronizer is shown in Fig. 22.9.

In the multiple clock domain designs, the data can be passed from slower clock domain to faster clock domain or from faster clock domain to slow clock domain depending on the design architecture and requirement. In both the cases, the synchronizers need to be incorporated in the design. The synchronizers need to be incorporated in the data and control path for the design.

Passing of the control signal from the slower clock domain to the faster clock domain is not a problem as the signal launched by the slower clock domain can be sampled multiple times by the faster clock domain.

As discussed above, consider clk1 is of 100 MHz and clk2 is of 200 MHz. As second clock domain is faster as compared to the first clock domain, there is no any issue while sampling the control signals passed to the second clock domain. But in the practical design scenario, problem occurs when the control information need to

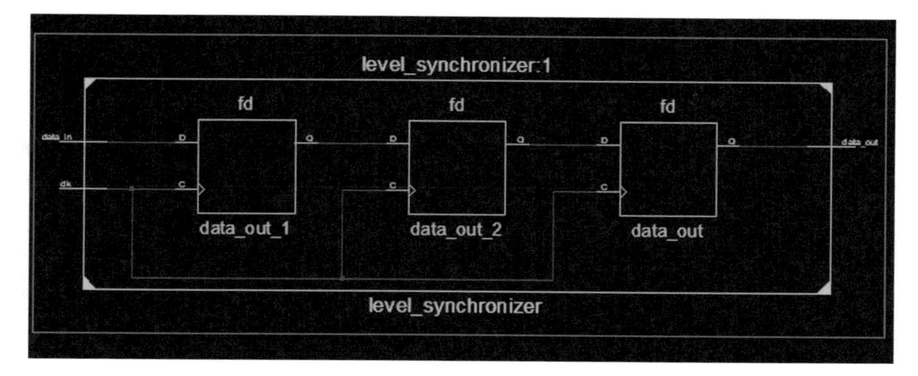

Fig. 22.9 Synthesis result for the three-stage level synchronizer

be passed from faster clock domain to the slower clock domain. The issue is due to non-converging of the legal states of the control signals passed from clock domain 1 to clock domain 2.

As shown in Fig. 22.10, due to slower clock clk2 in the clock domain 2, the data output data_out_1 is sampled on the active edge of clock clk2 but unable to produce the desired output. Due to that, both the registers in the second clock domain generate output as logic 0 and which is unintended output. Both data_out_2 and data_out outputs from are at logic 0 and shown in the timing sequence. The issue of sampling the data from faster clock domain to the slower clock domain can be

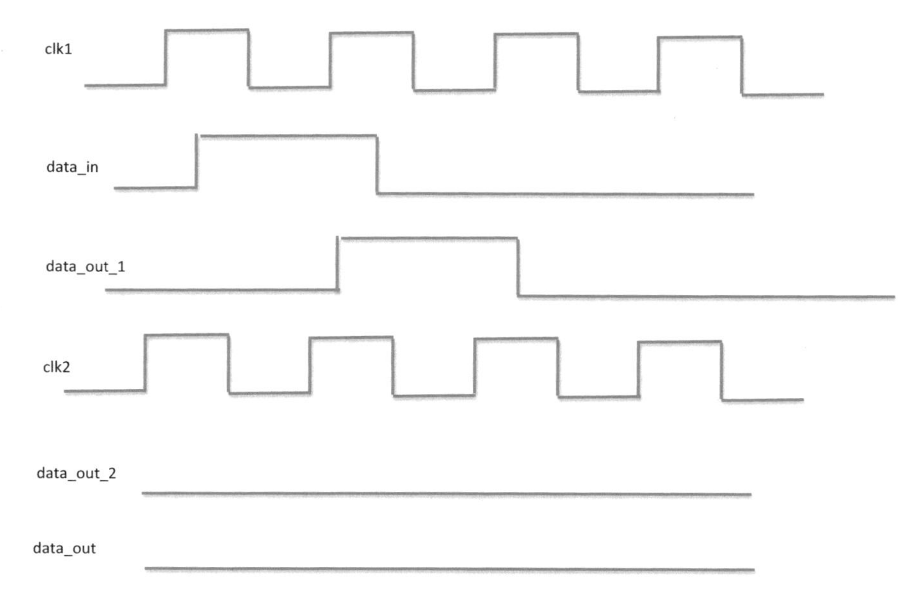

Fig. 22.10 Timing sequence for capturing the data in the slower clock domain

resolved by using pulse stretcher. The level to pulse generator on the positive clock edge is shown in Fig. 22.11.

Another mechanism to achieve the legal converging of the data is by using a handshaking mechanism by using the handshaking signals.

As shown in Fig. 22.12, the sampled signal in the clock domain 2 is reported as a handshaking signal to clock domain 1. This handshake mechanism is like acknowledgement or notification to the faster clock domain 1 that the control signal passed by the faster clock domain is successfully sampled by the slower clock domain. In most of the practical scenarios, this kind of mechanism is used and even the faster clock domain can send another control signal after receiving the valid notification or acknowledgement signal from the slower clock domain.

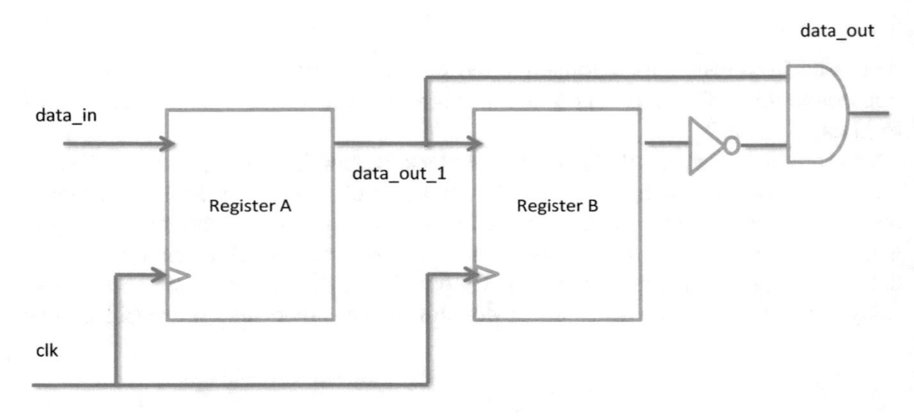

Fig. 22.11 Level-to-pulse conversion logic

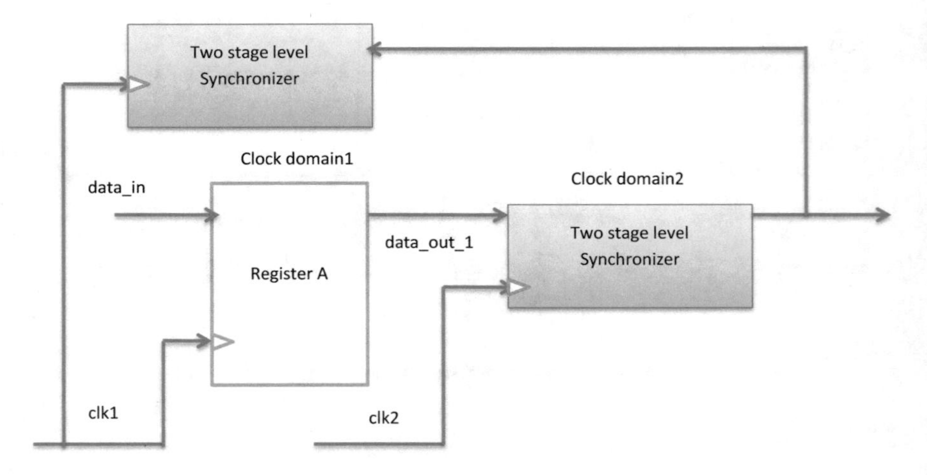

Fig. 22.12 Handshake control signal mechanism

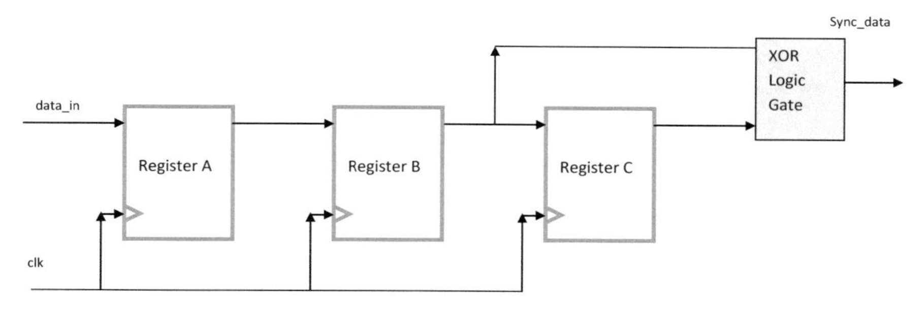

Fig. 22.13 Pulse synchronizer

22.4 Pulse Synchronizers

This type of synchronizer uses the two-stage level synchronizer with the additional register to sample the output of two-stage level synchronizer. The output synchronized data is generated by XOR of the output from the two-stage level synchronizer and the sampled output from the two-stage synchronizer. This kind of synchronizer is also named as toggle synchronizer and used to synchronize the pulse generated in the sending clock domain into the destination clock domain. While passing the data from faster clock domain to the slower clock domain, the pulse can be skipped if two-stage level synchronizer is used. In such scenarios, the pulse synchronizers are very efficient and useful. The pulse synchronizer diagram is shown in Fig. 22.13.

22.5 MUX Synchronizer

Use the pair of the data and control signals while sending the information from clock domain1 to clock domain2. Use the multiple bit data and use the single-bit control signal. At the receiving end depending on the ratio of the sending clock and receiving clock, use the level or pulse synchronizer to generate the control signal for the multiplexer. This technique is like the MCP and effective if the data is stable for multiple clock cycles across the clock boundaries. The diagram is shown in Fig. 22.14.

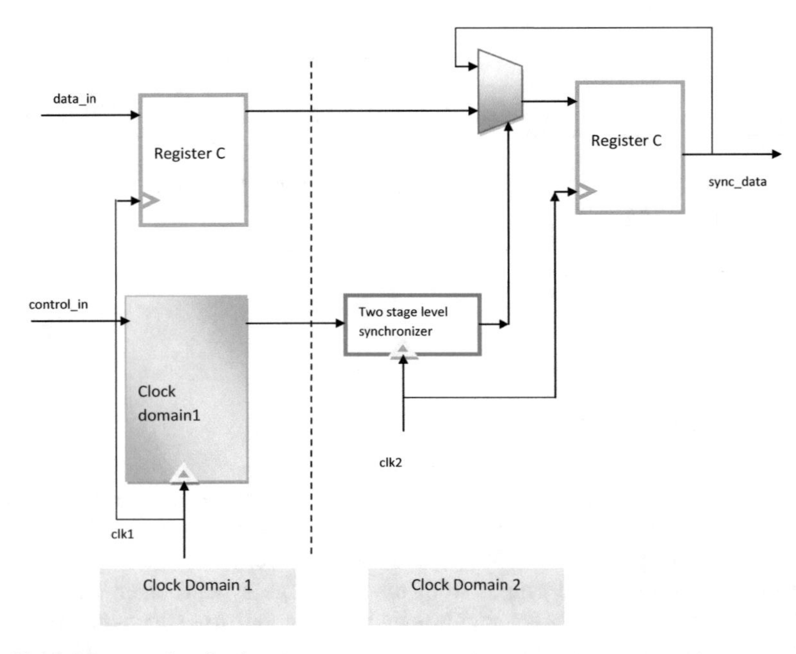

Fig. 22.14 Mux synchronization

22.6 Challenges in the Design of Synchronizers

Passing multiple control signals from one of the clock domain to another clock domain is one of the important challenges for an ASIC or SOC design engineer. When multiple signals are passed from one of the clock domain to another clock domain, then the arrival time of the entire control signals is especially important. If all the control signals are arrived at a time, then the skew is zero. Then there is no any issue while capturing these signals in another clock domain. But in the practical scenarios, there may be possibility that there may be skew between the multiple control signals due to different arrival time from clock domain 1 to clock domain 2. And this can be the cause of the synchronization failure. Consider the design scenario shown in Fig. 22.15, where enable, load_en, and ready need to be passed from one of the clock domain to another clock domain. In such scenario, if multiple-level synchronizers are used as shown then there might be synchronization failure at the receiving end due to skew (Fig. 22.15).

Consider the case where ready and load_en_c2 are arrived and sampled at a time but due to late arrival of enable input in the receiving clock domain2. The data output from the first register of synchronizer does not change, and it does not sample the new value; then again there will be synchronization failure. The sampling of multiple control signals is shown by using Example 6.

```
//////////////////////////////////////////////////////////////////////////////
// the sampling of multiple control signals
module sampling_multiple_control_signals
( input clk2,
  input data_in,
  input enable, ready, load_en,
  output reg data_out );
reg load_en_c2, load_en_c2_tmp;
reg enable_c2, enable_c2_tmp;
reg ready_c2, ready_c2_tmp;

// level synhronizers to sample the load_en
always@(posedge clk2)
begin
        load_en_c2_tmp<= load_en;
        load_en_c2 <= load_en_c2_tmp;
end
// level synchronizer to sample the enable
always@(posedge clk2)
begin
        enable_c2_tmp<=enable;
        enable_c2<=enable_c2_tmp;
end
// level synchronizer to sample the ready
```

Example 6 Verilog functionality to sample the multiple control signals

```
always@(posedge clk2)
begin
        ready_c2_tmp<= ready;
        ready_c2 <= ready_c2_tmp;
end

// Sequential design logic
always @ (posedge clk2)
begin
if ( load_en_c2 & enable_c2 & ready_c2)
        data_out <= data_in;
end
endmodule
/////////////////////////////////////////////////////////////////////////////
```

Example 6 (continued)

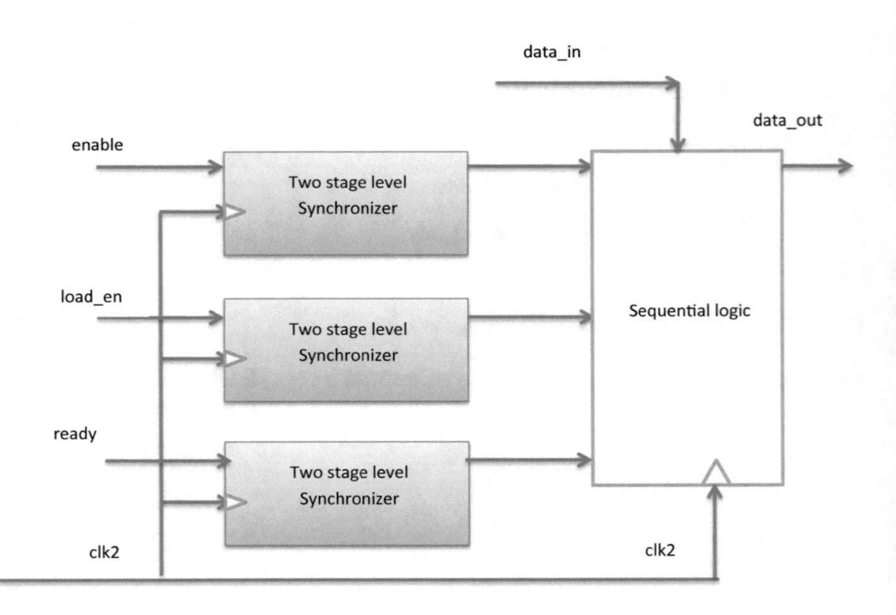

Fig. 22.15 Sampling multiple signals in the receiver clock domain

The synthesis result of Example 6 is shown in Fig. 22.16, and as shown the design uses the multiple level synchronizers in the control path.

The practical and feasible design solution to resolve the problem of data arrival at one of the input and to avoid the synchronization failure is discussed here! Tweak the RTL design in the clock domain 1 to generate the single control signal for enable, load_en, and ready. Pass this control signal from clock domain 1 to clock domain 2. The architecture is tweaked and is shown in Fig. 22.17.

The Verilog RTL is coded using multiple procedural blocks and shown in Example 7.

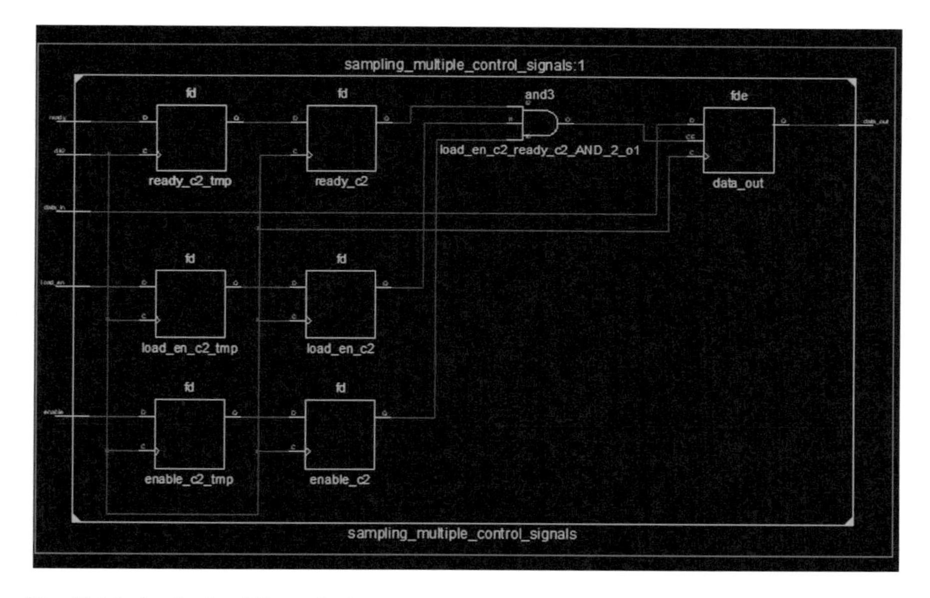

Fig. 22.16 Synthesis of Example 6

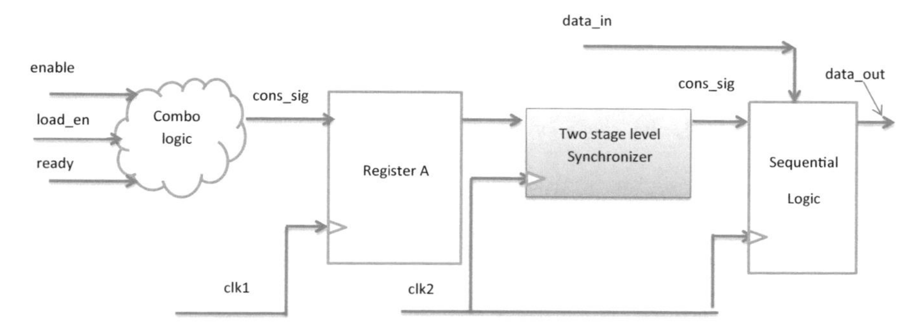

Fig. 22.17 Consolidated control signal passing in the multiple clock domain

```
/////////////////////////////////////////////////////////////////////////////
// sampling multiple control signals
module sampling_multiple_control_signals
( input clk1, clk2,
 input data_in,
 input enable, ready, load_en,
 output reg data_out);
reg [2:0] control_out;
reg [2:0] control_out_c2, control_out_c1;

// clock domain1 registered output
always@(posedge clk1)
begin
        control_out_c1<= {load_en, enable, ready };
end
// replicate the two-stage  level synhronizers to sample multiple signals
always@(posedge clk2)
begin
        control_out_c2<= control_out_c1;
        control_out <= control_out_c2;
end
// sequential design
always @ (posedge clk2)
begin
if ( &control_out )
        data_out <= data_in;
end
endmodule
/////////////////////////////////////////////////////////////////////////////
```

Example 7 Verilog RTL for consolidated control signal receiving

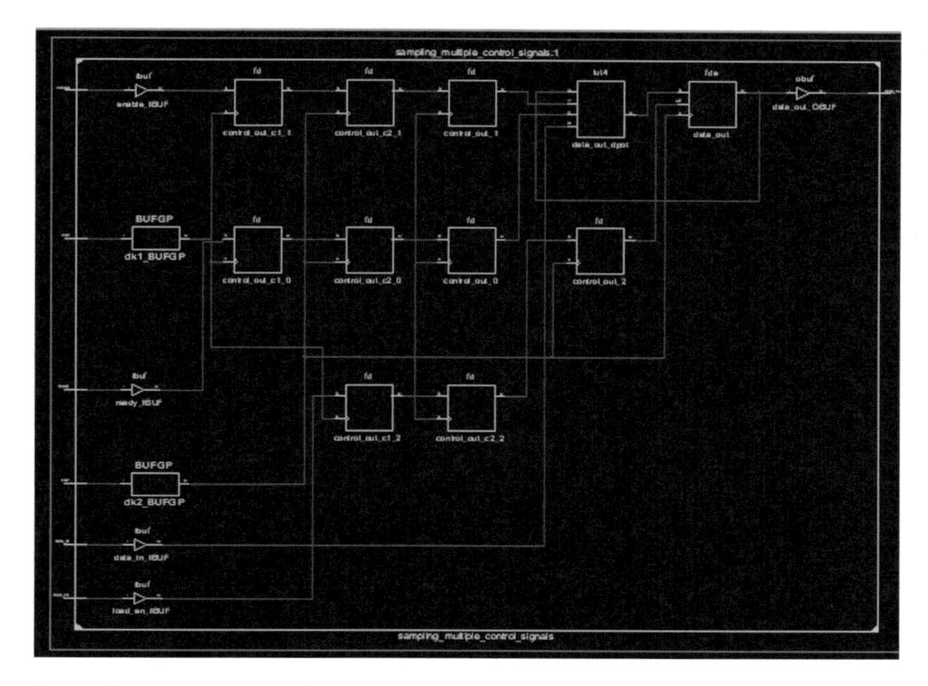

Fig. 22.18 Synthesis result of Example 7

The synthesis result of Example 7 is shown in Fig. 22.18, and as shown the multiple synchronizers are deployed at the output of clock domain 1 register.

Design scenario I

Consider the design scenario while passing the multiple signals from clock domain 1 to clock domain 2. If clock domain 1 has two output signals enable_1, enable_2 and the receiving clock domain 2 uses these two signals as shown in Example 8, then there may be chance of synchronization failure. The synthesis logic is shown in Fig. 22.19.

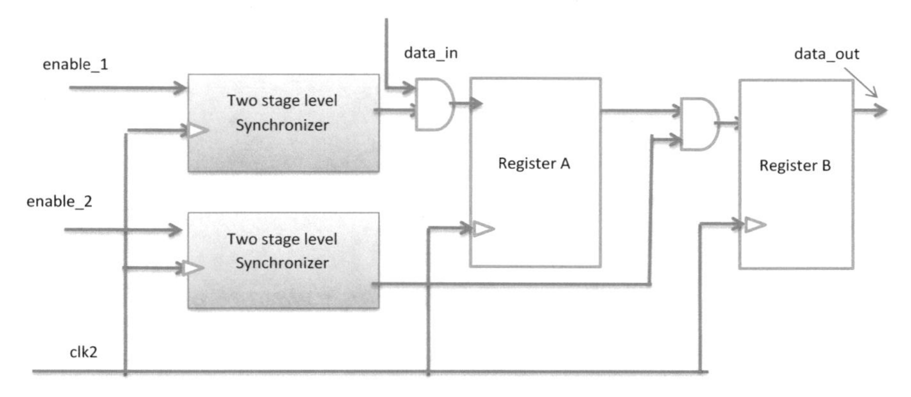

Fig. 22.19 Passing of multiple signals for the pipelined operation

```
/////////////////////////////////////////////////////////////////////////////
module multiple_signals_sampling_pipelined
(
input enable_1,enable_2,
input data_in,
input clk2,
output reg data_out);
reg data_out_2;
reg enable_1to2,  enable_1to2_1, enable_1to2_tmp, enable_1to2_tmp1,
enable_1to2_tmp2;
// two stage level synchronizer to sample enable_1
always@(posedge clk2)
begin
        enable_1to2_tmp <= enable_1;
        enable_1to2 <= enable_1to2_tmp;
end
// two stage level synchronizer to sample enable_2
always@(posedge clk2)
begin
        enable_1to2_tmp1 <= enable_2;
        enable_1to2_1 <= enable_1to2_tmp1;
end
// Pipelined design functionality
always@(posedge clk2)
begin
        data_out_2 <= data_in && enable_1to2;
        data_out  <= data_out_2 && enable_1to2_1;
end
endmodule
/////////////////////////////////////////////////////////////////////////////
```

Example 8 Verilog RTL for using the multiple signals for pipelined operation

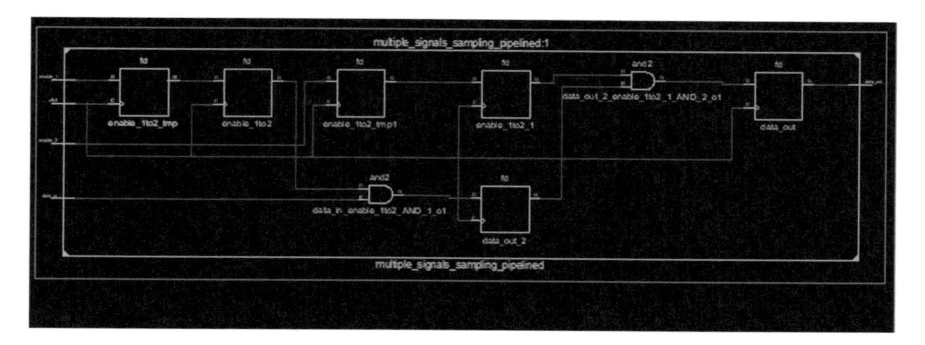

Fig. 22.20 Synthesis of Example 8

More precisely, the FPGA-based synthesis tool infers the result as shown in Fig. 22.20.

The issue for the Verilog RTL coded and shown in Example 8 is sampling of enable_1, enable_2 in the receiving clock domain2. Although the two-level synchronizers are used to sample enable_1 and enable_2, the small skew between arrivals of enable_1, enable_2 can cause the issue of synchronization failure. The pipelined stage shown in Fig. 22.19 can miss the data due to this issue and can result into the invalid output.

Figure 22.21 shows that the data_out is permanently zero and not loaded due to the small skew between the enable1_1 and enable2_1. If these two signals have skew, then there is gap of clock cycle while sampling these signals in the receiving clock domain.

Solution: The practical solution is to use the consolidated enable signal and sample enable_cons in the second clock domain to get the valid enable2_2 signal from the output of two-stage level synchronizer. Figure 22.22 shows the architecture tweak to generate the consolidated control signal.

Design Scenario II

Consider the design scenario of passing the multiple bit encoder output from one of the clock domain to another clock domain. Consider that an encoder output encoder_1, encoder_2 is to be passed from the clock domain1 to clock domain2. The output generated by the clock domain1 is sampled by the clock domain2 using the two-stage level synchronizer. The output of level synchronizer is used as an input to 2:4 decoder. There may be chance that the decoder output is error prone if there is some skew between the inputs of encoder_1 and encoder_2.

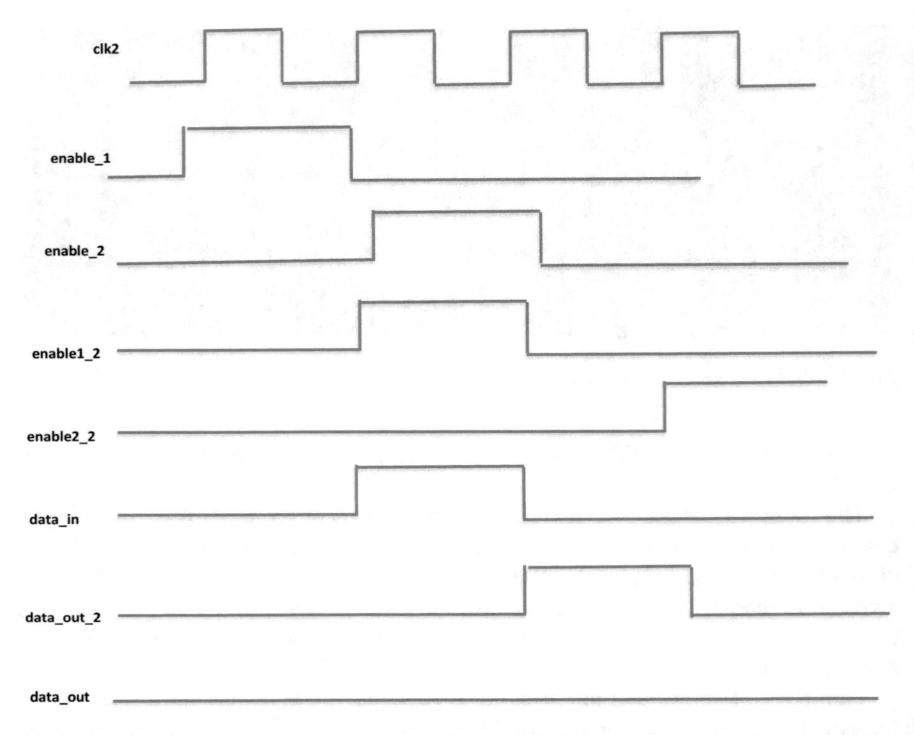

Fig. 22.21 Timing sequence for the use of multiple control signals for pipelined control logic

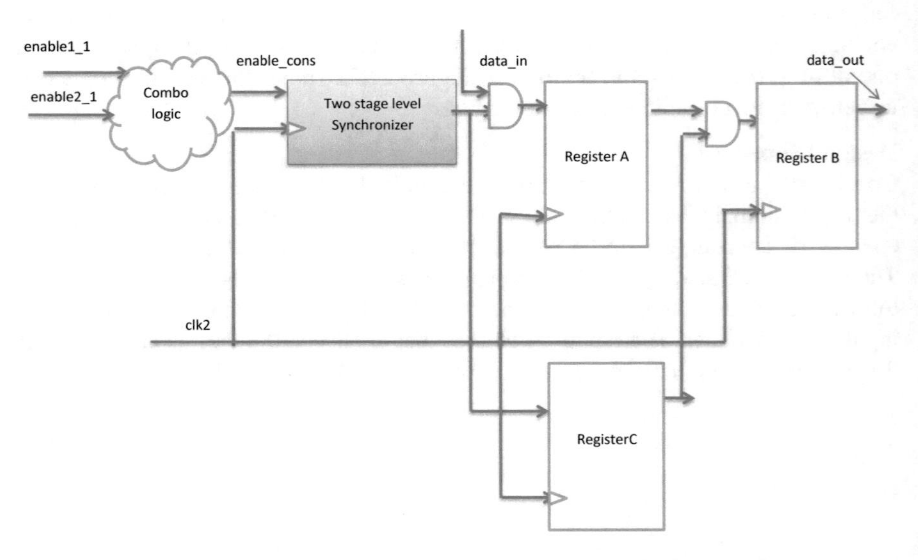

Fig. 22.22 Modified architecture to register the consolidated control signal for pipelined logic

```
//////////////////////////////////////////////////////////////////////////////

always@(posedge clk2)

begin

        {enable_2, enable_2_tmp} <= { enable_2_tmp, enable_cons};

        enable_2_2 <= enable_2;

end

always@(posedge clk2)

begin

        data_out_2 <= data_in && enable_2;

        data_out  <= data_out_2 && enable_2_2;

end

//////////////////////////////////////////////////////////////////////////////
```

Example 9 Partial Verilog RTL for the use of the consolidated control signals for pipelined logic

The Verilog RTL is shown in Example 10.

Consider the practical scenario with reference to the coding shown in Example 10; the issue in the output is due to skew between encoder_1, encoder_2. Due to the skew, the decoder output decoder_out[1] is permanently zero and never be asserted. This problem can be fixed by using the enable control signal while sampling the encoder_1 and encoder_2 signals from the clock doain1 by clock domain2. The enable control signal can be of one clock duration wide and can act as device ready or use the control signal to pass the control information when enable = 1. The enable signal can be asserted while asserting the encoder output or enable signal can be asserted after one clock cycle after assertion of the encoder output. Assertion and de-assertion logic can be designed separately for enable input.

Another important practical and viable approach is to generate the decoder output in the clock domain1 itself by repartition of the design and sample the decoder output in the clock domain2 by using the consolidated enable input and two-level synchronizers.

The Verilog RTL (Example 11) describes the sampling of the decoder output in the clock domain2.

```
////////////////////////////////////////////////////////////////////////////////

always@(posedge clk2)

begin

        {encoder1_2,encoder1_2_tmp} <= { encoder_1_tmp, encoder_1};

        {encoder2_2,encoder2_2_tmp} <= { encoder_2_tmp, encoder_2};

end

always@(*)

begin

case { encoder1_2,encoder2_2}

        2'b00 : decoder_out =4'b1110;

        2'b01 : decoder_out = 4'b1101;

        2'b10 : decoder_out = 4'b1011;

        2'b11 : decoder_out =4'b0111;

endcase

end

////////////////////////////////////////////////////////////////////////////////
```

Example 10 Partial Verilog RTL for sampling of the encoder output

```
///////////////////////////////////////////////////////////////////////////
always@(posedge clk1)
begin
case { encoder_1,encoder_2}
        2'b00 : decoder_out =4'b1110;
        2'b01 : decoder_out = 4'b1101;
        2'b10 : decoder_out = 4'b1011;
        2'b11 : decoder_out =4'b0111;
endcase
end
always@(posedge clk2)
begin
{decoder_out_2, decoder_out_tmp} <= ( decoder_out_tmp, decoder_out};
end
///////////////////////////////////////////////////////////////////////////
```

Example 11 Partial Verilog RTL for the pushing decoder in the single clock domain

22.7 Data Path Synchronizers

As discussed in the above section, to pass the multi-bit signals from one of the clock domain to another clock domain is difficult and error prone task. Although the multi-stage level synchronizers can be used but due to skew between the multiple clock signals, the synchronization cannot be achieved. So, for the multi-bit data, the other techniques are used to pass the data from one of the clock domain to another clock domain. There are two main techniques to pass multi-bit data, and these are used in the practical ASIC designs. These techniques are

(a) Handshaking mechanism
(b) FIFO memory buffers

22.7.1 Handshaking Mechanism

As discussed in the previous section, one or more than one handshake signals are required while passing the data from one of the clock domain to another clock domain. Consider the design scenario shown in Fig. 22.23, as shown the multi-bit data need to be passed from the transmitter to receiver. The transmitter is clocked in the clock domain1, and receiver is clocked by another clock in the second clock domain. So, the multi-bit data exchange by using only level synchronizer is not effective while passing data from transmitter to receiver. ASIC designer can think of the architecture by incorporating the handshake signals datavalid and deviceready. In most of the practical scenario where latency is not a bottleneck, this mechanism is effective to pass multi-bit data.

As shown in Figure 22.23, when transmitter passes multi-bit data from clock domain1, then the receiver receives the data in another clock domain at the edge of receive clock and generates active high datavalid signal to indicate the valid data has been received in the second clock domain. So, the transmitter uses the signal datavalid as handshaking signal. So, until datavalid signal is active high the transmitter cannot place the new data on the data lines. As two- or three-stage level synchronizers can be used to sample the data in the second clock domain, it is recommended that the datavalid signal should be active for at least two or three clock cycles. The overall latency while transferring the data is dependent upon the number of synchronizer stages and number of handshaking signals used. The poor latency is one of the biggest disadvantages of the handshake mechanism.

As required in most of the cases, another handshaking signal deviceready can be generated with the datavalid signal. The receiving clock domain can notify to the transmitter clock domain about the receiver status by asserting the deviceready signal. But while designing handshake mechanism, care needs to be taken for the generation of deviceready and datavalid signals. The deviceready handshake signal should go to logic 1 after de-assertion of datavalid signal.

FSM control uses the architecture shown in the Fig. 22.24.

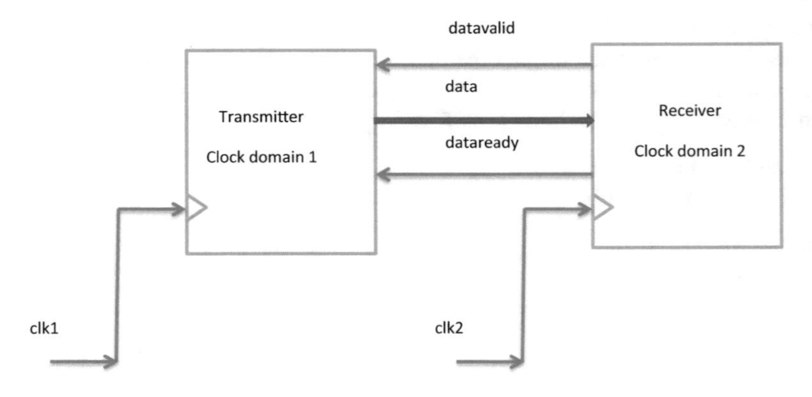

Fig. 22.23 Block diagram for handshake mechanism

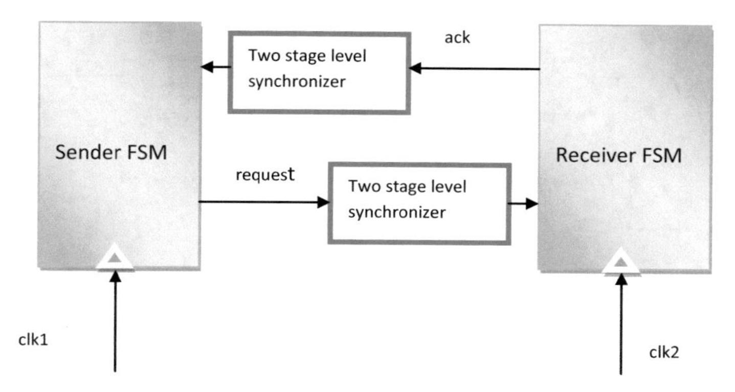

Fig. 22.24 FSM handshaking mechanism

22.7.2 FIFO Synchronizer

In the ASIC designs, the FIFO memory buffers are used as data path synchronizer to pass the data between multiple clock domains. The sender clock domain or transmitter clock domain can write the data into the FIFO memory buffer using write_clk, and receiver clock domain can read the data by using the read_clk.

So basically, FIFO consists of the memory buffer, write domain logic, read domain logic, and the empty and full flag generation logic. The FIFO having various functional blocks is shown in Fig. 22.25. The FIFO depth calculation and FIFO design are discussed in Chap. 23.

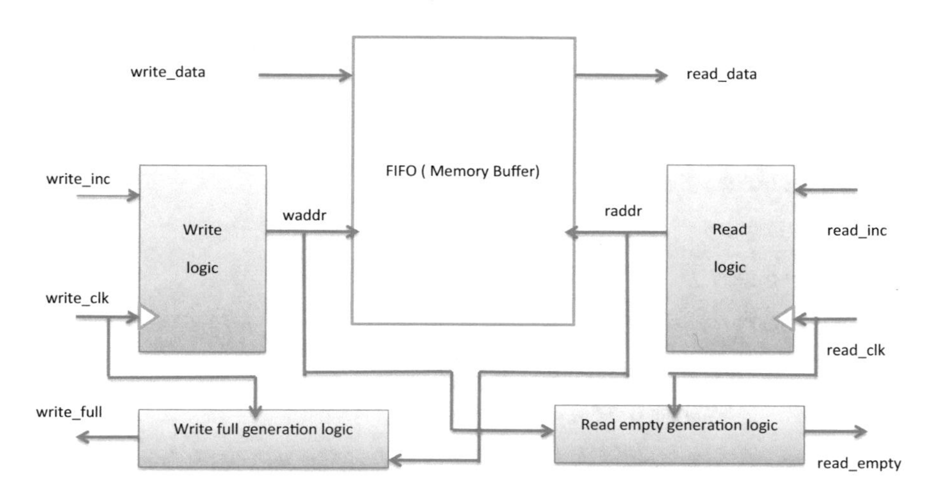

Fig. 22.25 Block diagram of FIFO

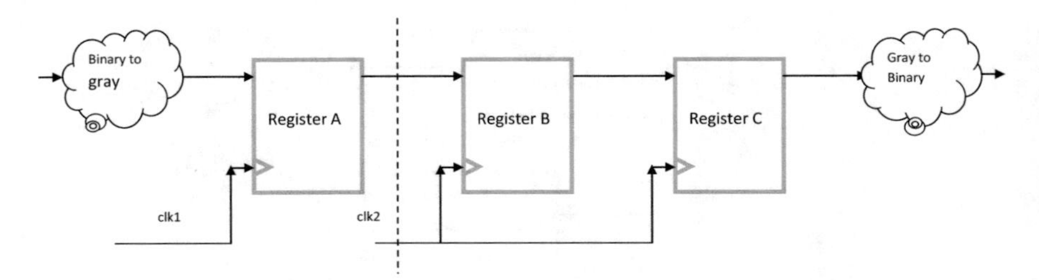

Fig. 22.26 Gray encoding technique

22.7.3 Gray Encoding

While passing the multiple bit of the data or control signals, it is essential to use the gray encoding technique as it is guaranteed to sample the single-bit change in the receiving clock domain. For example, if 4-bit binary data needs to be passed using the binary counter from sending clock domain to receiver clock domain, then use the binary-to-gray conversion logic in the sender clock domain. This guarantees only one bit change across the clocking boundary. After sampling of the gray counter value in the receiving clock domain, use gray-to-binary conversion logic to perform the operations on the binary numbers. The technique is shown in Fig. 22.26.

22.7.3.1 Gray to Binary Converter

Please refer Chap. 4 for the Verilog RTL of gray-to-binary converter. The gray-to-binary code conversion of 4-bit number is coded and shown in Example 12 (Fig. 22.27).

```
////////////////////////////////////////////////////////////////////////////////
module gray_to_binary_converter #(parameter data_size =4)
(
input [data_size-1 :0] gray,
output reg [data_size-1:0] binary);
integer m;
always@(*)
begin
        for (m=0; m <data_size; m=m+1)
        binary[m] = ^ (gray >> m);
end
endmodule
////////////////////////////////////////////////////////////////////////////////
```

Example 12 Verilog RTL for the gray-to-binary converter

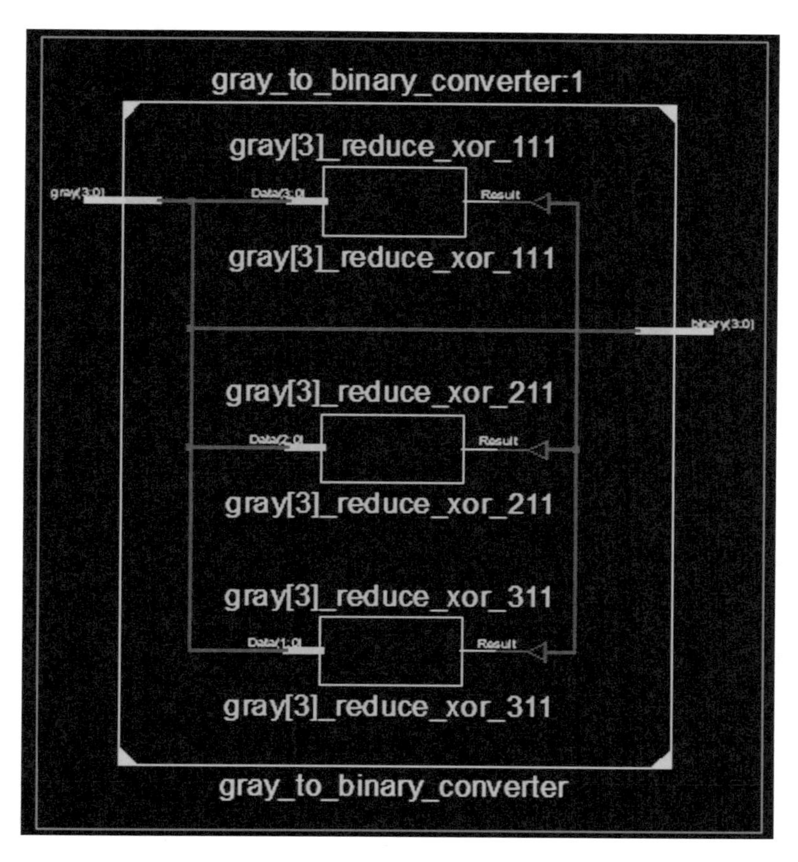

Fig. 22.27 4-bit gray-to-binary code converter

```
////////////////////////////////////////////////////////////////////////////////

module gray_to_binary_converter #(parameter data_size =4)

(

input [data_size-1 :0] gray,

output reg [data_size-1:0] binary);

integer m;

always@(*)

begin

        for (m=0; m <data_size; m=m+1)

        binary[m] = ^ (gray >> m);

end

endmodule

////////////////////////////////////////////////////////////////////////////////
```

Example 13 Verilog RTL for binary-to-gray converter

22.7.3.2 Binary to Gray Converter

Please refer Chap. 4 for the Verilog RTL of binary-to-gray converter. The binary-to-gray code conversion for 4-bit number is shown in Example 13 (Fig. 28.28).

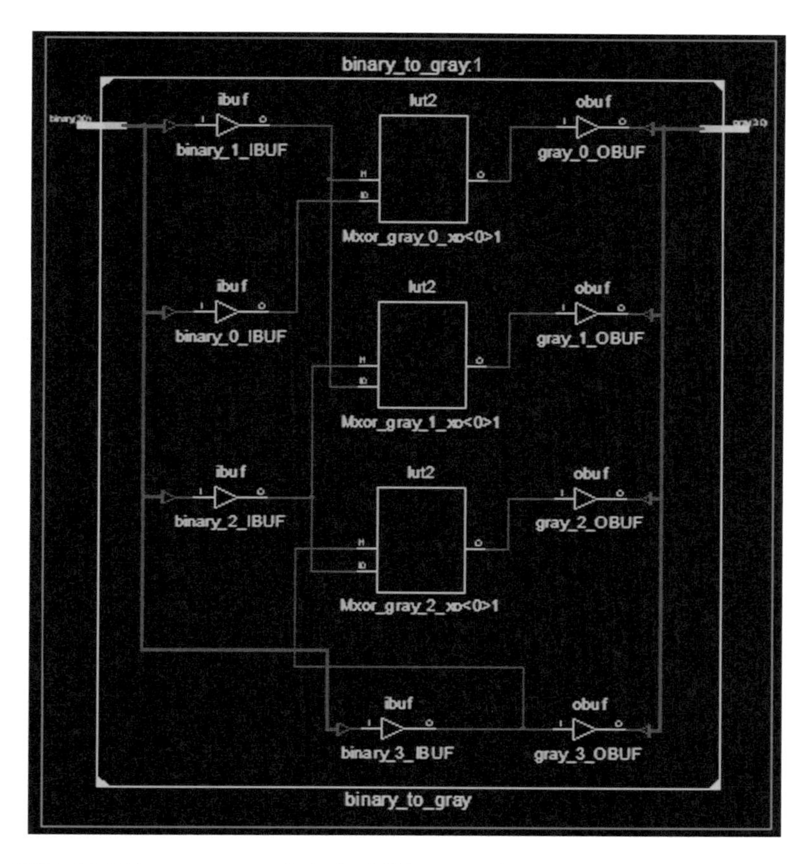

Fig. 22.28 4-bit binary to gray code logic (LUT structure)

22.7.4 Gray Counter

In the multiple clock domain designs, it is recommended to use the gray codes as in the two successive gray numbers only one-bit changes. The Verilog RTL of the gray counter is coded and shown in Example 14. The synthesis result of gray counter is shown in the (Fig. 22.29).

```
//////////////////////////////////////////////////////////////////////////////
module gray_counter #(parameter data_size =4)
(
input clk,
input reset_n,
input increment,
output reg [data_size-1:0] gray,
output reg [data_size-1:0] gray_next, binary_next, binary);
integer m;
always@(posedge clk or negedge reset_n)
if (!reset_n)
        gray <= 4'b0000;
else
gray <= gray_next;
always@(*)
begin
        for (m=0; m< data_size; m=m+1)
        binary[m] =^ (gray >>m);
        binary_next = binary +increment;
        gray_next = (binary_next >>1) ^ binary_next;
end
endmodule
//////////////////////////////////////////////////////////////////////////////
```

Example 14 Verilog RTL for the gray code counter

The gray counter FPGA synthesis result is shown in Fig. 22.30 and has the binary counters and gray counters.

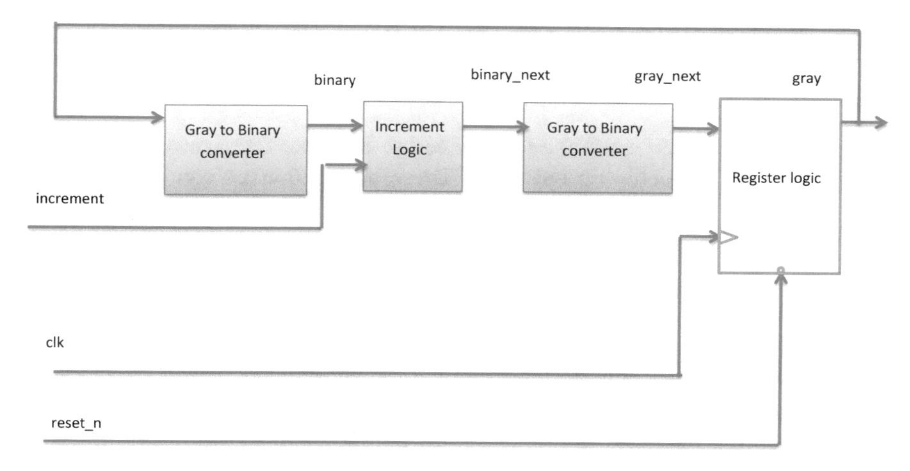

Fig. 22.29 Synthesis diagram for gray counter

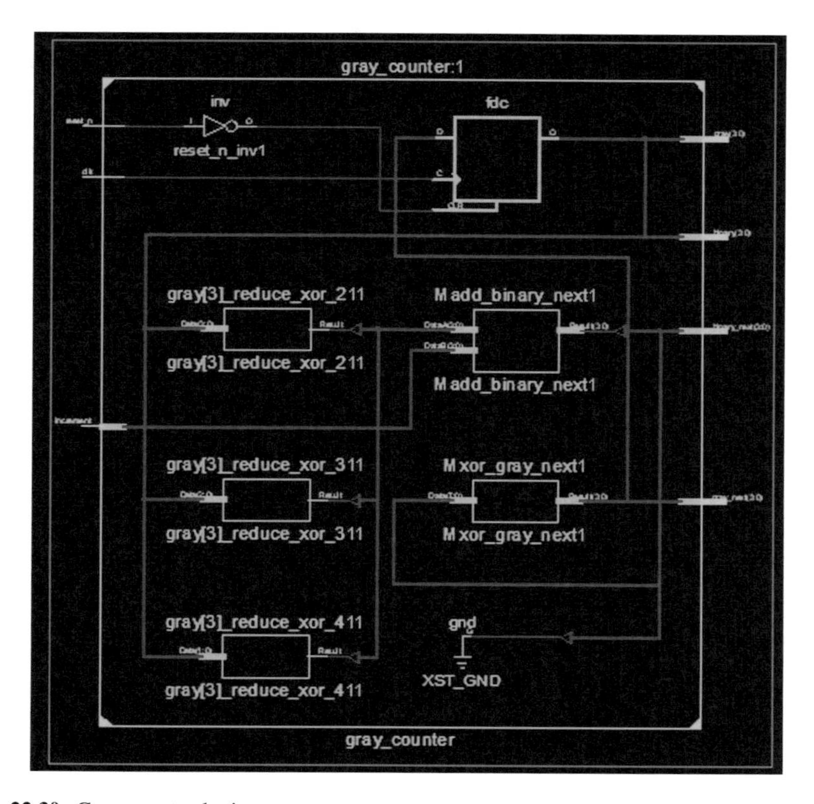

Fig. 22.30 Gray counter logic

22.8 Design Guidelines for the Multiple Clock Domain Designs

CDC design errors can cause the serious design failures. These design failures are expensive during the chip design cycle. These design failures can be avoided by using the following few guidelines during the design and verification phase.

1. **Metastability**: While passing the control signal information or data information, use the registered output logic in the sending clock domain. The reason being, if combinational logic is used at output to pass the data from the sending clock domain to the receiver clock domain, then there might be chances of glitches or hazards due to the multiple transitions in the single clock cycle. The multiple transitions during single clock cycle can be avoided by using the registered logic while passing the data. Metastability blocking logic is shown in Fig. 22.31.
2. **Use of MCP**: Multi-cycle path formulation is highly recommended to avoid the metastability issue while passing the data and control signal information across the clock domains. In the MCP, the strategy is to create the control and data pairs to pass the multi-bit data and single-bit control signal from sending clock domain to receiving clock domain. The control information can be sampled in the receiving clock domain by using the pulse synchronizer, and data can be passed to the receiving clock domain from sending clock domain with or without synchronizers. This technique is highly effective as the data can maintain the stable value for multiple cycles and can be sampled in the receiving clock domain by using the synchronized signal generated by using pulse synchronizer. Across the clock domain crossing boundaries, following are important points need to be considered:
    a. Control signals must be synchronized using the multi-stage synchronizers.
    b. Control signals should be free of hazards and glitches.
    c. There should be single transition across clock boundaries.
    d. Control signal should be stable for at least single clock cycle.

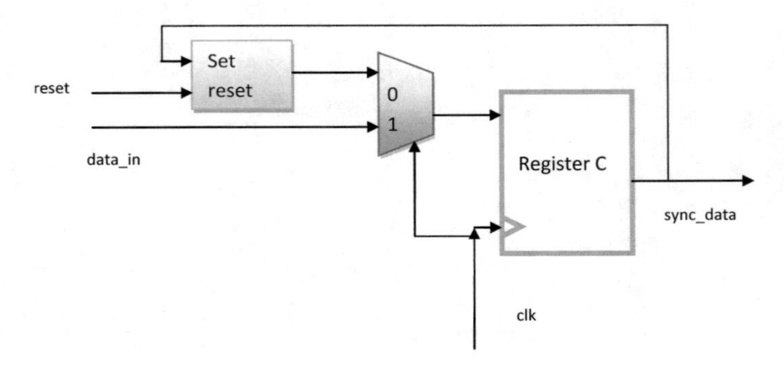

Fig. 22.31 Metastability blocking logic

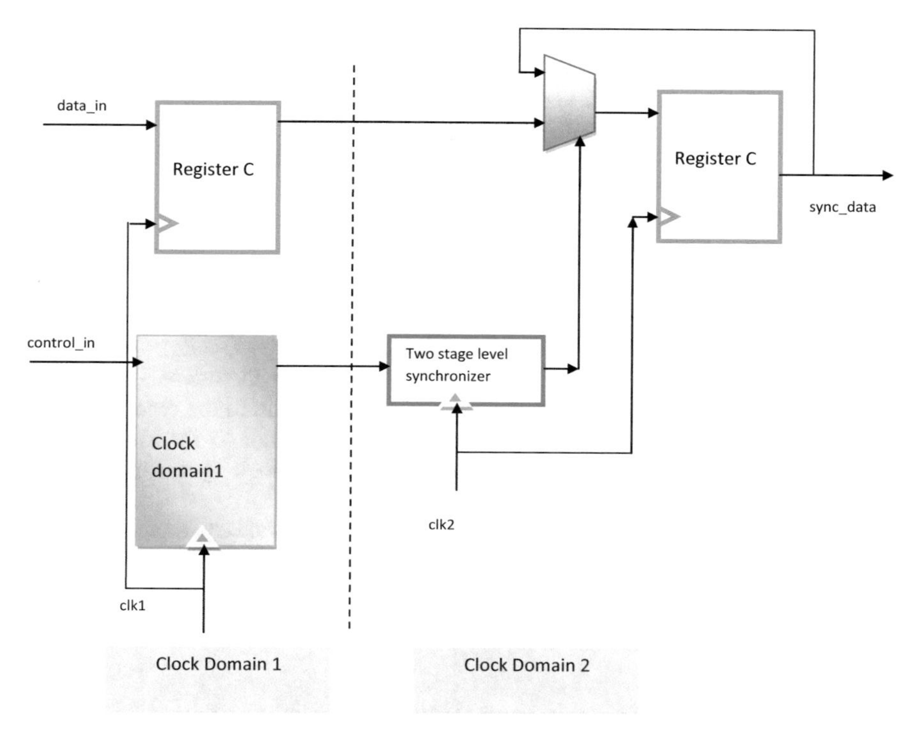

Fig. 22.32 MCP formulation

The MCP formulation is shown in Fig. 22.32.

3. **Use FIFO**: The common and effective technique to pass the multi-bit control or data information is use of asynchronous FIFO. In this technique, the sending clock domain writes the data into FIFO memory buffer and receiving clock domain reads the data from the FIFO buffer.
4. **Use gray code counters**: In most of the ASIC designs having CDC boundaries, it is essential to pass the counter values across the clock domains. If binary counters are used across the clock domain boundaries, then due to one or many bit change at a time, the sampling at the receiver clock domain is difficult and error prone due to the multiple transitions. In such scenarios, it is recommended to use the gray code counters to pass the data across the clock boundaries.
5. **Design partitioning**: While designing the logic for the multiple clock domain designs, partition the design by considering the single clock for every clock domain. This is highly recommended as the STA will be easier due to clean reg-to-reg paths. Even the design verification will be easier due to the design partitioning and by using the single clock.

6. **Clock naming conventions**: It is recommended to use the clock naming conventions to identify the clock domains and the clock sources. The naming conventions for the clock should be supported by the meaningful prefix. For example, for sending clock domain use clk_s, and for the receiving clock domain use clk_r.
7. **Reset synchronization**: For the ASIC and SOC designs, it is highly recommended to use the reset synchronizers while asserting the reset and even it is essential to incorporate the reset synchronizer to avoid the metastability during reset de-assertion. Every SOC has single reset, and either it is positive-level sensitive or negative-level sensitive. So, if synchronizers are not used, then there are chances of metastable states of flip-flops.
8. **Avoid hold time violations**: To avoid the hold time violations it is recommended to pass the stable from the sending clock domain to the receiving clock domain. If data is passed from the faster clock domain to the slower clock domain, then the data should be stable for multiple clock cycles to avoid the hold time violations.
9. **Avoid loss of correlation**: Across the clock domain boundary, there are several ways due to which loss of correlation can occur. Few of them are

   a. Multiple bits on the bus
   b. Multiple handshake signals
   c. Unrelated signals.

   To avoid this, use the clock intent verification technique; this technique will ensure the passing of multi-bit signal across the clock boundaries.

22.9 Summary

Following are important points to conclude this chapter:

1. Clock domain crossing (CDC) is critical to fix in the ASIC design, and these errors can cause the design failure.
2. For single-bit control signal transfer across the multiple clock boundaries, register the signal at the sending clock domain and avoid the glitches and hazard effect of the combinational logic.
3. Use multi-cycle path (MCP) formulation while passing the single-bit control signals across the clock boundaries.
4. Use the multi-stage synchronizers in the receiving clock domain while sampling the single-bit control signals at the receiver clock domain.
5. While passing the multiple control or data signals from one of the clock domain to another clock domain, use the consolidated control signal that is one-bit representation of the multiple signals in the sending clock domain.

6. Use the multi-stage synchronizer in the receiver clock domain while sampling the consolidated control signals.
7. To pass multiple control signals across the clock domains, use the MCP formulation.
8. Use the gray code counters instead of binary counters while passing the data across multiple clock domains.
9. Use FIFO in the data or control path while passing the multiple data bits or control bits.
10. Partition the design using the single clock at the receiving and transmitting (sending) ends.

Chapter 23
Case Study: FIFO Design

The First in First out (FIFO) is used in the data path to pass the data between multiple clock domains. The chapter is useful to understand the FIFO depth calculations and discusses about the FIFO design, simulation of FIFO, and synthesis.

As discussed in the previous few chapters, the FIFO is used in the data path as a data synchronizer. The understanding of the design techniques used in the multiple clock domain plays an important role. The chapter discusses about the FIFO depth calculations and the FIFO design case study.

23.1 FIFO Depth Calculations

FIFO is the storage buffers used to pass data in the multiple clock domain designs. The FIFO depth calculation is discussed in this section.

23.1.1 Asynchronous FIFO Depth Calculations

Scenario I: Clock domain I is faster as compared to clock domain 2 that is f1 is greater than f2 without any idle cycle between write and read.

> Consider the design where f1 = 100 MHz and f2 = 50 MHz, and the burst of data transfer from clock domain one to clock domain 2 is 100 without idle cycles that is consecutive write and read cycles.

V. Taraate, *Digital Logic Design Using Verilog*,
https://doi.org/10.1007/978-981-16-3199-3_23

The depth of FIFO can be calculated as:

1. **Find time required to write one data**:

$$\text{Twrite} = 1/100\,\text{MHz} = 10\,\text{ns}$$

2. **Find out time required to write burst of data**:

$$\text{Tburst\_write} = \text{Twrite} * \text{Burst length} = 10\,\text{ns} * 100 = 1\,\mu\text{s}$$

3. **Find time required to read one data:**

$$\text{Tread} = 1/50\,\text{MHz} = 20\,\text{ns}$$

4. **Find out number of data read in duration of Tburst_write:**

$$\text{No of reads} = 1000\,\text{ns}/20\,\text{ns} = 50$$

5. **The depth of FIFO:**

$$\text{Depth} = \text{Burst length} - \text{No of reads} = 100 - 50 = 50$$

Scenario II: Clock domain I is faster as compared to clock domain 2 that is f1 is greater than f2 with idle cycles between writes and reads.

Consider the design where f1 = 100 MHz and f2 = 50 MHz, and the burst of data transfer from clock domain one to clock domain 2 is 100 with idle cycles. Number of idle cycles between two successive writes = 1 and number of idle cycle between two successive reads = 3.

The depth of FIFO can be calculated as:

1. ***Find time required to write one data***:
   As between two successive writes the idle cycle is one therefore for every two cycles one data is written

$$\text{Twrite} = 2 * (1/100\,\text{MHz}) = 20\,\text{ns}$$

2. ***Find out time required to write burst of data***:

$$\text{Tburst\_write} = \text{Twrite} * \text{Burst length} = 20\,\text{ns} * 100 = 2\,\mu\text{s}$$

3. **Find time required to read one data**:
   As between two successive reads the idle cycle is three therefore for every four cycles one data is read

$$\text{Tread} = 4 * (1/50\,\text{MHz}) = 80\,\text{ns}$$

4. **Find out number of data read in duration of Tburst_write**:

$$\text{No of reads} = 2000\,\text{ns}/80\,\text{ns} = 25$$

5. **The depth of FIFO:**

$$\text{Depth} = \text{Burst length} - \text{No of reads} = 100 - 25 = 75$$

Scenario III: Clock domain I is slower as compared to clock domain 2 that is f1 is less than f2 with idle cycles between two successive writes and two successive reads.

Consider the design where f1 = 50 MHz and f2 = 80 MHz and the burst of data transfer from clock domain one to clock domain 2 is 100 with idle cycles. Number of idle cycles between two successive writes = 1 and number of idle cycle between two successive reads = 3.

The depth of FIFO can be calculated as:

1. ***Find time required to write one data:***
   As between two successive writes the idle cycle is one therefore for every two cycles one data is written

$$\text{Twrite} = 2 * (1/50\,\text{MHz}) = 40\,\text{ns}$$

2. ***Find out time required to write burst of data:***

$$\text{Tburst\_write} = \text{Twrite} * \text{Burst length} = 40\,\text{ns} * 100 = 4\,\mu\text{s}$$

3. ***Find time required to read one data:***
   As between two successive reads the idle cycle is three therefore for every four cycles one data is read

$$\text{Tread} = 4 * (1/80\,\text{MHz}) = 50\,\text{ns}$$

4. ***Find out number of data read in duration of Tburst_write:***

$$\text{No of reads} = 4000\,\text{ns}/50\,\text{ns} = 80$$

5. ***The depth of FIFO:***

$$\text{Depth} = \text{Burst length} - \text{No of reads} = 100 - 80 = 20$$

Scenario IV: Clock domain 1's frequency is equal to clock domain 2's that is f1 is equal to f2 and idle cycles between two successive reads and writes.

Consider the design where f1 = 100 MHz and f2 = 100 MHz and the burst of data transfer from clock domain 1 to clock domain 2 is 100 with idle cycles. Number of idle cycles between two successive writes = 1 and number of idle cycle between two successive reads = 3.

The depth of FIFO can be calculated as:

1. ***Find time required to write one data:***
   As between two successive writes the idle cycle is one therefore for every two cycles one data is written

$$\text{Twrite} = 2 * (1/100\,\text{MHz}) = 20\,\text{ns}$$

2. ***Find out time required to write burst of data:***

$$\text{Tburst\_write} = \text{Twrite} * \text{Burst length} = 20\,\text{ns} * 100 = 2\,\mu\text{s}$$

3. ***Find time required to read one data:***
   As between two successive reads the idle cycle is three therefore for every four cycles one data is read

$$\text{Tread} = 4 * (1/100\,\text{MHz}) = 40\,\text{ns}$$

4. ***Find out number of data read in duration of Tburst_write:***

$$\text{No of reads} = 2000\,\text{ns}/40\,\text{ns} = 50$$

5. ***The depth of FIFO:***

$$\text{Depth} = \text{Burst length} - \text{No of reads} = 100 - 50 = 50$$

23.2 FIFO Design Case Study

The FIFO design and case study are described by using the following Verilog RTL, and the important steps are documented in the following template

```
// FIFO Verilog RTL template
// Module instantiation and port definition
// define the intermediate signals using wire or reg
// Instantiation of FIFO memory buffer
// Instantiation of synchronizers for the write to read clock domain
//Instantiation of synchronizers for the read to write clock domain
// Instantiation of logic for the read empty flag generation
//Instantiation of logic for the write full flag generation
```

FIFO memory buffer top-level pin diagram is shown in Fig. 23.1 and has two different clock domains. Input clock domain or sender clock domain works on the

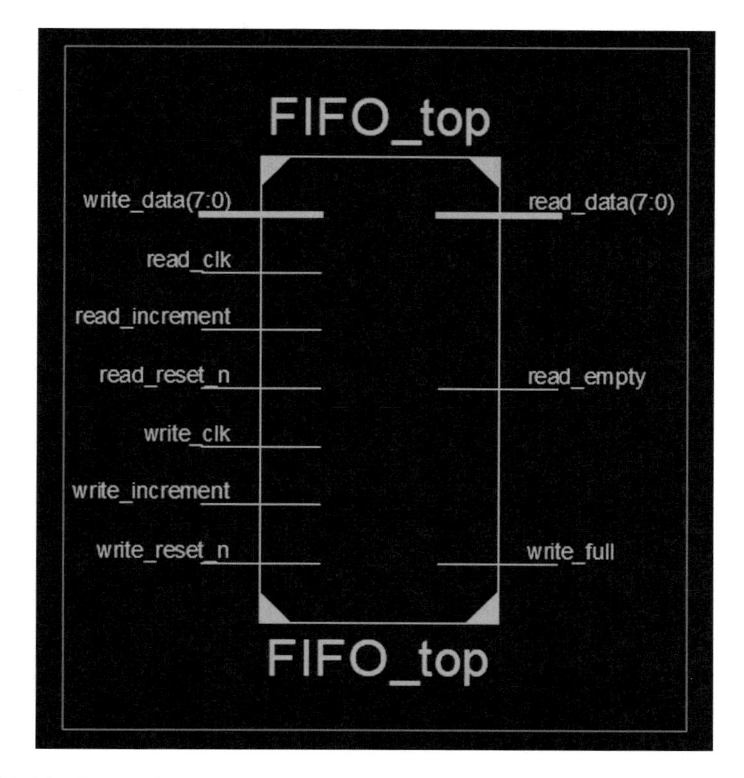

Fig. 23.1 Pin diagram for FIFO top-level module

write_clk, and another clock domain or receiver clock domain works on the read_clk.

Top-level signal description of the module FIFO is shown in Table 23.1.

FIFO Memory: FIFO memory buffer Verilog RTL is shown in Example 1.

Table 23.1 FIFO top-level inputs and outputs

Signal name	Direction	Width	Description
write_clk	Input	1-bit	The fast clock source for write clock domain
read_clk	Input	1-bit	The slow clock source for read clock domain
write_reset_n	Input	1-bit	The asynchronous active low reset for write clock domain
read_reset_n	Input	1-bit	The asynchronous active low reset for read clock domain
write_increment	Input	1-bit	The increment control input in the write clock domain
read_increment	Input	1-bit	The increment control input in the read clock domain
write_data	Input	[data_size-1:0]	The input data bus to carry the data to be written in the buffer
read_data	Output	[data_size-1:0]	The output data bus to get the data to be read from the buffer
write_full	Output	1-bit	The FIFO full flag as an output to indicate that do not write data now
read_empty	Output	1-bit	The FIFO empty output flag to indicate that do not read from FIFO buffer

```
//////////////////////////////////////////////////////////////////////////////

module FIFO_memory #(parameter data_size = 8, parameter ad-
dress_size = 3)

// let us define the data and address size to get 8 location X 8 bit memory

(

input [data_size-1:0] write_data,

input [address_size-1:0] write_address, read_address,

input write_clk_en, write_full, write_clk,

output [data_size-1:0] read_data);


localparam FIFO_depth = 1<<address_size;

reg [data_size-1:0] mem [0:FIFO_depth-1];

// read the data at the output of memory

assign read_data = mem[read_address];


//Write a data on rising edge of write clock at specific address location

always @(posedge write_clk)

if (write_clk_en && !write_full)

mem[write_address] <= write_data;


endmodule

//////////////////////////////////////////////////////////////////////////////
```

Example 1 Verilog RTL for the FIFO memory buffer

The top-level signal description is shown in Table 23.2, and the synthesis result is shown in Fig. 23.2.

Read synchronization logic: Table 23.3 is useful to understand about the inputs and outputs of this module. The logic is used to synchronize the pointers to avoid the metastability.

The RTL description is shown in Example 2.

Table 23.2 Pin description of FIFO memory buffer

Signal name	Direction	Width	Description
write_clk	Input	1-bit	The fast clock source for write clock domain
write_clk_en	Input	1-bit	Enable write when FIFO is not full
write_address	Input	[address_size-1:0]	The binary address of the memory to write the data
read_address	Input	[address_size-1:0]	The binary address of the memory to read the data
write_data	Input	[data_size-1:0]	The input data bus to carry the data to be written in the buffer
read_data	Output	[data_size-1:0]	The output data bus to get the data to be read from the buffer
write_full	Output	1-bit	The FIFO full flag as an output to indicate that do not write data now

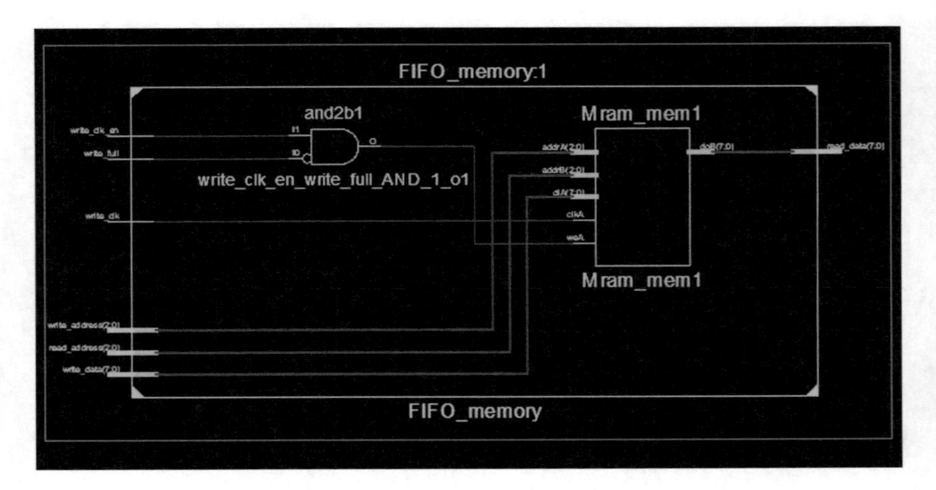

Fig. 23.2 FIFO memory buffer

Table 23.3 Pin description of the read synchronizer logic

Signal name	Direction	Width	Description
read_clk	Input	1-bit	The slow clock source for read clock domain
read_reset_n	Input	1-bit	The asynchronous active low reset for read clock domain
write_pointer	Input	[address_size:0]	The input as gray pointer which can be used during pointer comparison
read_to_write_pointer	Output	[address_size:0]	The pointer passing from read to write clock domain

```
////////////////////////////////////////////////////////////////////////////////
module sync_write_to_read #(parameter address_size = 3)
(
input [address_size:0] write_pointer,
input read_reset_n,read_clk,
output reg [address_size:0] read_to_write_pointer);
reg [address_size:0] tmp1_write_pointer;
//Multi-flop synchronizer logic for passing the control signals and pointers


always @(posedge read_clk , negedge read_reset_n)
if (~read_reset_n)
{read_to_write_pointer,tmp1_write_pointer} <= 0;
else
{read_to_write_pointer,tmp1_write_pointer} <= tmp1_write_pointer,write_pointer};
endmodule
////////////////////////////////////////////////////////////////////////////////
```

Example 2 Verilog RTL for the write to read synchronizer logic

As shown in the synthesis result, the write_pointer is synchronized in the read clock domain (Fig. 23.3).

Write synchronization logic: Table 23.4 is useful to understand about the inputs and outputs of this module. The logic is used to synchronize the pointers to avoid the metastability.

The RTL description is shown in Example 3.

As shown in the synthesis result the read_pointer is synchronized in the write clock domain (Fig. 23.4).

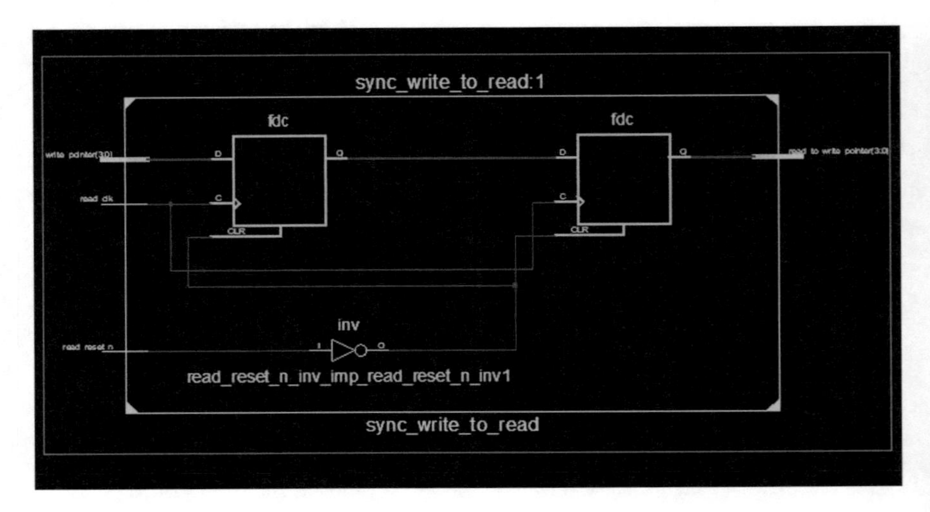

Fig. 23.3 RTL schematic of synchronizer for write to read clock domain

Table 23.4 Pin description of write synchronizer logic

Signal name	Direction	Width	Description
write_clk	Input	1-bit	The fast clock source for write clock domain
write_reset_n	Input	1-bit	The asynchronous active low reset for write clock domain
read_pointer	Input	[address_size:0]	The input as gray pointer which can be used during pointer comparison
write_to_read_pointer	Output	[address_size:0]	The pointer passing from write to read clock domain

```
//////////////////////////////////////////////////////////////////////////////
module sync_read_to_write #(parameter address_size = 3)
(
input write_reset_n,write_clk,
input [address_size:0] read_pointer,
output reg [address_size:0] write_to_read_pointer);

reg [address_size:0] tmp1_read_pointer;
//Multi-flop synchronizer logic for passing the control signals and pointers

always @(posedge write_clk , negedge write_reset_n)
if (~write_reset_n)
{write_to_read_pointer,tmp1_read_pointer} <= 0;
else
{write_to_read_pointer,tmp1_read_pointer} <= {tmp1_read_pointer,read_pointer};
endmodule
//////////////////////////////////////////////////////////////////////////////
```

Example 3 Verilog RTL for the read to write synchronizer logic

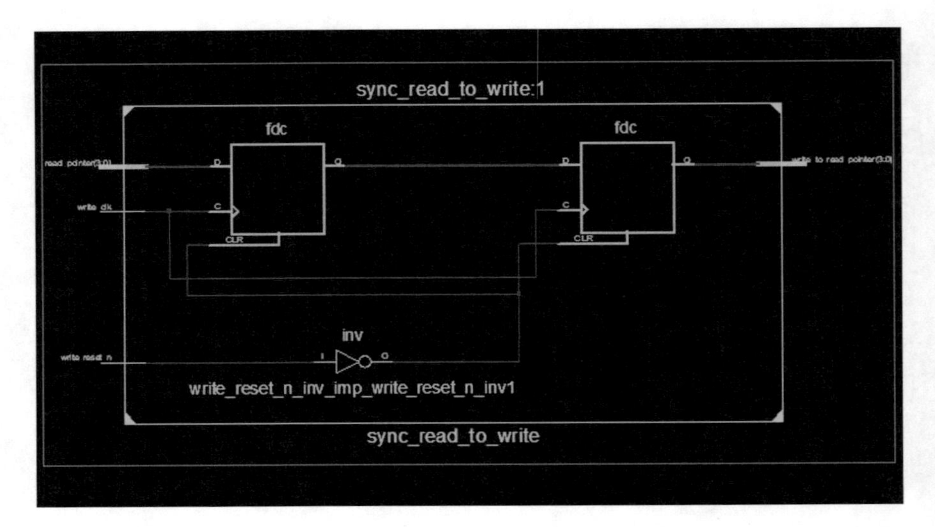

Fig. 23.4 RTL schematic of synchronizer for read to write clock domain

Read clock domain and empty flag generation: Table 23.5 is useful to understand about the inputs and outputs of this module. The logic is used to generate the empty flag after the pointer comparison that is read_address is equal to write_address the FIFO is empty. The assignment used to generate the empty flag is, read_empty <= (read_gray_next == read_to_write_pointer);

Example 4 is coded using the synthesizable Verilog constructs and used to generate the read_empty flag.

The synthesis result is shown in Fig. 23.5.

Table 23.5 Pin description of the read empty flag generation logic

Signal name	Direction	Width	Description
read_clk	Input	1-bit	The slow clock source for read clock domain
read_reset_n	Input	1-bit	The asynchronous active low reset for read clock domain
read_increment	Input	1-bit	The increment control input in the read clock domain
read_to_write_pointer	Input	[address_size:0]	The pointer passing from read to write clock domain
read_pointer	Output	[address_size:0]	The output as gray pointer which can be used during pointer comparison
read_address	Output	[address_size-1:0]	The binary address to memory buffer which is output from this module
read_empty	Output	1-bit	The FIFO empty flag as an output to indicate that do not read data now

```
read_empty <= (read_gray_next == read_to_write_pointer);
//////////////////////////////////////////////////////////////////////////
module read_pointer_empty #(parameter address_size = 3)
(
input  read_reset_n,read_increment,read_clk,
input [address_size :0] read_to_write_pointer,
output reg [address_size-1:0] read_address,
output reg [address_size :0] read_pointer,
output reg read_empty
);
// the temporary variable for binary data
reg [address_size:0] read_binary;
// let us declare the reg for the assign expression to get the gray and bi-
nary values
reg [address_size:0] read_gray_next, read_binary_next;


// Binary address pointer to memory
always@*
begin
 read_address = read_binary[address_size-1:0];
 read_binary_next = read_binary + (read_increment & ~read_empty);
 read_gray_next = (read_binary_next>>1) ^ read_binary_next;
 end
// let us incorporate gray pointers
```

Example 4 Verilog RTL for empty flag logic

```
always @(posedge read_clk , negedge read_reset_n)
if (~read_reset_n)
{read_binary, read_pointer} <= 0;
else
{read_binary, read_pointer} <= {read_binary_next, read_gray_next};

// FIFO empty logic

always @(posedge read_clk , negedge read_reset_n)
if (~read_reset_n)
        read_empty <= 1'b1;
else
        read_empty <= (read_gray_next == read_to_write_pointer);
endmodule

/////////////////////////////////////////////////////////////////////////////
```

Example 4 (continued)

Write clock domain and FIFO full flag generation: Table 23.6 is useful to understand about the inputs and outputs of this module. The logic is used to generate the FIFO full flag after the pointer comparison that is (Fig. 23.6).

```
write_full <= (write_gray_next=={~write_to_read_pointer[ad-
dress_size:address_size-1], write_to_read_pointer[address_size-2:0]});
```

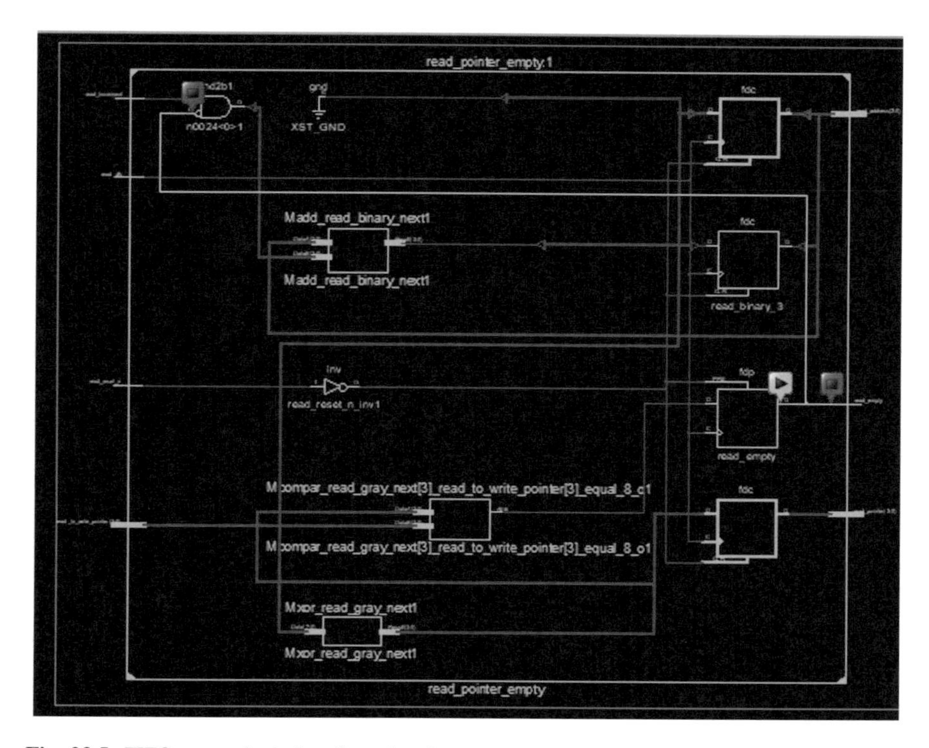

Fig. 23.5 FIFO empty logic interface signals

Table 23.6 Pin description of the write full flag generation logic

Signal name	Direction	Width	Description
write_clk	Input	1-bit	The fast clock source for write clock domain
write_reset_n	Input	1-bit	The asynchronous active low reset for write clock domain
write_increment	Input	1-bit	The increment control input in the write clock domain
write_to_read_pointer	Input	[address_size:0]	The pointer passing from write to read clock domain
write_pointer	Output	[address_size:0]	The output as gray pointer which can be used during pointer comparison
write_address	Output	[address_size-1:0]	The binary address to memory buffer which is output from this module
write_full	Output	1-bit	The FIFO full flag as an output to indicate that don't write data now

```
////////////////////////////////////////////////////////////////////////////////

module write_pointer_full #(parameter address_size = 4)

(

input write_reset_n, write_clk, write_increment,

input [address_size :0] write_to_read_pointer,

 output reg [address_size-1:0] write_address,

 output reg [address_size :0] write_pointer,

 output reg write_full);

 reg [address_size:0] write_gray_next, write_binary_next;

 reg [address_size:0] write_binary;


 // Binary pointers to the memory buffer

 always@*

 begin

  write_address = write_binary[address_size-1:0];

  write_binary_next = write_binary + (write_increment & ~write_full);

  write_gray_next = (write_binary_next>>1) ^ write_binary_next;


 end


 // let us use the gray pointers

 always @(posedge write_clk , negedge write_reset_n)

 if (~write_reset_n)

 {write_binary, write_pointer} <= 0;
```

Example 5 Verilog RTL for full flag logic

```
else

{write_binary, write_pointer} <= {write_binary_next, write_gray_next};


//Let us create FIFO_Full logic using the pointer comparison
always @(posedge write_clk , negedge write_reset_n)

if (!write_reset_n) write_full <= 1'b0;

else

write_full <= (write_gray_next=={~write_to_read_pointer[ad-
dress_size:address_size-1], write_to_read_pointer[address_size-2:0]});

endmodule

///////////////////////////////////////////////////////////////////////////
```

Example 5 (continued)

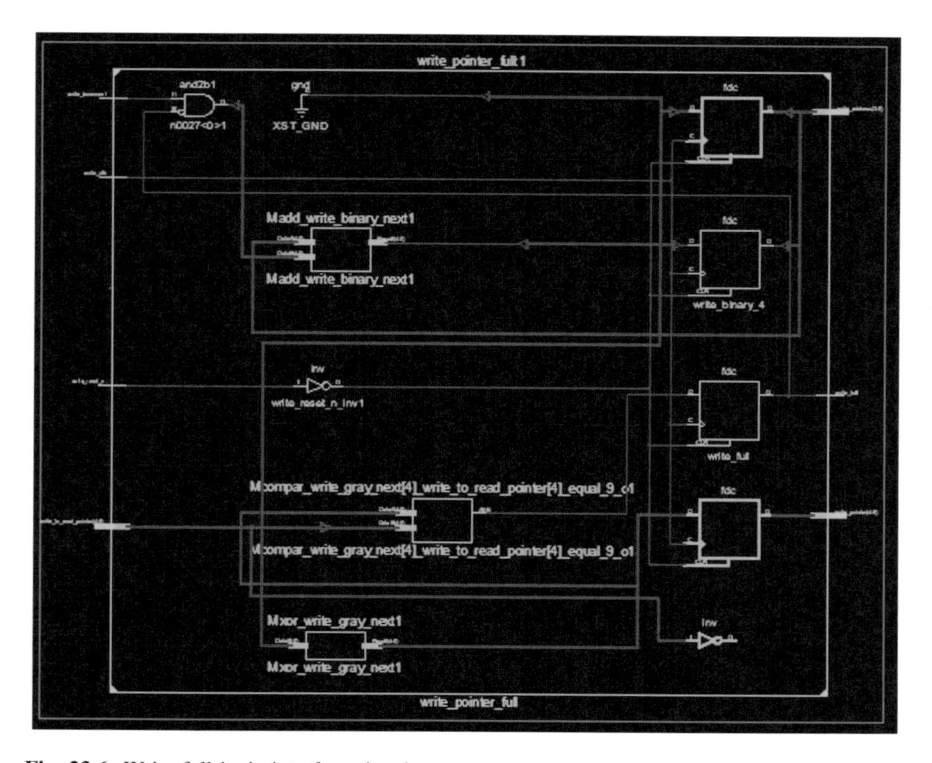

Fig. 23.6 Write full logic interface signals

Top-level instantiation: The FIFO top-level instantiation of FIFO memory, read_empty, write_full, synchronization from write to read and synchronization from write to read is shown in Example 6.

The top-level interfaces and the overall design functionality is shown in Fig. 23.7.

```
////////////////////////////////////////////////////////////////////////////////

module FIFO_top #(parameter data_size = 8,parameter address_size = 3)

(

input [data_size-1:0] write_data,

input write_increment, write_clk, write_reset_n,

input read_increment, read_clk, read_reset_n,

output [data_size-1:0] read_data,

output write_full,

output read_empty

);

wire [address_size-1:0] write_address, read_address;

wire [address_size:0] write_pointer, read_pointer, write_to_read_pointer,
read_to_write_pointer;

FIFO_memory #(data_size, address_size) fifomem

                         (

                         .write_clk(write_clk),

                         .write_clk_en(write_increment),

                         .write_data(write_data),

                         .write_address(write_address),

                         .read_data(read_data),

                         .read_address(read_address),

                         .write_full(write_full)

                         );
```

Example 6 FIFIO TOP level module and instantiation

```
read_pointer_empty #(address_size) read_pointer_empty
                              (
                              .read_clk(read_clk),
                              .read_reset_n(read_reset_n),
                              .read_increment(read_increment) ,
                              .read_address(read_address),
                              .read_pointer(read_pointer),
                              .read_empty(read_empty),
                              .read_to_write_pointer(read_to_write_pointer)
                              );
write_pointer_full #(address_size) write_pointer_full
                              (
                              .write_clk(write_clk),
                              .write_reset_n(write_reset_n),
                              .write_increment(write_increment),
                              .write_address(write_address),
                              .write_pointer(write_pointer),
                              .write_full(write_full)
                              .write_to_read_pointer(write_to_read_pointer)
                              );

sync_read_to_write  sync_read_to_write
                    (
                    .write_clk(write_clk),
```

Example 6 (continued)

```
                  .write_reset_n(write_reset_n),
                  .read_pointer(read_pointer),
                  .write_to_read_pointer(write_to_read_pointer)
                  );
sync_write_to_read  sync_write_to_read
                  (
                  .read_clk(read_clk),
                   .read_reset_n(read_reset_n),
                  .write_pointer(write_pointer),
                  .read_to_write_pointer(read_to_write_pointer)
                  );
endmodule
////////////////////////////////////////////////////////////////////////////////
```

Example 6 (continued)

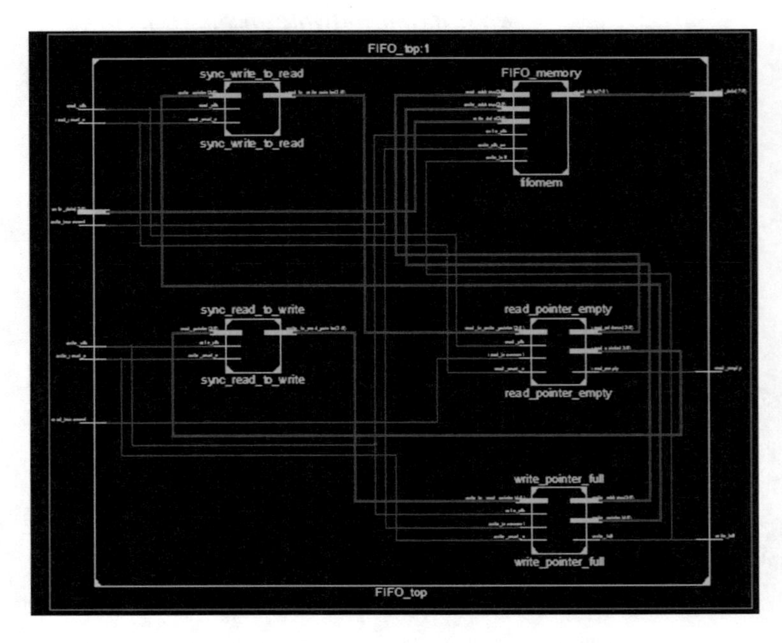

Fig. 23.7 FIFO Top-module having instantiation of the functional blocks

23.3 Testbench for FIFO

Using the non-synthesizable Verilog constructs, the testbench is coded for FIFO to check for the write and read transactions and shown in Example 7.

```
////////////////////////////////////////////////////////////////////////////////
module Test_FIFO_top;


        // Inputs

        reg [7:0] write_data;

        reg write_increment;

        reg write_clk;

        reg write_reset_n;

        reg read_increment;

        reg read_clk;

        reg read_reset_n;

        // Outputs

        wire [7:0] read_data;

        wire write_full;

        wire read_empty;


        // Instantiate the Unit Under Test (UUT)

        FIFO_top uut (

                .write_data(write_data),

                .write_increment(write_increment),

                .write_clk(write_clk),

                .write_reset_n(write_reset_n),
```

Example 7 Testbench of FIFO design

```
                .read_increment(read_increment),
                .read_clk(read_clk),
                .read_reset_n(read_reset_n),
                .read_data(read_data),
                .write_full(write_full),
                .read_empty(read_empty)
        );
    always #5 write_clk= ~write_clk;
    always #10 read_clk= ~read_clk;
    always #80 write_increment= ~write_increment;
    always #80 read_increment = ~ read_increment;
    always #200 write_reset_n = ~write_reset_n;
    always #200 read_reset_n = ~ read_reset_n;
        initial begin
                // Initialize Inputs
                write_data = 0;
                write_increment = 0;
                write_clk = 0;
                write_reset_n = 0;
                read_increment = 0;
                read_clk = 0;
                read_reset_n = 0;
                #100
```

Example 7 (continued)

```
// Wait 100 ns for global reset to finish

#10 write_data = 1;
#10 write_data = 2;
#10 write_data = 3;
#10 write_data = 4;
#10 write_data = 5;
#10 write_data = 6;
#10 write_data = 7;
#10 write_data = 8;
end
endmodule
////////////////////////////////////////////////////////////////////////////////
```

Example 7 (continued)

The simulation result is shown in Fig. 23.8 and as shown during FIFO full the write is not allowed and during FIFO empty read is not allowed.

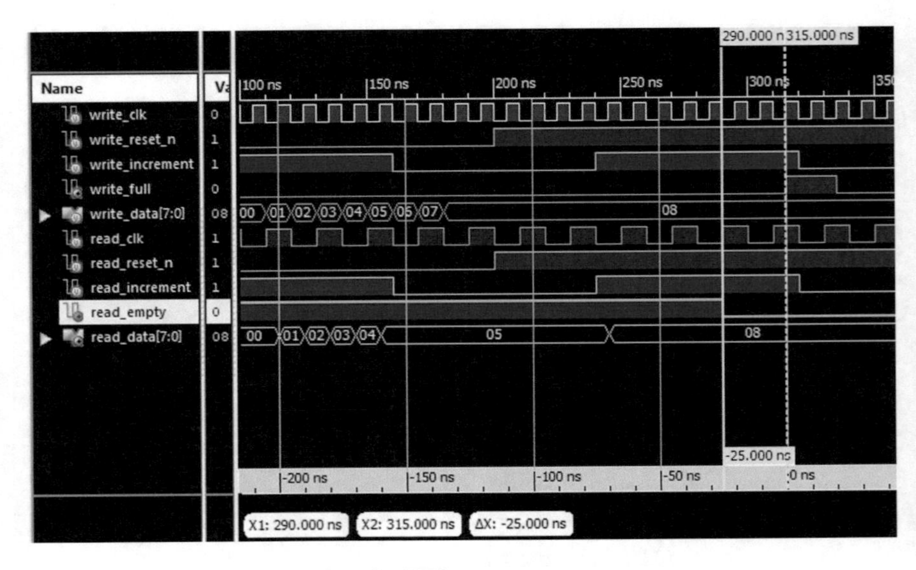

Fig. 23.8 Read and write operations for FIFO

23.4 Summary

Following are important points to conclude this chapter

1. Use the level synchronizers during sampling the data from different clock domain.
2. Use the gray code counters instead of binary counters while passing the data across multiple clock domains.
3. Use FIFO in the data or control path while passing the multiple data bits or control bits.
4. Partition the design using the single clock at the receiving and transmitting (sending) end.
5. Use the parameterized design while coding the RTL for the FIFO.
6. Include the reset logic and full, empty logic.

Chapter 24
Low Power Design

Low power design is one of the important requirements for ASIC designs. The power can be minimized using the different techniques and using the consistent power format. This chapter discusses about the low power design techniques and the need of Unified Power Format. This chapter also discusses about the important UPF commands.

In the modern low process node ASIC designs, the power is considered as the major factor. The low power design chips are required in many applications like mobile, computing, processing, and video and audio controller designs. Most of the SOC design needs the low power design architectures. This chapter discusses about the low power design techniques at the RTL level and the use of the consistent format UPF during the logical design. This chapter is useful for the RTL design engineers to understand the UPF terminology and the important commands to use the level shifter, retention, and isolation cells. Even this chapter describes about the multiple power domain designs using the UPF commands.

24.1 Introduction to Low Power Design

In the modern ASIC and SOC designs, the power optimization is very crucial. The power requirements for the ASIC or SOC designs play an important role in the overall design planning. The overall power estimation for the chip and the design of low power architecture and micro-architecture are decisive factors in the ASIC design flow. The goal of the chip architect is to design the architecture and micro-architecture for low power-aware designs. As process node has shrunk from 90 to 10 nm in the past decade, the voltage levels are dropped substantially.

As power is one of the crucial factors in the design of SOC, it has become the main problem in every category of the design. The power density is measured as watt per square millimeter and it rises with the alarming rate in the SOC designs. So, in the SOC design perspective, the power or energy management needs to be

V. Taraate, *Digital Logic Design Using Verilog*,
https://doi.org/10.1007/978-981-16-3199-3_24

used in the design from the architecture stage itself. The low power design techniques are essential to be used during the RTL to GDSII.

Power management is required for all the designs below the process node of 90 nm. As size of the chip has shrunk below 90 nm at the smaller geometry, it requires the aggressive management of the leakage current. The primary source of power dissipation in CMOS is leakage current. The leakage current is summation of all the cell leakage current and is state-dependent.

The dynamic power is defined as addition or the summation of the internal cell dynamic power and summation of power dissipated due to wires. The following are the few equations which describes the leakage and dynamic power.

$$\text{Pleakage} = \sum \text{Cell Leakage}$$

where cell leakage can be computed by using the library cell leakage and it is state-dependent.

$$\text{Pdynamic} = \sum \text{Cell dynamic power} + \sum \frac{1}{2} * C_l * V * V * T_r$$

where the C_l is the capacitive load at pin or net, V is voltage level, and T_r is toggle rate.

During the past decade, the main interest of design team was to improve the design performance that is throughput, latency and frequency and even to improve the silicon area. But below 90 nm the power management has become the important for the SOC designs. In the present scenario, due to the exponential growth in the field of the wireless and mobile communications and other home electronics intelligent applications, the demand is for the complex functionalities and high-speed computations. Even the low power management and the long battery life are key for such kind of applications in the competitive market. It is expected that such kind of devices should be of lightweight, small, cool, and even they should have the long battery life.

24.2 Power Dissipation in CMOS NOT Gate

Consider the ASIC standard cell as NOT gate and shown in Fig. 24.1. As shown in the figure, the NOT gate is designed by using PMOS and NMOS. PMOS passes strong 1 and NMOS passes strong 0. At a time either PMOS is ON or NMOS is ON. But practically the NOT gate cell is represented as On resistance of one of the ON transistors and the equivalent parasitic capacitance seen at the output port y_out. Energy stored in the capacitor is dependent on the capacitor value in nano or pico farad and the voltage applied to the NOT gate.

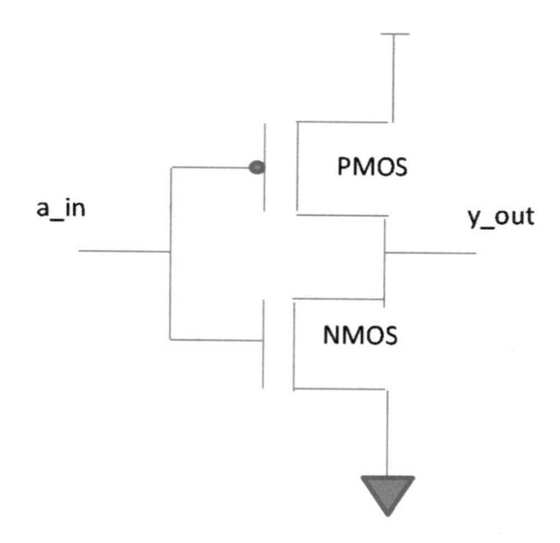

Fig. 24.1 CMOS NOT gate

The power dissipation for the CMOS NOT gate is computed by using the formula $p = (1/2) * C_s * V^2 * f$. So the power dissipation for any standard cell is directly proportional to the stray capacitance (C_s), applied voltage (V), and the frequency. To reduce the overall power for the chip, it is essential to minimize the applied voltage and essential to choose the process technology which can give minimum load capacitance at the output and input ports. Due to the high-speed design requirements, it is not possible to minimize the speed of the design by reducing the frequency, so there is always trade-off between the power requirements and speed of the design.

Sources of power consumption in CMOS are shown in Fig. 24.2.

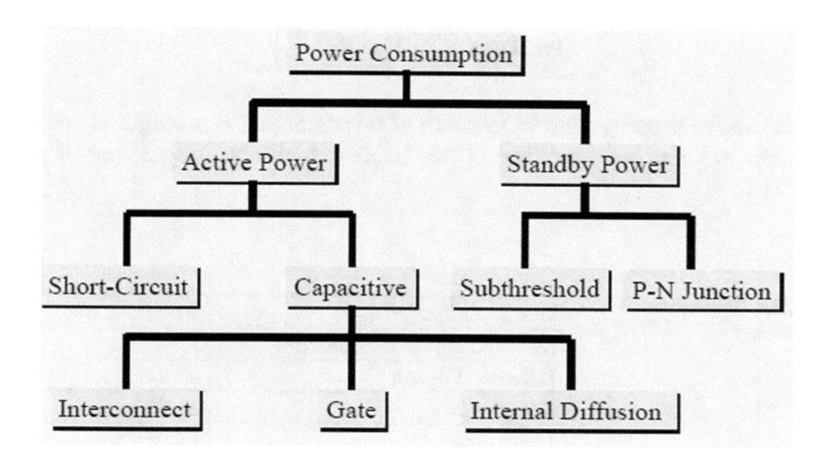

Fig. 24.2 Power consumption sources

So the power dissipation for any CMOS cell is function of the switching activity, capacitance, voltage, and the structure of transistor. So power is described as

$$\text{Power} = \text{Pswitching} + \text{Pshort} - \text{circuit} + \text{Pleakage}$$

The total power for any CMOS cell is summation of the dynamic and leakage power.

Dynamic power is summation of the switching power and short-circuit power.

The power dissipation is due to the charging and discharging of net capacitances. The short-circuit power dissipation is due to the gate switching, and it is due to the short circuit between the supply voltage and ground. The following equation describes the switching and short-circuit power

$$\text{Pswitching} = \alpha * f * C_{\text{eff}} * V_{\text{dd}} * V_{\text{dd}}$$

where α is switching activity, f is switching frequency, C_{eff} is effective capacitance, and V_{dd} is supply voltage.

$$\text{Pshort} - \text{circuit} = I_{\text{sc}} * V_{\text{dd}} * f$$

where I_{sc} is short-circuit current during switching, f is switching frequency, and V_{dd} is supply voltage.

Dynamic power can be reduced by reducing the switching activity, clock frequency (it reduces the design performance), also by using the capacitance and the supply voltage. If faster slew cells are used, then it consumes the less dynamic power and hence cell selection is important in reduction of the dynamic power.

Leakage power is given by the following equation, and it is function of the supply voltage V_{dd}, the switching threshold voltage V_{th}, and size of transistor.

$$\text{Pleakage} = f\left(V_{\text{dd}}, V_{\text{th}}, \frac{W}{L}\right)$$

In the above equation, the W is width of transistor and L is length of transistor.

Powers saving opportunities at the different design stages are listed in Table 24.1.

Table 24.1 Percentage power saving

Design abstraction stage	%power saving (%)
System design and architecture	70–80
Behavioral design	40–70
RTL design	25–40
Logic design	15–25
Physical design	10–15

24.3 Switching and Leakage Power Reduction Techniques

There are several techniques used to reduce the power, and few of the commonly used power management techniques are listed in Table 24.2.

Other few important techniques used in the power management at the various abstraction levels are listed in Table 24.3.

24.3.1 Clock Gating and Clock Tree Optimizations

This technique is very efficient and used to improve the dynamic power. In most of the application, the power is wasted due to unnecessary toggling of the clock signal. If the register input is changing or not, the clock signal toggles on every clock cycle. This is the major reason for the more dynamic power. Even the clock trees are the major sources for the larger dynamic power as they have the larger capacitive load and the switching requires the maximum rate. So if the data is not loaded in the register frequently, then significant amount of power is wasted and this can be saved by using clock gating technique. The clock gating is at the register level or leaf level, and if it is done at the block level, then the entire functional block can be disabled by disabling the clock tree. This reduces the switching and hence reduces the dynamic power.

Table 24.2 Power management techniques

Power management technique	Description
Clock gating and clock tree optimizations	In this technique, the portions of the clock tree which are not used at the instance of time are disabled
Logic restructuring	Use the cone structure to minimize the power. Move the low switching operations back in the logic cone and high switching operations up in the logic cone. This technique is used to reduce the dynamic power at gate-level optimizations
Operand isolations	This technique is effective in reducing the power dissipation in the data path of any blocks by using the enable signals
Logic and transistor resizing	Use the downsizing to reduce the leakage current and use upsizing to reduce the dynamic current by improving the slew
Pin swapping	Use the swapping gate pins to reduce the power. If the capacitance is lower, then the switching can be fast at the gate or pin

Table 24.3 Efficient power management techniques

Power management technique	Description
Multi-V_{th}	Use the multi-threshold libraries for the power reduction. Use the high switching threshold for lesser leakage power but it reduces the design performance. Use the low switching thresholds for the higher performance but it has higher leakage
Multiple supply voltage (MSV islands)	Use the multiple supply voltages for the different design blocks
Dynamic voltage scaling (DSV)	In this technique, the selected blocks can run at different supply voltages according to the design requirements
Dynamic voltage and frequency scaling (DVFS)	This is used to reduce the dynamic power. In this method, the selected blocks of design use the different supply voltages and frequencies on fly
Adaptive voltage and frequency scaling (AVFS)	This can be accomplished by using analog circuits, and in this technique based on the control loop feedback, the wide range of voltages is set dynamically
Power gating or power shut off (PSO)	If the functional blocks are not used, then the selected functional blocks are powered off
Splitting memories	If the memories are controlled by software or the data, then the portions of memories can be spitted into more number of portions. This is effectively used to save the power by shutting off the portion of memories which are not used

24.3.2 Operand Isolations

This technique is effective in reducing the dynamic power dissipation in the data path of any blocks by using the enable signals. Most of the times in the data path signals are sampled periodically, and hence, this sampling can be controlled by using the enable inputs. During inactive state of enable signal, the data path inputs can be forced to the constant value, and hence, it reduces the dynamic power due to lesser switching.

24.3.3 Multiple V_{th}

This technique is very effective while optimizing for area, power, and speed by using the different threshold voltage. Most of the libraries have the different switching threshold voltages. The efficient EDA tool used during synthesis can be able to use the different library cells of different switching threshold voltages to meet the area and speed constraints with the lowest power dissipation.

24.3.4 Multiple Supply Voltages (MSV)

In this technique, the different functional blocks operate at the different voltage levels. As the voltage level reduces, the active power is reduced as it is function of the square of the supply voltage. But it can degrade the design performance. While using this technique, it is required to use the level shifter to transfer the signals from one voltage domain to another voltage domain. If level shifters are not used, then the sampling of the signal is an issue!

24.3.5 Dynamic Voltage and Frequency Scaling (DVSF)

Dynamic voltage and frequency scaling is very efficient technique to reduce the active power consumption. As discussed in the earlier sections, the power dissipation is proportional to the voltage square so lowering the voltage has squared effect on the power consumption. In this type of technique, depending on the performance and power requirements, the frequency and voltage can be scaled down on the fly and hence it can reduce the power dissipation. This technique is very effective to optimize the static and dynamic power due to optimization of the frequency and voltage levels.

24.3.6 Power Gating (Power Shut Off)

Power gating or power shut off (PSO) is one of the effective techniques, and in this technique, the design modules which are not used are switched off using switches. This is one of the powerful techniques used to reduce the leakage power. In most of the industrial applications, the leakage power can be reduced by, more than 90% by using the power gating switches. To design this technique, it needs the clear understanding of the power-down sequence and isolation cells. It is essential to use the isolation logic with the state retention elements and even level shifters to implement the power gating.

24.3.7 Isolation Logic

This is used at the output of powered down block to prevent unpowered signals and floating signals from power-down block. In the simulation, these signals can be denoted by X. Isolation cells are used between the two power domains and connected between the power-off and the power-on domain. The reason for isolation cell in the two power domain is to isolate the output of blocks before the power switch off state and need to remain isolated until the power is switched on. In few design scenarios, isolation cells can be used to block level to prevent the connection to power down logic. Consider the block logic as driving power domain and it is in the off state then isolation cell can be located in the driving domain to isolate the signals from the driving power domain to the receiving on power domain.

24.3.8 State Retention

During the power-off mode, most of the time it is essential to retain the state of registers. The state of the registers is useful during the power recovery. In most of the low power designs, the state retention power gating flip-flops are used and these flip-flops are called as SRPG. Most of the EDA tool cell libraries are having the SRPG cell.

24.4 Low Power Design Techniques at the RTL Level

In the present scenario, there are many low power design techniques at the RTL and gate level. It is essential for the design team to understand about the low power design goals to use these techniques uniformly by ensuring the consistency and predictability in the overall design cycle. Most of the SOC design uses the low power design techniques using power analysis and optimization issues. This section focuses on the low power design technique.

1. Modeling and power estimation: For the low power design and the management of power for any SOC, it is essential to prepare the library models with the required power data. It is required to develop the transistor-level models for the custom blocks. The common practice in the SOC design at the RTL level is use of power compiler to understand the power consumption based on the switching activity information from the RTL simulation data. This technique is useful for estimation of the power consumption at early stage of the design. Another important point to be considered at the gate level is to develop the glitch-free low power designs. As gate-level analysis is more accurate as compared to the RTL-level analysis, it is essential to use the time-based analysis based on the peak power and hot spots.
2. Clock gating: Use the clock gating technique using the clock gating cells to minimize the power during the RTL design. Clock gating can be implemented by identifying the synchronous load enable register banks. Clock gating can be implemented by using the gating of clock with the enables instead of recirculating of the data when enable is inactive. If power compiler is used at the RTL level, then it automatically optimizes the static, dynamic power dissipation with the delay and area to meet the design constraints.

3. Clock gating stops the clock and forces the original circuit in the zone of no transition. In the practical scenario, if we consider the functional block as

```
always@(posedge clk)
begin
if(enable)
        data_out<=data_in;
end
```

The above block can generate the synthesis result shown in Fig. 24.3.

The above logic is without clock gating and has the higher power dissipation. To reduce the power consumption, the clock gating logic needs to be used and can be designed by eliminating the multiplexers at the input thus it is useful to avoid the recirculation of data. This results in the area and power savings and reduces the power consumption in the clock network. The synthesized logic using clock gating is shown in Fig. 24.4. The timing sequence is shown in Fig. 24.5.

The use of clock gating has drawback that the logic used to implement the clock gating technique is redundant, and hence, there can be issues in the testing and verification. Another important point needed to keep in mind is that, it is essential to stop the glitches and hazards on enable signal, and this is achieved by using the transparent latch between the enable and the AND logic gate.

Clock gating can be efficiently implemented by using the power compiler from Synopsys. Use the command ***set_clock_gating_signals***. Figure 24.6 shows the inputs and outputs used by the power compiler.

The outcome of the power compiler is the elaborated unmapped design. Power compiler uses the inputs as source RTL code and library to optimize for the low power.

The following are few of the important points need to be considered while implementing the clock gating using the power compiler.

1. General clock gating can be included or excluded from the design having the hierarchical modules. The command use is ***set_clock_gating_signals***. The care need to be taken by the design team while using the power compiler for the

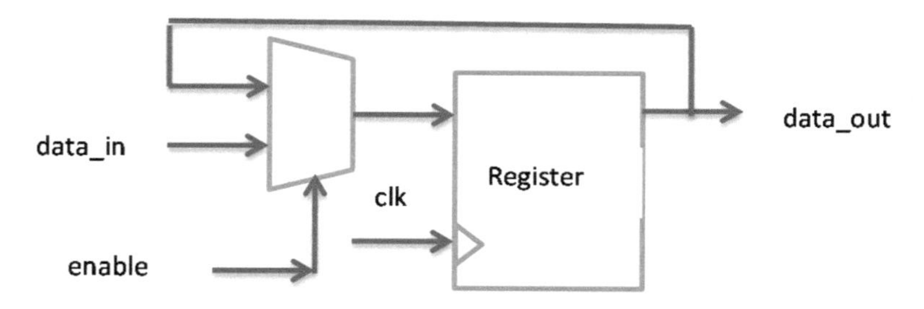

Fig. 24.3 Logic without clock gating

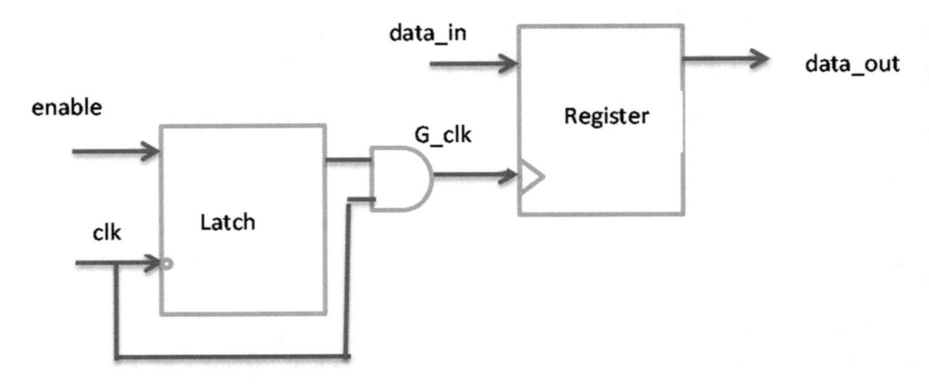

Fig. 24.4 Logic with clock gating

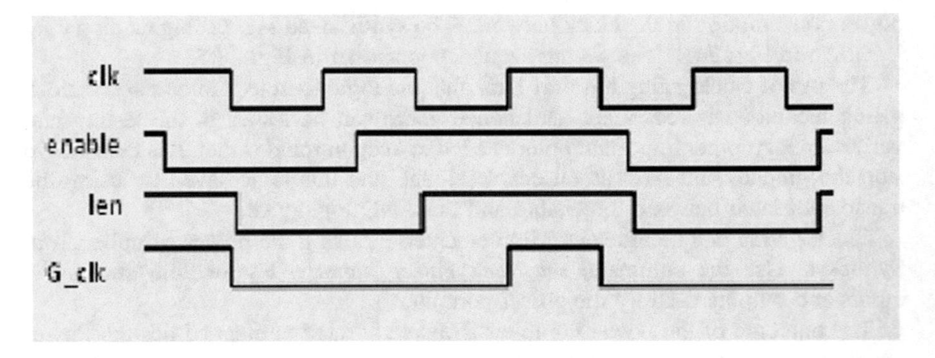

Fig. 24.5 Timing sequence for the clock gating

same. Each design should have the single command line for both the inclusion and exclusion of the clock gating.

2. If the design has multiple registers and few of the registers need to be excluded from the clock gating strategy, then they should have the separate enable signal. If same enable signal is used, then it generates the same clock gating for the entire register bank. For example, if the data bus is declared as data_in[7:0] with the registered inputs and if the lower nibble data_in[3:0] needed to be excluded from clock gating, then it should use the different enable and data_in[7:4] should use different enable.
3. Clock gating signals as single bit or multiple bits have added advantage as it avoids the recirculation of the data by removing the multiplexers. But it can consume more area and additional power due to the clock gating logic.
4. Do not use clock gating for the master–slave flip-flops. Generally, it is normal practice that clock gating logic is used at the slave flip-flop if the clock gating conditions are met. Such design may not perform the desired operation. Use the command ***set_clock_gating_exclude*** to exclude the master–slave flip-flops.

Fig. 24.6 Power compiler inputs and outputs

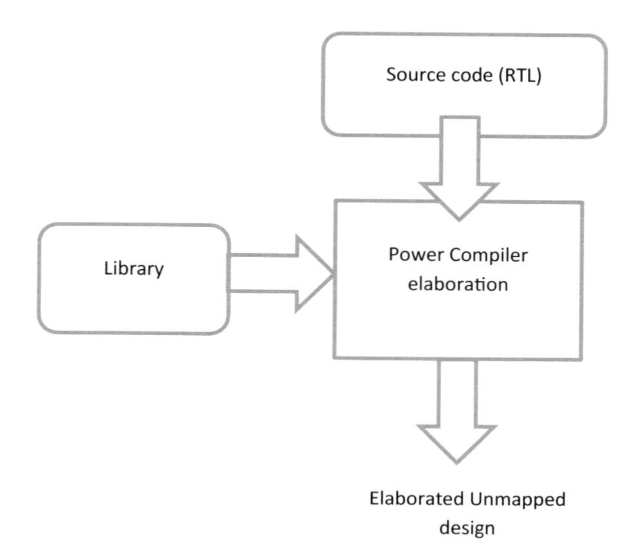

5. While using the clock gating, it is common practice to use the minimum bus width. The minimum bus width can be of 5 or more. Use the command ***set_clock_gating_style_minimum_bitwidth***.
6. In most of the design practices at the RTL level, if the procedural always blocks are used and if it consists of case with the default condition or conditional expressions like if-else, then tweak the RTL by including the default condition in every if-else statement. Example 1 describes the modification of the procedural block using default.
7. If same enable is shared by the multiple register banks, then the power compiler feature can be used to share the clock and enable signal to multiple register banks. This is used to save the overall area. Consider Example 2 shown below and it has two different procedural blocks, then the same clock gating logic can be used for both procedural blocks.
8. Use the simple clocking strategies for the automatic clock gating insertion. If the number of clock domains is minimum, then it gives simplifies timing analysis and clock tree synthesis. The lower down modules can have enable signals instead of dividing the clock. Use the ***set-don't_touch_network*** command to avoid the compilation changes on the clock network. During the multiple step compilation process, this avoids the changes on the clock gating logic.

Example 1 RTL tweaks for the power saving

```
case(a_in)
 2'b00: if (b1_in) c_in =d1_in;
 2'b01: if (b2_in) c_in =d2_in;
 default : c_in = e1_in;
endcase

//The above Verilog RTL can be modified as
case(a_in)
 2'b00: begin
         if (b1_in) c_in =d1_in;
         else c_in=e1_in;
      end
 2'b01: begin
         if (b2_in) c_in =d2_in;
         else c_in =e1_in;
endcase
```

9. Use the simple set and reset strategies. Complex set and reset strategies may result in the design logic which is prone to issues at the gate level and during functional debugging. The care need to be taken by the designer to have the better logic partition for synthesis while using the internal set and reset signals.
10. Clock balancing and the clock buffer signal insertion need to be used efficiently to have efficient clock tree synthesis (CTS). CTS tools work by adding or moving the buffers, resizing of cells along the clock tree network to manage the required skew and the insertion delay.

Example 2 Common clock enable for multiple procedural blocks

```
always @ ( posedge clk , negedge reset_n)
begin
    if (~reset_n)
            data_out <= 1'b0;
    else if (enable)
            data_out<=data_in;
end
always @ ( posedge clk , negedge reset_n)
begin
    if (~reset_n)
            data_out _1<= 1'b0;
    else if (enable)
            data_out_1<=data_in_1;
end
```

24.5 Low Power Design Architecture and UPF: Case Study [7, 8]

Unified Power Format (UPF) is the standard used to design electronic systems by considering the power as the feature. The standard is used for low power ASIC designs. The reasons for using UPF are

1. There is no any method which can support accurate management and distribution of low power at the RTL-level abstraction.
2. Vendor-specific power formats are inconsistent and are prone to bugs due to inconsistent specifications.

3. UPF provides the following and can be used consistently in low power ASIC designs

   a. Power distribution architecture

      i. Define the power domains
      ii. Define power switches
      iii. Define power rails

   b. Power strategy

      i. Creation of power state tables

   c. Set and map

      i. Isolation
      ii. Retention
      iii. Level shifter
      iv. Switches

UPF is IEEE 1801 standard and can be used throughout the design flow for power-aware design intent. Example 3 shows the use of UPF at various stages.

Example 3 UPF at various design stages

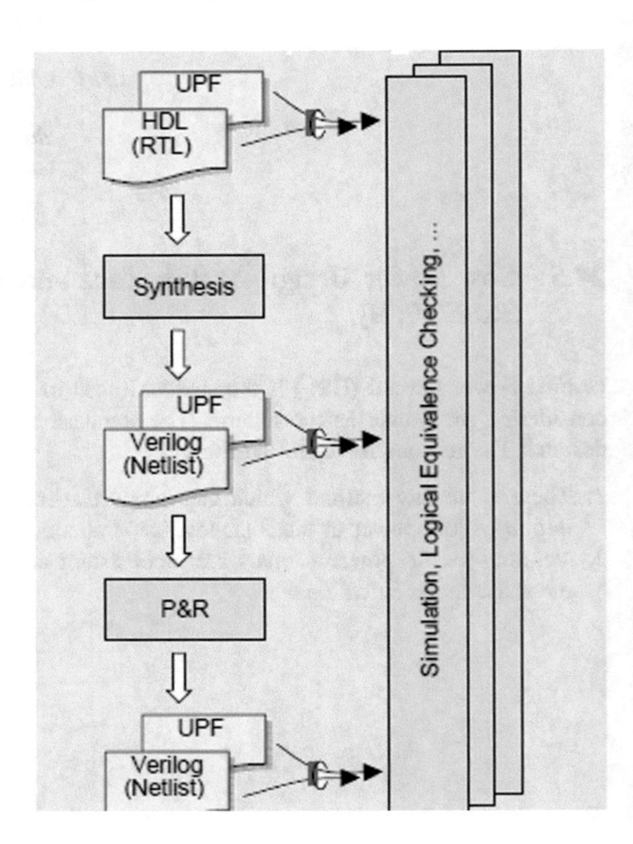

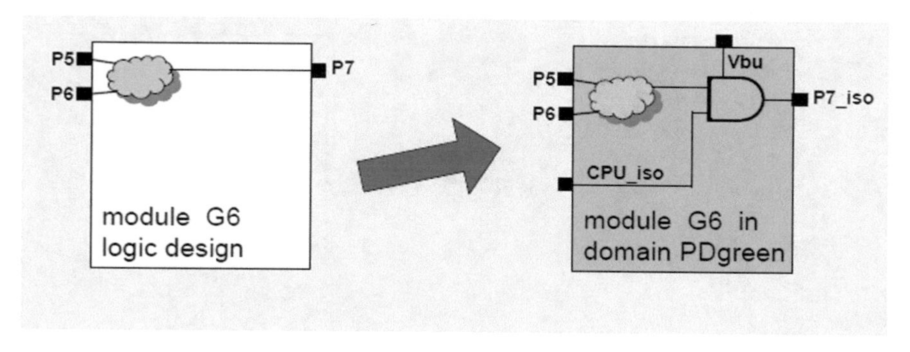

Fig. 24.7 Setting of isolation cell in logic design

24.5.1 Isolation cells

As discussed already, the isolation cells are used at the output of power-down block. The isolation cell can be set by using the UPF command. Figure 24.7 describes the design using isolation cell.

Set Isolation cell

```
set_isolation iso3
-domain PDgreen
-isolation_power_net Vbu
-clamp_value 0
-applies_to outputs
```

Set isolation control

```
set_isolation_control
iso3
-domain PDgreen
-isolation_signal CPU_iso
-location self
```

24.5.2 Retention Cells

As discussed already in the above section, the retention cells are used to retain the state of registers during power-off state. Figure 24.8 defines the setting of the retention cell in the design.

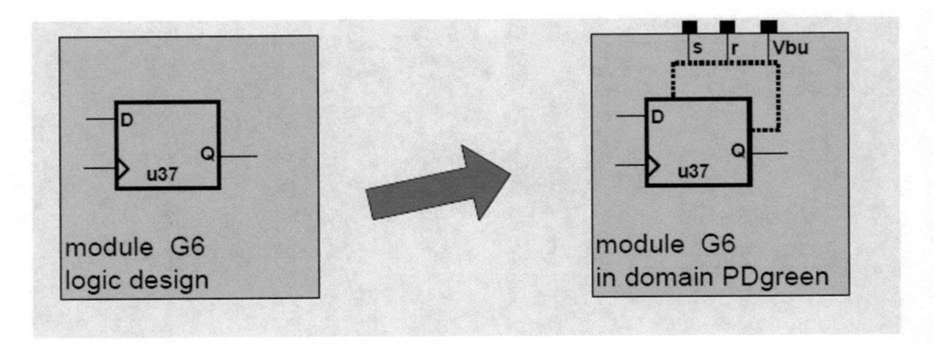

Fig. 24.8 Setting of retention cell in logic design

Set retention cell

set_retention ret3

–domain PDgreen

–retention_power_net Vbu

–elements { u37 }

Set Retention control

set_retention_control

ret3

–domain PDgreen

–save_signal s

–restore_signal r

24.5.3 Level Shifters

Level shifters are used to translate from one voltage level to another voltage level. The translation can be from low to high voltage level or high to low voltage level. Set and map level shifter can be achieved by using the following UPF commands. Figure 24.9 shows the use of command to set and map the level shifter.

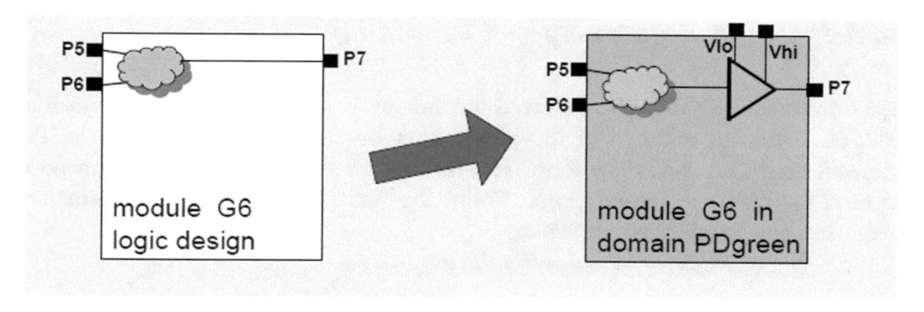

Fig. 24.9 Setting and mapping of level shifter in design

The important points to consider for the same are

1. Pick the correct power domain
2. Select input or output ports or both
3. Use upshift or downshift rule
4. Define the location.

Set level shifter

set_level_shifter my_ls

–domain PDgreen

–rule low_to_high

–location self

–applies_to outputs

Map level shifter

map_level_shifter_cell

ls_L2H

–domain PDgreen

–lib_cells { /lib/ls_123 }

24.5.4 Power Sequencing and Scheduling

Specific sequence should be followed for the power down. The sequence includes isolation, state retention, and the power shut off. For the power-up cycle, the opposite sequence need to be followed. During power-up cycle, it is recommended to have the specific reset sequence. Following timing sequence gives information about the power-up/down sequence.

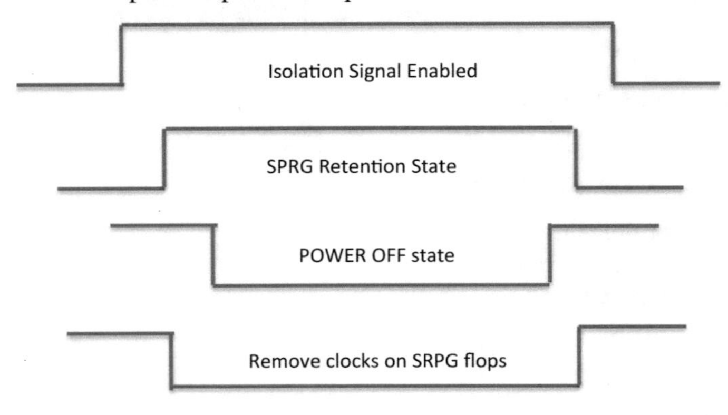

For the multiple clock domains with the different power sequence and the multiple clock gating with few common power control signals, it requires the higher verification efforts to ensure the correct sequencing during the power on and power off.

The UPF can be used from the RTL to GDSII and the basic UPF use at various stages is shown in Fig. 24.10. During the verification using the UPF, the functional and power intent should be analyzed and need the robust verification using the advanced verification techniques.

24.5.4.1 Creation of Power Domains

The power domains can be created by using the UPF command.

create_power_domain *domain_name*
*[**-elements** list]*
*[**-include_scope**]*
*[**-scope** instance_name]*

For example, creating the power domain having name pdA, the UPF command used is given below and the outcome is shown in Fig. 24.11.

create_power_domain pdA -include_scope A

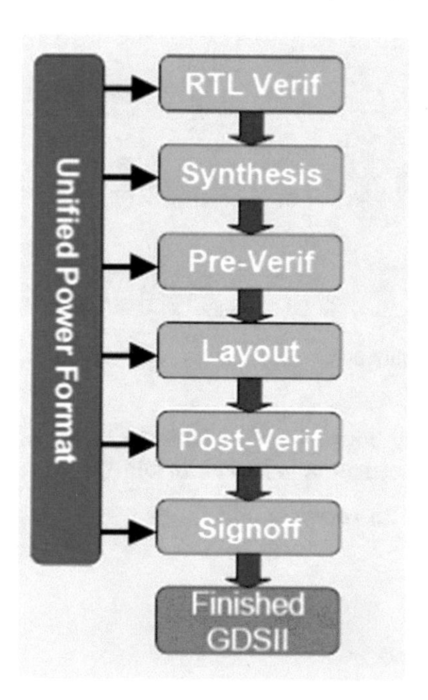

Fig. 24.10 UPF use at various stages from RTL design to GDSII

Fig. 24.11 Creation of power domain

24.5.4.2 Create Supply Port

The supply port can be created by using the UPF command.

create_supply_port *port_name*
 -domain *domain_name*
 [-direction <***in*** | ***out***>***]***

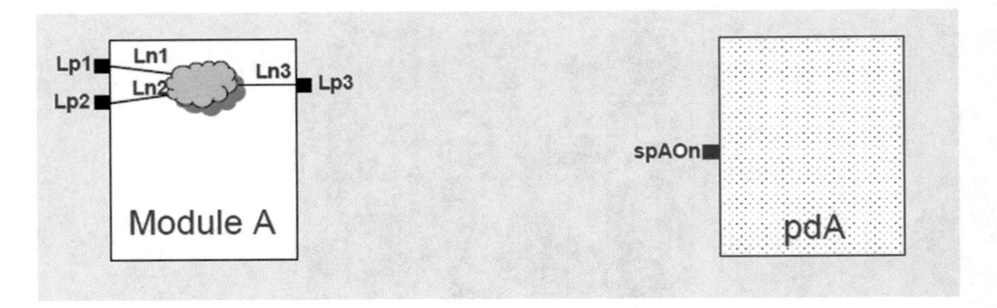

Fig. 24.12 Creation of supply port

For example, creating the supply port with the name spAOn, the command used is given below and the outcome is shown in Fig. 24.12.

create_supply_port spAOn - domain pdA

24.5.4.3 Create Supply Net

The supply net can be created by using the UPF command.

create_supply_net *net_name*
-domain *domain_name*
[-reuse]
[-resolve < unresolved
| one_hot
| parallel >]

For example, creating supply net named as RET, the UPF command used is given below and the outcome is Fig. 24.13.

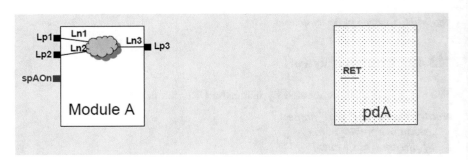

Fig. 24.13 Creation of supply net RET

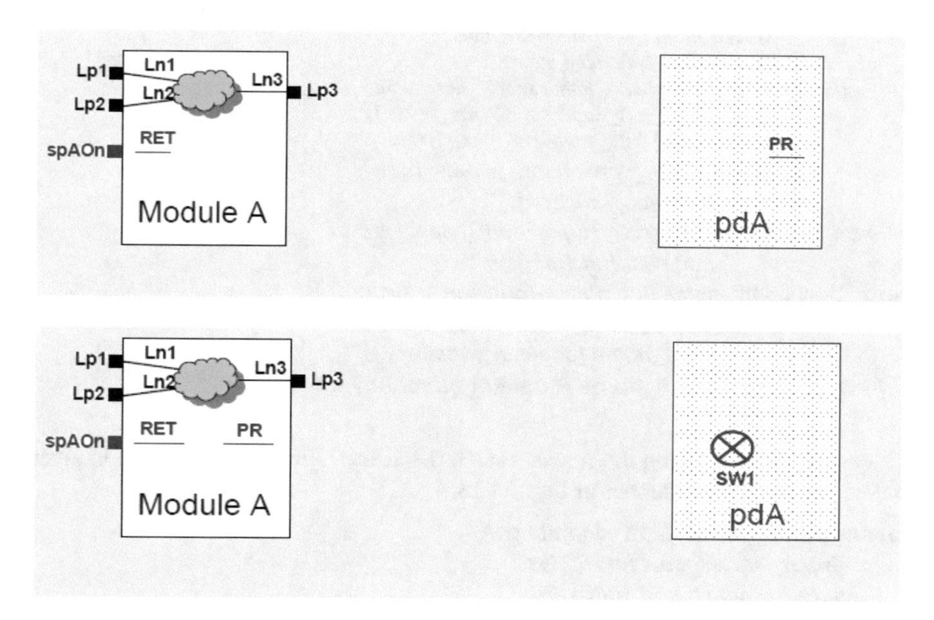

Fig. 24.14 Creation of supply net PR

create_supply_net RET -domain pdA

For example, creating supply net named as PR, the UPF command is given blow and the outcome is shown in Fig. 24.14.

create_supply_net PR -domain pdA

24.5.4.4 Create Power Switch

The power switch can be created by using the UPF command.

```
create_power_switch switch_name
    -domain domain_name
    -output_supply_port { port_name supply_net_name }
    {-input_supply_port { port_name supply_net_name }}*
    {-control_port { port_name net_name }}*
    {-on_state {state_name input_supply_port
        {boolean_function}}}*
    [-on_partial_state { state_name input_supply_port {
        boolean_function }}]*
    [-ack_port { port_name net_name [{boolean_function}] }]*
    [-ack_delay { port_name delay}]*
    [-off_state { state_name {boolean_function} }]*
    [-error_state { state_name {boolean_function} }]*
```

For example, creating the power switch SW1, the UPF command used is given below with the net outcome in Fig. 24.15.

```
create_power_switch SW1 -domain pdA
    -input_supply_port {inp RET}
    -output_supply_port {outp PR}
```

Fig. 24.15 Power switch creation

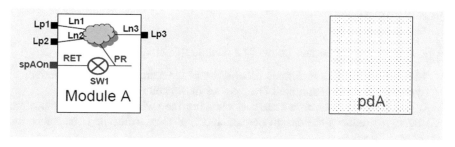

Fig. 24.16 Connecting the supply net

24.5.4.5 Connect Supply Net

The connection for supply net can be created by using the UPF command.

```
connect_supply_net net_name
          [-ports list]
          [-pins list]
        [< -cells list |
  -domain domain_name >]
[< -rail_connection rail_type |
    -pg_type pg_type >]*
      [-vct vct_name]
```

For example, connecting the power supply net, the command used is given below and the net outcome is shown in Fig. 24.16.

```
connect_supply_net RET -ports {spAOn}

    set_domain_supply_net pdA
      -primary_power_net PR
     -primary_ground_net VSS
```

24.6 Summary

Following are the important points to conclude this chapter

1. The dynamic power is defined as addition of the summation of the internal cell dynamic power and summation of power dissipated due to wires.
2. Dynamic power can be reduced by reducing the switching activity, clock frequency (it reduces the design performance), also by using the capacitance and the supply voltage.
3. Operand isolation is effective in reducing the dynamic power dissipation in the data path of any blocks by using the enable signals.
4. Dynamic voltage and frequency scaling is very efficient technique to reduce the active power consumption.
5. The retention cells are used to retain the state of key registers during power-off state.
6. Level shifters are used to translate from one voltage level to another voltage level. The translation can be from low-to-high voltage level or high to low voltage level.
7. Unified Power Format (UPF) is the standard used to design electronic systems by considering the power as the feature. The standard is used for low power ASIC designs.
8. This is used at the output of powered down block to prevent unpowered signals and floating signals from power-down block.
9. Power gating or power shut off (PSO) is one of the effective techniques, and in this technique, the design modules which are not used are switched off using switches. This is one of the powerful techniques used to reduce the leakage power.

Chapter 25
System-On-Chip (SOC) Design

The System on Chip can be realized and prototyped by using FPGAs. The SOC consists of many complex blocks like processors, arbiters, memories, peripherals. This chapter focuses on the generalized SOC architecture and discusses about the SOC design flow.

SOCs are complex density ASICs and need to be prototyped using the FPGAs. In the present scenario, there is more demand for the FPGA prototyping. Single or multiple FPGA can be used to prototype the desired SOC functionality. This chapter focuses on the discussion on the SOC components, challenges, and the SOC design flow. Even few of the important SOC design blocks RTL design strategies are discussed in this chapter.

25.1 What Is System on Chip (SOC)?

System on chip (SOC) is designed by using ASIC design flow, and for proof of concepts, FPGAs are used. In the present scenario, the designs are complex in nature and consist of multiple functional blocks to perform the desired operations and the requirement is higher design performance. The main important SOC design challenge is to have lower power, high performance, and lesser area.

As SOC complexity has increased during the past decade, it has become extremely important to detect the defects in the SOCs during early stage of design cycle. The best and affordable way is to use the modern FPGAs to prototype the design or to check the feasibility of the idea. In the present scenario, most of the complex designs are prototyped by using modern FPGAs having the high-speed capability and the complex architectures.

It is essential to understand about, why the FPGA prototyping has become popular during this decade? The main reason is the less non-recurring investment and the availability of the high-performance computing and reprogrammable features in the FPGA devices. SOCs consist of processor, IO interfaces such as

V. Taraate, *Digital Logic Design Using Verilog*,
https://doi.org/10.1007/978-981-16-3199-3_25

Ethernet, USB, UART, SPI, I2C, high-performance DSP computational capabilities, video and audio codecs and high-speed memory controllers like DDR II or DDR III. Modern FPGAs are used for SOC prototyping as they have most of the capabilities listed above to achieve the high performance.

25.2 SOC Architecture

In the present decade, IP and SOC complexity has increased so much. There is demand for SOC design, and FPGAs with the high-density functional blocks are used for validation of SOC functionality. This is also called as ASIC or SOC prototyping. If we consider typical SOC, it has processor core, various memories, and clock source as PLL, multiple power domain functional blocks, peripherals, communication interfaces and analog-to-digital and digital-to-analog converters. The important point during the design of SOC is to have better partition of the hardware and software resources. In the present scenario, the FPGAs are used during SOC prototyping due to reconfigurable capabilities and to accelerate the performance of design due to the use of soft and hard IPs in it.

The different blocks of SOC are shown in Fig. 25.1. If we consider any complex SOC, then it consists of the different communication interface such as USBs, Bluetooth, and most of the SOCs support the standard bus protocols. For any SOC design, it is essential to have the better area, high speed, and low power. Achieving the required design functionality with the constraints is one of the challenges due to the availability of lesser time to design and market the product due to high demand of new features and functional requirements. SOC design always needs the realistic plan, resources, and availability of necessary validation testing and prototyping setup.

Fig. 25.1 SOC design blocks

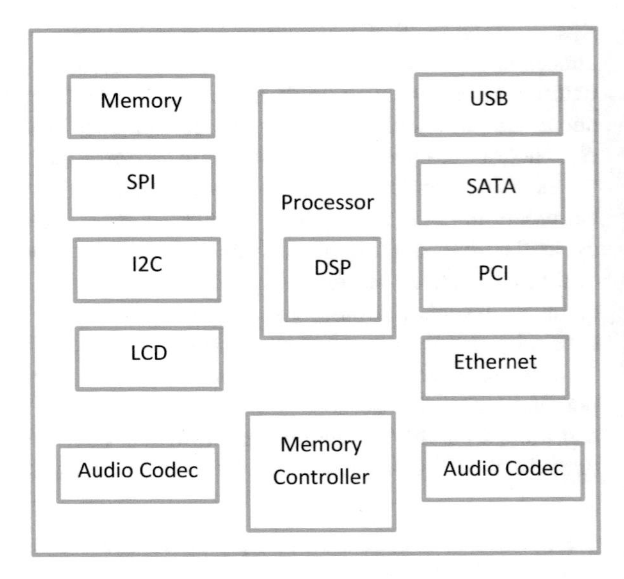

25.3 SOC Design Flow

The SOC design flow is shown in Fig. 25.2. As shown in the figure, it has multiple steps which include the design feasibility and implementation, FPGA prototyping and testing and ASIC porting. The important steps are discussed in the subsequent section.

25.3.1 IP Design and Reuse

Most of the SOC uses intellectual properties (IPs). But as a design team, it is important to validate the IPs in SOCs for the available features, timing requirements, and functionality. The important parameters in IP design are the overall functionality of the design. The IPs are sold in the semiconductor market due to its features, timing performance, and low-power requirements. If we consider simple tablet, then the tablet selling point in the market is the availability of features, interfaces, and the compatibility with the software and other communication devices. The IPs are not sold in the market due to only interfaces but need to have the overall all above-mentioned features.

Most of the time the SOC design team uses the third party functionally and timing has proven IPs. Instead of spending the time on design of IPs, most of the time SOC design team uses multiple IPs required according to the desired or functional requirements. All the required IPs can be integrated together according to the speed and power requirements. Although there is challenge in overall integration of IPs that challenge can be overcome by understanding the architecture details of IPs, timing and power details of IPs used. The IP can be soft IP or hard IP. The IP vendor companies can provide the synthesizable and process independent RTL, or netlist with the necessary timing information and having the high-performance user-friendly interfaces.

The IPs should exhibit the required functionality and should be delivered with the synthesizable RTL, synthesis scripts, design constraints, and interface details. Then it becomes easy during the IP integration and validation to realize the SOC in less time. The reason for growing complexity of SOCs is due to the following few factors.

1. Requirement of number of features with limitation on area that is size of SOC
2. Less time to market
3. High-speed requirements
4. The multiple power domain designs and requirements.
5. Multiple clock domain and clock tree structure

The complex SOC designs have challenges during the routing and during the power optimization. Even timing analysis and meeting the timing requirement is one of the important challenge!

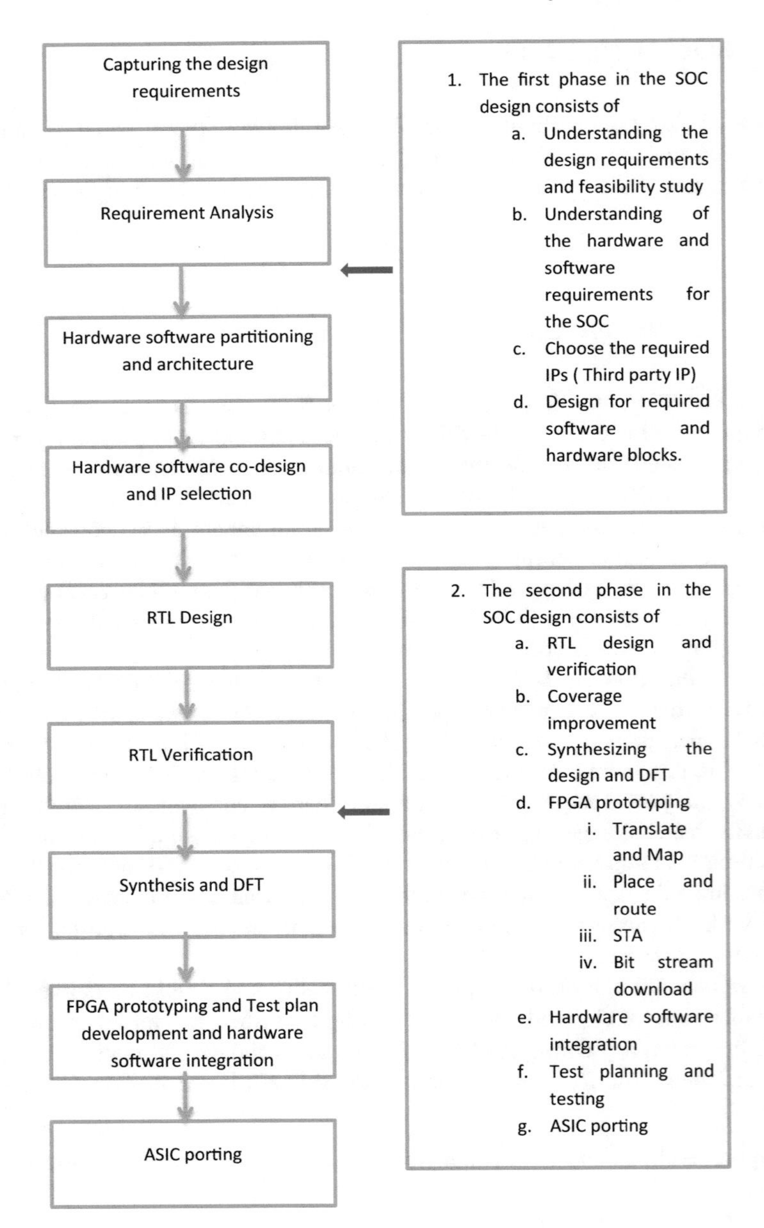

Fig. 25.2 SOC design flow

25.3.2 SOC Design Considerations

Capturing the design requirements and analysis of the design is the first important step during the SOC design. The input for this phase is the design or product specifications provided by the client or end user. The analysis involves the feasibility study of all the features provided. The feasibility study is an important phase as during this phase it can be easy to understand about the risk in the implementations as well as dependability during implementation. The feasibility study is needed and should be done for all the features by keeping in mind the time to market. This study gives roadmap and challenges involved during SOC design implementation cycle and is useful during various phases. The design specifications are analyzed and understood during this phase.

25.3.3 Hardware–Software Co-design

This is also called as design partitioning; the design must be partitioned into hardware and software. The important point of consideration is while partitioning the design; how parallel execution needs to be incorporated in the design? In the present scenario, as SOCs are complex the functionality can be implemented using the parallelism in the design which in turn can improve the design performance. The complex computational task or algorithms need to be partitioned during the design partitioning phase. Most of the complex computational blocks need to be implemented using hardware. Design partitioning is important and decisive phase to define what needs to be implemented using software? And what needs to be implemented using hardware?

For example, consider the design of video decoders which needs multiple frame support. The video decoder can be efficiently implemented using hardware, and even the parallelism can be incorporated for the few decoder features. The high computational DSP functional blocks which need filters like FFT, FIR, and IIR or high-speed multipliers can be effectively and efficiently implemented using hardware.

Let us consider the scenario of protocol implementation, most of the protocol like Ethernet, USB, and AHB can be efficiently implemented using hardware–software co-design. These algorithms should be functional and timing-proven. This can have advantage to overcome and to reduce latency in the design.

The major challenge in the hardware–software design portioning is the analysis of throughput and power requirements. For example, consider the scenario in SOC design where fixed length packets need to be transferred during the fixed time interval. If the design is implemented by using hardware, then care needs to be taken such that there should be minimum interaction between the hardware and software. To minimize the interaction between hardware and software, the strategy can be used by using FIFO buffers and timers.

25.3.4 Interface Timings

FIFO is First-in First-out Memory buffer and can be used to hold the packet information depending on the depth of FIFO. At the start of data transfer FIFO can interact with procedural calls defined in the software. To track the time duration, the timers can be implemented using hardware which can have communication interface with FIFO to indicate the end of timing intervals. Such type of handshaking mechanism can be implemented easily using hardware.

The design architecture for both the hardware and software activities can be created by considering power-aware design and throughput requirements.

25.3.4.1 Interface Details and Timing Requirements

For every SOC, it is essential to have the functional and timing-proven bus interfaces. In most of the applications, advanced high-speed bus protocols are used. These protocols need to be validated for the functional and timing correctness of the design. IO interfaces need to be targeted for the high-speed data transfer. There are many kinds of IO interfaces used in SOC designs. These IOs can be general purpose, differential IOs, multiplexed IOs, and high-speed IOs.

25.3.4.2 Reset Clock Requirements

Clock distribution network is used to provide the uniform clock skew to all the registers in the SOCs. The clocking policy plays the crucial role in overall design performance. The uniform clock skew can be achieved by using the suitable clock tree by using clock tree synthesis. Use of single clock structure or multiple clock domain structure needs to be decided at the architecture level. Also, the use of synchronous or asynchronous logic needs to be defined at the architecture level. Reset can be asynchronous or synchronous and needs to be defined at the architecture phase of SOC.

25.3.5 EDA Tool and License Requirements

Choose the desired and suitable EDA tools and licenses for FPGA prototyping of a SOC and during the ASIC porting. The most industry standard tools are.

Simulator: VCS.

Synthesis: Synplify pro and Synopsys DC.

STA: prime time (Synopsys PT).

25.3.6 Prototyping Development

For SOC validation, use the necessary prototyping and development platform. Prototyping platform can consist of the use of multiple FPGA boards to realize and validate SOCs, IP required, DSP functionality required, memories, and general-purpose processors required. The availability of desired prototyping boards with the necessary interfaces to realize SOC and use of debug or testing setup.

Most of the SOCs are tested by using the test setup consisting of available EDA tools and logic analyzers. At the start of the SOC design cycle, architect analyzes the design and functional requirements and according to the requirement of speed and estimation of gate count the prototyping platform can be built. Here the important factors are time to market, budget allocation, and design time requirements. If DSP capabilities are available in FPGA, then it is wise to implement the DSP functionality using FPGA dedicated resources.

25.3.7 Test Plan

For complex gate count SOCs, the necessary test cases need to be developed with the required test vectors. The features can be extracted using top-level functional specifications, and the required test cases can be documented in the test plan document. The test vectors developed can have significant impact on the quality of the verification to achieve the coverage goals. The test cases can be documented as basic, corner, and the random test cases. The constrained random verification with the required coverage goals should be targeted by using the required necessary test cases.

25.3.8 Verification Environment

Use the verification languages like Verilog and high-level verification languages like SystemVerilog or SystemC; for early detection of bugs and to achieve the coverage goals. The verification planning to improve the overall design quality by capturing the bugs during early design cycle is always crucial in the large gate count SOC designs. The overall objective is to achieve the required and designed functionality in less time. The verification environment needs to be built to achieve the coverage goals. The verification architecture can have the necessary bus functional models and the drivers, monitors, and scoreboards for robust checking of the design specifications. The overall verification planning and creation of environment with goal to achieve the automation and to minimize the time requirement to complete the functional checks in the lesser amount of time duration should be the main objective!

25.3.9 Prototyping Using FPGAs

At the architecture and micro-architecture level, the gate count estimation is done for the SOCs. As discussed already, the prototyping development can have multiple FPGAs with the desired high-speed interfaces. Depending on the complexity of design, FPGAs can be chosen. The main criteria are the use of FPGA to have the lesser power and more speed. The following are important points needs to be considered while prototyping using FPGAs.

1. Use of FPGA functional blocks to meet the desired area requirements. Choose the suitable FPGA platform and try to use the 70% of FPGA resources.
2. The area, speed, and power constraints need to be extracted at the chip level and at the block level.
3. Use the block-level constraints while synthesizing the blocks and use the chip-level constraints at the top level.
4. If high-performance DSP algorithms need to be coded, then use the DSP functional macroblocks to realize the high computational DSP filtering and the processing algorithms.
5. Try to choose the FPGA platform which has high-speed interfaces such as USB, Ethernet, PCI, and memory controllers.
6. Choose the mechanism to interact between software and hardware.
7. Choose the desired tool options for auto place and route of design to meet the design constraints.
8. FPGAs should have the capability to achieve the functionality at higher speed.

Most of the FPGA demands low power in the today's market scenario. SOCs can be designed to meet the desired power. Use the low-power design techniques to achieve the desired low-power requirements.

25.3.9.1 ASIC Porting

After performing the logic realization and validation of SOC using FPGA, the design needs to be migrated to an ASIC. For quick realization of ASIC, designer needs to do the following.

1. Replace the clock gating logic with the equivalent component from the ASIC library.
2. Insert DFT and check for the stuck at fault coverage.
3. Use the low-power intent design using UPF.
4. Use the block-level and chip-level constraints while migrating from FPGA to ASIC flow.
5. Synthesize the design for the required constraints.
6. Implement the physical design using the design flow for the required area, speed, and power.

25.4 SOC Design Challenges

While designing SOC, there might be many design challenges and few of them are listed below.

1. *Use of the modeling abstraction levels* In the practical scenario, different modeling levels are used from the design specifications to fabrication of chip. It is good decision to use the different levels of abstractions while design of SOC.

   a. *.Functional modeling* To describe the functionality and to get the valid and accurate output by using the simulators
   b. *Cycle accurate modeling* To understand the required number of cycles consumed while performing the operation.
   c. *Event-level modeling* To understand the number of events within a clock cycles are accurate or not?
   d. *Memory accurate modeling* To understand the memory contents and layout is accurate or not?
   e. *Transaction-level modeling* To understand for the number of transactions is accurate or not?

2. *RTL design* Efficient RTL design description and synthesizable design is one of the important challenges and RTL design team needs to take care of the following.

   a. Order of continuous assignments and loop-free design. The outcome is latch-free synthesis results.
   b. Defining hierarchy of design and having efficient design partition.
   c. Registering inputs and outputs for the module.
   d. Uses of register assignment in single clock domain.
   e. DFT friendly RTL design and low-power-aware RTL.
   f. Efficient use blocking and non-blocking assignments.

3. *RTL Verification* The goal is to detect the functional bugs during early design cycle and to achieve the coverage. So, the main challenge is to understand the usage of event-driven or cycle accurate simulators and use of their features. While creating the testbench architecture care need to be taken to have the self-checking testbench and to have the test automation for the higher coverage. Use of the transport and inertial delays during the verification and using zero delay models is another important challenge for the robust verification.
4. *Synthesis* The goal should be to meet the desired power, speed, and area requirements. For low-power designs use the isolation cells, retention cells, level shifters and clock gating logic. For speed improvement use the techniques like register balancing, pipelining, register retiming. For area optimization use the techniques like multiplex decoding, grouping, and constant data propagation.
5. *Hazard-free designs* For any efficient ASIC or SOC design, it is recommended not have the hazards. There are potential issues in the design due to hazards, for example, write after write hazard can create the potential issues in the design if

second write does not happen properly after first write of the data. Following are the few important points that need to be considered for the hazard-free design.

a. *Data Hazard* Can be potential problem if the data or address is not computed or arrived at the required time stamp.
b. *Structural Hazard* Can be potential problem due to the limited number of resources to perform the multiple activities at a time. To overcome these hazards, use the registers and sequence the operations using the pipelined structure. Following are few examples for the structural hazards.

    i. Memories with the limited number of ports and less latency.
    ii. Non-pipelined designs and limited number of processing units.
    iii. Implementation of multiplier algorithms without the pipelining or Booth multiplication.

c. *Control Hazards* Can be potential problem due to the late arrival of control signal or it is not clear when to perform the operation?
d. *Read and Write Hazard* Can be potential problem if the read and write operations are performed during the same time stamp.
e. *Timing estimation and analysis* The challenge is to meet the desired timing for the SOC and challenges are following

    i. Use of the pipelined design with the required pipelined stages.
    ii. Use of the grouping technique and logic duplications for the clean register to register paths.
    iii. se of the techniques to reduce the critical path timing delays.

6. *Interface and protocol implementations* Most of the SOC design uses the protocol and as discussed earlier meeting of the timing performance at the interface level is also important aspect for the efficient SOC. Following can be few points need to be considered while modeling the protocols and interfaces.

    a. Use of the handshaking mechanism for the transaction notification.
    b. Use of the general-purpose IOs and the special IOs for the interfaces.
    c. Understanding the timing details at the pin and signal level.
    d. Use of serializer, deserializer, and parallelism while modeling the protocols.

7. *SOC components* Selecting the required SOC components or describing the SOC RTL design is one of the important challenges. The main SOC components can be microprocessors or microcontrollers, IOs, arbiters, memories, general-purpose controllers, interrupt, and DMA controller. Describing the RTL for each individual component is one of the challenge as goal is to achieve the required area, speed, and power.
8. *Design Implementation and Testing* After completion of the hardware and software component design, the integration of hardware and software is the major challenge due to the interface synchronization requirements. The testing of the SOC needs the efficient verification and testing plan to test the features covered.

The SOC design case study for the moderate complex design is discussed in the following section. As discussed above, SOC consists of the microprocessor or microcontroller to perform the processing operation on the multiple operands, the memory banks RAM and ROM, general-purpose IO and control mechanism, counters, and timers and UART. For easy understanding of SOC, the complex modules like DSP controllers, DMA controllers, video controllers, and complex arbiters are not discussed in the case study. Readers are encouraged to use the logic design concepts to document the architectures and to code the RTL for the above complex modules.

25.5 SOC Design Blocks

The important SOC design blocks and the Verilog RTL for few of the blocks are discussed in this section. The important SOC design blocks which we use in most of the SOCs are.

1. Microprocessor or microcontroller
2. Counters and timers
3. General-purpose IO
4. UART
5. Bus arbitration logic

The memories are discussed in Chap. 10, and readers are requested to refer the memory section. The objective of this section is to discuss on the RTL design strategies for these blocks. Finally, these individual blocks can act as an IP and can be integrated together to achieve the desired functionality of the SOC.

The SOC with moderate gate complexity is shown in Fig. 25.3, and it consists of most of the blocks mentioned above.

25.5.1 Microprocessors or Microcontrollers

The generalized architecture for processor is shown in Fig. 25.4: Generalized Microprocessor block diagram below. As shown in the figure, it consists of ALU, instruction register and decoder logic, control and timing unit, program, and stack pointer. It also consists of bus arbitration logic. While designing the processor, it is essential to take care of the design partitioning and to have the RTL design of

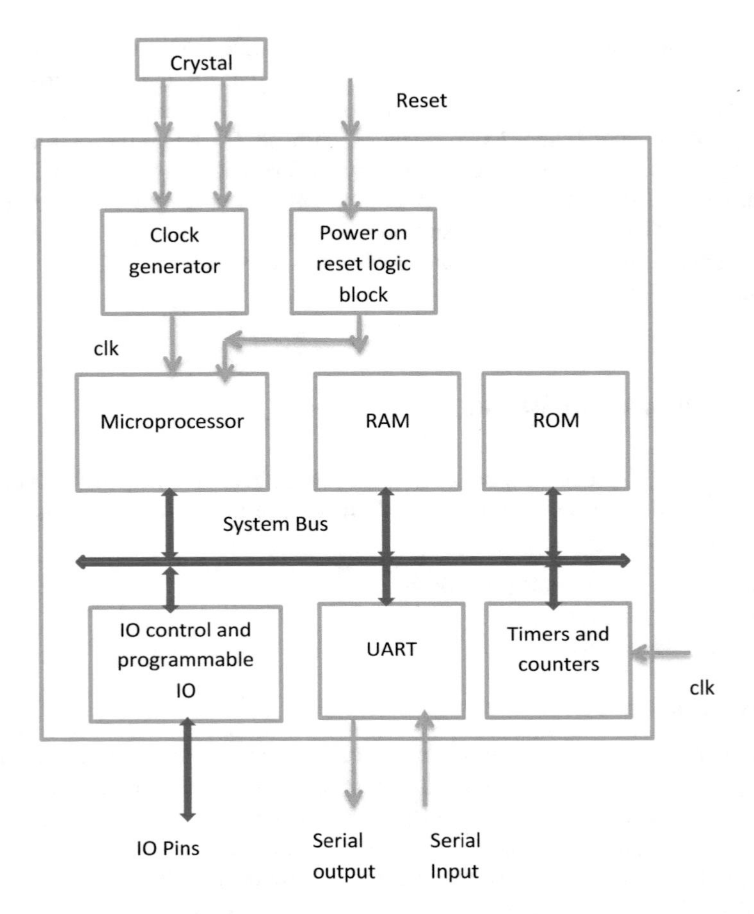

Fig. 25.3 Generalized SOC architecture

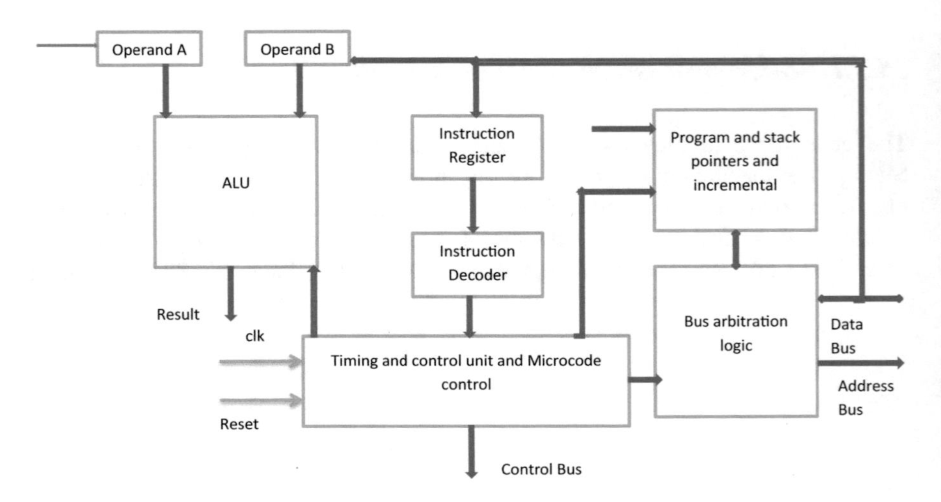

Fig. 25.4 Generalized microprocessor block diagram

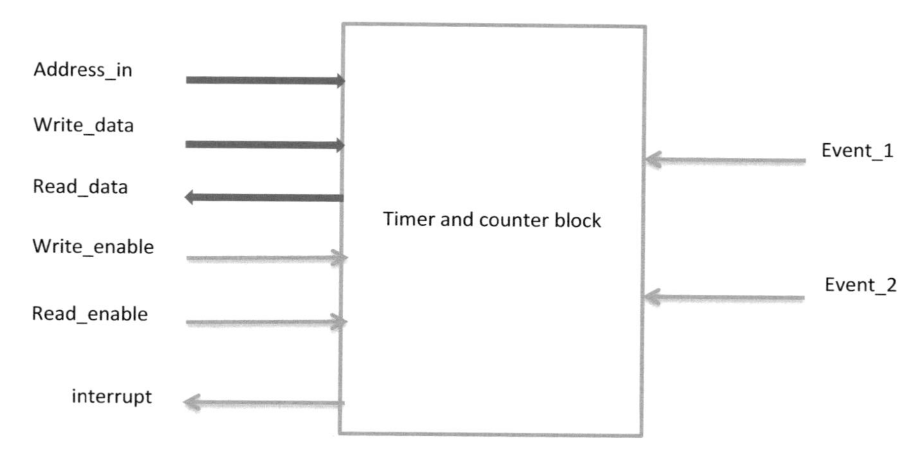

Fig. 25.5 Top-level signal diagram for the timer and counter block

individual modules using synthesizable constructs. Data path and control path logic need to be partitioned for better visibility and better timing and performance.

25.5.2 *Counters and Timers*

In most of the designs, the requirement is to count the predefined number of pulses depending on the external event by using active edge of the clock. An efficient RTL design having functional correctness of the design to achieve the desired performance is the major goal. Consider the block-level representation for the timer or counter block shown in Fig. 25.5. The RTL description for the block is shown in Example 1.

```
// define the module name
module counter_timer
 (
input clk,
input [3:0] Address_in,
input [7:0] Write_data,
input Event_1, Event_2,
input Write_enable,
input Read_enable,
output reg  [7:0] Read_data,
output interrupt
);
reg interrupt_pending, overflow, interrupt_enable;
reg [15:0] reload_counter;
reg [15:0] presale_count, presclae_count1;
reg [15:0] temp_count;
wire operation;
// Write operation functionality

always@(posedge clk)
begin
if (Write_enable  && Address_in==0)
        interrupt_enable <= Write_data[0];
if (Write_enable  && Address_in==4)
        prescale_count <= Write_data;  <----
if (Write_enable  && Address_in==8)
        temp_count <= Write_data;
end
// generation of operation signal high
operation = (Write_enable && Read_enable ==12);

//Read operation
always@ ( *)
begin
if ( Address_in ==0)
        Read_data =  { interrupt_enable, interrupt_pending};
else if ( Address_in==4)
        Read_data = prescale_count;
```

Consider Event_1=1 and Event_2=0, the write operation functionality initiated by processor is described with always block.

To clear the interrupt write to the Address 12 of the counter_timer

Example 1 Verilog RTL for timer-counter block

```
else if ( Address_in=8)
        Read_data = temp_count;
else
        Read_data = 0;

end

//Interrupt generation logic
assign interrupt = interrupt_enable && interrupt_pending;

//timer functionality
always@(posedge clk)
begin
        overflow <= (prescale_count ==prescale_count1);
        prescale_count <= (overflow) ? (0): (presclae_count+1);
end

always@(posedge clk)
begin
if (overflow)
        temp_count<= temp_count-1;
if (temp_count==0)
begin
        interrupt_pending<=1'b0;
        temp_count<= reload_counter;
end

if (operation)
        interrupt_pending<=0;

endmodule
```

The read operation functionality initiated by processor is described with always procedural block by using blocking assignment.

Depending on the channel address Address_in the respective required intermediate value is outputted on output line Read_data.

Interrupt generation logic is combinational logic and controlled by interrupt_enable and interrupt_pending flags.

The Timer functionality is described by using the procedural block and sensitive to the active edge of clock.

Depending on the status of overflow flag the temp_count and interupt_pending is assigned.

Example 1 (continued)

25.5.3 General-Purpose IO Block

In most of the ASIC or SOC design, the general-purpose bidirectional IOs are used. Multiple IOs are required depending on the required interface inputs and outputs. IOs are used to communicate with the outside world. The generalized structure for bidirectional IO is shown in Figs. 25.6.

The partial Verilog RTL is described in Example 2.

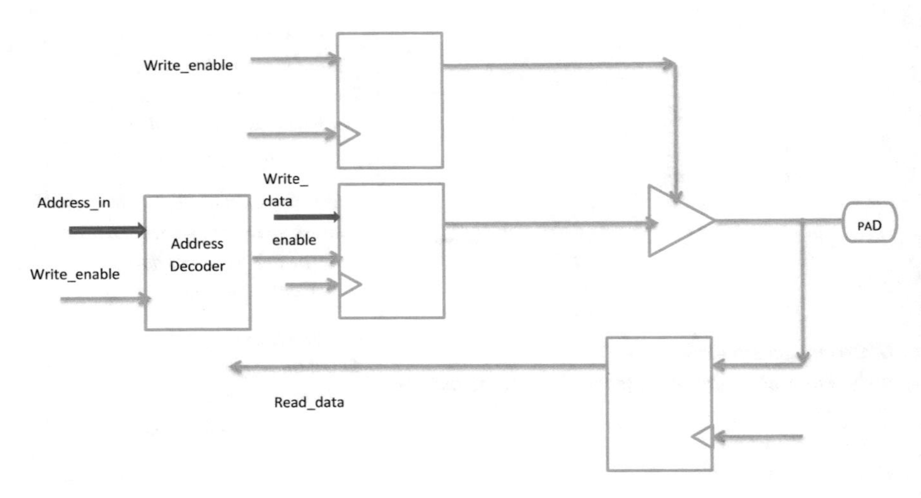

Fig. 25.6 General purpose IO block diagram

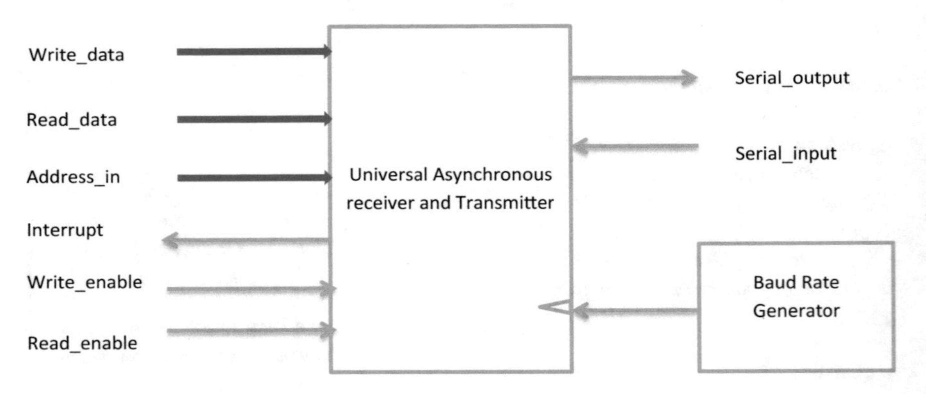

Fig. 25.7 Top level block diagram for UART

```
reg[15:0] ddr_out;
reg[15:0] data_out;
always@(posedge clk)
begin
if (Write_enable && Address_in=0)
        ddr_out<=Write_data;
if (Write_enable && Address_in=4)
        data_out<=Write_data;
end

//tri-state instantiation

tri_buf U1 ( d_datain[0], data_out[0], ddr_out[0]);
//for the 16 bit IOs there can be 16 more instantiation of the same

tri_buf U16 ( d_datain[15], data_out[15], ddr_out[15]);

//data read

always@(posedge clk)
begin

        read_data <= d_datain;

end
```

Double data rate IO structure and the read and write data operations are described in this code.

The Verilog RTL is not complete but can be used as a reference to describe the IO functionality.

Example 2 Verilog RTL for general purpose IO

25.5.4 *Universal Asynchronous Receiver and Transmitter (UART)*

These kinds of blocks can be used in the serial data transfer. The basic protocol is to use the active low start bit and then 8 bit of serial data and finally active high start bit. The data rate can be adjusted by generating the baud clock by using baud rate generator (Fig. 25.7).

The UART consists of transmitter to transmit the serial data using serial_output pin and receiver to receive the serial data using serial_input pin. The data rate is controlled by the baud rate control block. The control logic block can be designed using the multiple data buffers and FIFOs.

The block-level architecture for the UART is shown in Fig. 25.8.

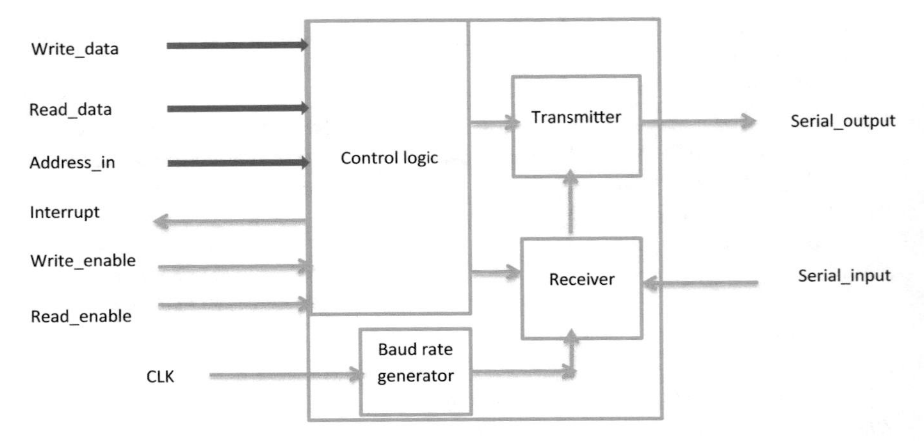

Fig. 25.8 Architecture diagram for UART

25.5.5 *Bus Arbitration Logic*

The bus arbiters are used to share the same resource by the multiple masters. In the practical scenario, typical shared resources are memories, multipliers, and buses. The arbiter decides to which master the service needs to be given, and the property can be static or round-robin. The arbiter is shown in Fig. 25.9, and the partial RTL is coded and is shown in Example 3.

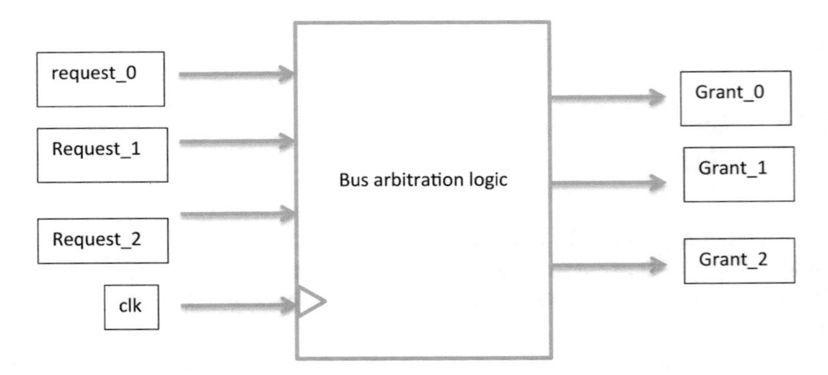

Fig. 25.9 Top-level diagram for bus arbitration logic

```
always@(posedge clk)
begin

 if (reset)
{ Grant_0, Grant_1,Grant_2} <=3'b000;
else
begin
Grant_0 <= Request_0;
Grant_1 <= (Request_1 && (! Resuest_0);
Grant_2 <= (Request_2 && (! (Resuest_0 || Request_1));
end
```

The static bus arbitration logic is shown in example and has three requests. The Request_0 has highest priority and Request_2 has the lowest priority.

Example 3 Verilog RTL for static arbitration scheme

In this manuscript, I have documented most of the important design scenarios and RTL design concepts. The Verilog constructs and their use during the RTL design and verifications. Readers can refer to the SOC prototyping books and other ASIC design books to understand the design concepts.

25.6 Summary

The following are important points to conclude this chapter.

1. A SOC is system on chip and consists of multiple processors, IPs, arbitration logic, peripheral interfaces, and protocols.
2. SOC can be prototyped by using single or multiple FPGA.
3. Third-party validated and functional accurate IPs can reduce the overall design cycle time during the SOC design.
4. Hardware and software co-design decides the overall turnaround time for the complex SOC design. The important challenge is the use of the handshake mechanism.
5. Interface timing is one of the critical challenges during the SOC design.
6. An efficient test plan and verification plan can boost the overall coverage during the SOC design.
7. For complex SOC prototyping multiple FPGA boards can be used to validate the design functionality.
8. ASIC migration and porting require the standard cell ASIC libraries with other power-aware RTL design support.

Appendix I: Important Verilog Keywords

The important Verilog keywords used in this book are listed below.

module	assign	reg	wire
input	output	parameter	endmodule
begin	end	fork	join
always	posedge	negedge	bit
function	endfunction	case	endcase
if	else	casex	casez
for	while	initial	
`define	localpar	`timescale	
inout	task	endtask	

V. Taraate, *Digital Logic Design Using Verilog*,
https://doi.org/10.1007/978-981-16-3199-3

Appendix II: Frequently Used Verilog Constructs

Verilog is case sensitive, and few of the important Verilog-2005 constructs are listed below.

V. Taraate, *Digital Logic Design Using Verilog*,
https://doi.org/10.1007/978-981-16-3199-3

1. module declaration

```
/////////////////////////////////////////////////////////////////////////
module comb_design ( input wire a_in, b_in, output wire
y1_out,y2_out, output reg [7:0] y3_out );

//Concurrent and sequential statements and assignments

endmodule
/////////////////////////////////////////////////////////////////////////
```

2. Continous assignment (neither blocking nor non-blocking)

```
assign y1_out = a_in ^ b_in;// net type is wire
```

3. ***always@**** // Combinational procedural blocks

```
/////////////////////////////////////////////////////////////////////////

always @*
begin
// blocking assignments or sequential constructs and net type reg
end
/////////////////////////////////////////////////////////////////////////
```

4. ***always@ (posedge clk)*** // sequnetial procedural block sensitive to positive edge of clock

```
/////////////////////////////////////////////////////////////////////////

always @(posedge clk)
begin
//synchronous reset and assignments
```

```
//non-blocking assignments or sequential constructs and net type reg
end
////////////////////////////////////////////////////////////////////////////////
```

5. ***always@ (posedge clk or negedge reset_n)*** // sequnetial procedural block sensitive to positive edge of clock

```
////////////////////////////////////////////////////////////////////////////////

always @(posedge clk or negedge reset_n)
begin
//asynchronous reset and assignments
//non-blocking assignments or sequential constructs and net type reg
end
////////////////////////////////////////////////////////////////////////////////
```

6. ***always@ (negedge clk)*** // sequential procedural block sensitive to negative edge of the clock

```
////////////////////////////////////////////////////////////////////////////////

always @(negedge  clk)
begin
// non-blocking assignments or sequential constructs and net type reg
end
////////////////////////////////////////////////////////////////////////////////
```

7. Multiple blocking (=) assignments in the procedural block

```
////////////////////////////////////////////////////////////////////////////////

begin
     tmp_1 = data_in;
     tmp_2 = tmp_1;
      tmp_3 = tmp_2;
     q_out = tmp_3;
end

////////////////////////////////////////////////////////////////////////////////
```

8. Multiple non-blocking (<=) assignments in the procedural block

```
//////////////////////////////////////////////////////////////////////////////

begin
    tmp_1 <= data_in;
    tmp_2 <= tmp_1;
     tmp_3 <= tmp_2;
    q_out <= tmp_3;
end

//////////////////////////////////////////////////////////////////////////////
```

9. Sequntial construct ***if –else*** within always procedural block

```
//////////////////////////////////////////////////////////////////////////////

    if(condition)
    //assignment or expression
       else
    //assignment or expression
    end

//////////////////////////////////////////////////////////////////////////////
```

10. Sequntial construct ***case--endcase*** within always procedural block

```
//////////////////////////////////////////////////////////////////////////////

  case (sel_in)

// conditions and expressions

  endcase
//////////////////////////////////////////////////////////////////////////////
```

11. Sequntial construct ***casex--endcase*** within always procedural block

```
///////////////////////////////////////////////////////////////////////////////

    casex (sel_in)

            // conditions and expressions

    endcase
///////////////////////////////////////////////////////////////////////////////
```

12. Sequntial construct ***casez--endcase*** within always procedural block

```
///////////////////////////////////////////////////////////////////////////////
    casez (sel_in)

            // conditions and expressions

    endcase
///////////////////////////////////////////////////////////////////////////////
```

13. Procedural block ***initial***

```
///////////////////////////////////////////////////////////////////////////////
initial
begin
//assignments with non-synthesizable intent
end

///////////////////////////////////////////////////////////////////////////////
```

For the syntax of other constructs, please refer Verilog-2005 language reference manual!

Appendix III: Xilinx Spartan Devices

The comparison of the Spartan 3 FPGA devices and the architecture is documented in this document.

- Xilinx Spartan 3 Devices

V. Taraate, *Digital Logic Design Using Verilog*,
https://doi.org/10.1007/978-981-16-3199-3

Device	System gates	Equivalent logic cells[1]	CLB array (One CLB = four slices)			Distributed RAM bits (K = 1024)	Block RAM bits (K = 1024)	Dedicated multipliers	DCMs	Max. user I/O	Maximum differential I/O pairs
			Rows	Columns	Total CLBs						
XC3S50W[2]	50K	1,728	16	12	192	12K	72K	4	2	124	56
XC3S200[2]	200K	4,320	24	20	480	30K	216K	12	4	173	76
XC3S400[2]	400K	8,064	32	28	896	56 K	288K	16	4	264	116
XC3S1000[2]	1M	17,280	48	40	1,920	120K	432K	24	4	391	175
XC3S1500	1.5M	29,952	64	52	3,328	208K	576K	32	4	487	221
XC3S2000	2M	46,080	80	64	5,120	320K	720K	40	4	565	270
XC3S4000	4M	62,208	96	72	6,912	432K	1,728K	96	4	633	300
XC3S5000	5M	74,880	1104	80	8,320	520K	1,872K	104	4	633	300

Notes:
1. Logic Cell = 4-input lookup table (LUT) plus a 'D' flip-flop. 'Equivalent Logic Cells' equal 'Total CLBs' × 8 Logic Cells/CLB × 1.125 effectiveness.
2. These devices are available in Xilinx Automotive versions as described in DS314: *Spartan 3 Automotive XA FPGA Family*.

• Spartan 3 Family Architecture

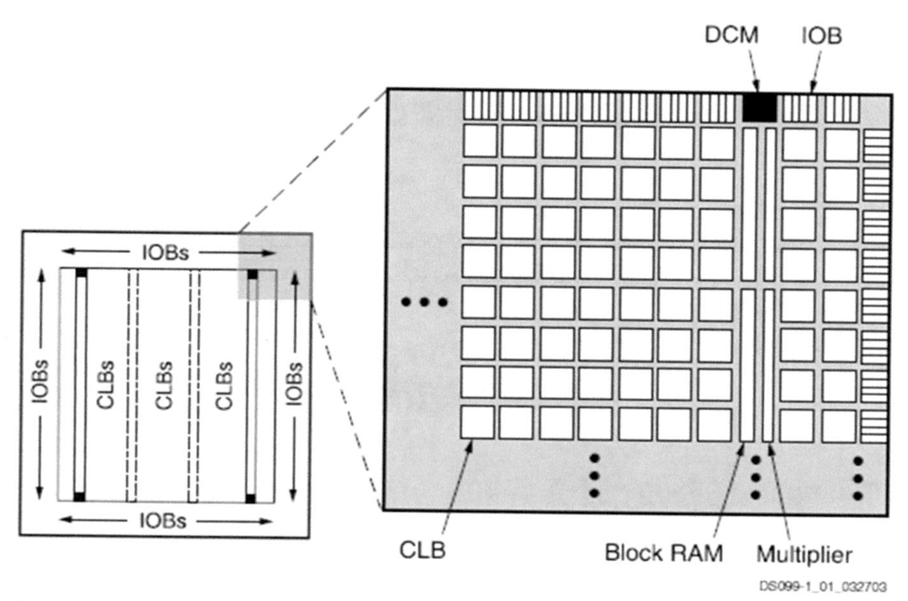

Notes:

1. The two additional block RAM columns of the XC3S4000 and XC3S5000 devices are shown with dashed lines. The XC3S50 has only the block RAM column on the far left.

- Xilinx Spartan 3 package information for Part no XC3S400-4PQ208C

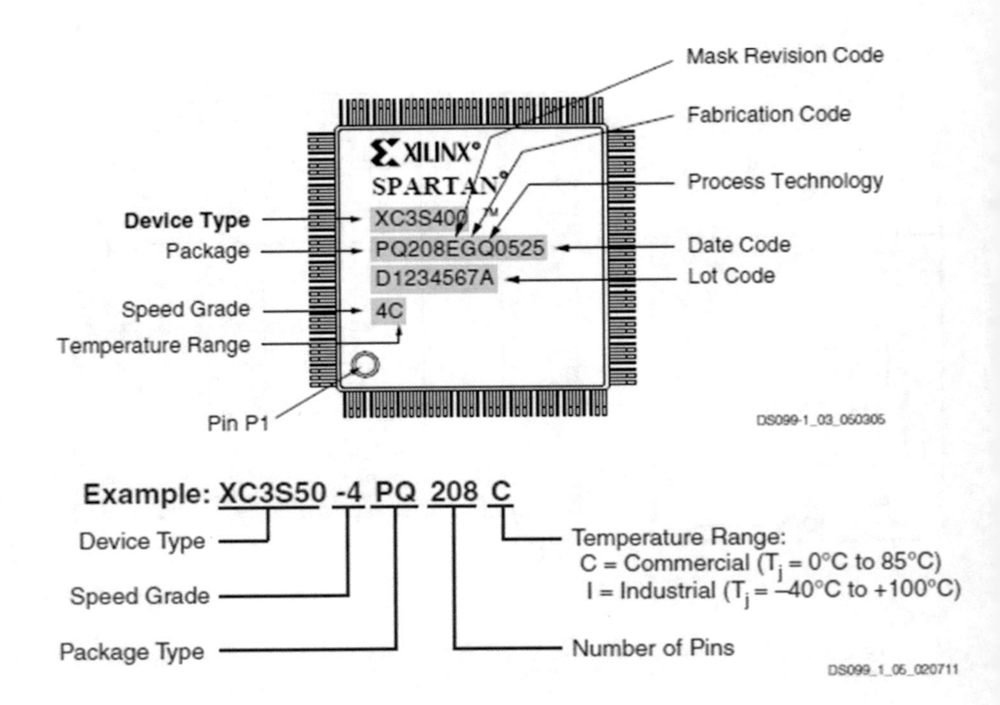

For more information please use the following link.
http://www.xilinx.com/support/documentation/data_sheets/ds099.pdf.

- Xilinx FPGA Spartan 3E Devices

Device	System gates	Equivalent logic cells	CLB array (one CLB = four slices)				Distributed RAM bits[1]	Block RAM bits[1]	Dedicated multipliers	DC Ms	Maximum user I/O	Maximum differential I/O pairs
			Rows	Columns	Total CLBs	Total slices						
XC3S100E	100K	2,160	22	16	240	960	15K	72K	4	2	108	40
XC3S250E	250K	5,508	34	26	612	2,448	38K	216K	12	4	172	68
XC3S500E	500K	10,476	46	34	1,164	4,656	73K	360 K	20	4	232	92
XC3S1200E	1200K	19,512	60	46	2,168	8,672	136K	504K	28	8	304	124
XC3S1600E	1600K	33,192	76	58	3,688	14,752	231K	648 K	36	8	376	156

Notes:
1. By convention, one Kb is equivalent to 1,024 bits.

- XILINX Spartan 3E Architecture

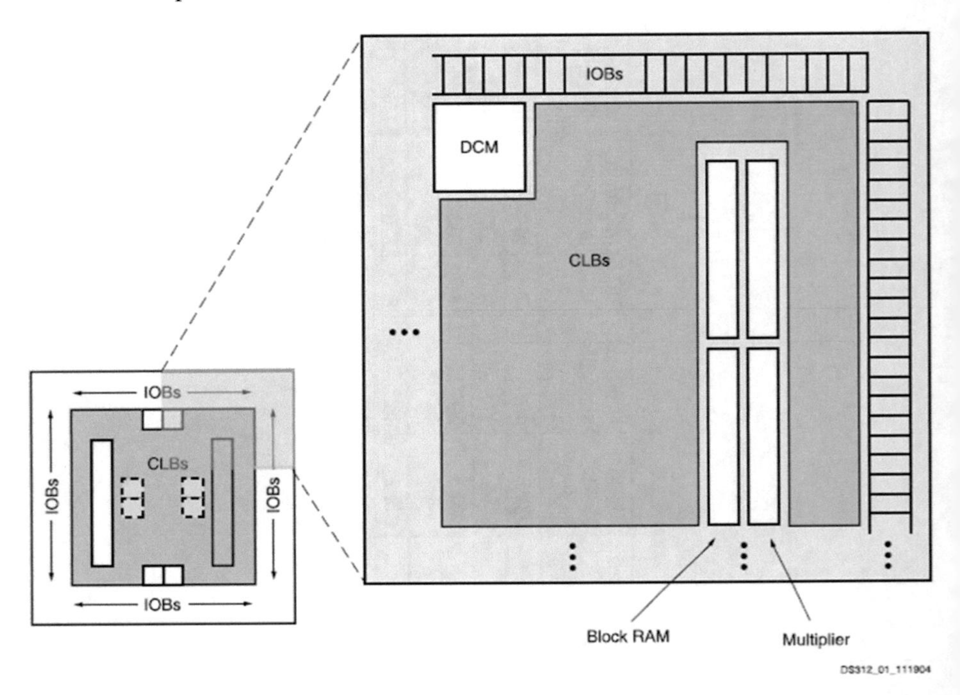

- XILINX Spartan 3E package information

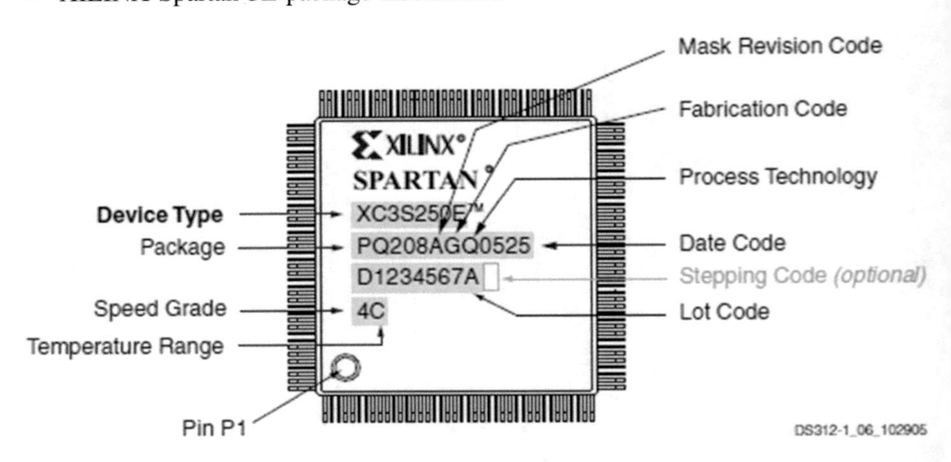

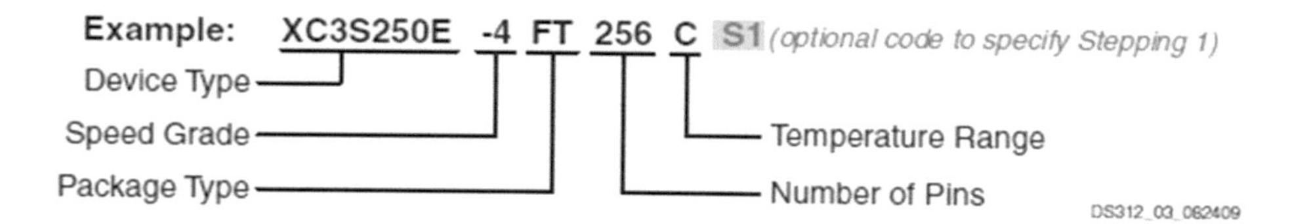

For more information please use the following link.
http://www.xilinx.com/support/documentation/data_sheets/ds312.pdf.

Bibliography

1. IEEE Standard. www.ieee.org, http://standards.ieee.org/getieee/1800/download/1800-2012.pdf
2. Guidelines and practices for successful logic synthesis version 1998.08, Aug 1998. Synopsys timing constraints and optimization user guide, version D-2010.03. www.synopsys.com
3. Guidelines and practices for successful logic synthesis version 1998.08, Aug 1998. www.synopsys.com
4. Design compiler® user guide version D-2010.03-SP2, June 2010. www.synopsys.com
5. Cummings CE. Synthesis and scripting techniques for designing multi asynchronous clock designs
6. Closing the loop on clock domain functional implementation problems. www.cadence.com
7. Power Compiler Reference Manual: Synopsys Inc. www.synopsys.com
8. IEEE 1801 low power design standard. www.ieee.org
9. RTL schematic capture using Xilinx Vivado tool: www.xilinx.com

V. Taraate, *Digital Logic Design Using Verilog*,
https://doi.org/10.1007/978-981-16-3199-3

Index

V. Taraate, *Digital Logic Design Using Verilog*,
https://doi.org/10.1007/978-981-16-3199-3

Printed in the United States
by Baker & Taylor Publisher Services